THE SENSORY EVALUATION
OF DAIRY PRODUCTS

THE SENSORY EVALUATION OF DAIRY PRODUCTS

F. W. Bodyfelt, M.S.

Professor of Food Science and Technology
Extension Dairy Processing Specialist
Oregon State University
Corvallis, OR

J. Tobias, Ph.D.

Professor Emeritus of Dairy Technology
University of Illinois
Champaign-Urbana, IL

G. M. Trout, Ph.D.

Professor Emeritus of Dairy Science
Michigan State University
E. Lansing, MI

An **avi** Book
Published by Van Nostrand Reinhold
New York

An AVI Book
(AVI is an imprint of Van Nostrand Reinhold)

Library of Congress Catalog Card Number 87–28078

ISBN 0–442–22685–3

Printed in the United States of America

Van Nostrand Reinhold
115 Fifth Avenue
New York, New York 10003

Van Nostrand Reinhold (International) Limited
11 New Fetter Lane
London EC4P 4EE, England

Van Nostrand Reinhold
480 La Trobe Street
Melbourne, Victoria 3000, Australia

Macmillan of Canada
Division of Canada Publishing Corporation
164 Commander Boulevard
Agincourt, Ontario MIS 3C7, Canada

16 15 14 13 12 11 10 9 8 7 6 5 4 3 2 1

Library of Congress Cataloging-in-Publication Data

Bodyfelt, F. W. (Floyd W.), 1937–
    The sensory evaluation of dairy products / F. W. Bodyfelt, J.
Tobias, G. M. Trout.
        p.   cm.
    Bibliography: p.
    ISBN 0–442–22685–3
    1. Dairy products—Sensory evaluation. 2. Dairy products—Flavor
and odor. I. Tobias, J. (Joseph) II. Trout, G. Malcolm (George
Malcolm), 1896–   . III. Title.
TX556.M5B63   1988
637—dc19   87–28078

Contents

Preface		vii
Chapter 1.	Development of Dairy Products Evaluation	1
Chapter 2.	Principles of Sensory Perception: An Overview	8
Chapter 3.	Practical Aspects of Dairy Products Evaluation	36
Chapter 4.	Sensory Defects of Dairy Products: An Overview	59
Chapter 5.	Sensory Evaluation of Fluid Milk and Cream Products	107
Chapter 6.	Sensory Evaluation of Ice Cream and Related Products	166
Chapter 7.	Sensory Evaluation of Cultured Milk Products	227
Chapter 8.	Sensory Evaluation of Cheese	300
Chapter 9.	Sensory Evaluation of Butter	376
Chapter 10.	Sensory Evaluation of Concentrated and Dry Milk	418
Chapter 11.	Preparation of Samples for Sensory Training	473
Chapter 12.	Sensory Testing Panels: An Overview	489
Appendices		527
Index		569

Preface

Three different methods are available for tracing causes of sensory problems in dairy foods: (1) chemical procedures, (2) microbiological tests, and (3) sensory evaluation. The simplest, most rapid, and direct approach is sensory evaluation. A food technologist trained and experienced in flavor evaluation of dairy products has an "edge" on someone who is competent only in performing the chemical and/or microbiological methods. Correct diagnosis of the type and cause(s) of a serious sensory defect is a prerequisite to application of remedial measures in production, processing, and distribution steps.

For dairy processors, the most important requirement of a comprehensive quality assurance program is careful flavor evaluation of all dairy ingredients. Based on sensory judgments, occasionally some milk, cream, or other dairy ingredients may require rejection. An important premise of the dairy industry is: "dairy products quality can be only as good as the raw materials from which they are made."

In this book, the authors have attempted to present a reasonably complete overview of the sensory evaluation of most of the major commercial dairy products in the United States. Furthermore, the authors have de-emphasized the terms "judging" and "scoring" in favor of the more contemporary terms "flavor" or "sensory evaluation." The latter terminology is more reflective of the marked progress made in relating flavor perception to the areas of sensory panel methodology, statistics, human behavior, psychology, and the psychophysics of human sensory perception.

This book is intended to serve as both a text and a general reference for students, production and quality assurance personnel in industry, and others interested in the sensory characteristics of the principal dairy products of the United States and Canada. The early chapters review the historical basis of relying on "flavor experts" or "judges" to critique the sensory characteristics of various dairy products, the fundamentals of human sensory perception and an overview of the sensory characteristics of dairy products. Subsequent chapters provide a description of various sensory defects, their causes, and remedial steps to minimize or eliminate their occurrence in fluid milk, frozen dairy desserts (ice cream), cultured dairy products, cheese, butter, and concentrated milk products. The final two chapters guide coaches or instructors through the preparation of samples for instructional purposes and provide an overview of sensory panel methods.

In preparing this edition, the two senior authors have attempted to reflect their philosophy and instructional techniques in conveying the "knack of how to recognize" and describe the sensory shortcomings of dairy foods. The reader should recognize that a clear distinction exists between the concepts of "quality," "preference," and "acceptability." The primary aim of this book was to treat the subject of sensory quality.

Since publication of the previous edition of this book, under the title *Judging Dairy Products* (Nelson and Trout), the definitions of many dairy products have appeared in the U.S. Code of Federal Regulations (CFR). If product quality is perceived as the absence of sensory defects, the consequences of compositional changes of dairy foods (as introduced or changed by CFR specifications) need not be reflected in quality changes. However, certain product characteristics may change as the result of formulation alterations. For instance, reduction of the milkfat content of ice cream from 12% to 10% certainly could affect the product's sensory and hedonic characteristics without affecting quality. In defining various dairy products, reference has been made to the Code of Federal Regulations throughout the book. The reader is cautioned that since changes in the CFRs may occur at any time, only the latest edition of this official document should be consulted for purposes of legal compliance.

Technological progress has all but eliminated many of the sensory defects of milk products reviewed in previous editions; hence, those product shortcomings no longer require much "attention" in the current edition. Some flavor descriptors or terms have continued in use over the years more by habit than due to logic. A better understanding of the causality of certain defects suggests that a different or "advanced" terminology is appropriate. In this edition, an effort has been made to bridge the traditional terminology with more advanced knowledge of the defects. By necessity, this transition must be gradual, to preserve our ability to accurately communicate the sensory properties of dairy products.

For many of the dairy products discussed in this book, various quality standards have been cited in either the appropriate chapter or the appendices. The appendices also include information on milk sampling and grading, examples of some additional dairy products score cards, and selected tests for quality monitoring.

The authors would like to acknowledge the following individuals for their outstanding efforts and assistance in preparing this book for publication: Helen Richardson and Marilyn Tubbs (Corvallis, OR) for word processing numerous copies of the manuscript; Julie Bodyfelt (Corvallis, OR) for technical illustrations; Drs. Randall K. Thunell (Logan, UT), Robert C. Bradley, Jr. (Madison, WI), and Mina R. McDaniel (Corvallis, OR) for chapter reviews and critique. The willing and helpful assistance of the following individuals for review of various chapters is also acknowledged: Bill Daley, Erin McDonnell, David Lundahl, Cynthia Carr Rich (Corvallis, OR), Christy Nelson (Portland, OR), and Betty Milton (Albany, OR).

From 1934 to 1965, when the first four editions of this book were published under the title *Judging Dairy Products,* the first author was the late Dr. John A. Nelson, Professor of Dairy Industry, Montana State College, Bozeman, Montana. Although his name has been omitted from the present revision, the authors are profoundly aware of his pioneering contributions to the unique and very successful early treatment of this subject matter.

The Development of Dairy Products Evaluation

The senses of smell and taste have always been used for guidance in our selection of food and beverages. The ability to discriminate between desirable and undesirable foods is apparently as old as the human race.

The selection of dairy-based foods that possess desirable flavor, particularly milk, butter, and various cheeses, dates back to the early use of these products in the U.S. Early American agricultural writers apparently recognized that the consumption of dairy products depended primarily upon their flavor characteristics. These writers cautioned dairymen concerning certain feeding and milk handling practices if a high quality dairy product was to be obtained. For example, Deane (1797) advised: "In feeding milch cows, the flavour of the milk should be attended to, . . . Feeding them with turnips is said to give an ill taste to the butter made of the milk."

EARLY HISTORY OF DAIRY PRODUCT EVALUATION

Displays of butter and cheese at fairs, exhibitions, and agricultural society meetings played an important role in the development of a consciousness of the quality of dairy products in the United States. However, not until the latter part of the nineteenth century did the grading of dairy products receive national and international attention. The establishment of product grades (with their attendant score cards), as well as standards for various dairy products, has paralleled quite closely growth of the dairy industry and development of dairy product markets.

Although the early dairy industries departments of U.S. agricultural colleges emphasized and taught the merits of quality in dairy products, it was not until 1916 that the first Students' National Contest in Judging of Dairy Products was held. In the first contest, butter was the only product judged, but the following year both Cheddar cheese and milk judging were introduced. Vanilla ice cream was evaluated in collegiate judging competition for the first time in 1926. Cottage cheese was added in 1963 and Swiss-style strawberry yogurt was introduced to the contest in 1977.

The International Collegiate Dairy Products Evaluation Contest, its current name, has been held annually from 1916, with the exception of 1918 and 1942 to 1946, inclusive, due to World Wars I and II (Trout *et al.* 1939, 1981). As many as 33 teams of 3 people each have participated in this international contest in a given year. This program has been most effective in helping provide the dairy industry with better-qualified personnel throughout the years. These students enter the dairy and food industry with developed skill levels and a basic knowledge of what constitutes quality in dairy products.

ESTABLISHMENT OF BRANDS AND TRADEMARKS

Basic and applied food research continues to play an important role in development of the U.S. dairy industry. During the past three decades, attention has been focused on the palatability of dairy products, with

Fig. 1.1. The Danish "Lur Brand" has become a widely recognized benchmark of quality for various dairy products, especially butter.

particular research directed toward the improvement and stability of dairy products' flavors. This research has given significant impetus to the evaluation and grading of dairy products.

The beginning of the twentieth century marked the establishment of brands and trade names for dairy products, particularly butter and cheese. This development necessitated recognition of set standards of quality by the manufacturer and the subsequent need for grading of finished products by an experienced, competent judge. Some brands of dairy products have become widely known and touted for their high quality. For example, the Lur mark for high-quality Danish butter (Fig. 1.1), instituted in 1906, has become perhaps the most famous export butter trademark. The Iowa trademark for butter (Fig. 1.2), adopted in 1915, was based upon specific quality factors that were established for the product, and also upon sanitary conditions and manufacturing methods within the plant (Iverson, 1942). While the Lur brand is still prominent in the international and domestic trade of Danish butter, the Iowa trademark has lost its market significance, and has succumbed to federal consumer grades established by the United States Department of Agriculture (Fig. 1.3).

Land O' Lakes Creameries, Incorporated, of Minneapolis, MN, established the Land O' Lakes brand of butter which is also based upon high quality. Likewise, Sealtest, Inc., a subsidiary of Kraft, Inc., has established the Sealtest brand for high-quality dairy products, primarily ice cream and milk. Many regional and national firms offer individual brands of dairy products which are readily recognized by the public based on the high standards of sensory quality. Official USDA product grades, though attached to many private labels, enjoy prominent significance when seen on butter, cheese, and nonfat dry milk (Fig. 1.4).

Fig. 1.2. The "Iowa Butter" trademark, adopted in 1915, served as a vital factor in helping establish quality butter in the U.S.

A   B   C

Fig. 1.3.   Examples of the grading and inspection marks (shields) of the Food Safety and Quality Service, U.S. Department of Agriculture: A—Graded products packed under USDA inspection; B—Graded products processed and packed under USDA inspection; C—Inspected products processed and packed under USDA quality control service (when there are no U.S. grade standards for the product).

Fig. 1.4.   Examples of brands and trademarks of dairy products of regional or national significance in the U.S.

THE IMPORTANCE
OF DAIRY PRODUCTS EVALUATION

While dairy products can be analyzed for chemical composition, micro-organisms, vitamin content, enzymatic activity, color, physical properties, and so forth, these determinations do not measure the true or actual "eating quality" of a product. Two samples of butter may have identical chemical composition, color, firmness, and spreadability; however, one sample may be highly relished by consumers, while the other product may leave a poor impression. A dairy food that is liked or preferred by a majority of consumers is considered to have good "eating quality." Butter of good to excellent quality generally conveys the impression of being clean, creamy, aromatic in flavor, and seems distinctly fresh and appetizing, whereas the sample that left a poor impression may be stale, rancid, oxidized, fishy, or have some other objectionable off-flavor.

Establishing the "eating quality" of a dairy product requires the application and "correct" interpretation of such sensations as mouth-feel, taste, and smell. The alert consumer experiences these "components of flavor" when the product is taken into the mouth. Although the essential parameters that constitute the "eating quality" of dairy products cannot be easily measured, either chemically or physically, they can be determined by using sensory evaluation techniques, such as those used by competent judges or trained panelists (Bodyfelt 1981).

The judging and grading of dairy products has received continuous attention due to: (1) increased consumer interest; (2) the interest of processors who prefer to sell their products on the basis of grade; and (3) the purchase of certain dairy products by the federal government (Nelson and Trout, 1964). Anyone engaged in the production, manufacture, sale, and purchase of dairy products should have some interest in how the grades for these products are established, updated, or revised as technology and consumer preference may dictate.

THE SEARCH FOR EXCELLENCE

Milk producers, who are co-partners with dairy products manufacturers in establishing a demand for uniform quality dairy products, should recognize that *dairy products cannot be of higher quality than the raw material from which they are made.* Without definite knowledge as to what constitutes desirable and undesirable flavors in finished products, successful production of high-quality raw material is more difficult. Unfortunately, milk flavor quality receives too little attention from most producers. Dairy producers should have a better understanding or awareness of the flavor demands and preferences of consumers. A knowledge of the relative importance of certain off-flavors and the various desirable flavors, plus specific methods of minimizing or eliminating objectionable off-flavors, should enable the production of milk that can be made into high-quality finished products. Such efforts should enhance dairy product sales.

Every enterprising dairy processor has, or should have, the desire to produce products of high quality. The ability to prevent certain objectionable off-flavors and manufacturing defects, and to recognize desirable flavors and product acceptance characteristics, enables processors to manufacture products that better meet consumer demands. A manufacturer who sells dairy products on the wholesale market should know product grades and be familiar with the flavor properties and workmanship required to meet the various grades. The manager who understands market demands and who has the ability to consistently select the grade or quality level desired by certain markets, will discover that his or her products, because of uniformity, meet with ready sale. During production shortages, dairy plant managers must occasionally purchase dairy products on the open market or from nearby plants to be sold as their own product. In such instances, the ability to discern quality or detect certain undesirable flavors is indispensable.

Consumers are interested in obtaining knowledge which will enable them to buy dairy products more intelligently. This knowledge includes an awareness of the product defects that may occur, the desirable and undesirable qualities of dairy food flavors and the important points in careful selection of high-quality dairy products. This information enables consumers to more wisely and economically purchase the dairy products which comprise an important part of the daily diet in most U.S. homes.

The management of the U.S. dairy foods industry needs to appreciate that increased sales of dairy foods is highly dependent upon the production and distribution of high-quality dairy products. Such products impart a pleasant, delicate flavor sensation to the consumer's palate. For dairy products, *high quality implies a relative degree of excellence.*

REFERENCES

Bodyfelt, F. W. 1981. Dairy product score cards: Are they consistent with principles of sensory evaluation? *J. Dairy Sci. 64*:2303.

Deane, S. 1797. *The New England Farmer.* Thomas, Worcester, MA., 78.

Iverson, C. A. 1942. The Iowa butter trade-mark. *Creamery J. 53*(7):8.

Nelson, J. and Trout, G. M. 1964. *Judging Dairy Products.* AVI Publishing Co., Westport, CT, 1–8.

Trout, G. M., White, W., Mack, M. J., Downs, P. A. and Fouts, E. L. 1939. History and development of the Students' National Contest in the judging of dairy products. *J. Dairy Sci. 22:*375.

Trout, G. M. and Weigold, G. 1981. Building Careers in the Dairy Products Evaluation Contest (60 Years of Student Judging, 1916—1981). Dairy Food Ind. Sup. Ass'n., Washington, D.C. 16 pp.

2

Principles of Sensory Perception: An Overview

The evaluation of dairy products for flavor is primarily a matter of noting carefully and interpreting correctly a set of sensory reactions after each product is sampled. The ability to critically evaluate dairy products can be learned, if close attention is directed to the delicate senses of smell, taste, touch, and sight with which practically everyone is endowed.

Some observers consider the process of evaluating dairy products to be an art skill. On the contrary, sensory evaluation is more appropriately based upon science. However, attaining proficiency in sensory judgment might best be considered an art skill based upon scientific principles.

This chapter will discuss the sensory physiology (psychophysics) of the human senses and their applications in the sensory evaluation of dairy products.

FLAVOR IS THE "VOICE" OF FOOD AND BEVERAGES

Moncrieff most eloquently summarized the complexity of flavor sensation at an Oregon State University symposium on the chemistry and physiology of flavors (Moncrieff 1967b):

> The study of flavor is one of those subjects in which science has never caught up with everyday experience. Mainly, flavor is composed of taste and odor. Hold the nose or even hold the breath, and flavor vanishes in a second; breathe again and it reappears at once. Of the other qualities that enter into it, texture is probably the most important: smoothness or roughness, particle size, solubility, even a glutinous quality can modify flavor. Less usual modifiers of flavor are the hotness of spices such as ginger, the coolness of menthol. Then there are the metallic, alkaline, and meaty tastes. If we are to accept the orthodox view that there are only four true tastes;

sweet, bitter, sour, and salt, then the metallic and alkaline tastes must presumably be accepted as modalities of the common chemical sense.

SOME FUNDAMENTALS
OF SENSORY PHYSIOLOGY

The Human Senses. Psychologists generally recognize 22 special senses (or subdivisions) within human beings (Amerine *et al.* 1965). On the authority of no one less than Aristotle, humans supposedly possess five primary senses for perceiving stimuli. They are the familiar senses of sight, hearing, touch, smell, and taste. The latter two senses are considered to be the most primitive (Brown and Deffenbacher 1979, Coren *et al.* 1978). Other human senses include temperature sensation (heat and cold), pain, visceral hunger, thirst, fatigue, sex (drive), and equilibrium (balance). See Table 2.1 for a more complete listing of human senses.

In human beings, at least three different senses respond to specific chemical stimuli: taste, smell, and the so-called common chemical or pain sense. Humans are primarily sight-guided in their search for food, whereas other animals, such as dogs and pigs, are scent-guided. Within humans, smell has a great complexity of qualities and features; in fact, the olfactory membrane compares well in absolute sensitivity with the retina (sight) and the organ of Corti (hearing) (Amerine *et al.* 1965).

The Sensory Receptors. As organisms, we experience our environment and many events occurring within our bodies not by direct means, and not in their entirety, but rather through specialized sense organs or sensory receptors. The more familiar of these sense organs are the eye, the ear, the skin as an organ of touch or pressure, the tongue as the organ of taste, and the nose as the organ of smell. Each of these sensory receptor devices responds to a particular range of environmental influences (stimuli) and transmits corresponding information to the brain via the central nervous system (Dudel 1981). In turn, specific sites in the brain are stimulated or energized by the initial sensory input. Up to a certain point, the response of the sensory cells is proportional to the stimulus intensity. Objectively, the response of the nerve is a function of the frequency of the electrical discharge of the nerve; the higher the frequency, the stronger the sensation. Nearly all sensory receptors vary in their sensitivity to stimuli (Amerine *et al.* 1965; Schmidt 1981; Coren *et al.* 1978; and Brown and Deffenbacher 1979).

Modality, Quality, Stimuli, and Sensory Impression. A group of similar sensory impressions, mediated by a given organ, is referred to as a

Table 2.1. Human Sensory Reactions with Associated Stimuli and Receptors.

Sensory Reaction	Stimulus	Specific Receptor	Human Experience
1. *Chemical Receptors*			
Gustatory	Chemicals—water soluble	Taste buds	Tastes
Olfactory	Chemicals—gas soluble	Olfactory cells in uppermost portion of nasal cavity	Odors
2. *Somesthetic (Body) Receptors*			
Cutaneous	1. Temperature changes	Cells in skin	Warmth, Coldness
	2. Mechanical	Cells in skin	Touch (light pressing)
	3. Extreme energy (i.e., intense heat, laser, etc.)	Free nerve endings	Pain
Kinesthetic	Mechanical pressures	Cells in tendons, muscles, joints	Active movement, weight, deep pressure
Vestibular (static)	Head movement (rectilinear or rotary)	Cells in semicircular canals and vestibule	Equilibrium (balance)
Organic	Chemical or mechanical action	Cells in viscera	Pressure, visceral disturbance (e.g., hunger, nausea)
3. *Distance Receptors*			
Visual	Radiant energy wavelength 10^{-4} to 10^{-5} cm (light waves)	Rods and cones of retina	Color hue, brightness, contrast
Auditory	Mechanical vibrations of frequency of 20–20,000 eps (sound waves)	Hair cells of the organ of Corti	Pitch, loudness

Adapted from Schmidt, 1981.

sense. A more technically precise term for a sense is *modality.* Hence, modalities include the "classic five senses": sight, hearing, touch, taste, and smell, as well as the following senses: temperature, vibration, pain, equilibrium, thirst, hunger, shortness of breath, and visceral sensation (Dudel 1981).

Each sense organ mediates sensory perceptions that can vary in intensity, but resemble one another in quality. *Quality* refers to a further distinction or special dimension of the sensory impression within each modality (sense). For example, the modality of hearing has the different pitches of tone for quality, the modality of vision can be segmented into the qualities of lightness (grey scale), red, green, and blue. The *qualities of taste* are represented by *sweet, sour, salt,* and *bitter.* According to one view, aroma qualities may be interpreted as fragrant, floral, burnt, and caprylic (goaty). The four qualities of mechanoreception (or touch) are considered to be: (1) pressure, (2) touch, (3) vibration, and (4) tickle (Schmidt 1981).

Factors from the environment or from body biochemistry that elicit sensory impressions or perceptions of a certain quality are called *specific sensory stimuli,* or simply *stimuli* (Dudel 1981). A given stimulus acquires its special dimension (quality) by virtue of its reaction with the stimulus detecting cells of the sense organs, the receptors. The term *sensory impression* serves experience. The perceived odor "goaty," the taste "salty," and the mouthfeel "gritty" would be examples of a sensory impression. Such impressions are very seldom received in partial or total isolation; a combination of such sensory impressions is called a *sensation.* In most situations, a pure sensation is accompanied by an interpretation, with reference to what has been experienced and learned by the individual, and the resultant overall impression is called a *perception.* We express a perception when we say, "This milk is too sweet."

Perception has at least four basic dimensions:

1. Time   3. Quality
2. Space   4. Intensity (quantity)

Objective and Subjective Sensory Physiology. Sensory physiology can be divided into two parts. First, the various responses of the nervous system to a stimulus is referred to as *objective sensory physiology.* The relationship between a stimulus and its perception by a sensory component of the central nervous system can be principally described as a physical and/or chemical process within humans. Second, the analysis of the statements that the subject makes about his or her sensations and perceptions is referred to as *subjective sensory physiology.* The relationship between a stimulus, its subsequent response(s) within the nervous system, *and* a conscious sensation, cannot be easily described in terms of physical and chemical processes—rather, it is eventually manifested as a subjective statement. In summary, sensory stimuli are physical and/or chemical components, whereas the resultant sensations and perceptions to stimuli are subjective processes (Dudel 1981).

Scientists, in their study of human behavior, have exercised concerted efforts to treat subjective statements about sensations and perceptions in the same unbiased manner as they accommodate recordings of cell potential for other chemical or physical data. Aided by the discipline of statistics, investigators can now make very precise statements about the objects of subjective sensory physiology (Dudel 1981). Furthermore, quantitative mathematical relationships have been established to "bridge the gap" between the quantitative (objective) and the qualitative (subjective) phases. Scientists have developed a concept known as "mapping" (Dudel 1981) to reflect, in mathematical terms, the unique association between the members of two sets for the objective and subjective phases. This area of natural science, dealing with subjective sensory physiology, is often termed *psychophysics.* One of its major efforts is to measure more effectively and accurately the variety and intensity of sensations.

MEASUREMENT OF SENSATION INTENSITY

Threshold Values. Amerine *et al.* (1965) defined the term "threshold" as a "statistically determined point on the stimulus scale at which occurs a transition in a series of sensations or judgments." Several types of thresholds associated with sensory processes have served as important tools in developing a better understanding of the relationships between the magnitude of a given stimulus and perceived sensations.

The least energy capable of producing a sensation is referred to as an *absolute threshold.* Sensation changes in a predictable (or lawful) manner as stimulus energy is increased, but changes in a slightly different way for each of the senses.

Other types of sensory-related thresholds are: the detection (stimulus) threshold, the difference threshold, the recognition (identification) threshold, and the terminal threshold. Amerine *et al.* (1965) defined each of these thresholds as follows:

Detection (stimulus) threshold—that magnitude of stimulus at which a transition occurs from no perceived sensation to a perceived sensation, often designated as *RL*. This is similar to, or perhaps the same as, the absolute threshold defined above.

Difference threshold—the least amount of change of a given stimulus necessary to produce a change in sensation, often designated as *DL*, and the interval or unit of difference is designated as the *JND* (just noticeable difference).

Recognition (identification) threshold—the minimum concentration at which a substance is correctly identified.

Terminal threshold—that magnitude of a stimulus above which there is no increase in perceived intensity of the appropriate quality for the stimulus. Above this point, pain often occurs to the subject.

Within the field or study of the measurement of sensation intensity (psychophysics), both objective and subjective parameters are applied. The subjective aspect is primarily concerned with the verbal or written statements that a human being makes about stimuli. The system of measurement must be based on a defined elementary unit, and on a fixed procedure by which a particular number of such units can be assigned to the quantity to be measured. Hence, appropriate units of measure must be estimated and/or assigned for the various sensations. In objective approaches of sensory physiology, the intensity of both the stimulus and the sensation are measured.

The absolute threshold for a given sensation can be considered as the reference intensity of sensation (a subjective unit, usually), and other degrees of sensation are expressed as multiples of that threshold sensation (Dudel 1981). The absolute threshold, RL can be objectively measured in experimental animals (rats) by establishing the weakest stimulus intensity to which the cell responds with a change in the frequency of action potentials (Zimmerman 1981).

The difference threshold (*DL*) is objectively measured by determining the amount by which the intensity of a suprathreshold stimulus (above the absolute threshold) must be changed in order to produce a frequency change of the action potentials of a neuron. It changes in a predictable way with stimulus intensity in accordance with *Weber's rule,* which describes the difference threshold in subjective sensation:

$$\frac{\Delta S}{S} = \text{a constant,}$$

where S = the suprathreshold stimulus (or the average of two successive just distinguishable stimuli), and

ΔS = the difference between the absolute threshold *(RL)* and *(S)*, the suprathreshold stimulus (JND, or the difference between two successive just distinguishable stimuli)

Weber, whose studies involved perception of different weights, based his generalization on a careful analysis of psychophysical data within the last century. The *DL* may be visualized as the distinguishable percentage change from the original sensory intensity. For instance, when applied to pressure stimuli on the skin, Weber's rule shows that whenever the magnitude of the stimulus is changed, the difference threshold is about 3% of the starting pressure (Zimmerman 1981).

From a practical standpoint, a recognition threshold represents the minimum concentrations of sapid (flavorful) substances possessing flavor characteristics that can be recognized upon tasting or smelling. Various threshold values tend to vary with each individual (Amerine *el al.* 1965, Schmidt 1981). Some individuals exhibit lower thresholds than others (Moncrieff 1967a). For example, those subjects that have a comparatively low recognition threshold for salt solutions might be expected to more readily detect a salty taste in test food samples, whereas individuals having a high recognition threshold may be unable to perceive the salty taste. One must bear in mind that threshold values are not a fixed entity with a given individual, but may vary with the temperature of tasting or other conditions (Amerine *et al.* 1965, Moncrieff 1967a, 1970).

Threshold values are somewhat difficult to compare, inasmuch as various investigators use different presentation techniques with their subjects. Presentation techniques have included applying a sapid substance to the tongue by means of droppers, tablets, pencils, brushes, or sponges.

PHYSIOLOGY OF SMELL AND TASTE

Sensitivity to various chemicals is basic to all human and animal life. Within the body, each living cell must assimilate the chemicals it needs and eliminate other chemicals (Brown and Deffenbacher 1979). The human senses are the means, or the channels for information, through which everything we know is received into the consciousness. Chemical sense receptors for humans are primarily in the nose and mouth, but some receptors for pain (chemical) occur in other locations of the body.

As mentioned earlier, the sensations of smell, taste, and pain (to some extent) are considered to be *chemical senses* inasmuch as receptor sites for these senses respond to chemical stimuli or the "arrival of chemical particles" (Amoore 1970). This is in direct contrast to the *physical senses* of sight, hearing, touch, and temperature, which respond to physical stimuli.

The sense of smell is substantially more refined than the sense of taste. A typical person requires relatively concentrated solutions in order to perceive a taste sensation. By contrast, an odoriferous substance, such as a mercaptan, may be diluted to the extent of 0.43 × 10^{-15} mg/L of air, and yet be recognized as such by the sense of smell (Moncrieff 1967a). The human sense of smell is more sensitive than sophisticated chemical procedures, performed by a well-trained chemist, who has use of the most advanced equipment.

Taste Receptors. Taste receptors are groups of sensory epithelial cells innervated (stimulated) by sensory nerve endings (Altner 1981a, Farbman 1967). As early as 1868, an investigator who identified and described the histology of taste receptors likened their appearance to that of a flower bud; hence, the term "taste bud" was coined.

Human taste buds are located on moist surfaces (Altner 1981a), within the oral cavity and pharynx, that are lined with stratified squamous epithelium (Fig. 2.1). The tongue serves as the major organ of taste. The taste bud sites are in the epithelium of the tip, lateral, and dorsal surfaces where they are generally associated with certain papillae (see Fig. 2.2); fungiform papillae on the anterior tongue segment, and the foliate and vallate papillae on the posterior part. Figure 2.3 illustrates the locations of the various taste bud sites within the oral cavity. A few nonpapillae-associated taste buds may be found in such locations as the soft palate, pharynx, and larynx. These taste buds are simply embedded within the epithelium of the mucous membrane (Farbman 1967).

The innervation of taste buds is derived from the "sensory" component of three different cranial nerves: the facial, glossopharyngeal, and vagus. The facial nerve branch innervates the taste buds in the anterior part of the tongue and those of the soft palate. The glossopharyngeal and vagus nerves accommodate sensory innervation of all the remaining taste buds (Farbman 1967).

Odor Receptors. According to Moncrieff (1967b), "somewhere in our anatomical equipment there is built a device that sorts out thousands of different kinds of odors and enables us to distinguish one flavor from another, and to detect even a slight off-flavor."

The olfactory receptor area is located in the roof of the nasal cavity (Fig. 2.4) and is lined with so-called olfactory epithelium which is distinguishable by its yellowish color from the pink respiratory epithelium around it. The surface of the olfactory epithelium is coated by a

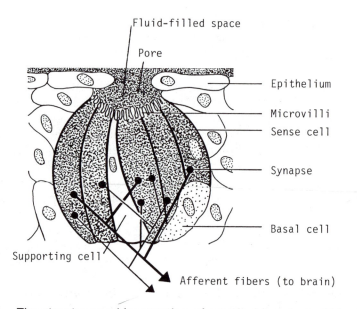

Fluid-filled space

Pore

Epithelium

Microvilli
Sense cell

Synapse

Basal cell

Supporting cell

Afferent fibers (to brain)

Fig. 2.1.   The structure and innervation of a typical human taste bud. The elements of the human taste organ are arranged in a fashion similar to the petals in a rose bud. The entire structure is located below the epithelial surface. About 50 afferent fibers enter and branch within a single taste bud (only two fibers are shown). From Altner, H., *Fundamentals of Sensory Physiology* (1981a). Courtesy of Springer-Verlag, Heidelberg, FRG.

layer of mucous (Fig. 2.5). Embedded in this mucous layer is a mat of fine hair-like appendages of the olfactory cells, called cilia.

Odor Perception. In order to perceive the odor of a chemical compound or substance, the chemical must be volatile. The volatility of a substance depends on its molecular weight and molecular bonding properties. The upper limit for "smellability" is usually a molecular weight (MW) of about 300, although one compound with a MW of 394 has demonstrated an odor (Moncrieff 1967a). This suggests that heavier-weight substances such as proteins, starches, fats, and many sugars are too heavy to become airborne under most circumstances. While molecular bonds vary with the chemical compounds and the temperature, more volatile molecules are derived from liquids than from solids; but the volatility of both states of matter increases with increasing temperature.

Odoriferous substances must be adsorbed or adhered to the chemoreceptor sites in the nose (Altner 1981b). To be adsorbed, a volatile sub-

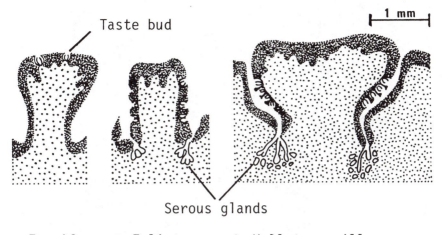

Taste bud

1 mm

Serous glands

A. Fungiform   **B.** Foliate   **C.** Vallate papilla
papilla   papilla

Fig. 2.2. The position of taste buds on the three different types of gustatory papillae: A—Fungiform, cover the anterior 2/3 of tongue surface; B—Foliate, cover the posterior 1/3 of lingual surface of the tongue, along with; C—Vallate papilla. From Altner, H., *Fundamentals of Sensory Physiology.* (1981a) Courtesy of Springer-Verlag, Heidelberg, FRG.

stance must be soluble to some extent in the receptor cell membrane, which consists of lipids, proteins and water. Most volatile organic compounds, are soluble to some extent in the membrane constituents.

Adaptation. It is a common experience that the perception of a constant odor diminishes with the elapse of time. This process is known as *adaptation.* Experimental results support the view that the olfactory organ slowly adapts to continuous and repetitive stimulation. When a stimulus has been applied to the olfactory system, its excitability or response is reduced. After the stimulus has been removed, odor sensitivity slowly recovers to the original level (Dravnieks 1967).

Theories of Olfaction. Many theories have been proposed regarding the generation of the neural signal by odorivector-chemoreceptor interactions. The most comprehensive theory that is compatible with common biochemical concepts, is that presented by Davies (Dravnieks 1967). Davies' theory holds that the polarized-neuron cell wall supplies energy, while the odorivectors trigger the release of this energy. However, at present, there is only circumstantial evidence for this theory (Dravnieks 1967). Some other common theories of olfaction can be listed as follows:

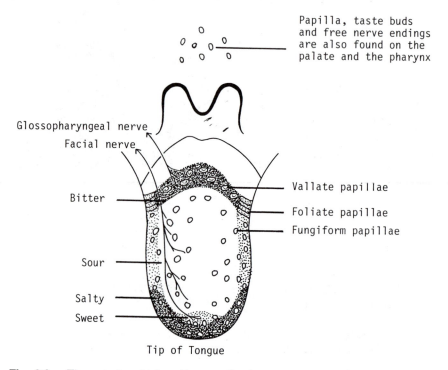

Fig. 2.3. The proximate location on the human tongue of the taste buds for each of the four primary taste (qualities), the distribution of the gustatory papillae and their innervation. From Altner, H., *Fundamentals of Sensory Physiology* (1981a). Courtesy of Springer-Verlag, Heidelberg, FRG.

1. *Vibrational theory*—based on correlations between infrared or Raman spectra and odor quality.
2. *Molecular size theory*—based on the ability of specific odoriferous molecules to fit or fill a corresponding receptor site.
3. *Molecular shape theory*—based on the degree of fit into postulated receptor sites.
4. *Intermolecular interaction theories*—dependent upon vapor pressure, solubility, and other bulk characteristics of the volatile compounds.

Physiological Aspects of Olfaction. The *centers of olfaction* are located chiefly in the uppermost regions of the nasal cavity, and not along the sides of the septum of the nose itself, as commonly supposed. The olfactory area, if spread out, would comprise only several square

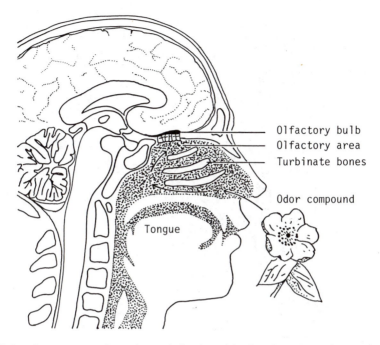

Fig. 2.4. A cross section view of the head indicating the olfactory area. From Amoore, J. E. et al. The stereochemical theory of odor. Copyright © 1964 by Scientific American, Inc. All rights reserved.

centimeters in area. From the light, yellowish-brown center-of-smell area are projected tiny cilia, or fine hairs, which are innervated (excited) by molecules of odoriferous substances.

Since this area of smell is reached chiefly by eddy currents, rather than direct passage of air during inhalation, the odoriferous substances must be "sniffed" or "whiffed" rather slowly, but strongly, while respiration is slowed or stopped. During exhalation there is no appreciable smell sensation.

Smell sensation may also be innervated by diffusion. However, diffusion of the odoriferous substance is so slow as to be of no practical value in ascertaining the odor of a substance. Even when the nose is filled with odor-laden air, an odor sensation is not noted when the breath is held. With these facts in mind, the evaluator must put more emphasis upon drawing in a full breath of air through the nose, rather positively and prolongedly.

Experienced tasters realize that during food mastication odors may

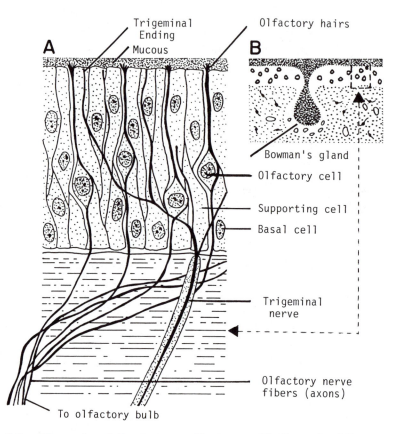

Fig. 2.5. Elements of the human olfactory epithelium: A—Microscopic diagram of the various cells that compose the olfactory mucosa (epithelium); B—Mucous membrane in the olfactory region.

be noted without consciously smelling the product. During mastication, molecules of aromatic substances pass to the olfactory area from the mouth. During the so-called "tasting" of dairy products the evaluator is actually sensing odor(s) and taste(s) of the product simultaneously. The authors feel that during the process of tasting, gentle exhalation through the nose is beneficial, because it serves to force volatile molecules toward the olfactory epithelium.

Taste Perception. The *taste receptors* are located primarily on the sides and on the base of the tongue, but may be found also on the soft palate and on the cheeks, particularly in young people. Papillae of various types can be noted chiefly at the tip, along the sides, and at the base of the tongue. The taste buds, with which the sapid substance (in liquid form) must make contact before a taste sensation occurs, are located in many of these papillae.

Taste buds differ somewhat in their response to stimuli. The sour taste may be noted chiefly along the sides of the tongue; saltiness along the sides and tip; sweetness generally at the tip; and the bitter taste at the base of the tongue. For this reason, a food or beverage sample being tasted should be manipulated about the mouth and rolled over the tongue to permit direct contact between the taste buds and the sapid substance.

The first requirement for tasting a material substance is water solubility of the food item. The two most primitive tastes, salt and sour, show good correlation with molecular structure. The slower-response sweet and bitter tastes are apparently less related to chemical formulation (Moncrieff 1967a). Practically all acids exhibit a sour taste. Common salt (NaCl) and other similar compounds taste salty, although as the constituent atoms of salts become larger (higher atomic weight), a bitter taste tends to accompany the initial salty taste. Sweetness is a characteristic of sugars and is apparently related to their possession of several hydroxyl groups. Bitterness, however, is not easily associated with any structural properties.

Role of Primary Senses in Dairy Products Evaluation. All five primary or "classic" senses; sight, smell, taste, touch, and sound are used in the sensory evaluation of dairy products. The extent to which each modality (sense) is used depends upon the product being evaluated. In the evaluation of dairy products, novices should be aware of the combined roles of all five of the senses. Greatest emphasis, however, should be placed on the senses of smell and taste.

1. *Sight.* Some of the desirable qualities and some defects of dairy products can be readily ascertained by careful observation. Among those factors which may be evaluated by sight are style, neatness, and cleanliness of the package exterior; attractiveness of product finish, package closures, body, and texture; color and overall appearance; meltdown characteristics (ice cream); and other items that determine the quality impression. Color and appearance aspects of milk products should not be slighted or overlooked in the evaluation process; occasionally these features may render a product unacceptable to consumers.

An experienced dairy product judge closely examines each product sample in an attempt to correlate possible deficiencies in visible items (clues) with flavor quality. A soiled butter package or an unattractive milk container, or a carelessly packaged cottage cheese carton can be readily evaluated by sight and may furnish the judge with a possible clue to flavor quality. Should the observer see evidence of carelessness in product workmanship, smelling and tasting the product should be undertaken carefully with the senses alert, in order to detect any possible associated quality defects. For example, if the judge examines a butter sample and notes a badly mottled color, then the sample should be tasted with greater concentration and effort than if the sample were found to be uniform in color. Not only must the judge know what features to observe in evaluating dairy products, but the eyes need to be trained to instantly detect shortcomings in product standards. Conversely, the judge should exercise some precautions to insure that judgment based on the senses of smell and taste are not unduly influenced by what is visible to the eye.

2. *Smell (Aroma)*. Generally, it is acknowledged that the sense of smell plays the paramount role in evaluating dairy products, but this does not exclude the need for tasting samples. As pointed out in previous discussion on the physiology of olfaction, it is not necessary to "whiff" each sample to perceive the odor sensation, since the odor substances may reach the olfactory receptor area through oral passages during tasting.

As a rule, the olfactory dimension of a food or beverage contributes substantially more to overall flavor perception than the taste dimension. On various dairy product score cards, flavor (a combination of odor, taste, and mouthfeel) is always assigned the greatest emphasis (higher numerical value) of any of the quality categories; since odor properties contribute so markedly to flavor, the sense of smell plays an especially important role in the evaluation of dairy products.

3. *Taste*. Taste is a "companion sense" with aroma in establishing the flavor characteristics of a food. In the evaluation of dairy products, with few exceptions, the product must be tasted. Not only will the taste sensation be noted upon mastication, but simultaneously, the odor sensation is perceived and the tactile (mouthfeel) properties of the food or beverage are noted. Thus, the role of taste as a part of flavor is more complex than the mere "act of tasting."

4. *Touch*. Not only are the senses of smell and taste closely related, but the sense of touch (or pressure) is involved also. Touch or mouthfeel plays an important part in the evaluation of dairy products. To illustrate: the tongue and palate register the peculiar feeling of mealy, or greasy (salvy) butter. They record, as well, the "sandy" defect in ice cream, where pressure between the teeth determines the presence of crystallized lactose. The "briny" defect in butter (due to incompletely

dissolved salt) or "mealy/grainy" in Cheddar cheese are also readily perceived by the sense of touch.

The finger tips and thumb may be brought into use to help substantiate "findings" of the tongue, roof, and other surfaces of the mouth. The fingers play an important role in examining the body and texture of cheese and butter. The tongue and the floor and roof of the mouth record the ease or difficulty with which the examined product goes into solution. The tongue, interior surfaces of the mouth, the finger tips, and the ball of the thumb are all important body-to-sample contact surfaces that help evaluate physical qualities which frequently lead the judge to other defects, which might otherwise be overlooked.

5. *Sound.* The sense of sound is sometimes used in evaluating dairy products. The judge can detect the relative size and distribution of the "holes" (or "eyes") in Swiss cheese by gently tapping the outside of the cheese with the fingers or a trier (sampling device) handle. The relative amount of free water in "leaky" butter can also be loosely determined by the "slushing" sound made when the sample plug (piece) is reinserted into the hole from which it was drawn.

The Concept of Flavor. Odor and taste, combined with mouthfeel (or tactile sensation), results in an overall concept of sensation referred to as "flavor." *Flavor* has been defined as the sum total of the sensory impressions (sensations) perceived when a food or beverage is placed into the mouth.

Classification of Tastes. Strictly speaking, we experience only four true taste reactions, namely, *sweet, sour, salt,* and *bitter.* Authorities agree on these four primary or basic taste reactions. Some sensory authorities believe, however, that there may be several other taste reactions, namely: alkaline, metallic, watery, and/or meaty. However, these may be, in part, tactual sensations and not true taste sensations, or modalities of the common chemical sense.

Truly, many different flavors are sensed instead of tasted (e.g., pepper, peppermint), as no doubt everyone has experienced. But these noted flavors are a composite of taste, smell, and touch reactions and not true tastes per se. In addition, such feelings in the mouth as the common chemical or pain sense, warmth, coolness, astringency, smoothness, numbness (anesthesia), and other feelings sometimes experienced in tasting are not taste reactions, but are sensations of touch or pressure (tactual). They are nevertheless important attributes of tasting and should be considered fully.

The true basic tastes (sweet, sour, salt, and bitter) may be sensed with the nose obstructed. In fact, when a person has a cold, the taste reactions, aside from the particular feel of the food in the mouth, are sometimes the only components of the flavor which may be detected by the individual.

Olfactory Stimulation. Moncrieff (1967a) indicated that approximately 50 different theories have evolved in the last century in an effort to explain what happens when an odoriferous substance is perceived by an organism. Most of these explanations are more appropriately labeled as hypotheses than theories, since most lack sufficient scientific evidence in the view of Amoore (1982). The convincing elements of these hypotheses have centered around either the chemical or physical attributes of the odorant molecule, and what causes a change or response in the olfactory receptor, which in turn leads to the experience of odor (Altner 1981b).

The earliest explanations of odor stimulation stemmed from a presumed analogy with vision and hearing, which are based on physical perception of electromagnetic or atmospheric vibrations, respectively. Historically, one of the earliest hypotheses was the *vibrational theory*, which suggested that the odor stimulus was conveyed at long range from the source to the nose (or antenna) by propagated electromagnetic radiation, as by ultraviolet or (more likely) infrared wavelengths. This theory was somewhat refined by several other researchers, who suggested that the effective stimulus was an intramolecular vibration of the odorant molecules, which was detected at extremely close range or by actual contact with the sensory nerves.

Many odor theorists ignored models of the vision and hearing senses and focused instead on a chemical sense, hence the so-called *chemical theories.* The odor stimulus was hypothesized to be related to some aspect of the chemical properties of the odorant molecules. Actual contact of the odor molecules with the olfactory receptor sites is usually postulated or implied. Furthermore, there are two distinct types/theories, as follows: (1) the relevant factor is some physicochemical property of the molecules, and (2) the causative agent for olfactory stimulus involves chemical reactivity.

The fact that nearly all chemical transformations in living cells are mediated by enzymes, due to either their catalytic power, proteinacious nature, or their capacity for allosteric transition, has resulted in a third category of theories for describing olfactory stimulation, namely the *enzyme theory.*

Amoore (1982) refers to the surviving theories—four theories that are noteworthy due to their relative longevity of existence and/or supporting evidence:

1. *Amoore's Stereochemical Theory.* Based on an idea expressed by Moncrieff in 1949, Amoore first published (1952) a theory which

proposed that the olfactory receptors are sensitive to the size, shape, and the electronic status of the odorant molecule (Amoore *et al.* 1966; Amoore 1970, 1982). Based somewhat on the lock and key concept of enzymology, Amoore expressed the viewpoint that all odor sensations are based on a combination of a limited number of primary odors and specific nerve receptor site cavities (Fig. 2.6). The internal dimensions and/or electrical affinities of each nerve cavity were complementary to the molecular morphology of the primary odorant.

2. *Davies' Penetration and Puncture Theory.* In 1953, Davies drew encouragement for his theory based on recent explanations of nerve impulse electrical conduction, which had been explored by Hodgkin and Katz (as discussed by Amoore 1982). The latter investigators had reported that nerve fiber (axon) membranes were composed of a lipid double layer with adsorbed protein on the inner and outer surfaces of the axon. In the resting state, there was an excess of Na and Cl ions on the outer surface and an excess of K ions on the inner surface of the axon. Davies suggested that the relatively bulky, awkward and rather rigid molecules of the odorant (upon contact of the axon) tended to penetrate and disorient the double-layer of lipid, if only temporarily. This, in turn, resulted in holes in the axon surface, which permitted ionic changes to occur, and thus initiated a nerve impulse to the brain (odor perception).

3. *Wright's Vibrational Theory.* In 1954, Wright (as discussed by

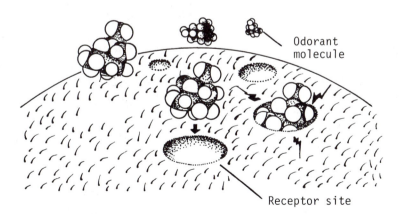

Fig. 2.6. The receptor site-fitting theory of odor perception. From Amoore, J. E., *The Molecular Basis of Odor* (1970). Courtesy of Charles C. Thomas, Publisher, Springfield, IL.

Moncrieff 1967a) suggested that the odors of given chemicals are a function of their intrinsic molecular vibrational frequencies, within the far infrared region of the electromagnetic spectrum $(100-700 \text{ cm}^{-1})$.
4. *Beet's Profile-Functional Group Theory.* Beets reported his theory in 1957 (as cited by Moncrieff 1967a), which stated that two molecular attributes were important in determining the characteristic odor: (1) the form and bulk of the molecule, and (2) the nature and disposition of the functional group(s) of the molecule.

Classification of Odors. Common experience indicates that the number of perceived odors are many and varied. Attempts have been made by numerous investigators to categorize these many odor sensations into basic or fundamental classifications, just as the various taste reactions were resolved into four basic categories.

As early as the mid-nineteenth century, Bain of Scotland (according to Moncreiff 1967a) developed a set of odor classifications in which the pleasant and the unpleasant aspects of four qualities formed the eight classes into which he divided odors. Bain's classification seemed to possess a logical basis, whereas later odor classifications were primarily empirical. Rimmel (according to Moncrieff 1967a) developed a list of 18 pleasant odors on the basis of empirical likenesses. However, this list was limited by the exclusion of foods and also repulsive or unpleasant odors. As early as 1895, Zwaardemaker offered a systematic classification of odors into nine classes, but it was clearly associative and subjective in Moncrieff's view (1967a).

In 1916, Henning offered a careful analysis of odors that basically outlined six odor-groupings (as cited by Moncrieff 1967a, 1970). To illustrate his concept, Henning diagrammed an arrangement for six fundamental odors in which interrelated or intermediate odors were shown as components of an Olfactory Prism (Fig. 2.7). On close examination of the prism, one can observe that a given odor can either be a fundamental odor by occupying a corner; when two odors are involved, then they would be located along an edge; or if three odors are involved, they would be located on a triangular surface.

Crocker and Henderson (1927) made notable contributions to this field by simplifying Hennings' six fundamental odor sensations. After careful study, they reclassified them into four groupings, arranged in order of preference by most subjects (Table 2.2). Furthermore, Crocker and Henderson postulated that four kinds of smell nerves existed in humans. Recognizing that any odor stimulates specific receptors for these odors, they set up standards for many odoriferous substances based upon the correct quantitative mixtures of the fundamental

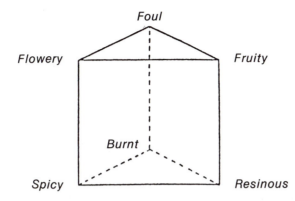

Foul

Flowery   Fruity

Burnt

Spicy   Resinous

Fig. 2.7   Henning's olfactory prism of six fundamental odors and the possible derivation of "intermediate odors."

odors. With a range of intensity of stimuli for each of the four basic odors numbered from 0 to 8, they reproduced odors simply by mixing certain intensities of the basic odors. Within this format, any given aromatic substance may contain all four fundamental odors; their relative degrees of stimulation determine the individuality of that odor. For example, 3803 and 6423 represent the four-digit symbols (Table 2.3) for acetic acid and rose aroma, respectively (Moncrieff 1967a).

In 1928, Boring studied the Crocker and Henderson approach to odor classification and (according to Moncrieff 1967a) reported good reproducibility. A convenient aspect of this method is that it readily enables an odor to be reproduced, or at least attempted, simply by interpretation of its code number. Boring projected that most persons could differentiate between 2,000–4,000 odors (Moncrieff 1967a). Amerine *et al.* (1965) speculated that some highly trained persons could probably differentiate as many as 10,000 different odors.

Moncrieff once attempted to classify common chemical stimuli, on a

Table 2.2. A Comparison of Odor Classifications by Several Investigators.

Henning (1927)	Crocker and Henderson (1927)
1. Spicy (cloves, cinnamon, fennel, anise)	1. Fragrant or sweet
2. Flowery (heliotrope, coumarin, geranium)	2. Acid or sour
3. Fruity (oil of orange, apple, citronella)	3. Burnt
4. Resinous (balsamic, turpentine, eucalyptus oil)	4. Caprylic or goaty
5. Burnt (pyridine, tar, scorched substances)	
6. Foul (hydrogen sulfide, carbon bisulfide)	

Table 2.3. Crocker and Henderson's (1927) Four-Digit Symbol Method of Describing a Given Odor.

Fundamental Odor	Simulated to a Degree of	
	Acetic Acid	Rose Smell
Fragrant	3	6
Acid	8	4
Burnt	0	2
Caprylic	3	1

Source: Crocker and Henderson (1927).

physiological basis, into seven classes. But he emphasized that odors cannot be readily classified by physiological reactions. As support for his argument, Moncrieff (1967a) offered the fact that "odorants are usually mild and undefined—a sniff, a drawing away or in extreme cases, disgust—and nausea. The number of types of odors seems large, and different judges do not agree as to the distinctions."

Another theory for odor classification was initially proposed by Amoore in 1952 (Amoore 1970). He initially suggested seven primary classes of odors as follows: ethereal, camphoraceous, musky, floral, minty, pungent, and putrid. Amoore eventually added an eighth primary odor to his set—sweaty. Amoore's basic concept was that different kinds of olfactory receptor sites existed for each of the primary classes of odors; locations whereon odorant molecules could fit when adsorbed to the olfactory sensitive area. The odorous molecules possessed shapes and sizes that were complementary to the shape and size of the olfactory receptors (Fig. 2.8). Two of Amoore's classes of odor, pungent and putrid, were exceptions, one requiring electrophilic and the other nucleophilic molecules (i.e., not shape and size fits). All other, complex, odorants were regarded as exhibiting molecular properties of two or more of the primary odors.

Moncrieff (1967a) felt that: "Molecular shape and size certainly play a part, perhaps a main part in determining odor. But little confidence can be felt in a restriction of odors to seven classes, and particularly to the seven suggested classes."

Inasmuch as evidence points to the fact that certain perceived odors might be considered to be a composite of two or more primary odor sensations, a dairy products judge should be alert to possible detection of individual components. While sense acuity by an individual may not be improved, *the power and value of concentration,* which is of primary importance in all tasting and smelling, can be materially improved.

Sensitivity of Taste and Smell. Much discussion has centered around a given individual's sensitivities of taste and smell and the effect of

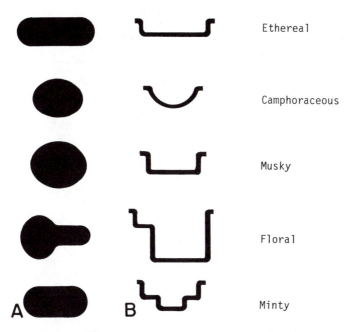

Ethereal

Camphoraceous

Musky

Floral

A   B   Minty

Fig. 2.8. Examples of Amoore's proposed receptor sites for odor perception (five of seven): A—as viewed from a flat plane. B—as viewed from an elevated profile. Adapted from Amoore, J. E., *The Molecular Basis of Odor* (1970). Courtesy of Charles C. Thomas, Publisher, Springfield, IL.

these sensitivities upon one's ability to evaluate dairy products. Studies show that threshold values (Table 2.4) of individuals vary (Moncrieff 1967a, Amerine *et al.* 1965) and that some individuals are taste-blind to certain substances (Blakeslee and Fox 1932, Richter and Clisby 1941). One investigator (Jacobs 1955) has emphasized that, surprisingly, few people make outstanding food product judges. It should be noted that other factors besides taste and smell sensitivity are involved in the development of capable dairy product judges. Judkins (1943) pointed out that errors in evaluating dairy product flavors are due to many considerations, chief of which are:

1. The individual lacks the ability to detect certain flavors.
2. The individual has a preconceived and incorrect terminology of certain flavors.
3. There are variations in the individual's threshold sensitivity (i.e., the amount of flavor required for a given individual to detect it).
4. The individual may have a poor memory, lack interest, or seem to have a poor mental attitude toward the activity.
5. Counteracting forces—such as the mood or personal feelings,

Table 2.4. A Summary of Detection Thresholds for Different Solutions Used to Represent the Four Primary Tastes as Determined by Five Groups of Investigators

Primary Taste	Investigator Group	Substance Used	% Solution Detected
Sweet	A	"Sugar"	0.5
	B	"Sugar"	1.0
	C	Sucrose	0.685
	D	Cane sugar	0.68
	E	Cane sugar	0.5
Acid	A	Sulfuric acid	0.05
	B	Acetic acid	0.25
	C	Hydrochloric acid	0.009
	D	Tartaric acid	0.019
	E	Hydrochloric acid	0.0067
Salt	A	"Salt"	0.045
	B	"Salt"	0.2
	C	Sodium chloride	0.234
	D	Sodium chloride (c.p.)	0.18
	E	Sodium chloride	0.25
Bitter	A	Quinine	0.00025
	B	Quinine sulfate	0.012
	C	Quinine hydrochloride	0.0016
	D	Caffeine (U.S.P.)	0.009
	E	Quinine	0.00005

A—Bailey and Nichols (1888).   D—Crocker and Henderson (1927).
B—Fontana (1902).   E—Best and Taylor (1943).
C—Parker (1922).

health, or general physical condition of the individual at the time of evaluation.

Judkins (1943) believed that individuals could be grouped, according to their abilities for judging dairy products' flavors, into the following three groups:

1. Those who always find the product satisfactory.
2. Those who may find the product unsatisfactory, but cannot tell why.
3. Those who are able to identify satisfactory products, and simultaneously be able to detect unsatisfactory products, and state the reasons(s) why they may be of lower quality.

Hollingworth and Poffenberger (1917, cited by Nelson and Trout 1964) pointed out that the connoisseur is sensitive to minute differences in the flavor of wines, tobaccos, and other products, and acquires a skill which is quite incomprehensible to the inexperienced person. However, they attributed this skill or art to the development of a special sensory acuity. They stated: "It is in large measure a matter of perception rather than one of sensation—a knowledge of what signs to look for and how to interpret these signs—rather than an increased sensitiveness to stimuli."

The experiences of the authors, of both the current and earlier editions, have led them to believe (despite marked differences in taste and smell sensitivities among tasters) that a lack of sense acuity by an individual does not necessarily mean that a person will not be a good evaluator. A person endowed with normal senses of taste and smell may, with proper training and diligent effort, become a capable judge of dairy products. Perhaps the greatest handicap of the beginner is not the inability to taste and smell, but the belief, which is purely psychological, that the ability to taste and smell is lacking. As previously emphasized, the sensitivity of taste and smell, important as it is, is not the limiting factor in determining whether a person will become proficient in the evaluation of dairy products.

Taste-Blindness. Blakeslee and Fox (1932) demonstrated that approximately three of every ten people are taste-blind to the bitter-tasting chemical, phenyl-thiolcarbamide, (P.T.C.), and that lack of taste acuity for this substance is an inherited trait. P.T.C.-treated paper prepared by the American Genetics Association lends itself well to classroom demonstrations in tasting. This treated paper is generally useful for illustrating that taste-blindness does occur with some individuals within a group; it also serves to stimulate interest in tasting. The above-mentioned investigators also reported taste-blindness for other substances.

Moncrieff (1967a) tends to deny that actual taste-blindness exists, but instead prefers to refer to "taste deficiencies." He reviewed Snyder's work with p-ethoxyphenyl thiourea, who found that 68.5% of tested subjects could taste bitterness, but 31.5% could not. Snyder (1931) noted that taste deficiency for bitter tastes is inherited by offspring according to Mendel's laws. Some investigators of taste deficiency towards P.T.C. concluded that lack of sensitivity to the bitter sensation was due to differences in the taste threshold, and not a real taste deficiency as such. Experiments definitely showed that the inability was inherited, but the work also demonstrated that the ability to taste P.T.C. did not run parallel with general acuteness of taste.

Moncrieff (1967a) summarized much of the research that has dealt with taste deficiency (taste-blindness), and in general concluded that: (1) differences in taste perception are not due to variations in substance solubility, as several investigators have suggested, or the saliva pH of individuals; and (2) all compounds that contain a $-NH-CH=S$ functional group do not necessarily show a dual taste reaction. However, those substances that seem bitter to some persons are tasteless for other people.

Jacobs (1955) concluded that individuals may exhibit varied responses to taste stimuli of certain chemicals, hence, simple classification into taste groups is not feasible. He directed attention to reconsideration of the universally accepted four basic tastes.

Being "blind" to a certain taste or off-flavor should not cause undue concern for the novice evaluator, since other factors play important roles in judging dairy products. Experience reveals that most expert food or beverage judges possessed no special taste acuity as beginners. These expert judges became proficient in their respective fields by applying mental concentration and good sensory techniques.

THE ROLE OF SALIVA
IN SENSORY EVALUATION

Saliva plays such an important role in tasting foods and beverages that the taster should be familiar with its origin, properties, and functions. In order to cause a taste reaction, the sapid substance must be in aqueous solution, so that it can be carried to the taste buds. Not all dairy products, upon placement into the mouth, are in an appropriate state for conduction to the taste nerve center. The food substance must be masticated, solubilized, and diluted before a taste reaction occurs. Toward this end, saliva performs a primary function.

Saliva is secreted mainly by three paired sets of glands: (1) the parotid, located in the cheeks with a duct (the duct of Stensen), opening from the inner cheek opposite the second molar tooth; (2) the submaxillary, the duct of which opens from the floor of the mouth to one side of the connecting membrane binding with the underside of the tongue; and (3) the sublingual, which have several duct openings beside the just-mentioned membrane. In addition, some small glands scattered over the mucous membrane of the oral cavity secrete a mucoid fluid (Best and Taylor 1943).

Tasters have frequently noted that saliva varies in viscosity and amount. Experiments have shown that the saliva secreted by the sublingual glands is usually thick and mucous, that of the submaxillary is either thin and watery or thick and viscid, depending upon the stimulus, whereas that of the parotid is thin and watery. When a copious flow of saliva occurs during normal mastication and tasting, the saliva is usually of the serous type (consists mainly of secretion from the parotid glands). Best and Taylor (1943) listed seven functions of saliva, four of which are of interest to tasters. These are: (1) preparation of the food for swallowing by altering its consistency; (2) solvent action; (3) cleansing action; and (4) moistening and lubricating action.

Pavlov's famous experiments on conditioned (stimulus) responses provide an insight into the nature of the secretion of saliva. Natural secretion of saliva is brought about in two ways, either: (1) through stimulation of the nerves in the mouth by the presence of food or other substances (unconditioned reflex); or (2) by stimulation of sense organs other than taste (psychic or conditioned reflex). Salivary secretion occurs readily when dairy products are taken into the mouth for tasting; the amount and nature of the saliva depending upon the product being tested. In general, ingestion of fluids such as milk, which need no dilution for tasting, stimulates a mucous, viscid, salivary secretion. Conversely, ingestion of semidry solids, such as cheese, results in the secretion of a thick, viscid, lubricating, submaxillary saliva, and copious quantities of diluting saliva from the parotid. Manipulating the sample about the mouth stimulates the flow of saliva. In fact, chewing, smoking, the presence of salts, acids, flavored substances, dry foods, paraffin, inert substances, or the introduction of dental tools serves to stimulate flow of saliva. These responses are inherent and should be used to the best advantage by tasters.

By contrast, Pavlov showed that the salivary glands could be conditioned to respond through the sight, smell, or thought of food, or sound associated with the preparation or dispensing of food. A dairy products judge should look with eagerness and pleasant anticipation toward the evaluation of dairy products in order to induce a generous flow of saliva.

Just as certain stimuli cause a secretion of saliva, other conditions, mainly of an emotional nature, may cause a temporary suppression of salivary secretions. It is often a common experience that an outstanding student judge may not perform well in a contest when surrounded by unusual working conditions and/or subjected to the emotional stress of competition. Other factors must necessarily enter into causes of reduced judging ability, but suppression of salivary secretion should not be overlooked.

Human saliva contains about 99.5% water and 0.5% total solids. The solids are composed of salts and organic substances; there is also a modest amount of gases in saliva. The pH of saliva varies, but it is usually slightly acidic, ranging from 5.75 to 7.05. The saliva of 86% of all persons falls within a narrower pH range of 6.35 to 6.85.

Harrow and Sherwin (1935) reported that the pH of saliva varied in a rather definite manner through the day. The pH rose sharply just after meals, but fell quickly to a slightly lower point than occurred just prior to meals, and approached neutrality between meals. These factors may have a bearing upon what constitutes the optimum time of day for conducting flavor evaluation sessions.

REFERENCES

Altner, H. 1981a. Physiology of taste. In: *Fundamentals of Sensory Physiology*. R. F. Schmidt (Editor). Springer-Verlag. New York. 220–227.

Altner, H. 1981b. Physiology of olfaction. In: *Fundamentals of Sensory Physiology*. R. F. Schmidt (Editor). Springer-Verlag. New York. 228–243.

American Society for Testing and Materials. 1968. Basic Principles of Sensory Evaluation. ASTM Special Tech. Pub. No. 433. Philadelphia. PA. 105 pp.

Amerine, M. A., Pangborn, R. M., and Roessler, E. B. 1965. *Principles of Sensory Evaluation of Food*. Academic Press. New York. 602 pp.

Amoore, J. E., Johnston, J. W., Jr., and Rubin, M. 1964. The stereochemical theory of odor. *Sci. Amer. 210*(1):42–49.

Amoore, J. E. and Venstrom, D. 1966. Sensory analysis of odor qualities in terms of the stereochemical theory. *J Food Sci. 31*(1):118.

Amoore, J. E. 1970. *Molecular Basis of Odor*. Charles C. Thomas Publishing Co., Springfield, IL. 236 pp.

Amoore, J. E. 1982. Odor theory and odor classification. In: *Fragrance Chemistry. The Science of the Sense of Smell*. E. T. Theimer (Editor). Academic Press. New York. 28.

Best, C. H. and Taylor, M. B. 1943. *The Physiological Basis of Medical Practice*. Williams and Wilkins Co. Baltimore, MD. 35.

Blakeslee, A. F. and Fox, A. L. 1932. Our different taste worlds. *J. Hered. 23*(3):96–110.

Brown, E. L. and Deffenbacher, K. 1979. *Perception and the Senses*. Oxford University Press. New York. 57.

Coren, S., Porac, C., and Ward, L. M. 1978. *Sensation and Perception*. Academic Press, New York. 112.

Crocker, E. C. and Henderson, L. F. 1927. Analysis and classification of odors. *Amer. Perfumer and Essential Oil Rev. 22*:325.

Dravnieks, A. 1967. Theories of olfaction. In: *Chemistry and Physiology of Flavors*. H. W. Schultz, E. A. Day, and L. M. Libbey (Editors). AVI Publishing Co. Westport, CT. 94–118.

Dudel, J. 1981. General sensory physiology, psychophysics. In: *Fundamentals of Sensory Physiology*. R. F. Schmidt (Editor). Springer-Verlag. New York. 1–30.

Farbman, S. I. 1967. Structure of chemoreceptors. In: *Chemistry and Physiology of Flavors*. H. W. Schultz, E. A. Day, and L. M. Libbey (Editors). AVI Publishing Co. Westport, CT. 25–51.

Fazzalari, F. A. (Editor). 1978. Compilation of Odor and Taste Threshold Values Data. American Soc. Test. Mater. Publ. DS 48A. Philadelphia, PA.

Fontana, A. 1902. Ueber die Wickung des Eucain B auf die Geschmacksorgane. *Ztchr. Psyect. u. Physiol. Sinnesorgane 28*:253.

Harrow, B. and Sherwin, P. L. 1935. *A Textbook of Biochemistry*. W. B. Saunders Co. Philadelphia. 797 pp.

Henning, H. 1927. Psychological studies on the sense of taste. *Handbook Biol. Methods. 6,A*:627.

Hollingworth, H. L. and Poffenberger, A. T., Jr. 1917. *The Sense of Taste*. D. Appleton Co. New York. 49 pp.

Jacobs, M. B. 1955. Variability in taste response. *Amer. Perfumer 66*:48.

Judkins, H. F. 1943. The judging of dairy products. *Sealtest News 9*(9):4–5.

Kramer, R. W. 1959. Glossary of some terms used in the sensory (panel) evaluation of foods and beverages. *Food Technol. 13*:730–736.

Moncrieff, R. W. 1967a. *The Chemical Senses*. Chemical Rubber Co. Press. Cleveland, OH.

Moncrieff, R. W. 1967b. Introduction to the symposium. In: *Chemistry and Physiology of Flavors*. H. W. Schultz, E. A. Day, and L. M. Libbey (Editors). AVI Publishing Co. Westport, CT. 1–22.

Moncrieff, R. W. 1970. *Odours*. Wm. Heinemann Medical Books Ltd. London, U.K. 26–189.

Nelson, J. A. and Trout, G. M. 1964. *Judging Dairy Products*. AVI Publishing Co., Westport, CT. 463 pp.

Parker, G. H. 1922. Smell, taste, and allied senses in the vertebrates, Philadelphia: J. P. Lippincott Co., pp 1–192.

Richter, C. P. and Clisby, K. H. 1941. Taste blindness. *Proc. Soc. Exp. Biol. Med. 48*:684–687.

Schmidt, R. F. 1981. Somatovisceral sensibility. In: *Fundamentals of Sensory Physiology*. R. F. Schmidt (Editor). Springer-Verlag. New York. 81–125.

Snyder, L. H. 1931. Inherited taste deficiency. *Science 74*:151.

Zimmerman, M. 1981. Neurophysiology of sensory systems. In: *Fundamentals of Sensory Physiology*. R. F. Schmidt (Editor). Springer-Verlag. New York. 31–80.

3

Practical Aspects of Dairy Products Evaluation

An understanding of sensory evaluation procedures and mastery of appropriate techniques should enable a person to become a good judge of dairy products. Efficient use of time in evaluating a number of samples is important; it results in having more time for rechecking and rescoring samples if necessary. Hence, knowing that time is available if needed, the evaluator is in a better position to concentrate and focus thoughts without being rushed. The application of mental poise and calmness in evaluating dairy products will generally result in more accurate judgment and superior performance, which may be lacking when the evaluator is hurried and/or confused. To enable the prospective judge to most effectively use time and energy, to heighten concentration and to gain self-confidence in evaluating (grading) dairy products, the following 16 guidelines (Nelson and Trout 1964) have proven helpful:

1. *Be in Physical and Mental Condition for Sensory Evaluation.* These conditions imply a state of good health, physical comfort, and mental poise for the evaluator. Consumption of a heavy meal just prior to evaluating foods tends to negate appetite, to dull much of the enthusiasm for tasting, and to lessen taste sensitivity. After eating certain foods, a pronounced "flavor aftertaste" may occur and confuse the evaluator's ability to ascertain more delicate flavors. Consequently, if it can be avoided, one should never judge after having consumed a large meal or after having consumed foods that have an intense flavor, such as hot pepper, garlic, onions, and many herbs and spices.

Any room used for sensory evaluation activity should be clean, well ventilated, adequately lit, and tempered for the physical comfort of the evaluators (Fig. 3.1). Restroom facilities should be located nearby.

For some individuals, it is often helpful to rinse the mouth with plain water prior to any tasting exercise. Use of a mild-flavored chewing gum prior to judging can be beneficial to saliva flow, and thus prepare the tongue and palate for sapid (having a taste or flavor) substances. In addition to stimulating the flow of saliva, the act of chewing seems

Fig. 3.1.   Dairy products should be evaluated in a clean, well-lit and ventilated room that is free from unwanted odors.

to have a nerve-quieting effect, which often is beneficial to student judges who are about to participate in a judging competition, perhaps for the first time.

The use of tobacco products just prior to any sensory discrimination activity is both questionable and debatable. While it is commonly believed that use of tobacco is generally detrimental to development and refinement of the sense of taste and smell, it should be mentioned that some dairy products judges, noted for their judging skills, are inveterate smokers.

2. *Know the Score Card and/or the Ideal Sensory Characteristics for each Product.* The score card is probably the most important tool available to the dairy products' judge. The skilled judge should learn and retain the various items and categories of the appropriate dairy product score card, and be able to instantly recall the assigned numerical value of each item.

3. *Learn the Important Sensory Characteristics of each Product and the Range of Defect Intensities.* This implies a thorough study of the flavors and many flavor defects applicable to each product and the relative desirability or undesirability of each in terms of the assigned

numerical values. Thus, to evaluate a given product, one must know what to expect in the way of sensory characteristics for each product. This information may be partially attained through hands-on experience in processing or manufacturing the product, by carefully studying the chapters in this text that deal with each product, and/or by working with a coach or an experienced judge of the sensory properties of each product.

4. *Have the Samples Properly Tempered.* Since the flavor, body, and texture characteristics can best be determined when the products are neither too cold nor too warm, each sample should be tempered to the optimum for that product. Thus, ice cream is tempered to around $-15\,°C$ to $-12.2\,°C$ ($5\,°F$ to $10\,°F$), and butter, cheese, and milk are tempered to about $15.5\,°C$ ($60\,°F$). Each respective temperature is most conducive to proper sampling and the subsequent study of the various sensory qualities of each dairy product. If product samples are too cold, the taste buds may be temporarily anesthesized; consequently, some of the delicate, more elusive flavors may go undetected. By contrast, if products are too warm, an accurate assessment of certain sensory qualities is more difficult, especially body and texture features.

5. *Secure a Representative Portion of the Sample to be Evaluated.* The sample must be taken accurately to provide a representative portion of the product. If a trier (sampling device) is used in sampling butter or cheese, it should be twisted one-half turn only, and the sample quickly removed by applying a slight back-pressure at the handle as the trier is withdrawn. A sample thus obtained should exhibit a clean-cut (smooth) surface. Regardless of the type of sampling tool used, the portion to be examined should always be cut-out, if possible, rather than obtained by scraping, compressing, or twisting. To insure representative portions of the product, avoid taking a sample from near the edge of the product. The judge should not secure a surface portion, or select a trier plug which touches an opening from where a previous sample had been removed. In the instance of liquid products, such as milk, cream, or buttermilk, the evaluator should be certain the product is well mixed before sampling.

6. *Observe the Aroma Immediately after Obtaining the Sample.* This is an important judging habit to form early in one's sensory evaluation experience (Fig. 3.2). Some aromas become less intense and disappear, at least in part, when exposed to the atmosphere. Thus, the best time to smell a sample is when the freshly cut surface is first exposed. If the aroma is not immediately observed, its true intensity may never be noticed. Other qualities of the product may be examined after initially noting the smell, since they remain fairly constant over time.

The importance of evaluating the aroma of the sample immediately

Fig. 3.2. The evaluator should observe the aroma immediately after obtaining the sample.

upon its removal cannot be overemphasized. An observation of experienced dairy products judges and graders is that, invariably, the most consistent evaluators carefully and conscientiously observe the aroma prior to tasting the sample. The relative sensitivities and features of the organs of smell and taste have been reviewed in Chapter 2. Evaluators should always bear in mind that the nose is substantially more sensitive than the tongue. As little as a fractional part per billion (ppb) of an aromatic vapor may be detectable by the human nose. In order to be able to taste the same substance, the tongue may require a million times as much material.

7. *Introduce into the Mouth a Sufficiently Large Volume of the Sample for Tasting.* There is a tendency among beginners to evade this requirement of tasting. Sometimes attempts are made to pass sensory judgment on a product without adequately tasting the product. The sample should be sufficiently large that delicate flavors may be detected, and yet small enough to permit easy manipulation of the warmed sample in the mouth. The taster should be in no hurry to ex-

pectorate the sample; it should be rendered completely liquid and warmed to body temperature before expectoration. Each sample should be held approximately the same length of time in the mouth, regardless of the quality of the product. Tasted samples are rarely swallowed and then only on specific occasions. The evaluator should also take note of the mouthfeel of the sample.

8. *Fix the Proper "Quality Ideal" in Mind.* Recognition of the so-called "quality ideal" can best be achieved by working closely with a sample or samples that are recognized as having superior quality. If one is working with an experienced judge, the novice should carefully note those qualities of the samples which merit a high score. The beginner should repeatedly smell and taste high-quality samples until the flavor is definitely fixed in mind. Learn to recognize when and in what respect(s) a sample fails to compare favorably with the "ideal." Without attainment of this mental guide or standard, the novice judge has no available "yardstick" by which to measure various products. The earlier the so-called "ideal qualities" of a given product are grasped, the sooner the beginner will become proficient in the judging or grading of dairy products.

9. *Observe the Sequence of Flavors.* The evaluator should, for a given product sample, observe particularly the first tastes and odors which are sensed, and note whether they change or are constant. If the flavors gradually disappear, one should note what other(s), if any, take their place(s). The sensory reactions to specific flavor stimuli remain the same; they can be expected to result in the same sensations the next time the flavor stimuli may make contact with the sense organs. The novice judge should remember these sensations and correlate them as early as possible with the specific flavor and the assigned flavor descriptor. After expectorating, the observer should note the relative time elapse before the taste sensation disappears or, by contrast, whether it persists. Some of the key indicators of flavor qualities have just been discussed; the skilled dairy products judge never overlooks these sensory techniques.

10. *Rinse the Mouth Occasionally.* The mouth should be cleansed or reconditioned at intervals of tasting, especially after having examined an intense, off-flavored sample. This may be done satisfactorily by rinsing the mouth with clean warm water or warm saline solution. Some prefer conditioning the mouth by eating portions of sound firm fruit, such as an apple, a pear, or grapes. Rinsing the mouth with water or with saline solution reconditions the mouth satisfactorily after having tasted milk and ice cream. The use of salt water or fruit seems best suited for reconditioning the mouth after having tasted butter or cheese. While some products such as buttermilk, cultures, cultured

sour cream, or yogurt appear to be "conditioners" in themselves, it is advisable to rinse the mouth occasionally with clean warm water to aid in preparing the mouth to detect mild, elusive flavors.

11. *Practice Introspection.* Introspection is accomplished by closing the eyes and mind to the world about you and practicing concentration of thought, solely about the effort of tasting (Fig. 3.3). Each evaluator looks back into their own mind and makes mental notes and determinations of the various taste and smell sensations perceived in each product sample. In other words, the novice judge should initially concentrate on the sample being examined, to the exclusion of everything else. The evaluator should practice a fixed degree of concentration during the scoring of a sample until the process becomes a fixed habit of judging. Judges should relax briefly after having finished the scoring of each sample, since sustained concentration can be tiring. Unless accompanied by alternate periods of relaxation, sustained concentration may eventually undermine the required mental poise that is so necessary to good judging performance.

12. *Do Not Be Too Critical.* Novice judges must guard against the initial tendency of trying to find objectionable flavors which may not be present. Such an overly critical approach may contribute toward the improvement of one's imagination, but does little to increase a person's judging ability. It often leads to the unfortunate habit of suspecting the presence of undesirable flavors in every sample (which may or may

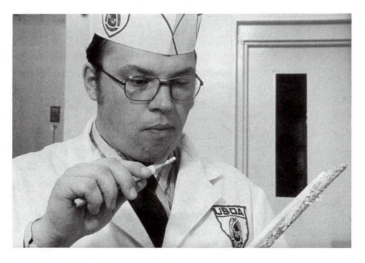

Fig. 3.3. A most important part of sensory evaluation is the practice of introspection. Courtesy USDA Grading Branch.

not be the case). However, the novice should be certain that the sample has been thoroughly examined to determine whether defects are present. Inexperienced judges are advised to give the sample the benefit of the doubt and above all, keep an open mind in judging.

13. *Once the Mind Is Made Up, Do Not Change It.* Vacillating judgment leads to guesswork. In fact, if a novice judge or contestant is guessing, experience has shown that the "first guess" is more likely to be correct than subsequent guesses.

14. *Check Your Own Scoring Occasionally.* This can best be done by closely comparing the flavors of two or more identical samples and observing whether the samples are scored consistently for flavor. A good check on scoring consistency can be made by rescoring several samples without knowing their identity. Rescoring a set of unknown samples identically with the first scoring can help develop self-confidence and mental poise.

15. *Be Honest with Yourself.* Make yourself use independent judgment; judge the sample itself. Do not be influenced by a name, a trademark on the package, or by the score previously assigned a similar product from a particular processor. The products made by some plants may have certain distinguishing characteristics which may reveal the commercial identity of the sample, despite all precautions to obscure sample identity. A judge should not attempt to identify the sources of samples available for evaluation. Concentrate on judging the samples at hand for the specific sensory characteristics they exhibit.

Student judges should particularly avoid trying to "judge the judge," and should score the product conscientiously, using their own independent judgment. Evaluators should keep a straight face—by so doing one avoids telegraphing observations intentionally or otherwise to another judge or contestant. A competent judge makes his or her own decisions, and after arriving at a conclusion, one must believe in their own judgment until shown or convinced otherwise.

16. *Recognize the Fact that Practice and Experience are Essential to the Development of Judging Ability.* Be assured that one must practice judging in order to develop the ability to taste, smell, and distinguish the delicate, often elusive flavors of dairy products. One should not become discouraged too easily. It is important to believe that each of us have adequate senses of taste and smell for judging dairy products. Usually, all that is needed to reveal skills of tasting and smelling is training and practice. Concentration, perseverance, and continued practice can yield astonishing results in terms of judging performance.

SCORE CARDS AND THEIR USE

Description. A score card is a tabulated list of the various factors used to assess the quality of a product with a numerical value assigned to each factor (Bodyfelt 1981).These factors are usually arranged on the score card either alphabetically or in the order of their relative importance. The score card also provides a numerical standard by which the quality of various dairy products can be measured.

In some instances of evaluating certain dairy products, use of laboratory equipment and facilities may be necessary or useful in determining a rating for some of the factors. However, this may prevent the complete evaluation and scoring of the product within a single time period. Consequently, in scoring such products at dairy products exhibits, clinics, or in student judging contests (where laboratory equipment is not available or when results are needed immediately), only those factors that can be readily determined are considered. A score card that lists the attainable factors (with their value) is known as an "abridged" or student score card. These abbreviated score cards are useful in comparing the sensory quality of different dairy products; they also provide a rapid set of results following sensory observation.

The educational and enforcement value of score cards in allied branches of the dairy industry, such as dairy farm inspection sheets and evaluation forms used for dairy plant inspection and laboratory certification, have long been appreciated.

Purpose. The score card furnishes a definite systematic means of arriving at a concise value of the product being evaluated. It is divided into component parts; each part has a numerical value. The score card reflects to some extent the comparative importance of the different terms that should be taken into consideration. The use of a score card enables the novice judge to establish better judging habits by facilitating a definite, orderly routine. This requires less time and effort to accomplish a desired result. Score cards that contain an orderly list (usually alphabetical) of criticisms under each item are very helpful in pointing out the possible defects that may be found in the product (Rainey and Jones 1985; Sjostrom 1969).

The score card in itself is usually of little value to the experienced judge as an aid in actual scoring. The capable evaluator has developed and mastered proper habits and perfected a procedure for the mechanical evaluation of the various quality factors. However, the score card conveniently serves the evaluator with a means of keeping an accurate, detailed record of the different samples examined. This also enables the judge to evaluate a greater number of samples at one time. Thus, it is evident that the use of a score card has many advantages, namely: (1) it is an educational tool; (2) it encourages formation of correct judging habits; (3) it permits a minimization of mechanical errors; (4) it is time saving; (5) it furnishes a permanent record; and (6) it serves as a guide for quality improvement.

Score Card Contents. With the advancement of judging techniques for dairy products, score cards have approached greater uniformity in design. The name of the product usually appears near the top center of the card. To one side there may be a blank space for the number or name of the contestant or judge. Below this, listed down the left side, usually in the order of decreasing numerical importance, are the different quality factors to be considered in scoring the product. Directly opposite the quality factors are placed the relative point values of each item. The sum of these values (often 25 points or 100 points total) is equivalent to a perfect score. Adjacent to the column for numerical values is a blank column in which the actual numerical evaluation is to be written. The score card is arranged so that listed defects may be marked as well as providing blank lines for the judge to write in comments. The use of different-colored score cards for each respective dairy product materially aids in sorting cards when several different products are scored in a judging contest.

Score Card Used. There are available score cards for nearly all the principal dairy products. Obviously, one standard score card can not suffice for satisfactory scoring all of the various dairy products, since each product differs materially in its characteristics from others. An illustration of each dairy product score card can be found in the chapter dealing with the particular product or in the appropriate appendix section of this book.

Some differences are often noted in the form or arrangement of the score card, depending on its use. However, the basic points of consideration and their relative importance remain unaltered. Score cards presented in this text are, in the main, the ones formulated and approved by the American Dairy Science Association and in general are used throughout the United States by the dairy industry. They are frequently used in class work in food science curriculums and in regional and national Collegiate Dairy Products Evaluation Contests.

Use of the Score Card as a Record. Since the score card is to be retained as a permanent record of the evaluation of the product, erasures, strike-outs, and untidy score cards should be minimized or eliminated by thinking through the process of judging and arriving at a definite conclusion before recording scores and criticisms (Fig. 3.4). Erasures and strike-outs reveal uncertainty and lack of confidence on the part of the judge. Unfortunately, the person addicted to changing decisions

CONTEST
ICE CREAM SCORE CARD

DATE _____ A.D.S.A.   CONTESTANT NO. _____

PERFECT SCORE	CRITICISMS		SAMPLE NO.										TOTAL GRADES
			1	2	3	4	5	6	7	8	9	10	
FLAVOR	CONTESTANT SCORE	▶	7	10	5	6	8	9	4	7			
	GRADE — SCORE / CRITICISM												
NO CRITICISM 10	COOKED		✓							✓			
	LACKS FLAVORING						✓						
	TOO HIGH FLAVOR												
NORMAL RANGE 1–10	UNNATURAL FLAVOR					✓							
	HIGH ACID												
	LACKS FINE FLAVOR						✓						
	LACKS FRESHNESS												
	METALLIC												
	OLD INGREDIENT					✓							
	OXIDIZED							✓					
	RANCID												
	SALTY								✓				
	STORAGE												
	LACKS SWEETNESS												
	SYRUP FLAVOR		✓										
	TOO SWEET												
	WHEY			✓									
BODY AND TEXTURE	CONTESTANT SCORE	▶	3	5	2	4	5	5	4	3			
	GRADE — SCORE / CRITICISM												
NO CRITICISM 5	COARSE/ICY		✓		✓	✓			✓	✓			
	CRUMBLY												
	FLUFFY												
NORMAL RANGE 1–5	GUMMY		✓										
	SANDY												
	SOGGY												
	WEAK				✓					✓			
COLOR 5	ALLOWED PERFECT IN CONTEST	▶	X	X	X	X	X	X	X	X	X	X	
MELTING QUALITY 5	ALLOWED PERFECT IN CONTEST		X	X	X	X	X	X	X	X	X	X	
BACTERIA 15	ALLOWED PERFECT IN CONTEST		X	X	X	X	X	X	X	X	X	X	
TOTAL	TOTAL SCORE OF EACH SAMPLE	▶											
	TOTAL GRADE PER SAMPLE												

Fig. 3.4.   A properly completed score card should reflect decisiveness and neatness—there should be few or no erasures.

indiscriminately in scoring dairy products never enjoys the inward satisfaction of believing that his or her judgment is correct. Such practice may tend to lead to an inferiority complex, which in itself handicaps the beginner from concentration of thought. Concentration is a necessary faculty in the correct identification of elusive, delicate flavors and aromas so often present in evaluated samples.

Denoting the Score. In denoting the score for a product sample, the general practice is to write in the points allowed or assigned by the judge, rather than designating point deductions. This method is both convenient and direct. Mental calculations should be undertaken and the final results written on the score card. By following this practice, associations are easier formed between the flavor and appropriate flavor score.

Writing in Criticisms. When the score is above a certain point, the practice is to not criticize the sample. The training and experience which comes from deciding definitely on the relative merits of an item and writing it down is a valuable experience for the novice judge. The evaluator who evades a definite decision and specifically avoids designating a characteristic flavor (whether the flavor is excellent, fair, or poor) is likely to slowly develop judging proficiency. Checking appropriate criticisms is most important in effective use of the score card.

General Methods for Using the Score Card. Details concerning the use of score cards for the various products are discussed in the respective chapters that discuss their sensory evaluation. However, general statements and recommendations regarding the use of score cards may be made. Of note, dairy products are seldom given a perfect or the highest attainable score by experienced judges.

In clinics, exhibitions, and fairs, the smallest deduction generally made on any one quality factor is one-half point. In the case of numerous high-scoring samples entered in keen competition, slight or subtle differences in flavor scores are sometimes differentiated by one-quarter or one-tenth points. The maximum deduction depends upon the quality factor considered, which varies with different products. For instance, in butter, the factors which deal with body and texture, color, package, and salt content are generally given a full score. Over the decades, improved processing and workmanship of dairy products manufacture have accounted for diminishing the occurrence of most of these defects. Furthermore, some tolerance must be made for regional characteristics and preferences when some dairy products are evaluated.

Use of the Official Score. To get the most value from the use of score cards, the beginner should keep in mind that experience has always been and will continue to be a great teacher. The judge should definitely decide on the appropriate critique for flavor, body, and texture, as well as the color and appearance of the samples. After determining appropriate scores for the several samples judged, the beginner should intelligently compare one's scoring results with those of the experienced judge. If the judge provides the "official" results orally, the beginner should record these findings on his or her own score card and later, when the opportunity presents itself, reexamine and rescore the samples and try to recognize the presence of the off-flavor(s) reported by the judge. Novice judges should work with any flavor defects long enough to obtain a lasting impression of them. Furthermore, one should associate any given off-flavor with the score assigned to that item. The beginner should bear in mind that an assigned score for a particular off-flavor will often vary somewhat from that noted by the official judge. This is primarily due to differences in perceived intensity by different individuals.

Satisfactory Scoring Performances. Beginning judges will observe that even the best judges will occasionally disagree among themselves as to what is the most appropriate final score for assignment to a given product. However, this should be no cause for alarm or discouragement about the scoring process. The fact remains that expert judges can independently score sample after sample, and either assign identical scores or remarkably similar scores to each given sample. Expert judges demonstrate perfect or near-perfect agreement on flavor criticisms of samples that are scored independently. We should expect some variation in assigned scores for a set of dairy product samples, since mental processes, perception abilities, keenness of smell and taste, and the ease or ability of perceiving certain flavors differs from person to person (Tobias 1976). It is indeed coincidental when contestants or participants assign identical scores for a majority of the samples, when compared to the experienced or official judge(s) of a contest or clinic. For individual samples, scores and the noted criticisms by participants or contestants may be identical with those of the official judge(s).

When a set of eight milk or butter samples is scored by a novice judge, an excellent "scoring performance" is attained if the differences between the score card sums of the contestant and the official judge does not exceed an average of 1.5 points per sample. For Cheddar cheese and ice cream judging, an average between contestant and "official scores" of 3 points per sample is considered an outstanding judging performance by the contestant. In scoring cottage cheese and yogurt, an outstanding scoring performance is achieved when the contestant scores at an average of 4 points lost per sample (for eight samples). Perhaps to the beginner, such scoring accuracy may seem impossible; as a person develops judging proficiency, close agreement between judges on denoted criticisms and scoring becomes more commonplace. See Appendix XVII for details on grading score cards.

Dairy Products Grading Outlines. Special forms are often employed on which to record observations in the routine examination of dairy products for sensory quality. These "quality report cards" differ according to the product examined and the policies of the processing plant in which they are used. In general, these outlines may have provisions for recording information on the various quality factors which appear on common score cards, but may have no provisions for recording numerical scores. In addition, space is often provided for recording additional information on quality which is not usually called for on the score card (Tobias 1976). For example, in examination of bulky (fruit, nuts, and/or candy incorporated) flavored ice creams, space may be provided for recording net weight of the sample, distribution of the fruit, nuts, or candy, and the package condition. The items listed on the given product evaluation (grading) outlines are usually those features about which the consumer might be concerned in terms of purchasing a safe, functional, and high-quality dairy product. Ample space is provided on such cards for recording the necessary or appropriate consumer acceptance parameters. Grading outlines and modified score cards are beneficial tools for quality assurance activities related to dairy products (Bodyfelt 1981).

CONTESTS AND CLINICS

The nature of dairy foods makes evaluation contests and clinics imperative, if a high quality level is to be maintained. The sensory characteristics of dairy products, not unlike other perishable foods, gradually but persistently change with age. Physical, chemical, and microbiological factors (or a combination of these factors) contribute to these changes. The rapidity with which these changes take place depends upon many conditions, such as storage temperature, exposure to light and air (oxygen), contact with certain container materials, chemical reactions, microorganisms and enzymes present, the initial quality of the raw materials used, and the skill and control exercised in processing. Processors and distributors recognize that dairy products must reach consumers in good condition and exhibit the intended desired qualities, if these products are to enjoy consumer acceptance.

Unfortunately, dairy processors often lack information on consumer reaction or impressions of their products (O'Mahony 1979). In order to assess and compare the sensory qualities of dairy products of competitive processors within a given market, quality assurance personnel may periodically conduct product clinics or flavor comparisons (sometimes referred to as "cuttings"). These dairy product evaluation sessions may be conducted in several different ways, depending on the desired objectives. Sometimes, state land grant universities, extension services, and/or state, regional, or national trade associations conduct product clinics for the educational or training benefit of industry personnel.

Objectives of Contests and Clinics

1. *Studying Competitive Samples.* A processor may desire to compare dairy products with other similar competitive dairy products sold within the marketing area. In this instance, the processor would purchase similar sample(s) of different brands of dairy products sold in the market area for the purpose of comparing them. The main objective of this form of clinic is to directly compare sensory characteristics and to study the relative qualities of competitive brands that may be enjoying good consumer acceptance.

2. *Trade Association Clinics.* A group of dairy processors may find it educationally beneficial to meet and compare the sensory qualities of their products. In this instance, a competent judge (or several judges) generally takes charge; the products are carefully evaluated and submitted samples are constructively criticized for the group of participants. To avoid brand bias or embarrassment, the product evaluation is best conducted in such a way that no participants (other than a clerk or coordinator) knows the identity of any of the samples until the evaluation is completed. After the sensory evaluation has been completed and the merits (or demerits) of each sample discussed, the official(s) may confide to each processor representative the brand identity of respective samples. Conduct of a product clinic in such an open manner permits a meaningful and unbiased comparison of product qualities by processor personnel.

3. *Consumer Demonstrations.* In order to acquaint consumers with the sensory properties of dairy products, consumers' educational clinics may be held. A group of consumers may be assembled and the desirable sensory properties and some of the more common defects of dairy products explained or demonstrated to them. A number of samples may be evaluated by sensory procedures and the quality factors pointed out to them by an experienced judge or instructor. The judge can also instruct consumers in what to observe and what to avoid in purchasing dairy products for home use.

4. *Training College Students.* Dairy products evaluation contests for students have proved most helpful in training future dairy industry personnel for many industry positions: quality assurance, product development, dairy product graders, production supervision and management. Graduates of food science or dairy technology who have had training in sensory evaluation are better prepared for numerous responsibilities. They should understand more clearly the problems involved in improving the quality of dairy products and have a better perception of quality standards for the different grades of dairy products.

5. *Youth Training and Contests.* Appropriately supervised contests for Future Farmers of American and/or 4-H Club members can do much to stimulate enthusiasm and interest in the improvement of dairy products. Contests that emphasize dairy foods' quality are particularly adaptable to training programs in vocational agriculture. U.S.D.A. Farmers Bulletin, No. 2259, "Judging and Scoring Milk and Cheese" is directed toward the Future Farmers of America National Contest; this pamphlet contains excellent information on the evaluation of dairy foods for high school students.

Evaluation Activities

Kinds of Contests or Competitions. Basically, there are two kinds of milk product-quality contests, in terms of how the samples are derived, which are termed "prepared" and "surprise" sampling. A contest in which the samples are specially selected and entered by the exhibitor for sensory evaluation is known as a "prepared" contest. A competition in which the participant organization does not know when the sample(s) are to be collected and entered into the competition (along with samples from competitors) is known as a surprise or "pick-up" contest.

A prepared contest has the limitation of only measuring the ability of the processor (or industry personnel) to select milk ingredients of good flavor quality and subsequently process and package a final product in such a manner as to merit a high score on the rating system employed. Unfortunately, this "submitted" contest sample may not be representative of the plant's daily output, and hence atypical of the quality of dairy product ordinarily presented to consumers. However, such a contest may educate or impress upon the industry worker the specific and exacting requirements for the production of high-quality products.

The surprise sampling form of contest serves to more appropriately measure the quality of the product which consumers receive on a day-to-day basis from the processor. The surprise contest is generally recommended over the prepared contest; it tends to provide more meaningful information and feedback for both the processor and consumers.

Collection and Preparation of Samples. Dairy product samples for surprise contests may be collected from warehouses, retail stores, or processing plants. The collected samples should be refrigerated ($\leq 4.2\,°C$ [$\leq 40\,°F$]) until evaluated. Each sample should be properly numbered or coded. When possible, brand identifications on the package should be obscured. The product may be transferred to another appropriate container so that no one, other than the contest superintendent or coordinator, knows the identity of the samples. After the samples have been arranged in order and the identification of the samples recorded by the superintendent, the products are ready for sensory examination.

An alternative method of handling milk samples is to decant the contents of the original containers into correspondingly numbered amber or clear glass bottles (if available). Not only will each milk sample appear identical, but milk in the amber containers will be protected during evaluation against light-activated off-flavor.

A contest provides the most educational value when the sensory properties of each sample are determined, compared, and discussed by an experienced product judge (or judges) at a meeting of the participants. This can be done without embarrassment if sample identities are not divulged, other than to a processor representative relative to their own entry. Each entry should be evaluated not only by the judge, but by all attendees of the session. After each sample has been evaluated, the sensory characteristics should be discussed, and possibly compared with some of the samples previously scored. If a flavor defect is present, its relative intensity, seriousness, and steps for possible prevention or improvement of the product should be considered. The lead judge designates the flavor criticism(s) and justifies or explains the assigned numerical score (see Chapter 5 for details).

Comparing the Scores. After each sample has been evaluated and the merits or demerits of various flavor characteristics have been discussed, the sensory evaluation remarks should be displayed on a score card summary, chart, or blackboard (by number), as indicated in Fig. 3.5. Display of sample criticisms and scores enables each clinic participant to evaluate their sample, in comparison with other samples. A discussion by the judge of the defects encountered and how to overcome them can prove helpful in improving product quality. In addition, the judge can give confidential advice to each processor on how to apply remedial steps.

Cheese Clinics and Contests

Cheese clinics and competition in which the various qualities of different cheese are critically examined have proven to be especially valuable in educating cheesemakers and improving cheese quality.

Due to the microbiological, chemical, and physical changes that occur in cheese when it is aged, and possible mold contamination which may result when it is stored, cheese clinics may be conducted differently from those of other dairy products. Since some cheese varieties are consumed as a fresh product, and others are consumed only after

ODI CONTEST

Date 10/2 FIRM SUNSHINE DAIRY 2% LOWFAT MILK SCORE CARD   Contestant No. 10

Criticisms		1	2	3	4	5	6	7	8	9	10	11	12	13	14	15
Flavor 10	Score	9	8	10	7	7	8	8	5	8	9	8	6			
	Astringent															
	Bitter															
No Criticism 10	Cooked	✓														
	Feed		✓				✓	✓		✓	✓	✓	✓			
	Flat															
	Foreign															
	Garlic/Onion															
Normal Range 1-10	High Acid															
	Lacks Freshness				✓	✓						✓	✓			
	Malty															
	Metallic								✓							
	Oxidized															
	Rancid															
	Unclean															

Criticisms		16	17	18	19	20	21	22	23	24	25	26	27	28	29	30
Flavor 10	Score															
	Astringent															
	Bitter															
No Criticism 10	Cooked															
	Feed															
	Flat															
	Foreign															
	Garlic/Onion															
Normal Range 1-10	High Acid															
	Lacks Freshness															
	Malty															
	Metallic															
	Oxidized															
	Rancid															
	Unclean															

Judges:

Fig. 3.5. An example of a dairy products clinic or contest form for summarizing results.

aging, the mechanics of conducting cheese clinics and contests vary widely according to the type of cheese.

Sometimes cheese is graded and branded or labeled when it is young. Some Cheddar cheese is quickly graded and sold on the basis of grade soon after manufacture. In order to properly evaluate cheese quality, the competent judge must be able to determine the quality of young cheese as well as that of the aged product. The judge must be able to evaluate cheese at various stages of development and determine or project the "quality potential" of the cheese after it has been aged for 3 to 18 months under prescribed conditions.

Some experience in cheesemaking is beneficial to becoming a cheese judge. By actually working with cheese, the judge learns the importance of the various steps in the manufacturing process and their impact on the finished cheese if these steps are not properly executed. An expert judge who has had some manufacturing experience can also provide effective advice to cheesemakers subsequent to judging and discussing the quality of cheese samples.

Size and Arrangement of Samples. The samples to be evaluated should be sufficiently large enough that the flavor will not be influenced by the surface area. For soft cheese which is generally consumed while fresh, a pint (455 g) sample will usually suffice. For cheese which is to be aged, an entire cheese, regardless of style or weight, should be secured for a sample. The samples should be numbered and placed in order as a class or group. All types of packaged cheese may be marked with a felt-tip ink marker, either by use of self-adhesive labels or directly on the package. Any method used for marking should not disrupt the outside covering of the cheese.

Examining Samples. Samples should be examined according to the methods for each kind of cheese, as described in Chapter 8. The report for each evaluated sample will depend on the use to be made of the obtained data. If cheese is evaluated to enlighten cheesemakers about cheese quality, the findings on each sample should be reported on score cards (illustrated in Chapter 8 and/or Appendix XII). After the sensory evaluation is completed, cheesemakers should be given the identity of their respective sample(s) so that comparisons of one's own product with others can be made. For the information of those participating in a clinic, a summary of the scores for all samples should be tabulated, in order of the quality ranking, on a blackboard or chart. After scoring, the cheesemakers in attendance will often desire information on probable causes of the noted defects and how to overcome or control them. An experienced cheese judge can do much to improve the future quality of cheese by explaining why different defects occur and how they might be prevented or remedied.

Ice Cream Clinics and Contests

Frequently, ice cream manufacturers wish to have their ice creams critically evaluated to determine how they compare with competitors' samples or measure up to standards of perfection. For this purpose, ice cream clinics and contests are usually held in connection with a dairy technology conference, where ice cream manufacturers may gather and compare numbered samples of ice cream. The results of this critical evaluation provide a guide for improvement or maintenance of product quality. Since the samples are numbered or handled in a manner to eliminate the possibility of brand identification, manufacturers usually welcome the opportunity to critically examine samples and study the quality information generated. This information, along with the manufacturer's identifying number (or code) indicated, can be sent to each manufacturer soon after the clinic. Usually, this information is summarized on an ice cream score card.

Clinics and Contests for Other Dairy Products

Conduction of clinics on dairy products other than milk, cheese, and ice cream can be highly educational for manufacturing personnel, distributors, sales personnel, suppliers, and others who work with these products. This is especially true of some of the more perishable dairy products which should be under constant surveillance when held in the plant and the market place awaiting sale to the consumer. Clinics similar to those previously described can be held on any dairy product, with the primary objectives being product quality improvement and enhanced consumer acceptance. Dairy products that are often evaluated within clinic or contest formats include butter, cottage cheese, yogurt, kefir, buttermilk, and sour cream.

Collegiate Dairy Products Evaluation Contests

An annual Collegiate Dairy Products Evaluation Contest, which entails evaluation of six different dairy products, is cosponsored annually by the American Dairy Science Association (ASDA), the Dairy and Food Industries Supply Association, Inc. (DFISA), and the United States Department of Agriculture (USDA). This international contest for U.S. and Canadian college students is featured at the time and place of DFISA's Dairy and Food Exposition (Expo), and in "off-show" years the contest is held in the city where the annual meetings of the Milk Industry Foundation and International Association of Ice Cream Manufacturers (both headquartered in Washington, D.C.) are conducted. Regional contests are usually held prior to the national contest, especially in the midwest and southern regions of the U.S. Contests on dairy foods' quality are also held for high school students who participate in vocational agricultural programs (Future Farmers of America—FFA) at state fairs or land grant universities. Winning teams of state contests are eligible to compete in the Annual National FFA Dairy Foods Quality Contest each fall.

Dairy products' evaluation contests are primarily sponsored to benefit interested students and help them become more proficient in the sensory evaluation of these products (Trout and Weigold 1981). The contests offer students an opportunity to put their training to a test and furnish an incentive for more intensive study. Students who are well-trained and skilled in product judging are in a position to be of greater service to the dairy industry. Persons skilled in sensory analysis are more ably prepared to help improve the quality of dairy products through the detection of flavor defects which often meet with consumers' disapproval. Contests furnish students with an opportunity to:

1. Assess one's training, knowledge, and experience in the evaluation of dairy products.
2. Observe the quality of dairy products in other sections of the country, and to note the standards of quality recognized by competent judges.
3. Put to a test a person's ability to arrive at conclusions within a definite time period and within an unfamiliar environment.
4. Meet similarly trained students from other schools in a challenging competition and in a friendly environment; and to gain the educational benefits of travel.
5. Obtain a beyond-the-classroom vision of the size and scope of the dairy and food industry; meet leaders and future leaders of the industry; and gain knowledge and inspiration by viewing the Dairy and Food Expo.
6. Kindle the desire to lead in a chosen field.

Organization of the Collegiate Contest. The Collegiate Contest is organized and supervised under the direction of a superintendent, (usually a representative of the U.S.D.A., Food Quality and Standards Branch, Washington, D.C.). The superintendent is guided by rules established by the Committee on Sensory Evaluation of Dairy Products of the American Dairy Science Association. The superintendent is responsible for the selection and arrangement of dairy products samples, the conduct of the students while they are judging, the correction of score cards, and tabulation of the final scores. When the judging contest is under way, the coaches of the teams are not allowed to converse with members of the team or enter the judging rooms until the contest is concluded. Codes are used to identify contestants' score cards so that card graders do not know the identity of any of the contestants. In 1983, computer tabulation of results was introduced for the purpose of calculating and summarizing contest results. In the near future, it is anticipated that electronic scanning or computer scoring and tabulation will be applied for grading score cards, thus eliminating the tedious and time-consuming manual approach of doing this operation.

Rules for the Contest. Rules for the Collegiate Dairy Products Evaluation Contest are revised from time to time. The current rules are the result of many years of trial and revision. Score cards for butter, Cheddar cheese, milk, ice cream, cottage cheese, and yogurt are constantly reviewed and modified as industry practices or changes in sensory characteristics occur. The A.D.S.A. Committee on Sensory Evaluation of Dairy Products (Champaign, IL) is responsible for approving these alterations of the score cards.

Current rules and other information for this annual contest may be obtained from the Contest Secretariat, Dairy and Food Industries Supply Association, Inc., Rockville, Md., or from the Contest Superintendent, U.S.D.A., Food Quality and Standards Branch, Washington, D.C. Contestant score cards may be obtained from the Executive Director of the American Dairy Science Association, Champaign, IL.

Selection, Judging, and Display of Products. The products judged in the annual Collegiate Contest are butter, Cheddar cheese, milk (Fig. 3.6), vanilla ice cream, cottage cheese, and strawberry (Swiss-style) yogurt. Samples are selected by official judges under the direction of the Superintendent of the Contest. An effort is made to obtain products that vary in quality, so that contestants encounter a range of defects.

The products, which have been previously evaluated and scored by official judges, are generally checked by a committee of coaches just before the beginning of the contest. All samples are marked, wrapped, or repacked to obscure the processors' identity. Thus, the student judge must consider each sample entirely on its sensory properties, and not on its identity or source.

Official Judges. Reliable "official judging" is the underlying strength of any successful contest. To this end, experienced dairy products' judges from industry are selected as official judges by the Superintendent of the contest. Judges are appointed for each product sufficiently early to enable them to "be on the lookout for," and to set aside samples which, in their estimation, would make a good sample for the subsequent contest.

Fig. 3.6. Retasting samples of butter after completion of a dairy products evaluation contest.

Fig. 3.7. The climax of the annual National Collegiate Dairy Products Evaluation Contest is the presentation of awards to the individual and team winners.

In addition to the official judges, two or more coach judges are designated by the committee chairman to judge the samples and ascertain if the official judgments are in accordance with the rules and scoring guidelines. New or less experienced coaches of college teams are appointed as "coach observers" in the official scoring process.

Awards. In order to stimulate interest in the contest, prizes are generally awarded to the three highest-ranking contestants and the three highest-ranking teams in evaluating all products, and for each of the six product categories. These awards are presented at an awards breakfast (Fig. 3.7) attended by hundreds of dairy and food industry personnel. The object of the awards is to recognize merit, arouse enthusiasm for quality products, and encourage students to pursue careers in the dairy and/or food industry.

FFA and Other Youth Contests. Many agricultural contests are featured as part of agricultural education programs in U.S. high schools. Dairy foods' quality contests contribute much to these programs. Teams trained at the local level may compete later at the state, regional, and national levels in vocational agriculture (FFA).

REFERENCES

Bodyfelt, F. W. 1981. Dairy product score cards: are they consistent with principles of sensory evaluation? *J. Dairy Sci. 64*:2303.

Nelson, J. and Trout, G. M. 1964. *Judging Dairy Products.* AVI Publishing Co. Westport, CT. 463 pp.

O'Mahony, M. 1979. Psychophysical aspects of sensory analysis of dairy products: a critique. *J. Dairy Sci. 62*:(12): 1954.

Rainey, B. and Jones, F. E. (Cochairmen). 1985. A symposium on in-plant sensory evaluation. *Food Tech. 39*(11):124.

Sjostrom, L. B. 1969. Flavor evaluation and numerical scoring. *Laboratory Section, Proc. Milk Ind. Found.* Washington, D.C. pp. 86–90.

Tobias, J. 1976. Organoleptic properties of dairy products. *In: Dairy Technology and Engineering.* W. J. Harper and C. W. Hall (Editors). AVI Publishing, Inc. Westport, CT. 75–140.

Trout, G. M. and Weigold, G. 1981. Building Careers in the Dairy Products Evaluation Contest (60 Years of Student Judging, 1916–1981). Dairy Food Ind. Sup. Ass'n. Washington, D.C. 16 pp.

4

Sensory Defects
of Dairy Products: An Overview

Flavor defects of milk and milk products may be divided into two basic categories: those already present when milk is harvested from the cow and those that develop later as a consequence of specific circumstances. The Committee on Flavor Nomenclature and Reference Standards of the American Dairy Science Association (Shipe *et al.* 1978) proposed seven categories of off-flavors in milk (Table 4.1): heated, light-induced, lipolyzed, microbial, oxidized, transmitted, and miscellaneous.

To the dairy or food technologist, a flavor defect represents a symptom which helps identify the cause of an existing or potential problem. Much of the published research on milk and dairy products has been focused on identifying various factors, conditions, and mechanisms that lead to off-flavor development, and with finding methods to prevent their occurrence (Badings and Neeter 1980; Badings 1984; Bradfield 1962; Forss 1969, 1979).

As is true for other experiences of the human senses, such as music, flavors are often difficult to describe by words. Usually, one has to experience a given flavor sensation to fully appreciate it. Even then, there needs to be a consensus reached by the persons involved as to the most appropriate verbal description. When available, reference standards of products (Shipe *et al.* 1978; Hammond and Seals 1972) that possess specific off-favors help immensely in promoting uniformity and a common understanding of flavor language. In practice, some off-flavors are labeled by their descriptive or associative term (e.g., acid), while the name for others, such as oxidized, is a generic term that relates to the defect cause. Table 4.1 lists a number of terms for describing milk off-flavors. Additional terms for flavor defects will appear later in this and subsequent chapters to deal with specific sensory problems of various dairy products.

Food flavors, as well as off-flavors, are primarily due to aromatic properties, although some occur as true tastes, that consist of one or more of the primary tastes of sweet, sour, salty, and bitter (Gould 1966;

Table 4.1. The Major Categories of Off-Flavors in Milk.

Causes	Descriptive or Associative Terms
Heated	cooked, caramelized, scorched
Light-induced	light, sunlight, activated
Lipolyzed	rancid, butyric, bitter, goaty
Microbial	acid, bitter, fruity, malty, putrid, unclean
Oxidized	papery, cardboardy, metallic, oily, fishy
Transmitted	feed, weedy, cowy, barny
Miscellaneous	adsorbed, astringent, bitter[a], chalky, chemical, flat, foreign, lacks freshness, salty

Source: Shipe et al. (1978).

[a]A bitter flavor may arise from a number of different causes. If the specific cause is not known, it should be classified under miscellaneous.

Crocker 1945; Moncrieff 1967). Odors are produced by volatile constituents that are capable of imparting characteristic odor sensations, often at extremely low concentrations. Typically, the concentration of flavor and off-flavor compounds is one or more parts per million (ppm) and in some cases, parts per billion (ppb). Even at this low level, an increase in the concentration of an odorant may change it from a desirable flavor constituent to an undesirable one. The volatile constituents of flavors or off-flavors are primarily molecules from several classes of organic chemicals, including carbonyls (i.e., aldehydes and ketones), alcohols, sulfur compounds, fatty acids, other carboxylic acids, lactones, and esters. More complex molecules, such as mono- and di-unsaturated compounds, keto acids, di-ketones, cyclic compounds, and amines may also be responsible for certain flavor notes in dairy products. Table 4.2 contains a partial listing of chemical compounds which have been found in the flavor and aroma fraction of dairy products by various investigators.

THE ROLE OF MILK CONSTITUENTS IN DETERMINING FLAVOR

Milkfat not only serves as a solvent for flavor compounds, but is itself the origin of many flavors (and off-flavors) frequently encountered in dairy products. The short-chain fatty acids have long been recognized as characteristic components of milkfat (butterfat) which, under conditions that bring about fat hydrolysis, produce characteristic flavors. Flavors that develop from free fatty acids are generally undesirable in fluid milk and cream, but are an essential flavor component in Blue and certain Italian cheeses. Milkfat is also a source of keto and hydroxy acids, which may, as a result of sufficient heat treatment, pro-

Table 4.2. A Partial List of Compounds Reported to Have a Flavor Significance in Dairy Products

Compounds	Flavor Significance
Dimethyl sulfide, $C_{2,4,6,8,10}$ n-alkanoic acids, $C_{8,10,12}$ δ-lactones, phenol, m- or p-cresol, o-methoxy-phenol, diacetyl	Butter
Lactic acid, acetic acid, acetaldehyde, dimethyl sulfide, diacetyl	Culture flavor
Acetic acid, butanoic acid, hexanoic acid, octanoic acid, acetone, 2-pentanone, 2-heptanone, 2-nonanone, 2-undecanone, 2-pentanol, 2-heptanol, 2-nonanol, 2-phenylethanol, ethyl butyrate, methyl caproate, methyl caprylate	Blue cheese
Methyl mercaptan, hydrogen sulfide, dimethyl sulfide, acetic acid, ethanol, 2-butanol, 1-propanol, 2-propanol, free fatty acids, methyl, ethyl, and n-butyl esters of fatty acids, $C_{3,4,5,7,9,11,13}$ n-alkanones, $C_{1,2,3}$ n-alkanals, 3-methyl butenal, methional, diacetyl, 1-butanol, 2-methyl-1-propanol	Cheddar cheese
Acetaldehyde, propionaldehyde, hydroxymethylfurfural, acetol, pyruvaldehyde, maltol, furfuryl alcohol, 1-furfural, 5-methylfurfural, acetylfuran, 2-acetylpyrrole, acetic acid, propionic acid, butyric acid	Caramel
Dimethyl sulfide, diacetyl, propionates, acetaldehyde, acetone, butanone, 2-methyl butyraldehyde, 2-pentanone, 2-heptanone, ethanol, 2-butanol, 1-propanol, 1-butanol, methyl hexanoate, ethyl butanoate, fatty acids	Swiss cheese
Ammonia, isovaleric acid, methanethiol, phenyl acetic acid, fatty acids	Limburger cheese
Oct-1-en-3-one	Metallic (oxidized flavor)
Oct-1-en-3-ol	Mushroom-like flavor
Isovaleraldehyde (3-methylbutanal)	Malty flavor
2-heptanone, δ-decalactone, acetone, 2-nonanone, γ-dodecalactone, 2-butanone, 2-undecanone, δ-dodecalactone, 2-pentanone, 2-tridecanone, δ-tetradecalactone, caprylic acid, capric acid, lauric acid, myristic acid	Aged evaporated milk flavor
Butyric acid, caproic acid, caprylic acid, capric acid, lauric acid (and salts of)	Rancid flavor
2-Hexenal, 2-heptenal, 2-octenal, 2-nonenal, 2-decenal, 2-undecenal, 2, 4-nonadienal, 2, 4-decadienal, 2, 4-undecadienal, 2, 4-dodecadienal, C_5–C_{16} n-alkanals	Oxidized flavor
Ethyl butyrate, ethyl caproate	Fruity flavor
Methyl sulfide	Fresh milk
Nonan-2-one, benzaldehyde, tridecan-2-one, acetophenone, naphthalene, δ-decalactone, benzothiozole, o-aminoacetophenone	Stale flavor
Hydrogen sulfide, dimethyl sulfide, δ-lactones, methyl ketones (odd numbered C_5–C_{15})	Heat-induced flavors

(*continued*)

Table 4.2. (*continued*)

Compounds	Flavor Significance
Acetone	Cowy flavor
4-*cis*-Heptenal	Cream flavor
Dimethyl sulfide, trimethylamine, indole, skatole, benzylmercaptan	Feed flavor
Methional, methanethiol, dimethyl sulfide (and disulfide)	Light-induced flavor
Acetaldehyde	Yogurt
4-mercapto-4-methyl-pentanone-2	Ribes or "catty" flavor
Pyrazines, pentanol, dimethyltrisulfide, 2-furfural, benzaldehyde, 2-furfuryl alcohol, and dimethylsulfone	Cheddar cheese whey powder

Source: Information in this table was compiled from a number of the references listed at the end of the chapter.

duce such substances as methyl ketones and lactones, respectively. Both of these classes of compounds contain many members which can produce characteristic flavors or off-flavors when dairy products have been subjected to particular processing or storage conditions.

Milkfat and the associated phospholipids contain enough unsaturated fatty acids to serve as substrates for lipid auto-oxidation. The auto-oxidation process is propagated by a chain reaction of free radicals which eventually leads to the formation of volatile carbonyl compounds. The identity of compounds formed from various unsaturated fatty acids may be predicted if the classic action mode for free radical formation is assumed. Indeed, many of the predicted end products of the free radical mechanism have been isolated and identified from oxidized milkfat.

Milk proteins serve as the precursors for several off-flavors. Whey proteins are a source of sulfur compounds, which are principally involved in flavors associated with the heat treatment of milk. Certain sulfur-containing amino acids are suspected participants in a reaction believed responsible for another form of oxidized off-flavor in milk. The reaction responsible for the light-induced off-flavor is dependent on the catalytic action of sunlight or fluorescent lighting, and the presence of riboflavin (Stull 1953; Patton 1954; Allen and Parks 1979). A bacterial or enzymatic action on milk proteins may result in hydrolysis of the peptide linkages, which usually leads to either a bitter or a "brothy" off-flavor. An accompanying chemical reaction may lead to formation of highly undesirable putrid or cheesy off-flavors.

The Maillard reaction, which occurs between reducing sugars and proteins or amino acids, is also known as the nonenzymatic browning reaction. Much research has been directed towards efforts to gain a better understanding of this reaction mechanism which leads to the simultaneous production of compounds responsible for an off-flavor and a brown pigment. Caramel-like (the most commonly encountered off-flavor that results from browning), is encountered in products that have been heated to high temperatures or stored at elevated temperatures. The "caramel" flavor does not appear to arise from a single organic compound, but rather from a mixture of several compounds. When reducing sugars and amino acids are heated together, particularly at neutral or alkaline pH, numerous degradation products are formed. Their numbers and concentration depend on the temperature, the time of exposure, and the chemical properties of the amino acids and sugars present. The compounds which cause the caramel off-flavor are intermediates in the browning reaction, as they may be isolated from a colorless concentrate from products which have browned. Many of the compounds have been identified by flavor chemists (Jenness and Patton 1959; Langner and Tobias 1967; Reynolds 1963, 1965).

Both sugars and amino acids vary in the rate and extent to which they participate in the browning reaction. The reaction is very active in the presence of lysine and fructose, but diminishes when glutamic acid is substituted, or lactose replaces fructose (Walstra and Jenness 1984). Amino acids such as lycine, alanine, and valine readily react with either fructose or glucose to undergo a browning reaction and produce caramel-like off-flavors. The major milk protein, casein, in the presence of reducing sugars and upon severe heating, is also susceptible to a browning reaction, but to a substantially lesser extent than the aforementioned amino acids.

Proteins and amino acids are also sources of potential flavor compounds. These compounds initially develop by hydrolysis, wherein an amino acid is liberated, followed by deamination, to produce an acid or by a combination of deamination and decarboxylation. Subsequent oxidation results in the formation of carbonyl compounds that have one less carbon than possessed by the original amino acid. Such chemical changes may be enzymatic; they most likely serve as prominent contributors to flavor development of certain cheese varieties.

Milk sugar or lactose, although considerably less sweet than sucrose, imparts a mildly sweet taste to milk at its normal concentration (approximately 4.9%). Lactose also contributes to the mouthfeel properties of milk and milk products. Furthermore, lactose is a key substrate for bacterial enzyme activity, which leads to flavorful fermentation end products. Similar flavor compounds may result from nonenzymatic browning reactions. Organic acids such as acetic, butyric, valeric, and lactic (surprisingly) have been isolated from heated sugar-amino acid mixtures.

The *minerals* in milk normally do not serve as flavor contributors, but under certain circumstances may impart a salty taste. Copper and iron, normally present in milk at very low concentrations, serve as oxidation catalysts. These metals, in particular, present a problem when they gain access to milk through external sources, such as contact with equipment.

Vitamins may also contribute to milk flavor in an indirect way. Vitamin E is an antioxidant and riboflavin is involved in the development of light-induced flavor. Depending on its concentration and whether present in the reduced or oxidized state, Vitamin C may serve either as an oxidizing agent or as an antioxidant.

OFF-FLAVORS IN MILK AND MILK PRODUCTS

Feed Off-Flavors

Certain bulk roughages consumed by cows just prior to milking tend to impart an off-flavor to milk which is characteristic of the feed. The most common offenders are certain grasses, hay, and silage. In different areas of the country, other offending feeds may include brewery and food processing wastes or commodities. The nearest State Agricultural Experiment Station can often be contacted for a list of offending feeds in a given milk production area. Although a slight degree of "feed" off-flavor is quite commonly encountered in milk, and is not considered a serious defect, a moderate or more pronounced feed off-flavor may frequently provoke consumer complaints. The type and intensity of the feed off-flavor may vary from season to season. This unfortunately tends to aggravate the problem in that consumers may notice a sudden change in milk flavor (as bulk feeds are altered) and hence become suspicious of its quality. For this reason, some processors use vacuum units in conjunction with their pasteurizers to eliminate feed off-flavors, or at least reduce it to an unobjectionable intensity. Vacuum treatment has proven to be effective for removing many of the volatile compounds of transmitted flavor, especially those related to feeds or weeds of slight to moderate intensity.

Control of feed off-flavors at the farm level is a problem primarily associated with milking management and feeding practices. Any feeds which are likely to "taint" milk should be fed immediately after milking (see Table 5.7). A Vermont University study (Bradfield 1962) indicates that silage off-flavor is detectable in milk when cows are allowed to consume it within 4.5 hrs prior to milking. To control or minimize grassy off-flavor in milk, pastures (when used) should provide sufficient forage to enable cows to consume their fill within just a few hours after milking. The quality of feed and forage is also important; moldy or musty feeds should be avoided. The forage should be relatively free from weeds to minimize the occurrence of off-flavors related to roughage consumption.

Experiments have shown (Dougherty *et al.* 1962, Shipe *et al.* 1962, 1978) that highly volatile materials from feeds (found in both the air which the cow breathes and in the eructated gas from the cow's rumen) rapidly pass from the lungs to the bloodstream and finally to the site of milk secretion in the udder. Flavor transfer through the digestive system is markedly slower. After the feed and the associated odor(s) are removed, a gradual reverse transfer of the volatiles from the udder through the bloodstream and back to the lungs can occur.

Feed off-flavors are encountered in milk and other fluid dairy products such as cream and skim milk, butter, cottage cheese, and occasionally some hard cheese varieties. Those products which are subjected to higher temperatures and vacuum processing are less likely to contain feed off-flavors.

Weed Off-flavors

Although related to feed off-flavors in the manner in which they are transmitted to milk, "weed" off-flavors are generally more serious and require immediate attention. The most common offenders are wild onions, garlic, and related plants, which impart a distinctive "flavor note" that is quite familiar to most people. These weeds grow particularly well during the rainy spring and fall seasons, though in some geographic regions they may be encountered at almost any time.

The control of onions and other weeds is primarily dependent on pasture management, including seeding, fertilization, and pasture rotation to ensure a good stand of forage. Good pasture management not only helps in crowding out undesirable weeds, but also makes it possible for cows to get all they want to eat in the shortest possible time. Due to the seriousness of certain weed off-flavors, both Grade A and manufacturing grade milk should be carefully checked for the possible presence of such off-flavors before acceptance of the milk at the plant. Generally, vacuum treatment is at least partially effective in removing volatile weed off-flavors; success is dependent on the initial intensity of the off-flavors, as well as on the efficiency of the vacuum treatment.

Weed off-flavors may be encountered in milk, cream, butter, and fermented products. The flavor-imparting constituents of some weeds, such as bitterweed, may be water soluble and relatively nonvolatile; hence, some weedy off-flavors may only be observed by tasting and not by smelling. If such weed-tainted milk were accepted for processing, the characteristic off-taste would most likely be transmitted to nearly all dairy products, with the possible exception of butter.

Unclean Off-flavors

"Unclean" off-flavors may also be described or referred to as "cowy" or "barny"; they are those objectionable off-flavors that are associated with unsanitary farm conditions and/or inadequate milk cooling, foul-smelling stable areas, or the growth of certain spoilage bacteria in milk, especially psychrotrophs or microorganisms of the coli-aerogenese group. Certain short-chain fatty acids have been reported to impart cowy-like and unclean off-flavors to milk (Al-Shabibi *et al.* 1964), which implies that, in certain occurrences of rancidity (or in its early stages), these off-flavors may be produced in milk by lipase activity. A cowy off-flavor (a result of an abnormally high acetone content of milk) has been reported due to ketosis, a fairly common disease of dairy cattle (Josephson and Keeney 1947). Unclean off-flavors may be encountered in many dairy products such as milk, cream, skim milk, cultured milk products, cottage cheese, Cheddar cheese, and ice cream.

As an aid in determining the cause of this more serious flavor problem (unclean), bacterial counts are useful in many cases. When an unclean off-flavor is accompanied by a high bacterial count (several million per milliliter), bacteria are undoubtedly implicated as the source of the difficulty. On the other hand, a low bacterial count does not prove conclusively that the off-flavor is not of bacterial origin. A high bacterial density milk may have become diluted with good quality milk, or the number of microorganisms may have been substantially reduced by pasteurization. It is possible, of course, that the unclean off-flavor was actually of nonbacterial origin, wherein odors were transmitted to the milk from ill-kept, poorly ventilated barns, as the result of cows inhaling foul-smelling air.

Unclean off-flavors are characterized by an unpleasant, lingering aftertaste. For this reason, this off-flavor must be considered a serious defect and procedures to eliminate the difficulty must be initiated without delay.

Off-flavors Produced by Microorganisms

Numerous flavor defects of dairy products may be caused by bacteria, yeasts, or molds. The most familiar example in this group, acid, high acid, or sour milk, is not as much of a problem today for fluid milk or cream, but it is by no means extinct. Currently, other bacterial off-flavors of milk, such as those produced by psychrotrophic bacteria, are generally more common. This is due to increases in the time between milk production and product consumption, improved but not ideal refrigeration conditions, and the trend toward larger-sized milk storage vessels. Collectively, these developments in milk handling and distribution result in milk being stored for a significantly longer period of time, and thus allow psychrotrophic microorganisms a chance to multiply. Some of the bacterial off-flavors encountered in dairy products are listed in Table 4.3, along with some pertinent comments.

Preventive measures center around thorough sanitation practices and maintaining a proper storage temperature ($\leq 4.4\,^{\circ}$C [$\leq 40\,^{\circ}$F]) for both raw and pasteurized milk and cream. Post-pasteurization contamination may come from poorly cleaned and unsanitized equipment and utensils; droplets of condensate; particulate matter in the air; water; people; and product containers. Carefully selecting milk and cream ingredients of high quality helps in producing butter, cheese, ice cream, and other dairy products that remain free of these defects. Except for certain spore-forming bacteria (Speck and Adams 1976; Adams *et al.* 1976; White and Marshall 1973; Anderson *et al.* 1981), undesirable microorganisms are generally sensitive to heat and are destroyed by pasteurization. Post-pasteurization contamination is commonly responsible for off-flavor production by psychrotrophic bacteria, which can grow steadily at borderline refrigeration temperatures ($\geq 4.4\,^{\circ}$C [$\geq 40\,^{\circ}$F]).

Oxidized Off-flavor

This flavor defect has undoubtedly received more research (e.g., Arnold *et al.* 1968; Bodyfelt *et al.* 1979; Bruhn *et al.* 1975; Chen and Tobias 1972; Dunkley *et al.* 1962, 1968; Forss 1964; Hammond and Seals 1972; Thurston 1937; Tracy *et al.* 1933) and quality control attention over the years than any other; yet, it still presents problems within many segments of the dairy industry. An oxidized off-flavor results from the action of oxygen on certain components of milkfat which produces flavor sensations which have been variously described as cardboardy, cappy, metallic, oily, fishy, painty, and tallowy. There is a likelihood that these may actually be different off-flavors imparted by different chemical compounds, but they all appear to be the result of auto-oxidation of lipids.

To cope with the ever-present threat of the development of an oxidized off-flavor in milk and nearly all other dairy products, some pertinent facts should be kept in mind (see Table 5.8 for details).

Since oxidation of lipids in food products requires the presence of

Table 4.3. Microbial Off-Flavors in Dairy Products.

Flavor Defect	Products Affected, Remarks, Causative Factors
Acid, high acid (sour)	Milk, cream, skim milk, flavored milk, cottage cheese, buttermilk, sour cream, Cheddar cheese, yogurt, ice cream, butter, etc. Usually due to the growth of lactic acid fermenting bacteria. The more perishable products at or near room temperature for an extended period of time.
Malty	May be encountered in nearly the same products as the high acid flavor. *S. lactis* var. *maltigenes*. The characteristic flavor apparently due to 3-methyl-butanal (isovaleraldehyde).
Fruity (apple-like)	Milk, cream, skim milk, and cottage cheese. Caused most commonly by psychrotrophic *Pseudomonas fragi*. The characteristic flavor apparently due to ethylbutyrate and ethylhexanoate (esters).
Fruity/Fermented	Cheddar cheese, cottage cheese, some other cheeses, and milk. Caused by growth of specific bacteria and yeasts that produce characteristic volatile fermentation end-products, especially certain esters.
Moldy (mildew)	Cheese, butter, and sweetened condensed milk. Mold growth may be observed on or near the surface of the product.
Yeasty (earthy)	Cheese, butter, sour cream, and yogurt. Yeast growth may also cause problems in other products that are not properly refrigerated.
Musty (swampy)	Milk, butter, and other products made from milk. Occasionally may be caused by bovine consumption of musty hay or silage, slough grass, or stagnant water. More commonly produced by specific bacterial contaminants (e.g. *Pseudomonas taetrolens*) and end products of their growth.
Old cream	Off-flavor term applied to butter made from old or aged cream that was not properly refrigerated prior to pasteurization.
Old ingredient	Off-flavor term applied to ice cream and yogurt made with mishandled dairy or other ingredients which may have developed microbial or other off-flavors.
Cheesy (Cheddar-like)	Cream and butter. Usually caused by growth of proteolytic bacteria, or possibly molds.
Putrid	Butter and milk; a type of cheesy/unclean off-flavor caused by excessive proteolysis, which results in a limburger-type cheese flavor. In milk, it is usually due to psychrotrophic organisms.
Utensil	An off-flavor term applied to butter and cheese when the milk or cream used in the manufacture of these products has come in contact with improperly cleaned/sanitized equipment or utensils; probably a form of unclean off-flavor.
Rancid	Milk, cream, cheese, and cultured milk products. Often caused by the growth of lipolytic microorganisms (possibly psychrotrophs). This off-flavor may also develop in milk or cream by mechanisms other than bacterial action, generally the result of physical abuse.
Unclean	May occur in nearly all dairy products. Growth of psychrotrophic or coliform organisms in milk or other ingredients used in the product.
Bitter	May occur in nearly all dairy products. Proteolysis by psychrotrophic bacteria. May also be due to nonbacterial origin. (Weeds or certain feeds.)

oxygen, any effort aimed at reducing the oxygen content or preventing the contact of the product with oxygen is beneficial. While complete product deaeration is not practically feasible, some examples of partial success may be cited: the vacuumizing and nitrogen-packing of dried milk and the use of packaging materials that are impervious to air.

From the practical standpoint of flavor control, the most significant fact about lipid oxidation is that it is catalyzed by divalent cations such as copper and iron—particularly copper. Occurrences of oxidized off-flavors are often traced to a source of metal contamination at various points between milk production on the farm and the processing facility (Hunziker et al. 1929; Hunziker 1923). The quantity of copper required to catalyze milkfat oxidation varies, depending on the oxidation susceptibility of a given milk, in the range from 0.01 to 1 ppm; therefore, serious contamination may be triggered by extremely small areas of exposed metal. For example, one of the authors has observed that a surface area as small as a brass chain (formerly chrome-plated), that was used to retain a sink stopper, can "trigger" an intense oxidized off-flavor in raw milk from a given herd. Both copper and metal alloys that contain copper must absolutely be avoided in all milk and milk-product handling and processing equipment. Stainless steel, glass, and approved plastics have been found to be very satisfactory surfaces for handling milk from the standpoint of protection against an oxidized flavor.

Homogenization of market milk affords protection against the development of an oxidized off-flavor. This factor, more than any other, has been responsible for the improvement in the flavor stability of milk. The mechanism by which homogenization protects the flavor of milk against fat auto-oxidation is not known, and therefore it is subject to speculation.

It is now generally accepted that the oxidized off-flavor of milk is caused by oxidation of the phospholipids, which are located on the surface membrane of fat globules, rather than of the milkfat itself, which primarily consists of triglycerides (Jenness and Patton 1959; Walstra and Jenness 1984). Since homogenization vastly increases the surface area of the fat globules, the composition and properties of the fat globule membrane are also expected to change. The newly formed, restructured membrane, which is assumed to consist primarily of protein, seems to protect the phospholipids from attack by oxygen in some as yet unknown way.

When and if homogenized milk has developed an oxidized off-flavor, there are usually some unsound milk handling/processing practices that occurred prior to homogenization and pasteurization of the milk. With the present trend toward every-other-day milk pick-up from the farm, raw milk may more readily develop an oxidized off-flavor which carries through into pasteurized milk and/or cream. In the usual case, there is substantial dilution of the oxidized milk with good flavor-quality milk, so that the presence of the off-flavor is hidden or not obvious. In small dairy plants or producer-distributor operations, the lack of dilution quite often leads to an oxidized off-flavor in milk and cream products. In other instances, the defect may be caused by improper or incomplete homogenization, in which case a cream layer forms that would be more susceptible to the development of an oxidized off-flavor. Certain other dairy products, particularly those stored over an extended period of time, may develop an oxidized off-flavor in spite of proper homogenization. Butter, ice cream, and powdered whole milk are examples of products that may be stored over an extended period of time. In these cases, a certain quantity of de-emulsified or liberated fat may be present, which would certainly be subject to oxidation. The observation that the phospholipids are responsible for an oxidized off-flavor in milk does not exclude the possibility that the unsaturated fatty acids in the milkfat may also be oxidized when the proper conditions present themselves, as may be the case in ice cream and powdered milk. Several types of oxidized off-flavors may be encountered in butter, which may be the result of oxidation of phospholipids or other unsaturated fatty acids that are present.

High heat treatment of milk effectively helps provide protection against the development of an oxidized off-flavor. This phenomenon has been attributed to liberation or activation of sulfhydryl groups from the whey proteins, which act as antioxidants. A temperature of at least 76.6°C (170°F) is required to initiate the protective action. Since the addition of chemical antioxidants to milk and dairy products is generally forbidden by statutes or regulations, the practice of adequately heat-treating milk to render its proteins antioxygenic has become widespread. Of course, such heat treatment produces a cooked flavor; however, this is not considered anywhere near as objectionable as an oxidized off-flavor, and the "heated" intensity actually diminishes during storage. In the case of sweet cream butter, the higher heat treatments of cream prior to churning is considered so desirable or essential, that the resultant cooked flavor note is not criticized; in fact, in most instances the resultant butter receives a perfect USDA Grade of "AA," assuming no other sensory shortcomings exist.

One source of development of an oxidized off-flavor is catalyzed by light. Milk should not be exposed to direct sunlight for even a short time (i.e., in excess of several minutes) or to artificial light any longer than necessary. Homogenized milk, when exposed to light, will not develop a typical oxidized off-flavor, but will develop an equally undesirable light-activated (or light-induced) off-flavor which will be discussed later.

While their use in dairy products is generally illegal, it is of interest to note that antioxidants could be most helpful in the control of an oxidized off-flavor. Some of the materials which have been shown to impart protection against milkfat oxidation are certain ortho- and para-substituted phenols, oat flour, tyrosine, pancreatic extract, and, at relatively high concentrations, ascorbic acid (vitamin C). Feeding increased levels of alpha-tocopherol or tocopherol acetate (a form of vitamin E) within grain concentrates to cows can provide improved resistance to oxidation, if incorporated at relatively high feed supplementation levels (Dunkley *et al.* 1968, Bruhn *et al.* 1975).

The role that ascorbic acid plays in connection with lipid oxidation in milk has been the object of considerable research. When present at normal concentration, 10 to 20 mg per liter, vitamin C does not act as an antioxidant; ironically, there is evidence that it actually encourages oxidation, possibly through its reducing action on copper which, in turn, becomes a strong oxidation catalyst. It may be demonstrated that when ascorbic acid is removed from milk by oxidation, the milk then becomes quite resistant to the development of an oxidized off-flavor, even in the presence of added copper. On the other hand, when the concentration of vitamin C is increased in milk, a degree of protection against lipid oxidation is observed. This seemingly anomalous behavior of ascorbic acid suggests that its action is rather complex and probably depends on the concentrations of both copper and ascorbic acid and the reduced or oxidized state of the latter compound.

Milks vary in their susceptibility to the development of an oxidized off-flavor. This fact led Thurston (1937) to propose the following classifications: (1) spontaneous milk—one which develops an oxidized off-flavor without copper or iron contamination; (2) susceptible milk—one which does not develop an oxidized off-flavor spontaneously, but will develop it in the presence of copper or iron contamination; and (3) non-susceptible milk—one which does not develop an oxidized off-flavor, even when copper or iron is added. (See Appendix VI for the application of a copper sensitivity test for classifying milk according to its oxidation susceptibility.)

Some additional factors may be of significance in the control of an oxidized off-flavor. The inclusion of green feed in cow rations has been observed to impart resistance to oxidized off-flavor development in milk. This probably explains why summer milk in some locales is usually more resistant to this defect than milk produced during the late fall or winter months. Also, milk from cows on dry-lot feeding tends to be more susceptible to the oxidized off-flavor defect. Sanitizing agents frequently contain chlorine, which is a strong oxidizing agent. The effect of chlorine on the development of an oxidized off-flavor has not been well documented, but it would seem prudent to avoid unnecessary contamination of milk with chlorine-type sanitizers. The use of iodine sanitizers seems to help minimize the occurrence of an oxidized off-flavor in raw milk supplies. This is probably due to the lower pH of this form of sanitizer, and its ability to solubilize and remove critical divalent cations from equipment surfaces.

Several studies have attempted to relate the oxidation-reduction potential of milk to an oxidized off-flavor, but the results have usually been difficult to interpret. Researchers observed that the addition of copper to milk typically raised the oxidation potential and the milk then became more susceptible to development of the oxidized defect. They also noted that the application of high heat treatments to milk reduced the oxidation potential and that the milk acquired an increased resistance to the development of an off-flavor. However, an oxidized off-flavor has been found to develop in milk regardless of whether the potential was high or low. Efforts have also been undertaken to implicate xanthine oxidase as the "triggering" enzyme in the development of an oxidized off-flavor. While this enzyme is present in milk, and possesses many characteristics which might *a priori* associate it with the oxidized defect, investigators have not agreed as to its actual role, if any, in producing this off-flavor.

Consideration should also be given to the metallic off-flavor, which is frequently differentiated from the oxidized defect. The metallic defect is characterized by an astringent, metallic sensation, which is similar to that observed when an iron nail or metal foil is placed in the mouth. Hunziker (1923) reported that the metals which showed definite corrosion when placed in milk also had a decided effect on the formation of this flavor defect. The metals that are particularly detrimental are iron, copper, and copper alloys. Hunziker *et al.* (1929) also reported that various metals were soluble in lactic acid solutions and that the resultant lactates had a bitter, puckery, astringent, and metallic taste. Copper lactate was the most serious offender. The metallic off-flavor is quite frequently encountered in fermented dairy products such as buttermilk, sour cream, yogurt, cottage cheese and cream cheese, but may also be found in other dairy products, including butter, certain types of cheese, and ice cream.

Off-flavors due to lipid oxidation may actually be encountered in so-called "fat-free" products which are made from skim milk or non-fat milk solids. Some phospholipids remain in the skim milk obtained by centrifugal separation. Since phospholipids contain unsaturated fatty acids which are prime targets for oxidation, the oxidized defect will frequently develop, under opportune conditions. Of the fluid dairy products, only evaporated milk appears to be immune to the development of an oxidized off-flavor, presumably because the extremely high heat treatment ($116°C$ [$240°F$] for 15 min. or more) used in its manufacture provides conditions that are unfavorable to lipid oxidation. An oxidized off-flavor is also seldom observed, if ever, in Cheddar cheese, presumably due to an unfavorable oxidation-reduction potential, which prevents its development.

Once lipid oxidation is initiated, it proceeds like a chain reaction. New free radicals are generated to carry on the reaction and the off-flavor becomes progressively stronger. In the reaction process, peroxide radicals give rise to hydroperoxides which undergo scission (breakdown). Among the scission products are the chemical compounds which are responsible for the oxidized off-flavor. Vinylamyl ketone imparts a metallic off-flavor. This compound, in combination with certain aldehydes, exhibits a cardboardy off-flavor (Hammond and Seals 1972; Stark and Forss 1964; Wilkinson and Stark 1967). Figure 4.1 shows the chain reaction involved in the auto-oxidation of milk lipids, and Table 4.2 lists some of the chemical compounds believed to play a role in the oxidized off-flavor of dairy products.

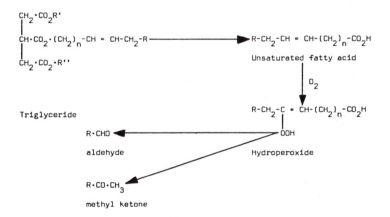

Fig. 4.1. The mechanism for auto-oxidation of lipids. Unsaturated fatty acids, in the presence of oxygen, form peroxides which decompose into aldehydes and ketones (carbonyls). The reaction is initiated by the catalytic action of certain metals (Cu^{++}, Fe^{++}) and proceeds by a free-radical mechanism. The carbonyls that are formed contribute to the sensation of the oxidized off-flavor (metal-induced).

Light-induced or Sunlight Off-flavor

Exposure of milk to sunlight, or extended exposure to fluorescent light initiates development of a type of oxidized off-flavor referred to as light-induced (sunlight) off-flavor. This light-activated off-flavor has been variously described as cabbage-like, chemical-like, burnt protein, burnt feathers, burnt plastic, or mushroom-like. The oxidized and light-induced defects may occur simultaneously, but usually one off-flavor will predominate over the other; however, which predominates usually depends upon the type of milk exposed to the light source. Homogenized milk is definitely more susceptible to a light-induced off-flavor, while unhomogenized milk is definitely more susceptible to the conventional auto-oxidation defect. Bradley (1980) published a review on the effect of light on the nutritional value of milk and milk flavor.

Substantial evidence has been presented by Patton (1954) to implicate the amino acid methionine as the target compound of milk for development of the light-induced off-flavor. Patton postulated that when methionine is acted upon by light in the presence of riboflavin (a vitamin), a degradation occurs that leads to the formation of methional. The addition of this compound to milk was demonstrated to impart a milk off-flavor that exhibited the same characteristics as those produced by the action of sunlight on milk. This implies that the development of a sunlight off-flavor is an oxidative process which involves the sulfur-containing amino acids of milk proteins. The oxidative nature of the reaction is further supported by research findings that indicate that the mechanism is retarded by certain antioxidants (Weinstein and Trout 1951a, b) and by the exclusion of air (oxygen). The involvement of copper as a catalyst is probable, but not entirely clear (Dunkley et al. 1962, Tracy et al. 1933). The role of sulfur-containing compounds has also been implicated. A study which used radioactive sulfur showed that mercaptans, sulfides, and disulfides increased in light-exposed milk (Samuelsson 1962). Figure 4.2 summarizes the "sunlight" reaction.

The literature also contains accounts which question the role of methional in the light-induced defect (Wishner and Keeney 1963). Even though the "burnt" character of this off-flavor calls to mind a sulfur compound, the possibility that other key chemical compounds may be involved should not be overlooked. The responsible compounds appear to lack stability; the sunlight off-flavor tends to diminish and undergo a change in its characteristics after storage for several days. Flake et al. (1938) reported that the combined oxidative actions of hydrogen peroxide, oxygen, and cupric ions dissipated or removed the off-flavor

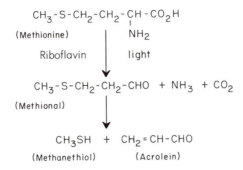

Fig. 4.2. By the action of light and in the presence of riboflavin, methionine is transformed into methional, and subsequently to methanethiol. Methional is believed to be responsible for the light-induced oxidized off-flavor.

from milk. Surprisingly, a prolonged exposure of milk to sunlight has been claimed to diminish the off-flavor.

When milk is exposed to sunlight, two flavor defects tend to develop (Dunkley *et al.* 1962; Finley *et al.* 1967). In unhomogenized milk, the "cardboardy" or lipid-oxidized off-flavor primarily develops, while in homogenized milk the "sunlight" or light-induced off-flavor always predominates. Inexperienced judges may find it difficult to differentiate a sunlight from an oxidized off-flavor, probably because some degree of both defects are present when milk is exposed to sunlight (or extended exposure to fluorescent lights).

Irradiation of milk by other than optical energy also leads to flavor difficulties. Irradiation of milk by gamma-particles leads to very unpleasant off-flavors, which suggest an attack and effect on both the milkfat and proteins of milk. Application of high-frequency sound can also produce off-flavors in milk.

Eliminating "sunlight" off-flavor requires minimizing the exposure of "unprotected" milk and other milk products to light. This off-flavor is catalyzed by either direct sunlight or certain artificial light, particularly the type of fluorescent lights described as "white" or "daylight" (Herreid *et al.* 1952). Experimental evidence indicates that practically all light below a wavelength of 620 mμ must be excluded to protect milk against the formation of this light-activated defect (Josephson 1946). Due to their higher opacity, paper milk cartons generally provide greater protection than clear or translucent glass or plastic containers. Some progress has been made in the development of opaque or pigment-modified containers by the addition of light-blocking agents to the resins (before forming single-service plastic bottles). Most fluid milk and cream products are vulnerable to the development of off-flavors as a result of exposure to light. Cottage cheese, ice cream, and many other dairy products may also develop a light-induced off-flavor when transparent or "see-through" lids or closures are used on product containers.

Rancid Off-flavor

Due to the unique fatty acid composition of milkfat, hydrolytic rancidity is a typical defect found in dairy products when careless milk handling occurs. Under certain conditions, the triglycerides of milkfat may be partially hydrolyzed as the result of the catalytic action of lipase, an enzyme indigenous to raw milk. If lipid hydrolysis occurs, free fatty acids are released in milk. These low molecular weight fatty acids (carbon length, C_4-C_{12}) impart an undesirable, soapy, bitter off-flavor, commonly referred to as "rancid" (Al-Shabibi et al. 1964; Scanlan et al. 1965).

Since lipase is a normal component of raw milk, the development of rancidity is a continuous threat. This enzyme generally remains inactive, but when raw milk is subjected to certain conditions of abuse or process treatments, lipase can be activated. Some current practices of milk production and raw milk handling are problematic from the standpoint of potentially triggering a rancid off-flavor. Precautions in processing must be taken to avoid rancidity problems in producing many of the various dairy products (refer to Fig. 4.3).

While the exact mechanism for the initiation of lipase activity is not known, it has been postulated that this enzyme is normally found in the aqueous portion of milk, where it is inactive. Before lipolysis can occur, the enzyme must come into physical contact with the fat. A number of treatments to which raw milk may be subjected, commonly known as activation, are thought to help in bringing about this physical contact. The most common known causes of lipase activation are: (1) excessive agitation; (2) alternate warming to about 32°C (90°F) and subsequent cooling; (3) foaming or excessive air incorporation; (4) homogenization of raw milk; and (5) the addition of raw milk to pasteurized milk, particularly homogenized milk. Unfortunately, some of these conditions are frequently encountered on dairy farms or in the transport of milk from the farm to the processing plant (International Dairy Federation 1974, 1975; Johnson and Von Gunten 1962). Milk may be stirred with a mechanical agitator whose action is more violent than necessary, or the agitator blades may be oversized or may not be com-

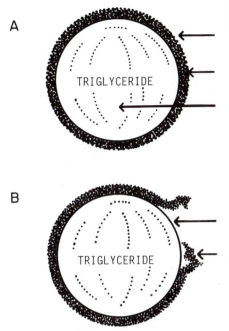

A

Outer layer of adsorbed material—lipase, cell material, agglutinins, etc.

Fat globule membrane—phospholipids and proteins

Inner core—primarily tri- and diglycerides

B

Newly exposed area of fat globule

Adsorbed material in direct contact with milkfat

Fig. 4.3.   A simplified visualization of a milkfat gobule: A—an intact normal membrane layer with adsorbed material; and B—an example of a globule in which the membrane is broken and the adsorbed material (including lipase), is in direct contact with exposed triglycerides.

pletely submerged in the milk. Excessive foaming may be observed in some pipeline installations, in the farm bulk tank, and/or the transport tanker. Entire milk tanker loads may become contaminated with "activated" milk from one or more producers.

Similarly, raw milk storage tanks at the processing plant (into which the milk is transferred) may become contaminated with "activated milk". Whether such contamination leads to a rancid flavor problem depends on the extent of "activation," the amount of dilution and the elapsed time before pasteurization. The longer raw milk is held unheated (unpasteurized) the greater are the chances for development of the rancid defect (Ludzinska *et al.* 1970).

Some factors that contribute to the development of a rancid off-flavor are not completely understood. Milk from various sources (different herds) seem to vary in susceptibility to the rancidity defect, but whether such variations are due to hereditary, environmental, and/or nutritional factors is not currently well-understood. Factors such as the quantity or type of lipase secreted, and the nature of the fat globule membrane (for which the enzyme may have a greater or lesser affinity), may explain the varied susceptibility of milk to rancidity. Observations have indicated that milk from cows in late stages of lactation is more susceptible, while milk produced from cows on pasture seems less susceptible to this defect. In most instances, rancid off-flavor in raw milk can usually be traced to one or more of the known conditions which bring about activation, although occasional incidents of rancidity arise for no readily apparent reason.

The extent of lipolysis may be followed by several tests, the most common of which is a determination of the formed free fatty acids within the milkfat. A test for the Acid Degree Value (ADV) was developed by Thomas *et al.* (1955), and is sufficiently simple and rapid enough to be suitable for routine quality control work (see Appendix IV). More recently, a copper-soap modification of the method has been proposed (Shipe *et al.* 1980). The value of the ADV test lies in the fact that a given milk may be checked for incipient lipolysis before a rancid off-flavor is actually detected by sensory evaluation. This may serve as a warning signal that the given milk should be pasteurized without delay. A high degree of incipient lipolysis should also alert quality-control or field personnel to initiate an investigation into the source of the difficulty. Small quantities of free fatty acids are indigenous to milk, regardless of how fresh or old the product is. At these low concentrations of free fatty acids, no undesirable flavor is imparted. As the state of lipolysis progresses, a gradual but significant increase in the concentration of free fatty acids occurs, until a quantitative level is reached when an off-flavor begins to appear. The initial rancid defect may be described as "cowy," "stale," or "lacks freshness," followed by an "unclean-like" off-flavor. Rancid will eventually taste bitter, soapy, and manifest a Blue cheese-like odor. The actual ADV at which a rancid flavor is detected by sensory evaluation will vary with the milk source (see Appendix IV) and with the person(s) performing the sensory evaluation.

Many individuals have great difficulty in detecting a rancid off-flavor or may not be able to detect it at all. Some persons finally learn to recognize rancidity through continued practice, but may not find this off-flavor that unpleasant in their perception process. A possible explanation is that these persons may have a relatively high flavor threshold for this type of off-flavor. In fact, their description of the rancid off-flavor (even when most intense) often is vague, such as simply indicating it is "rich" or "sour." An inability to perceive certain off-flavors may be analogous to a "blindness to certain colors" or a "deafness to certain tones"; Unfortunately, little can be done to alter this physiological fact.

Although there are no known studies to substantiate it, the experience of the authors and other coaches in training collegiate judges indicates that more students seem to have a problem recognizing rancidity than most other off-flavors. A further complication arises from the observations that not only do people perceive this off-flavor differently, but there may actually be different types of rancid off-flavors (Willey and Duthie 1969). Furthermore, Duthie *et al.* (1983) were surprised to learn that approximately 25% of the taster-attendees at a Vermont State Fair indicated a preference for rancid milk compared to nonranious fatty acids was reported as follows:

There is general agreement that short-chain fatty acids, which contain an even number of carbon atoms from 4 to 12, are responsible for the rancid off-flavor. However, efforts to duplicate the off-flavor by the addition of a mixture of fatty acids to milk have not been completely successful (Scanlan *et al.* 1965). The simulated off-flavor is close but not identical to actual samples of rancid milk. Whether this is due to a missing, unknown constituent or a difference in the distribution of the fatty acids between the fat and the aqueous phase, is not known. Each of the individual short-chain fatty acids seems to differ in the specific flavor sensation that it imparts to milk. Under the conditions of one study (Al-Shabibi *et al.* 1964), the individual contribution of various fatty acids was reported as follows:

Butyric acid—strong odor (characteristic of butyric acid)
Caproic acid—cowy, unclean
Caprylic acid—cowy, goaty, unclean
Capric acid—bitter, soapy, rancid
Lauric acid—bitter, soapy, rancid
Myristic acid—none.

One may speculate that people differ in their sensitivity to the individual fatty acids, and thus each person describes rancidity by what they perceive as the predominant component. All the various flavor terms that apply to the individual fatty acids (just previously listed), have been used, in turn, to describe rancidity by various investigators.

The short-chain fatty acids are lipid soluble, but to some extent they are also water soluble. At the normal pH of milk, the water-soluble fraction of the fatty acids is found largely in the form of salts, since the pK_a value of the fatty acids is approximately 4.8, which means that, at pH 4.8 they are half-neutralized. The flavor intensity imparted by salts of the free fatty acids is less than that of the acids. This helps explain why it is possible to have rancid off-flavored cultured products, even though they may have been made from milk ingredients which were not rancid. At the lower pH of cultured dairy foods, a greater proportion of the free fatty acids would exist in the un-ionized acid form than as the salt form.

Increases in the free fatty acid content of milk may also be observed as the result of certain microbial activity and high heat treatment, as well as the advanced age of certain dairy products (Andersson et al. 1981). The flavor note of "free fatty acids" is believed to be a component of the "caramelized" off-flavor so commonly observed in heat-sterilized dairy products, although this flavor cannot be described as "rancid." However, some commercially sterile dairy products may develop a rancid off-favor in storage, several months after manufacture. This is particularly true of those products sterilized by short exposure to extremely high temperatures and then aseptically packed. These incidents of rancidity are difficult to explain, because available evidence (Hetrick and Tracy 1948) indicates that lipase is effectively inactivated by relatively low process temperatures, such as those used for milk pasteurization. One possible explanation is that certain heat treatments do not irreversibly inactivate lipase, and a reactivation of the enzyme occurs following prolonged storage. Another possibility is that, even though the heat treatment may have destroyed the microorganisms, bacterial lipases may remain active because they are more heat resistant than milk lipase.

Surprisingly, rancidity is a desired flavor component in some foods. Rancid milkfat and rancid milk powders (produced by treatment with special lipase preparations) are sometimes used as ingredients for certain milk chocolates and other food products. In the manufacture of Blue cheese and some Italian cheese varieties (such as Romano and Parmesan), lipase activity helps develop the desired character; it is commonly described as the "picante" flavor note in some Italian cheeses.

A rancid off-flavor may feasibly be encountered in nearly any dairy product; but, in most cases, its origin may be traced to the original (raw) milk or cream from which it was made. Thus, control measures should include a close surveillance of farm practices to eliminate or minimize conditions that contribute to the activation of lipase. In the processing plant, extreme care must be taken to avoid mixing raw with pasteurized milk or the homogenization of raw milk (when it is not immediately followed by pasteurization). Raw milk temperatures should not be allowed to fluctuate widely and milk should not be unnecessarily or violently agitated.

Cooked Flavor

Readers may have noted that the authors chose to refer to "cooked" as a flavor, rather than classify it as an off-flavor per se. In many instances (with most dairy products in the U.S.) some degree of the cooked sensation is actually desired or preferred for a number of reasons (which will be reviewed later). Hence, the authors are reluctant to "downgrade" the flavor note of "cooked" to the lower status symbolized by the term "off-flavor."

When milk or cream is subjected to a heat treatment just beyond the minimum conditions for pasteurization, the flavor properties are generally altered. This flavor alteration is often described as "cooked." The type of flavor which results depends on such factors as: (1) the intensity of the heat treatment; (2) the time of heat exposure; (3) the composition of the product; (4) the rate of heating and cooling; and (5) the extent of "burn-on" that occurs on the heat exchanger surfaces. The term "cooked" is thus a generic name that is applied to a number of heat-induced flavor defects, also variously described as heated, nutty, scalded, burnt, scorched, custard-like, or caramel. The cooked flavor is believed to result from the action of heat on certain proteins, particularly the whey proteins and the fat globule membrane proteins. The more serious forms of cooked flavor, such as scorched or caramel, probably involve interactions of protein with lactose and the subsequent breakdown products, in addition to protein and lactose degradation.

In the U.S., a mild intensity of cooked flavor is typically observed in market milk (Thomas 1981), powdered milk, ice cream, and sweet cream butter. In the case of butter, an obvious, characteristic "nutty-cooked" flavor, is generally considered highly desirable, if not essential, for the overall flavor dimensions of the product. This flavor objective for sweet cream butter is quite simply attained by appropriately heating the sweet cream. The presence of a cooked note provides reliable assurance that adequate heat treatment was applied to the cream to definitely protect against the development of unwanted lipid oxidation in the butter.

Prior to the time of compulsory milk pasteurization, persistent consumer objections were raised against even the mildest forms of cooked flavor; to them it was considered an off-flavor. But over the years consumers have gradually become accustomed to this common milk product flavor. Hence, most U.S. dairy processors now pasteurize milk and cream at higher than the minimum times and temperatures required by public health standards. This is justified by the dairy industry on the basis of: (1) improved flavor characteristics and stability; (2) efforts to mask feed off-flavors; and (3) increased product shelf-life properties.

The cooked flavor is most obvious immediately after heating, but its intensity gradually decreases with storage time. The mild to moderate degree of cooked flavor developed in milk or sweet cream butter may not completely dissipate with storage time, but its intensity will become substantially reduced. However, this is not the case with the more serious scorched or caramel off-flavors, which may actually become more objectionable as storage time is prolonged. Though the intensity of a cooked flavor depends on a combination of the time and temperature parameters of heat treatment, the diversity of the encountered flavor sensations makes it rather difficult to predict the intensity of the flavor for any given heat treatment. Detection of a sulfide-like odor when milk is heated above 76.7°C (170°F), even for a short time, signals the appearance of what is considered the true "cooked flavor." Short-time exposure of milk to lower temperatures than that just mentioned commonly results in a milder form of the cooked flavor, generally referred to as a "pasteurized milk" or "heated" flavor. In the U.S., some version of this flavor note may be detected in practically all dairy products, with the exception of most ripened cheese varieties. In essence, this is a form of cooked flavor that is no longer considered objectionable by the majority of consumers in this country.

The more intense forms of cooked flavor (occasionally an objectionable level is reached), result from the exposure of milk to extremely high temperatures for prolonged periods of time, as in the manufacture of evaporated milk. In some products, such as ice cream, pasteurized dairy ingredients are blended and repasteurized, sometimes at higher temperatures (≥ 93.3°C [≥ 200°F]) than the initial thermal process. In ice cream manufacture, pasteurization of the mix is universally required by public health regulations and the use of previously pasteurized milk and cream ingredients does not exempt the manufacturer from the mix pasteurization requirement. When repeated heat treatments are of high intensity, a detectable cooked flavor is commonly produced in the final product. Fortunately, ice cream mixes can "tolerate" considerably more heat treatment than milk before objectionable levels of cooked flavor are attained in the finished product.

The chemical nature of cooked flavor is not fully understood, but hydrogen sulfide and other sulfides probably contribute to it (Swaisgood 1980). A number of other compounds (Scanlan et al. 1968) are known to be produced by the action of heat on milk constituents; these may also contribute to the overall cooked flavor sensation. In dairy products that receive more severe heat treatment(s), intermediates of the Maillard browning reaction serve as potential flavor compounds

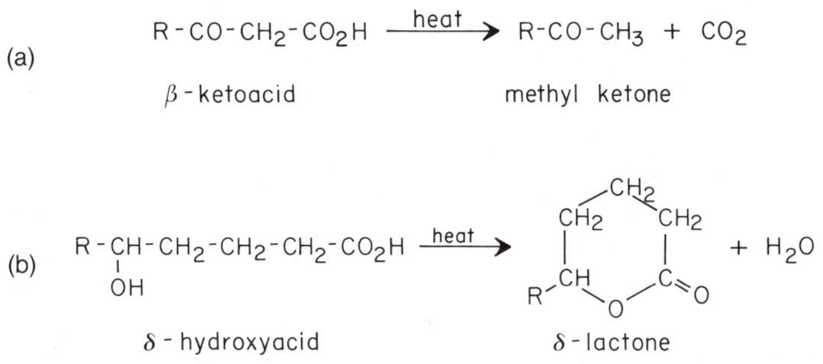

(a) $R-CO-CH_2-CO_2H \xrightarrow{\text{heat}} R-CO-CH_3 + CO_2$

β-ketoacid   methyl ketone

(b) $R-CH-CH_2-CH_2-CH_2-CO_2H \xrightarrow{\text{heat}}$

$\underset{OH}{|}$

+ H_2O

δ-hydroxyacid δ-lactone

Fig. 4.4. (a) β-ketoacids upon heating decarboxylate to form various methyl ketones; (b) δ-hydroxyacids and γ-hydroxyacids close to a ring form and produce the corresponding lactone.

(Forss 1979; Patton 1955; Badings 1984; Langner and Tobias 1967; Badings and Neeter 1980). Also, see Fig. 4.4.

Foreign Off-flavor

Since milk contains both an aqueous and a fat phase, it acts as a solvent for many (foreign) substances which may impart undesirable off-flavors. Possible contaminants are sanitizing agents, udder medications, insecticides, cleaning compounds, and boiler compounds; theoretically, the list is long. From a practical standpoint, the most common off-flavor in this category is best described as "medicinal," since it conveys an odor sensation which is reminiscent of a hospital. Some potential chemical contaminants of milk may be toxic, and hence precautions must be taken against their entry into milk.

The most common form of medicinal-like flavor defect results from the interaction of chlorine and phenol. Either of these chemicals are capable of imparting a medicinal off-flavor to milk, but to do so, their respective concentrations must be relatively high. However, when in combination, only a fraction of a part per million of these substances is required to form an objectionable off-flavor. Usually, the source of these contaminants is easily traced. Since all dairy equipment must be sanitized, there is the constant possibility that some disinfectant may be accidentally absorbed by the milk. If a given milk contains phenol as a minor contaminant, requirements for serious off-flavor formation are met. The phenol (carbolic acid) may derive from udder medications (milk from treated cows should always be withheld from the farm bulk tank for the prescribed time) or other sources, including the water supply. Chloro-phenol off-flavors sometimes develop in cottage cheese as a result of chlorinating wash water that contains phenol residues, which is commonly associated with leaf-fall during autumn. The leaves provide a source of organic material which combine with the chlorine to form the chloro-phenol.

Other examples of foreign off-flavors that may occur in fluid milk include one described as "rubbery," "hot paraffin," or "burnt plastic." Its point source can be quite difficult to confirm without prior experience. This particular flavor defect is commonly the result of an excessively high operating temperature of the "sealing-jaws" of paper milk carton fillers. In such an instance, the polyethylene plastic coating is partially incinerated or scorched; volatiles from the plastic are readily absorbed by the milk product. The defect suddenly disappears when the temperature of the machine sealing "jaws" is adjusted to the correct intensity, as recommended by the filler manufacturer.

Another foreign off-flavor has been observed at times in those milk products that are processed by direct steam injection. This flavor defect appears to be related to the condition of the boiler and the purity or cleanliness of the steam supply that may be incorporated into the product. To some observers, this flavor defect bears a vague resemblance to the "smokey flavor of roast ham." But regardless of what sensation this defect is associated with, its presence is readily noted and objected to by consumers.

Unfortunately, foreign off-flavors may be encountered in any dairy product, and their occurrence creates a serious problem which requires immediate attention. Undesirable conditions on the farm, in transport, in the plant, or a combination of abuses may be responsible for the difficulty. There are several possible modes of transmission of foreign off-flavors into milk and milk products, including absorption from the atmosphere, direct entry of the undesired substance into product and transfer of residues from contact surfaces of equipment and containers. Foreign off-flavors may be transferred to milk when multiuse plastic bottles have been misused (by consumers) as containers for storing home, garden, or farm chemicals, gasoline, fruit punch, or other beverages (Bodyfelt *et al.* 1975, Landsberg *et al.* 1977). When the nature and source of the contaminant cannot be immediately ascertained, line samples of the milk at all stages of processing and handling should be examined to pinpoint, if possible, where the contamination is occurring.

Storage Off-flavor

For many cheese varieties, storage under proper conditions of temperature and humidity constitutes an integral step of the overall manufacturing process. This processing phase is termed "aging," "ripening," or "maturation." By contrast (and unfortunately), those dairy products which do not require ripening generally undergo flavor deterioration during or following storage. The resultant off-flavors may be variously described according to their nature as: surface taint, stale, lacks freshness, old, or simply "storage." In all instances, these stored products just seem to lack the anticipated flavor luster and/or refreshing quality, which hopefully exists in comparable, freshly made dairy products.

The mechanism of storage flavor development is not the same in all cases. First, in its simplest form, the storage-like flavor defect can simply be due to the absorption of potent odors from the surroundings (environment) in which the product was processed or stored. This is a phenomenon experienced by many consumers who may have consumed "tainted" butter which had been stored in the refrigerator alongside other aromatic foods (i.e. onions, garlic, or cantaloupe). Other dairy products also have the capability of readily absorbing atypical or foreign odors. Secondly, some auto-oxidation occurs in many dairy products (depending on packaging conditions), which leads to an oxidized-like defect. Surface oxidation is quite common on such products as butter, ice cream, and some cheese; this usually results in a pronounced, undesirable surface taint or off-flavor. Thirdly, an off-flavor described as "stale" may develop by a mechanism which is not yet fully understood. Stale may simply be another form of oxidation, in which the end-products differ from those obtained when auto-oxidation occurs. All of the circumstances for these oxidative forms of deterioration are not fully understood. The stale off-flavor seems to occur even when conditions that do not favor auto-oxidation exist. Examples of this are sterile milk and other products that are heated to relatively high temperatures or milk powders that are stored in an atmosphere of nitrogen. An additional argument against the assumption of an oxidative mechanism is the observed ineffectiveness of antioxidants in preventing the stale flavor defect. On the other hand, it may be argued that "staling" may only develop when "limited oxidation" is possible, since the "normal pathway of auto-oxidation" is prevented or minimized by conditions which are unfavorable to it.

More research is needed to clarify the process of staling, and hopefully, to prevent the development of stale off-flavors in dairy products. This flavor defect, together with the oxidized off-flavor, are the principal stumbling blocks in industry efforts to develop sterile, nonperishable forms of milk and cream in which the flavor quality might approximate that of fresh pasteurized milk and cream (Arnold et al. 1968; Bodyfelt et al. 1979; Muck et al. 1963).

Often, a rather serious storage off-flavor is encountered in fluid milk when it is stored in paper cartons in storerooms where the odor of certain fruits and vegetables is particularly strong or offensive. These odors can permeate two separate plastic coatings as well as the paperboard layer, and are subsequently absorbed by the milk; this usually leads to numerous consumer complaints. This "environmental hazard" of milk distribution should be suspected when consumer complaints are reported from only one retail store, an isolated route, or a single distribution point.

Storage off-flavors may be encountered in fluid milk products, ice cream, butter, cottage cheese, and fermented milk products. Effective packaging provides a considerable degree of protection, but only a reasonable duration for shelf-life should be expected for any perishable food product. For this reason, a system of effective product rotation must be implemented by the processor and/or distributor. Storage off-flavors are not necessarily serious defects, although there are various degrees of intensity. This is sufficient reason for carefully determining the expected shelf-life for each product and marking containers with a pull date or an expiration date. The stale-type of storage off-flavor poses a problem because it is encountered in certain dairy products that are intended for storage or distribution through a prolonged period of time (several weeks to several months). Most unfortunately, no effective preventive measures for the staling of dairy products are known or available at this time.

In the discussion of individual dairy products (in subsequent chapters) "lacks freshness" and "storage" are, in some cases, treated as separate defects to distinguish between the absorption mechanism of storage off-flavors and the gradual quality deterioration associated with lacks freshness (or staleness, as a more serious case). In other instances, the various storage defects are grouped together under a single heading. However, the evaluator still has the option of choosing the appropriate term to most usefully identify the cause of the difficulty.

Salty Off-taste

Under typical conditions, the minerals present in milk do not cause milk to taste salty, but the normal mineral-lactose balance may be disturbed by bovine disease, in which case a salty off-taste may develop. Milk from cows in late stages of lactation also tends to be saltier than that obtained in the earlier stages of lactation. However, a salty off-taste is not frequently encountered in market milk, presumably because a given milk supply is made up of milk from many cows from many herds. This tends to normalize the flavor and cancel the effect of an occasional salty-tasting milk from an animal that may be producing milk of abnormal composition or be in late lactation.

Salty defects may be encountered in other dairy products when more than the required quantity of salt has been incorporated into the formulation (Bodyfelt 1982). This may be the occasional situation for buttermilk, butter, cottage cheese, and certain flavors of ice cream. A Cheddar cheesemaker generally guards against the use of too much salt, since it markedly slows down flavor development and breakdown of the cheese body during the aging process. The correct amount of salt, however, is required for proper flavor and rheological properties in cheese.

It is difficult, perhaps impossible, to make a general statement about the most appropriate concentration of salt in the various dairy products, due to the wide variations in consumer preference for the salty taste or its absence. However, as a general rule, the salt content should not be so high as to be objectionably sharp, or that "salty" be the first or only noted taste in the dairy product. Salt should not disturb the desirable flavor blend; and it should be completely dissolved so that product mouthfeel will not be perceived as "gritty" when consumed. Gritty salt (briny) is still occasionally encountered in poorly worked butter; hence, a typical indication of poor workmanship.

Bitter Off-taste

The bitter defect may stem from several causes which have already been identified under previous headings. To resummarize, bitterness can be either: (1) a component of rancid off-flavor; (2) or it may be imparted by certain weeds; and (3) it can be produced by some psychrotrophic bacteria. It is generally acknowledged that specific protein fragments (peptides and some amino acids) manifest a distinct bitterness. These protein fragments (bitter peptides) are products of proteolytic enzyme activity (Juffs 1973a, 1973b, 1975). Under just the right circumstances, bitterness may be encountered in practically all dairy products, although the off-taste may be more difficult to detect in many sweetened products. Proteolytic enzymes are generally of bacterial origin, either from microbial contaminants or lactic starter cultures. Rennin (chymosin) is also classified as a proteolytic enzyme. The preferred absence of bitterness in most cheese largely depends on the compatible behavior of rennin and starter peptidases (specific types of proteases) during cheese ripening. Currently, it is not clear whether there are bitter compounds, other than peptides and certain amino acids, formed in cheese.

Flat or Lacks Flavor

Interestingly, the term "flat" suggests different meanings for various dairy products. For milk, the descriptor "flat" simply denotes a watery sensation; a lack of richness and/or mouthfeel. Even though milk need not be diluted with water to exhibit this defect, flatness may be easily simulated by adding water to whole milk (approximately 15–20%). Either accidental or unintentional "watering" or a low milk solids content are the most common reasons for the flat defect.

A cheese that is described as flat, simply "lacks the flavor intensity" characteristic of the given variety. A young Cheddar cheese (less than three months of age) is typically expected to be somewhat flat, but for an aged cheese (more than nine months), flatness is generally considered to be a notable defect. When the flavor attributes are graded as flat, it does not necessarily mean that a cheese has no flavor. The cheese sample may actually possess other off-flavors, but simultaneously, it lacks the desired or expected flavor character (and/or intensity) sought in a cheese of this type or variety.

Buttermilk, cottage cheese, yogurt, sour cream, and other cultured milk products should exhibit a delicate, desirable aroma in addition to the perceived clean, lactic acid taste. The character of this aroma largely depends on the performance of the microorganisms used in the fermentation (Law 1981; Lindsey 1967; Morgan 1976). Absence of the characteristic or desired aroma causes the product to seem flat; however, an excessive aroma can be somewhat undesirable. A principal aroma component of buttermilk, sour cream, and similar products is diacetyl; the primary aromatic component of yogurt is acetaldehyde (Law 1981).

In flavored products such as yogurt and ice cream, the term "flat" is replaced by the descriptor, "lacks flavor," since the associated defect in these products is obviously due to an insufficient concentration of added flavoring substance(s).

Astringency Defect

This sensory defect is actually a tactual sensation, (i.e., perceived by the sense of touch), but since astringency is only detected when and if the sample is placed into the mouth, it is traditionally listed as a flavor defect. Other descriptions of astringency include: mouth coating, dry, puckery, chalky, and powdery. A green persimmon or alum are examples of substances that exhibit extreme astringency. All the possible causes of astringency are probably not known, although for milk this defect is usually associated with high heat treatment and it seems more often related to lowfat products. Possibly, astringency may be attributed to a certain particle size of milk proteins or other milk constituents. One theory of astringency suggests that a specific particle size may be required to activate the sensory response that leads to the perception of this defect; either smaller or larger particles may pass unnoticed by the taster.

Sweetening Agent Defects

The principal dairy products that require the addition of sweeteners are frozen dairy desserts, flavored milks, flavored yogurts, and sweetened condensed milk. In the case of ice cream, ice milk, and sherbets, the function of sweeteners is two-fold: (1) to provide the desired level of sweetness, and (2) to supply a portion of the needed solids for effecting desirable body and texture characteristics. In flavored milks and yogurt, the most important function of sugar is to supply the sweetness needed to achieve the proper flavor balance. In sweetened condensed milk, sugar acts primarily as a preservative and as a modifier of the product's body or viscosity.

The most obvious sensory defects that directly relate to sweetening agents are either a lack of sufficient sweetness or excessive sweetness. This situation is often complicated by the recognized range of variation in consumer preferences for sweetness intensity. This fact makes it rather difficult, if not impossible, to formulate frozen dairy desserts that will please everyone. For establishing the sweetness level in most sweetened dairy-based foods, the wisest approach (dictated by past experience) appears to be to "try to satisfy the *majority* of consumers." An excellent application of consumer panels is in the effort to obtain public reaction to sweetener intensity or characteristics for various products. This information can lead to formulation and reformulation of sweetened products.

Product sweetness as a sensory property can only be measured by taste. The carbohydrate, sucrose, is considered to possess a "true sweet" taste, basically free from other taste notes or odors. Honey is also sweet, but it usually contains other flavor notes as well; hence, a product sweetened with it generally acquires a recognizable "honey-flavor." Corn syrups that have varying degrees of hydrolysis (identified by a numerical value for the Dextrose Equivalent [DE]), and fructose-containing corn syrups are commonly used as a partial replacement for sucrose. Under typical conditions, an ice cream may be criticized for exhibiting an "unnatural sweetness" or a "syrupy" off-flavor. An important consideration in noting occurrences of this defect is the final flavor blend of the ice cream. For instance, when only a mild flavor (such as vanilla) is the objective for a frozen dessert, then sweetener ingredient "background" off-flavors may be detected and serve to detract from the anticipated or desired flavor attributes.

There are stimuli and response interactions between and amongst the basic tastes of sweet, sour, salty, and bitter. The consequence of this is that a variety or appropriate balance of primary taste elements frequently adds a sense of depth or desired richness to the final product. Some tastes and odors also nicely complement or "play off" (interact with) each other. Thus, the bitter flavor of chocolate may be pleasantly modified by salt, vanilla, and (especially) sugar; the excessively sweet flavor of fruit sherbet is made more acceptable by addition of an acid. The success of these combinations largely depends on achieving a perfect or balanced blend of the various flavor and/or sweetener elements.

Sweeteners may be the source of undesirable flavors which are not related to their primary sweetening function. In the transport and storage of sugar syrups, serious quality problems can arise from yeast contamination. The fermentation end-products that arise from these unsanitary conditions may actually render some syrups unusable for dairy products manufacture.

Browning and caramelization are typical difficulties encountered with reducing sugars under certain thermal conditions that must be minimized. Browning can proceed relatively quickly in syrups that are stored at elevated temperatures, particularly if the nitrogen content of a syrup exceeds a certain level.

Defects from Added Flavorings

Dairy product forms to which various types of flavorings may be added are constantly increasing in number. In addition to the familiar examples of flavored milks, ice cream, and yogurt, the list includes dips, cheese spreads, butter spreads, cream cheese, cottage cheese, and whipped toppings. Obviously, formulation and blending for each flavored product requires specialized knowledge about flavor chemistry, component compatibilities, and aspects of consumer acceptance or appeal. In assessing the sensory qualities of flavored products, there are some general guidelines which the evaluator should keep in mind.

Once it is determined that a high-quality flavoring has been selected for the product formulation (which is entirely judgmental), the possible shortcomings related to flavoring material are: (1) an insufficient quantity was used ("lacks flavor") or (2) an excessive quantity was used ("too high flavor"). The selected flavoring should be present in sufficient concentration to be pleasantly perceived and readily recognized, but not be so intense in flavor as to seem overwhelming to the evaluator. An insufficient flavor level may convey the impression that the processor is trying to over-economize, while an excessive flavoring level may cause the flavor to appear artificial or harsh, or it may impart a lingering, unpleasant aftertaste.

Generally, the flavoring system should impart a smooth, delicately balanced flavor to the product; a flavor which is perceived as being characteristic and fully representative of the named flavor that the product (label) bears. Usually, this may be best accomplished by the use of so-called "natural flavorings" (though at higher cost, as a rule), such as carefully selected pure vanilla extract, specially processed chocolate liquor and/or cocoa, and fresh or frozen fruits, fruit purees, or essences. However, these materials may differ widely in their per unit flavoring strength and effectiveness, depending on the source and quality of the raw material, the variety of fruit, and/or the method of processing it. In dairy products evaluation there seems to be an unwritten general premise, perhaps subconsciously practiced by most dairy product judges. It can be informally stated and loosely interpreted as follows: "The overall sensory quality of a flavored dairy product should be based on how closely its flavor (the taste, aroma, and mouthfeel) and the color and appearance approaches that which would have been obtained by the use of the most suitable, best quality, natural flavoring known to be available."

Answering the following questions can serve as a helpful set of guidelines for evaluating flavored dairy products:

1. Do the given flavoring materials and dairy ingredients blend in a pleasing manner?
2. Is the flavoring material present in the proper concentration?
3. Is the flavor quality the best that can be obtained? If not, how closely does it approach the best recognized flavor system?

The terms or descriptors which are commonly used to describe defects within the flavoring system of dairy products are: lacks flavor, too high flavor, lacks characteristic flavor, lacks fine flavor, and unnatural (artificial or imitation) flavor. Given the nature of unnatural flavor, it is probably best categorized as being an off-flavor.

TACTILE PROPERTIES OF DAIRY PRODUCTS

Each dairy product possesses its own characteristic properties of mouthfeel, which are commonly defined in terms of body, texture, or consistency. Tactile properties are an integral and important component of overall sensory quality of a product; they are detected by the sense of touch. In some instances, there is a probable correlation between tactile properties and taste and odor, since factors which affect one response create conditions which often influence the others. However, in other product samples, this interdependence may be completely absent. For example, a given product may have an excellent body and texture, but exhibit poor flavor, or vice versa.

The tactile properties of some food products can influence consumer acceptability to the same extent as taste and aroma (Little 1958; Prentice 1984; Schultz et al. 1967). For instance, the argument as to whether the flavor of an ice cream is as or more important than the body and texture in determining consumer acceptability, is generally inconclusive. This is apparent inasmuch as consumers may reject a given product for poor flavor attributes in one instance, or for poor body and texture characteristics in another case. The clear implication should be that flavor, body and texture, color, and appearance are all essential components of the sensory quality of dairy foods.

Fluid Dairy Products

The category of fluid milk products (formerly referred to as market milk) include whole milk, lowfat milk, skim milk (nonfat), various forms of cream, cultured or fermented milks, flavored milks (including chocolate and eggnog), reconstituted milk powder and concentrated milk, evaporated milk, and sweetened condensed milk. Ice cream, ice milk, and milk shake mixes are also evaluated for certain tactual features. The principal criteria related to tactual properties are viscosity, texture, and astringency. The desired viscosity varies from one product to the next. For flavored and cultured milks, the extent of viscosity is often varied to satisfy the range of consumer preferences in different regions of the country.

Differences in product viscosity may be detected subjectively by mouthfeel, or more objectively by observing how easily the product flows or resists flow. Physical methods are also available for the measurement of product viscosity. These procedures are objective and lend themselves to routine quality control testing. Normally, fluid whole milk varies in viscosity within narrow limits. These limited variations may be due to composition, the extent of heat treatment, and the applied temperature(s), pressure(s), and stage(s) of homogenization. Significant departure from expected viscosities may be caused by the action of microorganisms and their associated enzymes, which may, in extreme cases, lead to ropiness or protein coagulation. The viscosity of cultured milks may be varied or controlled by a combination of heat treatment, final product acidity, and/or selection of starter culture strains (proteolytic activity). Dairy products such as flavored milks, yogurts, and frozen dairy dessert mixes generally contain water-binding agents (stabilizers) which are added to influence the final product viscosity.

The texture (consistency) of fluid dairy products should be homogenous and smooth. Principal defects are grainy, curdy, chalky, and gritty or sandy, which may be easily detected on the tongue and other parts of the mouth. They are usually detected as numerous, extremely fine, undissolved particles within the masticated product. These defects may result from partially denatured milk proteins due to severe heat treatment, freezing, or acidity development. A similar sensation may be produced by the presence of finely divided, churned fat particles within the product. In highly concentrated milks, such as sweetened condensed milk, a chalky or sandy texture may develop from partial precipitation of lactose (as crystals).

An astringent or puckery sensation is sometimes associated with chalkiness and is considered most undesirable in dairy products. This defect, apparently related to the particle size of certain milk components, generally results from severe heat treatment of the product. Astringency is sometimes encountered in reconstituted powdered milks and sterilized milk products.

Butter, Cheese, and Ice Cream

The rheological properties of butter, cheese, and ice cream are commonly referred to by the terms "body" and "texture," although there does not always seem to be a distinct differentiation between these two descriptors. Especially, with butter and cottage cheese, the terms "body" and "texture" are apparently used interchangeably. With Cheddar and certain other cheese varieties, the term "body" generally denotes the consistency of the product mass, while "texture" refers to the relative number, type, and size of openings, as observed by the sense of sight. In ice cream, body also denotes the product consistency and the resistance to bite, while texture refers to the relative size of ice crystals, size and distribution of air cells, the presence of churned fat agglomerates, and/or undissolved lactose. These variations in the meaning of texture are more likely due to different rheological characteristics for each of these products than to disagreements per se between experts about definitions of terminology. From the standpoint of reflecting on a sensory attribute for most dairy products, general use of the combined terms "body" and "texture" seems satisfactory or appropriate.

Considerable effort has been expended in developing instrumental techniques for assessing the rheological properties of butter and cheese (Baron 1952; Prentice 1984). Instrumental techniques have also been developed for determining the proper time to perform critical steps in the manufacture of cheese. Since the final body and texture characteristics of most dairy products are so dependent on variations in the manufacturing process, any methods which serve to standardize existing procedures would seem worthwhile.

Surprisingly, instrumentation for measuring or monitoring rheological properties or physical-chemical changes is not currently in general use by the U.S. dairy industry for determining critical endpoints or stages of process. Consequently, many of the operations in dairy manufacturing remain somewhat of an art form. Hopefully, more advances in objective methods of assessing rheological properties of dairy foods will be implemented in the near future. The ultimate aim should be to better assess rheological properties (i.e., properties relating to the flow and deformation of matter) in terms of numerical instrument readings, in an effort to replace the present subjective judgments, which are less reliable and nonprecise. The properties measured instrumentally may include elastic modulus, crushing strength, elasticity, plasticity, shear modulus, superficial density, and other physical parameters.

Tactile Properties of Butter

Since the body of butter is affected markedly by temperature, it is necessary to define the temperature at which tactile properties are evaluated. Butter body properties should be evaluated at a product temperature between 7.2°C and 12.5°C (45°F–55°F). Within this temperature range, the body of butter should be firm, waxy, and consist of such closely knit granules that it appears as a uniform mass. Water and air, in proper amounts, should be uniformly distributed and closely bound by the triglyceride superstructures. The "ideal" butter should cut easily and evenly when sliced and be readily spreadable. When placed in the mouth, butter should not appear greasy, salve-like (salvy), nor contain undissolved salt particles (gritty).

Variations in butter body and texture may be caused by changes in the composition or balance of triglycerides and/or modifications of the manufacturing procedure. The triglycerides of milkfat consist of numerous fatty acids, some of which cause the triglycerides to have either a relatively high or low melting point. The relative proportion of these acids varies seasonally, depending on the type and amount of feed consumed by cows. To overcome these differences in the firmness or softness of milkfat, the buttermaker can compensate by varying the heat treatment of the churning cream, controlling the cooling and holding temperatures of pasteurized cream, adjusting the temperature(s) of churning and/or washing of butter granules, and changing the "working time" of butter within the churn.

Samples of butter for evaluation are generally obtained by a cheese or butter trier, a tool which facilitates collection of a long, cylindrical plug (sample) of butter. Some body and texture characteristics may be assessed by carefully observing the plug and the backside of the metal trier. Butter should not adhere to the back of the trier, nor should there be any visible droplets of water. The plug should be well rounded (symmetrical), have a smooth, waxy surface, and be void of breaks or openings.

Tactile Properties of Cheddar Cheese

The consistency of Cheddar cheese should be determined at 7°C–9°C (45°F–50°F), since body characteristics are quite dependent on sample temperature. The body should be firm, smooth, and reasonably pliable as determined by actual handling or manipulation of the cheese by the evaluator's hand. A plug of cheese removed by means of a trier (similar to that used for butter) should be full and free from openings (closed appearance). When bent, the plug should not break sharply but exhibit some flexibility and plasticity and tear or shred like the breast meat of thoroughly cooked chicken. When worked between the thumb and first two fingers, a well-developed cheese should gradually break down with moderate resistance into a smooth, cohesive mass.

Since Cheddar cheese may be marketed after various stages of aging, from mild (young) to extra sharp (aged), the body characteristics would be expected to vary. The body of a young cheese generally possesses considerably more springiness (rubberiness) than that of an older or aged cheese; it may fail to break down to a uniform consistency when worked between the thumb and forefingers (curdy). Young cheese requires sufficient time for the curd to undergo enzyme-induced changes (proteolysis), which occurs during ripening.

The body and texture properties of a given Cheddar cheese are usually closely correlated with the flavor, since common factors tend to affect both of these sensory categories (Day 1967; Marth 1963; Morris et al. 1966; Ohren and Tuckey 1969; Scott-Blair 1967). For instance, the excessive development of acidity in the cheesemaking process generally produces an acid flavor; simultaneously, high acidity often causes a short (brittle) and mealy body. The presence of gas holes (symmetrical) frequently signifies an atypical or undesirable fermentation, which may produce an associated fruity/fermented, yeasty, or unclean off-flavor. This condition may also be linked to a high moisture content, which may give the cheese a weak and/or pasty (sticky) body. A hard, firm body, on the other hand, may be associated with a lack of typical flavor development (flat). Experienced cheese judges definitely rely on observations of body and texture for a possible link or association with a particular off-flavor(s) or vice versa.

Tactile Properties of Cottage Cheese

The body and texture characteristics of cottage cheese are controlled by such factors as the solids content of the milk, pasteurization temperature, developed acidity (pH), enzyme concentration (if used), rate of heat application in cooking, and final cooking temperature of the curd and whey. There is no general agreement as to the degree of curd firmness that is considered the most desirable, due to variations in consumer preference. However, the minimum requirements for cottage cheese appear to be that the body and texture be smooth, uniformly "meaty" throughout the curd, and individual curd particles should retain their distinct identity and be free from pockets of whey.

Some experienced cottage cheese evaluators describe a well-produced cottage cheese as being "tender—but firm," to applied pressure or contact between the teeth, tongue, and the roof of the mouth. Unlike the procedure used in judging Cheddar cheese for body and texture, the most appropriate way to evaluate the tactile properties of cottage cheese is by determining its various mouthfeel characteristics.

Tactile Properties of Ice Cream and Related Products

The body and texture of ice cream is affected by many factors, such as: the milkfat content, milk-solids-not-fat content, types and relative amounts of sweetening agents, total food solids content, type and amount of stabilizer and/or emulsifier, temperature history of the ingredients, heat treatment of the mix, the temperature, pressure, and stages of homogenization, rate of freezing and hardening, amount of overrun, and temperature fluctuations that occur during storage, delivery, and distribution. Certain other subtle factors, related to the state or condition of milk proteins, may serve to influence the selection of ingredients and the subsequent composition of the mix, or can dictate modification of processing parameters. These factors are believed to be related to minute changes in the salt balance of milk and/or the relative concentration(s) of various milk protein fractions.

When ice cream and related products such as ice milk, sherbet, frozen yogurt, and mellorine (imitation ice cream or ice milk) reach the consumer, their texture should be relatively smooth and homogenous. For the smoothest product, the ice crystals should be extremely small and the air cells should also be small and uniformly distributed. There should be no discernible particles of churned fat. The product should be completely free from any crystallized sugars or lactose, which would tend to impart a "sandy" texture. The body may vary in resistance to bite (or applied tongue pressure) or chewiness, depending on consumer preference. Generally, though, the product should possess some degree of substance and exhibit a defined minimum level of resistance to applied mouth pressure. For a frozen dairy dessert to retain its refreshing and pleasing characteristics, the mouthfeel should not be excessively cold, dry, or astringent. It would be convenient and desirable to have available a practical physical measurement of the body (and texture) of ice cream, since a preferred degree of chewiness is most difficult to define by descriptive terminology. However, in the absence of such objective measurements, the needed steps appear to be: (1) to determine consumers' preferences; (2) to develop and maintain suggested product formulation and processing specifications; and (3) to have the product(s) regularly monitored by competent sensory evaluators.

The freezing point of a given ice cream or related product (which is dependent on both the sugar content and the types of sweetening agents), affects product consistency at any given temperature. Obviously, the temperature of the frozen product, at which the body and texture is evaluated, is of considerable importance. Since the typical serving temperature of ice cream (and related products) varies between −18°C and −12°C (0°F and 10°F), it appears reasonable to select −15°C (5°F) as the standard temperature for presenting these products for evaluation of body and texture.

Tactile Properties of Other Dairy Products

Unfortunately, to enumerate the desirable tactual properties of all dairy products would be a task of considerable proportions and beyond the scope of this book. In many instances, there are no well-established standards, and in other cases guidelines for overall consumer acceptance may be divided or unknown. For example, the body of buttermilk may be light, medium, heavy, or "ropy"; but there seems to be a wide difference in opinion (regional) as to which viscosity level is the most desirable. The viscosity of eggnog is a matter of individual preference, although most persons will agree that it should be more viscous than milk. A similar argument may be made for the viscosity of chocolate milk.

We can make a general statement about the tactile properties of various cultured products, such as sour cream or yogurt, by stipulating that each product should be smooth and homogenous, free of curdiness, and contain no visible "free whey." The preferred consistency of these products is difficult to describe without resorting to quantitative physical measurement, but typically, a moderate degree of firmness, without being excessively gel-like, is considered desirable in these products.

As stated earlier, the tactual features of some dairy products are closely related to other sensory properties. Rheological characteristics cannot be modified without either compromising the overall quality or, indeed, changing the identity of the product, which may, in effect, amount to the development of a new product. This may be desirable, provided that the new product enjoys the necessary measure of consumer acceptance. In addition to Cheddar cheese, some of the more common and popular cheese consumed in this country are creamed cottage, cream, Colby, Monterey Jack, Swiss, Blue, Italian varieties, Edam, Gouda, and Brie, to name a few. Each cheese type has its own characteristic tactile properties which can often be closely correlated with the typical, desired flavor. A Brick cheese, for instance, whose body may have become too soft may taste similar to Limburger cheese, but when the body of Brick cheese seems too firm, it may exhibit a flavor more similar to Cheddar cheese. A firm-bodied Limburger cheese is likely to lack characteristic flavor (Singh and Tuckey 1968), and a soft-bodied Romano cheese would hardly qualify as a grating cheese; in this condition it would probably also be subject to the development of an atypical flavor.

Generally speaking, the consumers' acceptance of a food is influenced to a great extent by the tactile properties exhibited by the product. *The consumer is the final and decisive judge of eating quality.* This is an important tenet that all quality assurance personnel and experienced judges of food products must remember and continuously bear in mind.

COLOR AND APPEARANCE

The importance of color and appearance of dairy products is not just limited to the obvious and important aesthetic reasons, but must be considered in relation to other consumer acceptance criteria. For instance, a faded color in Cheddar cheese may not in itself be highly objectionable, except for its likely relation to excessive acidity, which is usually the original cause of this problem. On the other hand, a most unattractive container, which otherwise satisfies all of the needed functional requirements of packaging, would be objected to primarily on aesthetic grounds.

Color

With respect to color, dairy products fall into two categories: (1) those to which no color is added, and (2) those to which a food coloring agent may be added. Heat treatment or extended storage of some products, such as evaporated milk or powdered milk, can lead to a darkening of color, which may carry over into finished dairy products that use the former as ingredients. In assessing sensory quality, an undesirable color change brought about in this manner should be differentiated from that change in appearance produced by the addition of a flavoring agent or a food colorant.

Unflavored fluid milk and cream and unflavored cultured products as a rule have no color added to them. Milk from individual cows of different breeds may exhibit different shades of cream color, depending on the milkfat content and the carotene content of the fat. Beyond these natural variations, the appearance of other colors should be guarded against, since it may suggest or signify an abnormal condition. An atypical color could be the result of chemical contamination, the growth of certain microorganisms, or (in the case of raw milk on the farm) the presence of blood from diseased or injured animals.

Since both the white color and opacity of milk are due to the colloidal dispersion of casein and the milkfat emulsion, physical changes, such as heating, homogenization, and freezing (which have an effect on the colloidal properties), may produce changes in the product's appearance. No pigment formation is involved here; instead it involves a change in the reflectance properties of some of the product's milk components.

Cheddar cheese, butter, and vanilla ice cream are examples of products to which color may be added at the option of the manufacturer. The only requirement is that Food and Drug Administration-approved colors be used. Processing guidelines are: (1) color to a shade that is natural for the given product; (2) the intensity should be neither "anemic" nor overwhelming; (3) the color should be uniformly distributed throughout the product; and (4) the shade (or color hue) and intensity should be uniform or consistent from day to day. Sudden changes in product color should be avoided, since consumers often interpret this as actually a more severe change in product properties than is the case. A different color, however, may be used to advantage to help introduce a new line of products (Walford 1984).

Flavored dairy products usually have approved food colors added. The product color should be representative of the given flavor. Chocolate flavoring ingredients generally provide their own color; the color intensity is usually a function of the manner in which the chocolate was processed. Since the specific method of processing also determines or affects the flavor, it may occasionally be difficult to obtain the desired combination of color and flavor. In instances where artificial color is added to help simulate fruits and other flavors, the same general requirements for food colorants should be adhered to as were discussed in the preceding paragraph.

Appearance

If possible, the term "appearance" should be carefully defined as to its scope and application for each product to which it is applied. Frequently, the evaluation of "product appearance" may be difficult to distinguish from the entities of "color" and "package."

For unflavored dairy products, the overall appearance is probably best described in terms of the color and package, except for a few possible undesirable characteristics such as a cream plug, a cream line, or a cellular sedimentation in homogenized milk. The chance discovery of foreign objects in a dairy product is a particularly discomforting observation to anybody, especially consumers.

Usually, flavored dairy products require a more careful visual examination of appearance than unflavored products. In low-viscosity products there should be little or no stratification or settling (precipitation) of the flavoring material. Other products, such as ice cream flavored with fruit, candy, nuts, or a variegating syrup, require a "good showing" or "ingredient display" as well as an even, pleasing distribution of the flavoring substance(s).

In Cheddar cheese, the appearance and product "finish" are customarily considered jointly. In this case, these terms refer to an assessment or evaluation of the cheesemaker's demonstrated skill and workmanship as applied to the outer covering of the cheese. As a rule, the exterior surfaces of most cheese should be sound, smooth, clean, and attractive. A damaged or faulty cheese exterior may lead to economic loss by encouraging deterioration at the cheese surface. Subsequently, poor workmanship on the surface finish may lead to a penetrating, off-flavor development within the cheese interior. In Swiss cheese, the extreme importance directed to the size and distribution of the "eyes" could conceivably be evaluated under the heading of appearance. The magnitude and visual impact of *Penicillium* mold growth in Blue and related cheese varieties is certainly an observation for product appearance.

Appearance is an important sensory observation for cultured dairy products. For fluid products, a general objective is the absence of curdiness and free whey. Preferably, cottage cheese particles should not be shattered, matted (clumped), uneven in size or shape, or gelatinous in appearance. There should be little or no evidence of either under- or over-creaming, or a failure of the curd particles to absorb some of the cream (free cream), or the appearance of "free whey." Sour cream and plain yogurt should present a slight to moderate degree of surface gloss (sheen) and convey an overall appearance of having a smooth, homogenous consistency. No shrinkage of the milk gel (clot) and/or associated "free whey" should be evident.

Package

Food product packages must be both functional and aesthetic; well-designed dairy food containers should simultaneously satisfy both criteria. The desirability for attaining an attractive container that has lasting consumer appeal is obvious. However, it is not necessary for elaboration here, since this is more appropriately a focus and concern of marketing activities. In the interest of quality assurance, a functional package should protect the given dairy product against all known applicable modes of quality deterioration, including sensory properties.

Assuming that a package of appropriate material(s) and the proper design has been selected, the attention of the product evaluator should be directed to the following points: (1) the overall cleanliness of the container; (2) the soundness of the seal; (3) the correct level of fill; (4) the exterior or interior surfaces for possible damage to the container (including deformation or stress cracking); and (5) legibility of the product manufacturing expiration (sell-by) date or lot code. The significance of each of these points should be self-evident. An improperly filled or sealed container, or a defective or damaged one, is likely to subject the product to the harmful effects of contamination, or at least unwanted contact with air and/or odors from the storage environment. For example, both the flavor and the tactile properties of ice cream may be affected by an incompletely filled container. The air (oxygen) in the space above the product level promotes surface dehydration, possible development of a "surface taint" and under certain conditions, simultaneously, facilitates product shrinkage. This is considered to be a somewhat serious body/texture and appearance deterioration of ice cream.

This chapter has attempted to provide a general discussion of the more common flavor, body/texture, and color/appearance defects (or characteristics in some instances) of the more popular dairy products in the U.S. Subsequent chapters, through a product-by-product approach, will attempt to deal with specifics of the "what, when, why, and how" of recognizing, "labeling," and controlling (or eliminating) many of these and other sensory shortcomings of dairy products.

REFERENCES AND BIBLIOGRAPHY

Adams, D. M., Bausch, J. T., and Speck, M. L. 1976. Effect of psychrotrophic bacteria from raw milk on milk proteins and stability of milk proteins to ultra-high temperature treatment. *J. Dairy Sci. 59*:823.

Allen, C. and Parks, O. W. 1979. Photodegradation of riboflavin in milks exposed to fluorescent light. *J. Dairy Sci. 62*:1377.

Allen, J. C. and Hamilton, R. J. (Editors). 1983. *Rancidity in Foods.* Elsevier Sci. Publ. Co., Inc. New York.

Al-Shabibi, M. M. A., Langner, E. H., Tobias, J. and Tuckey, S. L. 1964. Effect of added fatty acids on flavor of milk. *J. Dairy Sci. 47*:295.

Anderson, D. F. and Day, E. A. 1966. Quantitative evaluation and effect of certain microorganisms on flavor components of blue cheese. *J. Agr. Food Chem. 14*:241.

Andersson, R. E., Danielsson, G., Hedlund, C. B., and Svensson, S. G. 1981. Effect of heat-resistant microbial lipase on flavor of ultra-high temperature sterilized milk. *J. Dairy Sci. 64*:375.

Arnold, R. G., Libbey, L. M., and Day, E. A. 1968. Identification of components in the stale flavor fraction of sterilized concentrated milk. *J. Food Sci. 31*:566.

Badings, H. T. 1984. Flavors and Off-flavors. *In: Dairy Chemistry and Physics.* P. Walstra and R. Jenness. John Wiley and Sons. New York. 336.

Badings, H. T. and Neeter, R. 1980. Recent advances in the study of aroma compounds of milk and dairy products. *Neth. Milk Dairy J. 34*:9.

Baron, M. 1952. *The Mechanical Properties of Cheese and Butter.* Dairy Industries. London, U.K.

Begemann, P. H. and Koster, J. C. 1964. Components of butterfat. *4-cis*-heptenal: a cream flavored component of butter. *Nature 202*:552.

Bodyfelt, F. W. 1982. Processors need to put some pinch on salt. *Dairy Record 83*(4):83.

Bodyfelt, F. W., Andrews, M. V., and Morgan, M. E. 1979. Flavors associated with the use of Cheddar cheese whey powder in ice cream mix. *J. Dairy Sci. 62*:51.

Bodyfelt, F. W., Morgan, M. E., Scanlan, R. A., and Bills, D. D. 1975. A critical study of the multiuse polyethylene plastic milk container system. *J. Milk Food Tech. 39*:481.

Bradfield, A. 1962. Causes and prevention of some undesirable flavors in milk. *Vermont Univ. Agr. Expt. Sta. Bull.* 624.

Bradley, Jr., R. L. 1980. Effect of light on alteration of nutritional value and flavor of milk: A review. *J. Food Prot. 43*(4):314.

Bruhn, J. C., Franke, A. A., and Goble, G. S. 1975. Factors relating to the development of spontaneous oxidized flavor in raw milk. *J. Dairy Sci. 59*:828.

Chen, C. C. W. and Tobias, J. 1972. Migration of copper between different fractions of milk. *J. Dairy Sci. 55*:759.

Crocker, E. C. 1945. *Flavor.* McGraw-Hill Book Co., Inc. New York.

Day, E. A. 1967. Cheese Flavor. *In: The Chemistry and Physiology of Flavors*, E. A. Day, H. W. Schultz, and L. M. Libbey (Editors), AVI Publ. Co. Westport, CT. 331.

Dougherty, R. W., Shipe, W. F., Gudnason, G. V., Ledford, R. A., Peterson, R. D., and Scarpellino, R. 1962. Physiological mechanisms involved in transmitting flavors and odors to milk. I. Contribution of eructated gases to milk flavor. *J. Dairy Sci. 45*:472.

Dunkley, W. L., Franklin, J. D., and Pangborn, R. M. 1962. Influence of homogenization, copper, and ascorbic acid on light-activated flavor in milk. *J. Dairy Sci. 45*:1040.

Dunkley, W. L., Franke, A. A., and Robb, J. 1968. Tocopherol concentration and oxidative stability of milk from cows fed supplements of d- or dl-α-tocopheryl acetate. *J. Dairy Sci. 51*:531.

Duthie, A. H., Hosmer, S. P., Aleon, J., Wulff, S., Fox, J., Jensen, L. A., Ryan, J. J., and Atherton, H. V. 1983. Vermont fair data suggest market potential for lipolyzed milks. (Abstract) *J. Dairy Sci. 66*:88 Suppl. 1.

Finley, J. W., Shipe, W. F., and O'Sullivan, A. C. 1967. Observations relative to light induced off-flavors in milk. *J. Dairy Sci. 50*:983.

Flake, J. C., Jackson, H. C. and Weckel, K. G. 1938. Studies on the activated flavor of milk. *J. Dairy Sci. 21*:A145.

Forss, D. A. 1964. Fishy flavor in dairy products. *J. Dairy Sci. 47*:245.

Forss, D. A. 1969. Flavors of dairy products. *J. Dairy Sci. 52*:832.

Forss, D. A. 1979. Mechanism of formation of aroma compounds in milk and milk products. *J. Dairy Res. 46*:691.

Gould, R. F. 1966. Flavor Chemistry. Advances in Chemistry Series 56. American Chemical Society, Washington, D.C.

Hammond, E. G. and Seals, R. G. 1972. Oxidized flavor in milk and its simulation. *J. Dairy Sci. 55*:1567.

Herreid, E. O., Ruskin, B., Clark, G. L., and Parks, T. B. 1952. Ascorbic acid and riboflavin destruction and flavor development in milk exposed to the sun in amber, clear, paper, and ruby bottles. *J. Dairy Sci. 35*:772.

Hetrick, J. H. and Tracy, P. H. 1948. Effect of high-temperature short-time treatments on some properties of milk. II. Inactivation of the lipase enzyme. *J. Dairy Sci. 31*:881.

Hunziker, O. F. 1923. Selection of metals in the construction of dairy equipment. *Proc. World's Dairy Cong. 2*, 1189.

Hunziker, O. F., Cordes, W. A., and Nissen, B. H. 1929. Metals in dairy equipment— metallic corrosion in milk products and its effect on flavor. *J. Dairy Sci. 12*:140.

International Dairy Federation. 1974. Lipolysis in cooled bulk milk. IDF Document No. 82. Rome, Italy.

International Dairy Federation. 1975. Proceedings of lipolysis symposium. IDF Document No. 86. Rome, Italy.

Jenness, R. and Patton, S. 1959. *Principles of Dairy Chemistry.* John Wiley and Sons, New York.

Jeon, I. J., Thomas, E. L., and Reineccius, G. A. 1978. Production of volatile flavor compounds in ultra-high temperature processed milk during aseptic storage. *J. Agric. Food Chem. 26*:1183.

Johnson, P. E. and Von Gunten, R. L. 1962. A study of factors involved in the rancid flavor of milk. *Oklahoma Agr. Expt. Sta. Bull.* B-593.

Josephson, D. V. 1946. Some obversations regarding the effect of various wave lengths of light on riboflavin content and flavor of milk. *J. Dairy Sci. 29*:508.

Josephson, D. V. and Keeney, P. G. 1947. Relation of acetone bodies to "cowy" flavor in milk. *Milk Dealer 36*(10):40.

Juffs, H. S. 1973a. Proteolysis detection in milk, I. *J. Dairy Res. 40*:371.

Juffs, H. S. 1973b. Proteolysis detection in milk, II. *J. Dairy Res. 40*:383.

Juffs, H. S. 1975. Proteolysis detection in milk, III. *J. Dairy Res. 42*:31.

Landsberg, J. D., Bodyfelt, F. W., and Morgan, M. E. 1977. Retention of chemical contaminants by glass, polyethylene and polycarbonate multiuse milk containers. *J. Food Prot. 40*:772.

Langler, J. E., Libbey, L. M., and Day, E. A. 1966. Volatile constituents of Swiss cheese. *J. Dairy Sci. 49*:709.

Langer, E. G. and Tobias, J. 1967. Isolation and characterization of ether soluble sugar-amino acid interaction products. *J. Food Sci. 32*:495.

Law, B. A. 1981. The formation of aroma and flavor compounds in fermented dairy products. *Dairy Sci. Abstracts 43*:143.

Lindsay, R. C. 1967. Cultured dairy products. In: *Chemistry and Physiology of Flavors*. H.W. Schultz, E. A. Day, and L. M. Libbey (Editors). AVI Publ. Co. Westport, CT. 315.

Little, Arthur D., Inc. 1958. *Flavor Research and Food Acceptance.* Reinhold Publ. New York, NY.

Ludzinska, D., Pijanowski, E., and Zmailicki, S. 1970. Proteolytic and lipolytic changes in raw milk stored at different temperatures. *Roczn. Technol. Chem. Zywn. 18*:45.

Marth, E. H. 1963. Microbiological and chemical aspects of Cheddar cheese ripening: A review. *J. Dairy Sci. 46*:869.

Mehta, R. S., Bassette, R., and Ward, G. 1974. Trimethylamine responsible for fishy flavor in milk from cows on wheat pasture. *J. Dairy Sci. 57*:285.

Moncrieff, R. W. 1967. *The Chemical Senses.* Chemical Rubber Co. Cleveland, OH. 760 pp.

Morgan, M. E. 1976. The chemistry of some microbiologically induced flavor defects in milk and dairy foods. *Biotechnol. and Bioeng. 18*:953.

Morris, H. A., Angelini, P., McAdoo, D. J., and Merritt, C. J. 1966. Identification of volatile components of Cheddar cheese. *J. Dairy Sci. 49*:710.

Muck, G. A., Tobias, J., and Whitney, R. M. 1963. Flavor of evaporated milk. I. Identification of some compounds obtained by the petroleum ether solvent partitioning technique from aged evaporated milk. *J. Dairy Sci. 46*:774.

Ohren, J. A. and Tuckey, S. L. 1969. Relation of flavor development in cheddar cheese to chemical changes in the fat of the cheese. *J. Dairy Sci. 52*:598.

Parks, O. W., Keeney, M., and Schwartz, D. P. 1963. Carbonyl compounds associated with the off-flavor in spontaneously oxidized milk. *J. Dairy Sci. 46*:295.

Patton, S. 1954. The mechanism of sunlight flavor formation in milk with special reference to methionine and riboflavin. *J. Dairy Sci. 37*:446.

Patton, S. 1955. Browning and associated changes in milk and its products: A review. *J. Dairy Sci. 38*:457.

Patton, S., Forss, D. A., and Day, E. A. 1956. Methyl sulfide and the flavor of milk. *J. Dairy Sci. 39*:1469.

Prentice, J. H. (Editor). 1984. *Measurements in the Rheology of Foodstuffs.* Elsevier Sci. Publ. Co., Inc. New York.

Reynolds, T. M. 1963. Chemistry nonenzymatic browning. I. The reaction between aldoses and amines. *Adv. Food Res. 12*:1.

Reynolds, T. M. 1965. Chemistry of nonenzymatic browning. II. *Adv. Food Res. 14*:167.

Samuelsson, E. G. 1962. Experiments on sunlight flavor in S[35] labelled milk. *Milchwissenschaft 17*:401.

Scanlan, R. A., Sather, L. A., and Day, E. A. 1965. Contribution of free fatty acids to the flavor of rancid milk. *J. Dairy Sci.* 48:1582.

Scanlan, R. A., Lindsay, R. C., Libbey, L. M., and Day, E. A. 1968. Heat induced volatile compounds in milk. *J. Dairy Sci.* 51:1001.

Schultz, H. W., Day, E. A., and Libbey, L. M. (Editors). 1967. *The Chemistry and Physiology of Flavors.* AVI Publ. Co. Westport, CT. 552 pp.

Scott-Blair, G. W. 1967. Rheological properties of dairy products. *Rheol. Acta* 6:(3):201.

Shipe, W. F., Ledford, R. A., Peterson, R. D., Scanlan, R. A., Geerken, H. F., Dougherty, R. W., and Morgan, M. E. 1962. Physiological mechanisms involved in transmiting flavors and odors to milk. II. Transmission of some flavor components of silage. *J. Dairy Sci.* 45:477.

Shipe, W. F., Bassette, R., Deane, D. D., Dunkley, W. L., Hammond, E. G., Harper, W. J., Kleyn, D. H., Morgan, M. E., Nelson, J. H., and Scanlan, R. A. 1978. Off-flavors in milk: Nomenclature, standards, and bibliography. *J. Dairy Sci.* 61:855.

Shipe, W. F., Senyk, G. F., and Fountain, K. B. 1980. A modified copper soap solvent extraction method for measuring the free fatty acids in milk. *J. Dairy Sci.* 63:193.

Singh, S. and Tuckey, S. L. 1968. Chemical changes in the fat and protein of limburger cheese during ripening. *J. Dairy Sci.* 52:942.

Speck, M. L. and Adams, D. M. 1976. Symposium: Impact of heat stable microbial enzymes in food processing. Heat-resistant proteolytic enzymes from bacterial sources. *J. Dairy Sci.* 59:786.

Stark, W. and Forss, D. A. 1964. A compound responsible for mushroom flavour in dairy products. *J. Dairy Res.* 31:253.

Stark, W. and Forss, D. A. 1966. n-Alkan-1-ols in oxidized butters. *J. Dairy Res.* 33:31.

Stark, W., Smith, J. F., and Forss, D. A. 1967. n-Pent-1-en-3-ol and n-Pent-1-en-3-one in oxidized dairy products. *J. Dairy Res.* 34:123.

Strobel, D. R., Bryan, W. G., and Babcock, C. J. 1953. Flavors of milk. A review of literature. *U.S. Department of Agriculture Bull.* Washington, D.C. 91 pp.

Stull, J. W. 1953. The effect of light on activated flavor development and on the constituents of milk and its products: A review. *J. Dairy Sci.* 36:1153.

Swaisgood, H. E. 1980. Sulfhydryl oxidase: Properties and applications. *Enzyme Microb. Technol.* 2:265.

Thomas, E. L., Nielson, D. J., and Olson, Jr., J. C. 1955. Hydrolytic rancidity in milk—a simplified method for estimating the extent of its development. *Am. Milk. Rev.* 17(1):50.

Thomas, E. L. 1981. Trends in milk flavors. *J. Dairy Sci.* 64:1023.

Thurston, L. M. 1937. Theoretical aspects of the causes of oxidized flavor, particularly from the lecithin angle. *Proc. 30th Ann. Conv. Intern. Assoc. Milk Dealers, Lab. Sect.* 30:143.

Tracy, P. H., Ramsey, R. J., and Ruehe, H. A. 1933. Certain biological factors related to talloviness in milk and cream. *Illinois Univ. Agr. Expt. Sta. Bull.* 389.

Walford, J. (Editor). 1984. *Developments in Food Colours—2.* Elsevier Sci. Publ. Co., Inc. New York.

Walstra, P. and Jenness, R. 1984. *Dairy Chemistry and Physics.* John Wiley and Sons, New York, NY.

Weinstein, B. R. and Trout, G. M. 1951a. The solar-activated flavor of homogenized milk. II. The role of oxidation and the effectiveness of certain treatments. *J. Dairy Sci.* 34:559.

Weinstein, B. R. and Trout, G. M. 1951b. The solar activated flavor of homogenized milk. III. Effect of deaeration, surface area of fat globules, and relation of the Kreis test. *J. Dairy Sci.* 34:568.

White, C. H. and Marshall, R. T. 1973. Reduction of shelf-life of dairy products by a heat-stable protease from *Pseudomonas fluoresens* P26. *J. Dairy Sci. 56*:8.

Wilkinson, R. A. and Stark, W. 1967. A compound responsible for metallic flavor in dairy products. II. Theoretical consideration of mechanism of formation of oct-1-ene-3-one. *J. Dairy Res. 34*:89.

Willey, H. A. and Duthie, A. H. 1969. Evidence for existence of more than one type of rancid flavor. *J. Dairy Sci. 52*:277.

Wishner, I. A. and Keeney, M. 1963. Carbonyl pattern of sunlight-exposed milk. *J. Dairy Sci. 46*:785.

5

Sensory Evaluation
of Fluid Milk
and Cream Products

The sensory evaluation of milk, in both the bulk and packaged form, is of utmost importance to the market (fluid or beverage) milk industry. The sale of packaged fresh milk comprises a major share of the U.S. dairy industry. Since fluid milk is consumed regularly by people of all ages and most ethnic groups, this product is constantly being assessed for quality by consumers. If the flavor of milk is not appealing or appetizing, less of it will be consumed. Furthermore, off-flavored milk may cast an unfavorable reflection on other dairy products that are sold or distributed under the same brand name and thus unfavorably affect sales of those products as well.

The sensory characteristics of any dairy product is most dependent on the quality attributes of the milk ingredient(s) used to produce them. An important truism of the dairy industry is that "finished milk products can be no better than the ingredients from which they are made." The quality and freshness of the various milk and cream components is most critical to product sales. Most flavor defects of finished dairy products could be substantially minimized, or perhaps eliminated, if all dairy manufacturers would more critically assess the essential quality parameters of all ingredients, especially the milk-based ones.

It is generally conceded among dairy product judges that the scoring or differentiation of milk into different quality classes (known as grading) demands keener, more fully developed senses of smell and taste than does the sensory evaluation of other dairy products. Many of the off-flavors present in fluid milk are more delicate, less volatile, or otherwise more elusive than those encountered in other dairy products.

Since milk (or cream) is the basic material from which all dairy products are made, it behooves milk producers, dairy processors, distributors, and other personnel involved with dairy products to be aware of how various flavor defects of milk affect the quality of manufactured products. Processing personnel should have the ability to detect off-flavors in milk and be able to assess or project the impact of these on the flavor quality of finished dairy products.

CLASSES OF MILK

Milk may be divided into two general classes, namely, *market milk* (Grade A) and *manufacturing grade milk*.

Market Milk. "Market" or "beverage" milk is typically consumed in the fluid form. It is processed, packaged, and retailed or distributed to the homemaker, restaurant, hotel, or other food service institution, where it is used for either beverage or culinary purposes. This product form reaches the consumer in the natural, fluid state, as contrasted to milk forms that may be converted into frozen dairy desserts, cheese, butter, fermented milk foods, concentrated milk, or other types of dairy products.

In the U.S., market milk is nearly always "Grade A pasteurized" milk. The Grade A Pasteurized Milk Ordinance (P.M.O.) specifies requirements for the production of *Grade A raw milk for pasteurization* and regulations that pertain to pasteurization equipment and procedures, physical facilities, containers, packaging, sealing, and storage of finished products. The P.M.O. is recommended by the U.S. Department of Health and Human Services, Public Health Service/Food and Drug Administration, Washington, D.C. 20204. At this time, the latest available document that specifies Grade A milk requirements is the 1978 P.M.O. (Fig. 5.1.). The pasteurization ordinances adopted by individual states and communities may differ in some respects, but the 1978 P.M.O. recommends that only Grade A pasteurized milk (or Certified pasteurized milk) and milk products be sold to consumers, restaurants, food service operators, grocery stores, or similar establishments. A provision in the P.M.O. is made for the sale of ungraded pasteurized milk in an emergency situation, but it must be distinctly labeled "ungraded."

Market milk is used primarily for consumption as whole milk or may be centrifugally separated to produce either lowfat milk, skim milk, light cream, whipping cream, and/or half-and-half. Some of the aforementioned products may be flavored or fermented. This class of milk may be grouped or further categorized with respect to the particular heat treatment to which the milk is subjected in processing, namely as pasteurized or ultrapasteurized.

Manufacturing Grade Milk. "Manufacturing grade milk" is basically any milk intended for processing into dairy products other than

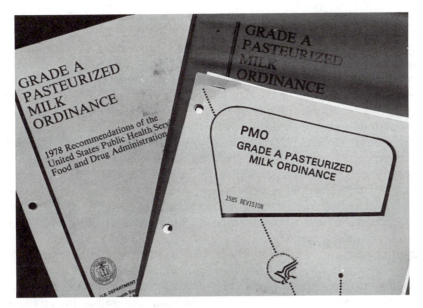

Fig. 5.1. The USPHS/FDA Grade A Pasteurized Milk Ordinance (PMO)—1978 recommendations (with appropriate revisions) serves as an important model code for most states and fluid milk and cream products in interstate commerce.

market (fluid or beverage) milk. Such milk generally does not comply with the specific sanitation and production facilities standards established for producing Grade A raw milk. Recommended requirements for manufacturing grade milk have been issued by the U.S. Department of Agriculture, Consumer and Marketing Service, under the title "Milk for Manufacturing Purposes and Its Production and Processing (1972)."

PRECAUTIONS FOR EVALUATING RAW MILK

The 1978 P.M.O. contains the following statement: "Compilation of outbreaks of milkborne diseases by the U.S. Public Health Service, over many years, indicates that the risk of contracting disease from raw milk is approximately 50 times as great as from milk labeled 'pasteurized'." This implies that, even though raw milk samples should not be swallowed, there is an inherent risk in tasting them. Smelling raw milk samples, rather than tasting them, is substantially less objectionable, especially if none of the milk comes in contact with the mouth of the person performing the sensory evaluation (for possible off-odor(s)).

If tasting of the given samples of milk is imperative, then small milk quantities should be "laboratory pasteurized." There is no standard procedure for performing this. Hence, good technique must be employed to insure that every particle of the milk sample has been subjected to the minimum pasteurization temperature for the required time period to render it pathogen-free. Some possible heating combinations are: (1) 65.5°C (150°F) for 30 minutes; (2) 70°C (158°F) for 10 minutes; or (3) 74°C (165°F) for 2 minutes. The authors stress that the aforementioned temperatures are intended to be actual, correctly measured milk sample temperatures, not temperatures of the water bath or other heating media. Timing should not begin until the sample has reached the required temperature. Some agitation of "heated" milk samples is advised, since all milk particles within any portion of the sample vessel must be properly heated to insure "complete pasteurization" of the milk sample(s). One approach is placement of raw milk samples into appropriate-sized test tubes (identity labeled), and insertion into plastic or metal racks for subsequent immersion into a heated water bath. The tubes must be clean (and sterile) so as not to impart off-flavors to the samples. There must be no milk residue on the upper portion of the test tubes, or the entire sample will not receive the required heat treatment. Sensory detection of serious off-flavors in raw milk is not affected by any of the above-listed laboratory pasteurization conditions (Bodyfelt 1983).

GRADES OF MARKET MILK

More and more, health officials and dairy processors are recognizing the practicality and economic reality of a "single grade" of milk for human consumption. This is particularly true for market milk. The 1978 P.M.O. refers to the Code of Federal Regulations (CFR), Title 21, Section 131.110, for the following legal definition of milk:

> Milk is the lacteal secretion, practically free from colostrum, obtained by the complete milking of one or more cows. Milk that is in final package form for beverage use shall have been pasteurized or ultrapasteurized, and shall contain not less than 8.25% milk solids-not-fat and not less than 3.25% milk-fat. Milk may have been adjusted by separating part of the milkfat therefrom or by adding thereto cream, concentrated milk, dry whole milk, skim milk, concentrated skim milk or nonfat dry milk. Milk may be homogenized.

The various whole milk products which may require sensory evaluation include those listed below.

Raw Milk

Certified. "Certified raw milk for pasteurization" is that milk which conforms to the latest requirements of the American Association of Medical Milk Commissions, Inc., Rt. 3, Box 34, Alpharetta, Georgia 30201.

Grade A. "Grade A raw milk for pasteurization" is that milk which conforms to the latest regulations and the highest standards established by the United States Public Health Service, Pasteurized Milk Ordinance (P.M.O.) and Code (1978). Grade A milk may also be produced under a given state's regulations, which usually conform closely to the federal standards, but may be slightly more rigorous for certain criteria.

Pasteurized Milk

Certified. "Certified pasteurized milk" is certified raw milk which has been pasteurized, cooled, and bottled in a milk plant which conforms with the regulations for Grade A pasteurized milk, as well as the Medical Milk Commission requirements.

Grade A Pasteurized Milk. This is Grade A raw milk which has been pasteurized in accordance with the regulations of the United States Public Health Service Pasteurized Milk Ordinance and Code. Such milk must meet all the regulations, pasteurization confirmation tests, and sanitary requirements for this grade of milk.

Kinds of Market Milk and Associated Quality Characteristics

Whole Milk. "Whole milk," or simply "milk" may be pasteurized, ultrapasteurized, or commercially sterilized. Pasteurized milk is milk which has been subjected to pasteurization temperatures for a prescribed period of time, in equipment which complies with the requirements of the P.M.O. With respect to times and temperatures of pasteurization, the 1978 P.M.O. states:

> Every particle of milk or milk product is heated in properly designed and operated equipment to one of the temperatures specified in the following table and held continuously at or above that temperature for at least the time specified:

Temperature		Time
63°C*	(145°F)	30 minutes
72°C*	(161°F)	15 seconds
89°C	(191°F)	1 second
90°C	(194°F)	0.5 second
94°C	(201°F)	0.1 second
96°C	(204°F)	0.05 second
100°C	(212°F)	0.01 second

*If the milkfat content of a milk product is 10% or more, or if it contains added sweeteners, the specified temperature shall be increased by 3°C (5°F), *provided* that eggnog shall be heated to at least the following temperature and time combinations:

Temperature		Time
69°C	(155°F)	30 minutes
80°C	(175°F)	25 seconds
83°C	(180°F)	15 seconds

When a minimum 30 minute holding time is required, the pasteurization process is known as the "batch" or "holding" method; with holding times less than this, but greater than 1 second, the process is termed "high temperature-short time pasteurization" (HTST); and with holding times of 1 second or less, the designation is "higher heat-short time pasteurization" (HHST). Ultrapasteurization requires heating to 138°C (280°F) for at least 2 seconds, either before or after product packaging. The term "ultra-high temperature" (UHT) designates a process for "commercially sterilizing" milk at a temperature of about 149°C (300°F) or higher, with a holding time of a few seconds. The sterile product is then aseptically packaged in sterile containers. The equipment used for milk pasteurization or sterilization (Hall and Trout 1967; Harper and Hall 1976; Henderson 1971; Herreid and Tobias 1959) varies widely in design (see Fig. 5.2). Some plants employ vats, while others use plate heat exchangers, tubular heaters or other forms of heat exchangers. Most commonly, heating is achieved by an *indirect* approach through heated metal surfaces, but there are processing units which heat by introducing steam into the product (*direct*). A vacuum chamber removes the water added to milk due to the condensed steam.

Pasteurized milk commonly possesses some degree of a heated or cooked flavor, especially immediately after processing, but the intensity of the cooked flavor diminishes during storage (Gould and Sommer 1939; Gould 1940; Patton *et al.* 1956; Sliwkowski and Swaisgood 1980; Swaisgood 1980). The initial flavor intensity depends on the tem-

Fig. 5.2.   Examples of typical vat (A), HTST (B), and UHT (C) pasteurization systems for fluid milk products. ((C) Courtesy of Dasi Industries, Inc., Chevy Chase, MD.)

Fig. 5.2. (C) *(continued)*

perature and holding time employed, as well as the method of heating. The factors which may influence milk flavor include: (1) heating-up and cooling time; (2) temperature difference between the product and the heating medium; (3) velocity of the product in a continuous system; (4) occurrence of product "burn-on"; and (5) direct versus indirect heating methods.

Currently, practically all milk marketed in the U.S. is both pasteurized and homogenized. "Homogenized" is defined in the 1978 P.M.O. as follows:

> The term 'homogenized' means that milk or the milk product has been treated to insure breakup of the fat globules to such an extent that, after 48 hours of quiescent storage at 7°C (45°F), no visible cream separation occurs on the milk; and the fat percentage of the top 100 milliliters of milk in a quart, or of proportionate volumes in containers of other sizes, does not differ by more than 10% from the fat percentage of the remaining milk as determined after thorough mixing.

As pointed out previously, the definition of milk in Title 21 of the CFR ends with the simple statement, "Milk may be homogenized." Except for being homogenized, homogenized milk does not differ in composition or any other provision of the definition from unhomogenized milk. However, there are some differences between the two products in their susceptibility to development of certain off-flavors (Doan 1933; Halloran and Trout 1932; Trout 1940, 1941, 1950); for this reason we shall examine them separately.

Unhomogenized Milk. Since pasteurization standards represent the minimal time and temperature requirements, milk is frequently heated in excess of the minimum. However, it is less likely that unhomogenized milk would be heated much above the minimum requirements because the cream line, which is the unique characteristic of this product, is progressively reduced in volume by increasing the heat treatment. Therefore, it is also less likely that this product will ever exhibit a strong "cooked" flavor. Immediately after pasteurization, milk may manifest a definite "heated" flavor erroneously described as a "cooked" flavor. During storage, the "heated" flavor diminishes in intensity and may entirely disappear, especially if significant levels of divalent cations are present in the milk (from water or equipment), as pointed out by Gould (1940). The flavor of pasteurized, unhomogenized milk has a marked tendency to undergo flavor changes during storage as follows:

Heated→ normal→ flat→ metallic→ oxidized.

The extent of the flavor deterioration depends on the storage time, season of the year, the type of roughage fed to cows, and the relative levels of cupric or ferric ions.

Unhomogenized milk is particularly susceptible to the cardboard-like oxidized off-flavor which results from the oxidation of lipids. This tendency toward oxidation is usually greater in the winter months and/ or when pasture or green feeds are not available. The bitter, rancid (soapy-like) off-flavor that is encountered in raw milk, due to the hydrolysis of triglycerides, should not develop in properly pasteurized milk. If a lipolytic defect is noted, either this off-flavor was present when the milk was pasteurized, or the milk was contaminated after pasteurization with psychrotrophic bacteria (that possess lipase activity).

Homogenized Milk. There are several properties and flavor characteristics of homogenized milk that differentiate it from unhomogenized milk. First, since there is little or no concern about a cream line in homogenized milk, higher processing temperatures may be employed at the option of the manufacturer, with a resultant higher incidence and/or greater intensity of the cooked flavor. This occurs not only in ultrapasteurized and sterilized milks (or cream products), but frequently with pasteurized products as well.

Second, if milk is homogenized while still raw, before lipase enzymes are destroyed, or if it is not pasteurized immediately after homogenization, or if the pasteurized milk is contaminated with raw milk, rancidity rapidly develops. Such milk exhibits distinct hydrolytic rancidity within a few hours after processing and becomes quite bitter and soapy within 24 hours when stored at a low temperature. Dorner and Widmer (1931) and Halloran and Trout (1932), who worked independently, showed that all cows' milk is subject to the development of rancidity upon homogenization, unless adequately heat-treated to inactivate the indigenous lipase of milk. Doan (1933) found that the critical temperature for inhibiting rancidity development in homogenized milk by flash heating was $\geq 63.9\,°C$ ($\geq 147\,°F$). Furthermore, it must be emphasized that raw milk must never be mixed with homogenized milk in the course of processing, or a rancid off-flavor is almost certain to occur. Hence, if a rancid off-flavor is encountered in homogenized milk, one should assume that either: (1) all the milk ingredients were not adequately heat-treated, or (2) rancidity existed within the milk prior to the pasteurization process.

Third, homogenized milk is distinctly less susceptible to development of the metal-induced, cardboardy, or oxidized off-flavor. This was first noted in studies by Tracy et al. (1933), and later substantiated by other researchers. If homogenized milk products are properly pasteurized, properly refrigerated, and not unduly exposed to light, then the pleasant, rich flavor of these homogenized products should remain fixed and stable for a considerable time. This period of flavor stability is in excess of that within which unhomogenized, pasteurized milk might be expected to exhibit some degree of flavor deterioration.

Fourth, homogenized milk is more susceptible to the development of the light-activated or light-induced off-flavor (sometimes also referred to as "sunshine flavor") when exposed to light than unhomogenized milk, as pointed out by Hood and White (1934). This off-flavor has a burnt-protein (or burnt-feathers) character and should not be confused with the cardboardy taste and puckery mouthfeel sensation of the generic oxidized flavor.

Sedimentation. In homogenized milk, the absence of milkfat separation may prompt any destabilized protein, colloidal form of soil, or any yeast and somatic (body) cells to readily precipitate and form a yellowish to smokey-grey layer on the bottom of the container. When the milk container is agitated slightly, or the milk is heated moderately, this deposit may clump into feathery, wooly, or oily-appearing masses that resemble soil, oil, or extraneous material in milk. A milk judge should be familiar with the possibility of sedimentation in homogenized milk as well as with its characteristic behavior upon handling. Freshly packaged homogenized milk subjected to proper refrigeration and little or no agitation, generally shows no sediment formation when evaluated six to eight hours later. However, the same milk examined after the elapse of 24 hours, or after some agitation, might show considerable sediment. Obviously, sedimentation is more readily noted in transparent or translucent containers. Clarification (centrifugation) of milk is routinely practiced in the dairy industry to preclude objectionable sedimentation problems.

Watery Appearance. If homogenized milk is allowed to freeze and then slowly defrost, the upper portion usually appears watery due to precipitation of some of the milk solids, including milkfat (Hood and White 1934 and Trout 1940, 1941). A competent milk judge will have become familiar with the behavior of homogenized milk under some of these unfavorable conditions of environment and storage so that "suspect" milk samples are not unduly criticized for possible water adulteration.

Cream Layer, Cream Plug, or Fat Ring. If homogenized milk is inadequately processed, temperature abused, agitated severely or held for an extended time at room temperature, it may form objectionable cream layers, cream plugs, or fat rings (sometimes referred to as "spaghetti") of varied intensity.

Vitamin-Fortified Whole Milk. The 1978 P.M.O. and Title 21 of the Federal CFR do not contain a separate definition for vitamin-fortified whole milk. Vitamin addition is recognized as optional within the definition of milk, but specific provisions are given only for vitamin A (2000 International Units) and vitamin D (400 International Units) per quart. Safe and suitable carriers (fat solvents) for vitamins A and D are also permitted. The added vitamins themselves apparently do not impair the flavor of fortified milk, but industry experience has shown that occasionally the vitamin carriers may be suspected of introducing some degree of off-flavor. Certain preparations of vitamin A concentrate have been known to impart a detectable, objectionable off-flavor, particularly to skim milk and lowfat milks, and occasionally to whole milk products. Quality control procedures that include actual flavor trials in milk (in the manufacture of vitamin concentrates) should minimize defective batches of vitamin concentrate. A "hay-like" off-flavor, associated with the presence of added vitamin A (or carriers) in milk and subsequent exposure to light, has been reported in the literature (Weckel and Chicoye 1954) and observed by the authors as an ever-increasing problem for milk processors.

Since vitamin-fortified milk is also homogenized, it is expected to behave the same as homogenized milk with respect to flavor and other sensory characteristics. Though vitamin fortification of whole milk is optional, the practice is near-universal among U.S. milk processors.

Lowfat Milk. According to Title 21 CFR, Part 131.135, lowfat milk is milk from which sufficient milkfat has been removed to produce a food having, within limits of good manufacturing practice, one of the following milkfat contents: 0.5, 1, 1.5, or 2%. Lowfat milk is pasteurized or ultrapasteurized, must contain added vitamin A (not less than 2000 I.U. per quart), and contains not less than 8.25% milk solids—not fat—and may be homogenized. The addition of vitamin D is optional, but if the vitamin is added, the finished product must contain 400 I.U. per quart.

Although lowfat milk may lack the typical richness and mouthfeel of whole milk, this is a natural consequence of a lower milkfat content, and is not considered a defect per se. The product is evaluated in the same manner as whole milk, and may potentially possess the same off-flavors. Thus, a perfect flavor score, if deserved, may be assigned to either a lowfat or whole milk based solely on the absence of off-flavors. Obviously, individual taste preferences may or may not be the same for whole and lowfat milks; preferences will vary with the individual.

In addition to vitamins (according to 21 CFR 131.135), other optional ingredients in lowfat milk include concentrated skim milk, nonfat dry milk, or other milk-derived ingredients to increase the nonfat solids content, *provided* that the ratio of protein to total nonfat solids of the food and the protein efficiency ratio of all protein present shall not be decreased as a result of adding such ingredients. Stabilizers and emulsifiers are also permitted in an amount of not more than 2% by weight of the solids in the optional ingredients actually used. According to the CFR, the lowfat milk may be labeled "protein-fortified" if it contains not less than 10% of milk-derived nonfat solids.

When some of these optional ingredients are used, their relative freshness and quality will impact on the finished product. The processing history and age of these optional ingredients may affect flavor. Long shelf-life products may develop a "stale" flavor following storage, or possibly an oxidized off-flavor. A history of high heat treatment may be responsible for cooked or caramel off-flavors. By exercising thorough quality control of the added ingredients, any significant incidence of the aforementioned problems is probably avoidable.

Skim Milk. The definition of skim milk (21 CFR 131.145) differs from lowfat milk only in the requirement that its fat content be less than 0.5%. All provisions regarding optional ingredients are the same. Most comments relative to the flavor of lowfat milk are also applicable to skim milk. An off-flavor most commonly described as "lacks freshness," "stale," "chalky," or "storage flavor" is frequently encountered by judges in the sensory evaluation of skim milk samples. The composition of skim milk appears to favor occurrence of this off-flavor; it probably stems from the ratio of proteins to milkfat found in skim milk.

Concentrated Milks. "Concentrated milk" is defined in 21 CFR 131.115 as the liquid food obtained by the partial removal of water from milk; the milkfat and total milk solids content must be not less than 7.5% and 25.5%, respectively. This product must be pasteurized, will generally be homogenized, and may have vitamin D added (25 I.U./ fluid ounce). Water is removed under partial vacuum; as much as 3 parts of the milk may be concentrated to 1 part of concentrated milk. Water is added back by the consumer, and savings are realized in transportation and packaging costs, although processing costs are higher.

Frozen concentrated milk and *commercially sterile concentrated milk* are different and more complex product forms. They are intended for longer storage, which unfortunately provides opportunities for physical and chemical factors to influence sensory properties. Flavor is a function of the processing temperature, storage temperatures, and age of the product. On prolonged storage, the flavor may become stale, oxidized, or caramelized. Even a fresh concentrate may taste somewhat flat upon reconstitution, although the flatness sensation is generally lessened upon storage. Reconstituted concentrated milk is usually evaluated from the standpoint of utilization as a beverage or fluid milk.

Evaporated milk is a special type of sterile concentrated milk with its own definition in 21 CFR 131.130. Although this product could be made by a combination of UHT processing and aseptic packaging, evaporated milk is commonly sterilized in the final container at a lower temperature, but a much longer holding time. The addition of vitamin D (25 I.U./ounce) is mandatory, and the use of emulsifiers and stabilizers is permitted. The flavor characteristics of this product are influenced by the heat treatment applied, storage temperature, and age. Off-flavors such as cooked, caramel, and stale are frequently observed. This product may display varying degrees of browning and excessive viscosity. Curdiness and fat separation are undesirable characteristics. Evaporated milk is discussed in greater detail in Chapter 10.

Reconstituted Milk. Reconstituted milk is the product resulting from either: (1) recombining milkfat and nonfat dry milk or (2) dry whole milk with water in appropriate proportions, to yield the milk constituent percentages that typically occur in fluid milk. For this purpose, various forms of milkfat such as butter, anhydrous milk fat, and fresh or frozen cream and nonfat dry milk, dry milk, or concentrated milk may be used as ingredients. Any form of reconstituted milk is practically always homogenized. Despite the fact that homogenization (an integral part of the process), inhibits the development of an oxidized off-flavor in milk, an oxidized defect of slight to moderate intensity may be present in reconstituted milks with some degree of frequency. This off-flavor is generally derived from any one of several susceptible dairy ingredients prior to their reconstitution. Other types of off-flavors associated with reconstituted milk are flat, heated, cooked, and stale.

Half-and-Half and Cream. Title 21 of CFR gives definitions for heavy cream (36% milkfat), light whipping cream (30% to less than 36% milkfat), light cream (18% to less than 30% milkfat), and half-and-half (10.5% to less than 18% milkfat). All of these cream-based products are either pasteurized or ultrapasteurized, and may be homogenized. Although not normally consumed as beverages, cream products are listed here since their flavor characteristics are evaluated in basically the same way as milk; they are subject to essentially the same off-flavors. Due to their higher fat content, and the optional presence of stabilizers and emulsifiers, the mouthfeel of these products differs markedly from that of milk. In addition to sensory qualities, important functional properties such as whippability (see Appendix VII) and coffee-whitening properties should also be tested by recommended or standardized procedures.

Miscellaneous Products. The 1978 P.M.O. describes low-sodium milks, i.e. whole milk, lowfat milk, and skim milk; and lactose-reduced milks, i.e. whole milk, lowfat milk, and skim milk. Other dietary products may also be encountered where permitted by local ordinances, in the form of mineral- and/or vitamin-fortified milk. The flavor properties of such products should be evaluated in a manner similar to milk. These "low-sodium milks" must contain less than 10 milligrams of sodium per 100 milliliters to be so labeled. Lactose-reduced products must have sufficient lactose converted to glucose and galactose (a mixture which is sweeter than lactose) by the addition of safe and suitable enzymes to cause the remaining lactose to be less than 30% of its original concentration. Hence, some effect on flavor (taste) would be expected, but these products should be relatively free of other flavor defects common to milk.

Commercially Sterile and UHT Milks. The commercial availability of this product has already been alluded to previously under several headings in this text. A 1981 amendment to the 1978 P.M.O. permitted a Grade A designation for aseptically packaged UHT-milk. From a microbiological standpoint, a "sterile" label implies the absolute absence of all microorganisms (both pathogenic and spoilage types) in milk products. Commercially sterile milk products can be successfully stored without benefit of refrigeration for extended time (up to 9 months). By contrast, the label "ultrapasteurized" connotes extended shelf-life under refrigerated conditions.

Depending on the method of sterilization or heat treatment, both of these product forms are generally expected to exhibit varying intensities of cooked flavor. If intense, the flavor defect may be described as scorched, scalded, burnt, or caramel. With the advent of improved sterilization systems, only the more subtle cooked, sulfide-like flavor predominates. During storage, the intensity of the cooked flavor gradually diminishes, so that under the most favorable circumstances, a sterilized product may taste similar to pasteurized milk. The discovery that addition of the enzyme, sulfhydryl oxidase (Swaisgood 1980), can reduce the cooked flavor in commercially sterilized milk may have significant future implications for UHT-processed milks. It has been suggested that a commercial process could be developed for treating heat-processed milk with an immobilized form of this enzyme. In one experiment and subsequent flavor panels, the enzyme-treated UHT-milk could not be distinguished from pasteurized milk (Sliwkowski and Swaisgood 1980). During prolonged storage, particularly when not refrigerated, various storage flavors may be encountered which result from lactose and protein interaction, protein and/or fat degradation, and staling.

THE MILK SCORE CARD

The original score card developed by the United States Department of Agriculture for milk and cream is shown in Fig. 5.3. Though it has been extensively modified over the years, both by changes in legal requirements and revised scoring methods, this score card still provides a significant historical document which has served as a most effective tool for evaluating and improving milk quality.

An expanded and modified version of the milk score card and an associated scoring guide (patterned after the American Dairy Science Association "official" score card) is reproduced in Fig. 5.4 and Table 5.1, respectively. Frequently, a score card is only used for recording flavor observations, although the importance of other quality factors that were included in the original score card should not be ignored. Bacterial counts, milk sample temperatures, and sediment tests can be important data provided by the laboratory; they continue to be components of the overall quality profile for a given milk product. Evaluating the container and the closure is also a valid quality criterion; they

United States Department of Agriculture
Bureau of Dairy Industry

Score Card for Milk and Cream
(Approved By The American Dairy Science Association)

Place ....

Class ....   *Exhibit No*...............

	Perfect Score	Score Allowed	Remarks
Flavor and odor	45	....	{ "Flavor defects" listed on other side } ......
Bacteria	35	....	{ Bacteria found per milliliter } ....
Sediment	10	....	....
Temperature	5	....	Degrees ....
Container and closure..	5	....	{ Container .... Closure ....
Total	100	....	

Exhibitor ....

Address ....

   (Signed) ....

   ....

   ....
                                                   Judges

Date...................

B. D. I-64
(Rev. 1941)   16—24004-1

Fig. 5.3.   The original USDA score card for milk and cream.

SCORE CARD FOR MILK QUALITY

Product: _____   Date: _____

SAMPLE NO.

		1	2	3	4	5	6	7	8
FLAVOR 10	CRITICISM SCORE ➡								
	ACID								
	ASTRINGENT								
NO CRITICISM	BARNY								
10	BITTER								
	COOKED								
	COWY								
	FEED								
UNSALABLE	FERMENTED/FRUITY								
0	FLAT								
	FOREIGN								
	GARLIC/ONION								
NORMAL RANGE	LACKS FRESHNESS								
1-10	MALTY								
	OXIDIZED LIGHT INDUCED								
	OXIDIZED METAL INDUCED								
	RANCID								
	SALTY								
	UNCLEAN								
SEDIMENT 3	SCORE ➡								
PACKAGE 5	SCORE ➡								
	CONTAINER BULGING/DISTORTED								
	DENTED/DEFECTIVE								
	DIRTY INSIDE								
NO CRITICISM	DIRTY OUTSIDE								
5	LEAKY								
	NOT FULL								
UNSALABLE	CLOSURE DEFECTIVE								
0	COATING FLAKY/CRACKED								
	HEAT SEAL DEFECTIVE								
NORMAL RANGE	ILLEGIBLE PRINTING								
1-5	LABELING/CODE INCORRECT								
	LIP CHIPPED								
	COVER NOT WATERPROOF								
	UNPROTECTED								
BACTERIA 5	SCORE ➡								
	STANDARD PLATE COUNT								
	COLIFORM COUNT								
	KEEPING QUALITY								
TEMPERATURE 2	SCORE ➡								
	TEMPERATURE (°F or °C)								
	TOTAL SCORE OF								
	EACH SAMPLE SCORE ➡								
DESIRED __%	FAT CONTENT (%)								
DESIRED __%	SOLIDS NOT FAT (%)								
	UNDER/OVER FILLED								
FUNCTIONAL	TITRATABLE ACIDITY								
AND OTHER									
TESTS PERFORMED									
ON SAMPLES									

SIGNATURES OF EVALUATORS: _____ _____ _____

Fig. 5.4.   A modified and expanded version of the ADSA milk score card.

Table 5.1 A Suggested Scoring Guide for Flavor of Milk (Suggested Flavor Scores for Designated Defect Intensities).

Flavor Defect[a]	Intensity of Defect				
	Slight[b]	Moderate	Definite	Strong	Pronounced[c]
Astringent	8	7	6	—[d]	—
Barny	5	4	3	2	0–1
Bitter	5	4	3	2	0–1
Cooked	9	8	7	6	0–5
Cowy	6	5	4	3	0–2
Feed	9	8	7	6	0–5
Fermented/					
Fruity	3	2	1	0[e]	0
Flat	9	8	7	—	—
Foreign[f]	3	2	1	0	0
Garlic/Onion	5	4	3	2	0–1
High acid	3	2	1	0	0
Lacks freshness	8	7	6	—	—
Malty	5	4	3	2	0–1
Metallic	5	4	3	2	0–1
Oxidized					
Light-induced	6	5	4	3	0–2
Metal-					
induced	5	4	3	2	0–1
Rancid	4	3	2	1	0
Salty	8	7	6	5	0–4
Unclean	3	2	1	0	0

[a] "No criticism" is assigned a numerical score of "10." Normal range is 1–10 for salable products.
[b] Highest assignable score for defect of slight intensity.
[c] Highest assignable score for defect of pronounced intensity. However, a sample may be assigned a score of zero "0" (unsalable product).
[d] A dash (—) indicates that this level of intensity for the defect is unlikely to be observed.
[e] When a product is determined to be unsalable for a given flavor defect, a "0" (zero) numerical score is assigned for flavor.
[f] Due to the variety of possible foreign off-flavors, suggesting a fixed scoring range is inappropriate. Some foreign off-flavors warrant a zero score even when their intensity is slight (e.g., gasoline, pesticides, lubricating oil).

should be evaluated when appropriate or required. Flavor on the new score card is evaluated on a 10-point scale according to the scoring guide (Table 5.1). A 100-point score card similar to the original U.S. Department of Agriculture card (but which allows a maximum of 20,000 bacteria per milliliter and a maximum temperature of 7.2°C (45°F)) may still be used by industry and in some clinics, competitions and state fair judging. Other instruments for recording scores derived from sensory observations may be in use by individual companies or have been developed for specific purposes. In 1987, the ADSA Committee on Evaluation of Dairy Products approved a Collegiate Contest milk score card which accommodates electronic grading of student contestants' performance (see Appendix X).

Familiarity with the score card and use of the scoring guide is important for the milk product judge. The scoring guide provides a standard yardstick to be applied from day to day for quality assurance activities and making comparisons of different samples or brands of a given product.

Some Milk Scoring Techniques

Preparation of Samples for Evaluation. The preparation of packaged milk samples for sensory evaluation will depend upon the purpose or objectives of the activity, the number of participants, and the quality criteria to be assessed. If several persons are to score the milk samples for flavor, container and closure, sediment, and other criteria, then several containers of each individual lot of milk must be provided. If the presence of sediment in raw milk is to be noted and evaluated, sediment discs for each lot of milk should be available (preferably mounted on a card or enclosed in a covered petri dish). With these provisions, each judge may score the container and closure, and the sediment (if applicable), with assurance that samples are representative of the lot. Each container and each sediment disc needs to be correctly labeled. Sediment discs are generally placed directly in front of the corresponding container. The milk containers used for flavor evaluation should be clearly marked so that the sample identity is easily noted; this identification must be fastened to the container in a secure manner (Fig. 5.5). A paper or plastic tag fastened to the container with a rubber band or fiber cord or use of self-adhesive labels has proven quite satisfactory

Fig. 5.5. Numbered tags serve as a convenient method for identifying milk samples in flavor evaluations.

for facilitating sample identification. However, the tag and the tie need to be odorless. Some precautions must also be taken against using certain marking pens which contain strong aromatic inks that markedly interfere with odor detection of milk samples.

For routine sensory evaluation of packaged milk, where only several judges score the samples simultaneously within the milk plant, preparation of the samples as previously described seems unnecessary. However, systematic arrangement of the samples before the actual scoring begins is an aid to proficiency. Tagging of the containers or preparation of sediment discs may not be required. Identification of the sample can be made by labeling (or numbering) the bottle closure and/or the container. The sediment rating may be determined by carefully observing the bottom of unshaken and undisturbed transparent containers. Obviously, potential sediment problems in milk in paper or opaque containers can only be evaluated by means of a sediment disc prepared from the milk sample.

Order of Examination and Scoring. A scoring routine should be followed which enables the evaluator to make efficient use of time and which enhances "concentration of thought." Furthermore, this routine should enable the judge to make direct comparisons between different samples, with respect to the various categories listed on the score card. Before beginning, the name (or other identification) of the evaluator should be placed in the space provided on the score card. If not already indicated on the card, the numbers or identity of the samples should be placed consecutively thereon. A basic order of examination might be as listed in the following paragraphs.

Sediment. If appropriate or conducted, sediment scoring should be performed first. The kind, the amount, and the size of the sediment particles should be carefully observed and scored. In scoring sediment discs, visual examinations and scoring may be compared with standard charts or photographs of standard discs. However, a mental image of this chart or photograph should become a part of the evaluator's skill, so that continued comparisons of sediment discs with actual visual standards is not always necessary. Photographs of standard sediment discs are shown in Figs. 5.9 and 5.10.

Closure. After having evaluated the milk for sediment, the closure (if evaluated) should be carefully observed and scored. A perfect closure has three main functions, namely: (1) to contain the milk in the package or bottle; (2) to protect the pouring surface against contamination; and (3) to seal the container against tampering without some visible detection. In order to fulfill the protection requirements for bottles, the cap (if employed) must cover the pouring lip at its greatest diameter. When appropriate, the evaluator should observe whether the cap

Fig. 5.6. Example of an electronic thermometer and thermistor probes for monitoring storage temperatures of milk and other perishable dairy products.

is properly seated, so that there is no leakage which might cause microbial contamination. If a cap is covered, this covering should be tight, waterproof, and tamperproof. If possible, it should be determined whether the closure was inserted by hand or by machine. Handcapping is generally prohibited by milk ordinances, due to the greater risk of contaminating milk through associated human contact. Thus, certain observations and judgments should be made relative to the closure itself; namely, whether it fully protects the pouring lip, whether it is properly seated, whether it is leaky, and (should the closure be covered) whether the covering is fastened securely and made of waterproof material, and whether the closure adequately seals the container. The 1978 P.M.O. states:

All caps and closures are designed and applied in such a manner that the pouring lip is protected to at least its largest diameter and, with respect to fluid product containers, removal cannot be made without detection. Single service containers are so constructed that the product and the pouring and opening areas are protected from contamination during handling, storage and when the containers are initially opened.

In principle, the same criteria apply to closures for paper containers. An examination of the heat seal of the carton is appropriate. It must be adequate to prevent contamination of the milk, but it should not be so rigid or tenacious as to make opening of the carton unduly difficult. Also, excessive heat from the "sealing jaws" of the carton filler may burn or scorch the polyethylene coating. This may lead to an unattractive carton appearance at best and a "burnt-plastic" off-flavor at worst; the latter (flavor) defect is most objectionable to consumers.

Container. Multiuse containers should be examined for the extent of fullness, cleanliness, and freedom from cracks or chips, especially on or near the pouring lip. Any condition of the container that may interfere with contents safety and wholesomeness should be carefully observed and noted. With practice, this observation may be made quickly and accurately.

Single-service plastic containers have exactly the same requirements for cleanliness and freedom from leakage and damage, but they generally lack the sidewall rigidity to readily determine the precise level of fill. The 1978 P.M.O. contains sanitation guidelines for the manufacture of single service containers for milk and milk products. Single service paper containers are examined for cleanliness, rigidity, freedom from leakage, smoothness, and adherence of coating. The correct fill level can best be determined by actual measurement of milk volume per container by pouring contents into a graduated cylinder.

Flavor. The evaluation of milk for flavor is generally done after the other items of sediment, container and closures have been considered. At the time of scoring, the milk should be adequately tempered to optimize the detection of any possible odor(s) in the sample(s). Simultaneously, the milk sample should be sufficiently low in temperature that it will increase appreciably when the sample is placed into the mouth. A temperature range of 12.8°C to 18.3°C (55°F–65°F) for the sample has been found to be most satisfactory for scoring milk.

Occasionally, when appropriate or a problem is suspected, the evaluator should remove the cap before mixing the milk and closely inspect the underside of the closure for possible adherance of cream or foam; and then examine the milk sample for the possible presence of a cream plug.

Milk samples for tasting should be poured into clean, odorless drinking containers (i.e., sanitary and nontoxic) that are made of glass (preferably), plastic, or paper. Any size between 30–120 ml (1 to 4 ounces) is appropriate. As soon as the sample (10–15 ml [½oz]) is poured, the judge should take a generous sip, roll it about the mouth, note the flavor sensation, and then expectorate. Sometimes, any aftertaste may be enhanced by drawing a breath of fresh air very slowly through the mouth, and then exhaling slowly through the nose. Swallowing the milk as a means of detecting off-flavors is an inadvisable practice. The milk judge should make certain that the milk is well mixed by gently swirling the container contents in a circular pattern just before sampling. By placing the nose directly over the container immediately after the milk has been swirled in the container, and taking a full "whiff" of air, any off-odor that may be present can be more readily noted.

Agitation (or swirling) of the milk leaves a thin film of milk on the inner surface which tends to evaporate, thus readily optimizing the opportunity to detect any odor(s) that may be present. If the evaluator is perceptive, even the faintest odors may be detected in this way. If several judges participate in the sensory evaluation, the containers, when temporarily uncapped and whiffed, should always be handled in a sanitary manner.

REQUIREMENTS OF HIGH-QUALITY FLUID MILK

Evaluating Container and Closure

Multiuse (glass and plastic) containers should have an attractive appearance, be clean, and contain the full volume of milk (as indicated by the label). The bottle contents should be protected from contamination (Bodyfelt et al. 1976; Gasaway and Lindsey 1979; Landsberg et al. 1977) by a well-made, properly seated, waterproof cap which protects the pouring lip. Attractive milk bottles should appear bright and shiny, be free from dirt and dust and should exhibit no case wear and/or caustic etching (surface abrasions). A chipped bottle lip often results in a leaky or poorly seated cap, and may harbor microorganisms due to roughened surfaces.

Single-service paper and plastic containers should reflect cleanliness, recent filling, and freshness and should possess a dry, firm, rigid, and milk solids-free surface. A weakening of the packaging material, as indicated by pronounced bulging of the container sidewalls, should not be evident. There should be no evident leakage of unopened containers.

Fullness of the Container. There is a legal requirement that milk containers must be filled with the full volume of milk, as indicated by the size of the container and/or label statement. Tolerances and the methods of measurement may vary from state-to-state, but certain compliance requirements are inescapable. Some containers may have an indicated fill-line and can be assessed for fullness by visual observation. These would have to imply rigid containers, such as those made of glass. When more flexible packaging materials are used, or when the container is opaque so that the level of fill cannot be seen, a volumetric measurement of the contents at a predetermined temperature is necessary. It should be remembered that the density of a liquid varies with temperature; the volume increases with temperature rise.

Bottle Closures. As previously stated, the closure has three basic functions: (1) to retain the milk within the container; (2) to protect the pouring lip from contamination; and (3) to seal the container against tampering. The closure is assessed on the completeness with which it fulfills these three functions. The cap is intended primarily to retain the milk within the bottle. In addition, a cap that meets the U.S.P.H.S. requirements for Grade A milk protects the pouring lip of the bottle from contamination; it also protects the filled container against tampering without some evidence of detection.

In the past, more kinds of milk bottle closures or caps were used than is the present case. As the recommendations of the P.M.O. were more widely adopted, many of the then-existing closures simply did not comply. Current container closures generally meet all of the requirements regarding protection of the pouring lip and provide some safeguards against tampering. Table 5.2 lists possible defects which apply to containers and closures of both multiuse and single-service containers.

The term "unsealed" is used to mean "not tamperproof." Closures that meet the requirements of the 1978 P.M.O. satisfy the "sealed" criterion. The term "tamperproof" may be subject to legal interpretation which cannot be adequately addressed here. Approval of specific containers and closures by appropriate public health enforcement agencies is a necessary requirement, as possible tampering with milk would be a serious matter. When evaluating closures, the presumption that a closure is sealed when it cannot be removed and replaced without obvious detection appears to be a practical one. Unfortunately, to make a container absolutely tamperproof would require extreme measures and perhaps prohibitive expense.

Scoring Containers and Closures. Since there is no recently accepted system for scoring containers and closures, the following may be used as a suggestion in developing a scoring guide (Table 5.3). A so-called

Table 5.2. Possible Defects of Milk Containers and Closures of the Multiuse and Single-Service Types.

Container closure unsealed	Flaky or cracked coating
Container not full	Closure poorly seated or leaky
Container dirty on the outside	Defective heat seal
Container dirty on the inside	Lip chipped
Container dented or defective	Lip unprotected
Container leaky	Lip cover not waterproof
Container bulging or distorted	Torn closure cover
Illegible printing on container	Lack of, or incorrect, code or labeling

"perfect" container could be assigned a score of "5.0." At the other extreme, any milk container which does not meet the 1978 P.M.O. recommendations should be disqualified and assigned a score of "zero." Containers that are dirty inside, leaky, or have closures that are defective or leaky should also be disqualified and receive a score of "zero." Most other defects might carry a penalty of 1 point for slight, 2 to 3 points for moderate, and 4 or 5 points for pronounced intensity. In

Table 5.3. A Suggested Scoring Guide for the Appearance and Integrity of Milk Containers.

	Intensity of Defect				
Defect[a]	Slight[b]	Moderate	Definite	Strong	Pronounced[c]
Container:					
bulging/distorted	4	3	2	1	0[d]
dented/defective	3	2	1	0	0
dirty inside	0	0	0	0	0
dirty outside	2	1	0	0	0
leaky	0	0	0	0	0
not full	4	3	2	1	0
Closure defective	0	0	0	0	0
Coating cracked/flaky	4	3	2	1	0
Heat seal defective	4	3	2	1	0
Illegible printing	4	3	2	1	0
Incorrect label/code	3	2	1	0	0
Pouring lip:					
chipped	4	3	2	1	0
cover not waterproof	3	2	1	0	0
unprotected	3	2	1	0	0

[a] "No criticism" is assigned a score of "5." Normal range is 1–5 for a salable product.
[b] Highest assignable score for a slight intensity of the given defect.
[c] Highest assignable score for a pronounced intensity of the given defect.
[d] An assigned score of zero ("0") is indicative of an unsalable product.

this scoring scheme, if several defects are encountered, the deductions should be additive.

Evaluating Temperature

The temperature at which pasteurized market milk and other fluid products are held is very important in determining the keeping quality, and for retention of good flavor characteristics. Even commercially sterile milk, which is microbiologically stable at room temperature, may actually suffer more rapid flavor deterioration at higher storage temperatures.

The 1978 P.M.O. recommendations for storage temperature of Grade A pasteurized milk sets 7.2°C (45°F) as the maximum acceptable temperature. In view of the longer keeping-quality demands placed on milk, 7°C (45°F) should be considered the highest milk storage temperature permissible; however, temperatures *below* 4.4°C (40°F) are definitely preferable for helping extend shelf-life. Frequent line temperature checks should be made of milk coming from the cooling section of the pasteurizer, surge tanks, and filler, and the product when packaged, in cold storage, in transport, and in retail store coolers and display cases (Bodyfelt 1974, 1980; Bodyfelt and Davidson 1975, 1976). See Figs. 5.6 and 5.7 for examples of electronic thermometers and probes and techniques useful for assessing critical temperature checkpoints. These can be readily charted to produce a temperature profile of processing and distribution operations (Fig. 5.8).

There is no generally accepted scoring system for temperature. What follows is only a suggested approach which may be applied for scoring the temperature of milk products. For in-house quality assurance program purposes, it seems more logical to record or graph the actual temperature(s) (Bodyfelt 1978) than to assign a score. If a score is more appropriate, such as in competitions (when samples are picked up at the plant or from a retail establishment), a 2-point scale may be employed. A sample that is above 7.2°C (45°F) is not in compliance and should conceivably receive a score of "zero." At the other extreme, samples at a temperature of 4.4°C (40°F) or lower could be assigned a perfect score of "2." When the sample temperature is between 4°C and 7.2°C (40°F and 45°F), a score of "1" would be assigned. Sample temperatures of 10°C (50°F) or higher should probably be disqualified from competition, since both quality and public health concerns may be at stake.

Requirements for Grade A raw milk for pasteurization as specified by the 1978 P.M.O. are as follows: "Cooled to 7°C (45°F) or less within two hours after milking, provided that the blend temperature after the first and subsequent milkings does not exceed 10°C (50°F)." Thus, the "temperature scoring" of raw milk would depend upon the time

Fig. 5.7. A technique for using an electronic thermometer for monitoring product temperatures.

elapsed after milking that the temperature was measured. After two hours, the scoring system would be the same as that used for pasteurized milk, since the requirements are identical. The milk should be disqualified from competition whenever its temperature is above 10°C (50°F).

Evaluating Sediment in Milk

Milk can be scored for sediment either by observing the particles of sediment which may have settled to the bottom of a bottle or by observing the sediment collected on a cotton disc. Obviously, direct observation for sediment is only possible when transparent containers are used. When several samples are compared, the container size or the sample size (from which the sediment is obtained) should be standard-

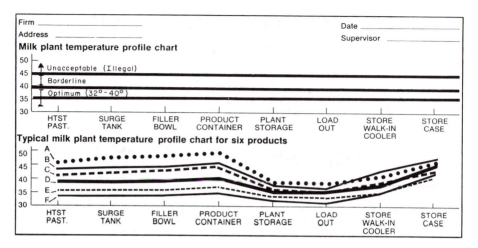

Fig. 5.8. An example of a profile chart for critical temperature checkpoints. From Bodyfelt, F. W. (1980a). Courtesy of *Dairy Foods* magazine, Chicago, IL.

ized. For the cotton disc method, one-pint samples are used under standardized conditions of temperature and aspiration. The comparisons with a chart or standard photograph should be made on the potential sediment found in one pint of tempered milk (35°C–38°C[95°F–100°F]).

The visual assessment for sediment particles on the bottom surface of bottles (when held above the eyes) is somewhat tedious and inaccurate. When several evaluators are handling the same milk samples, some of the sediment particles are likely to be remixed with the milk, which makes them invisible. In the absence of good light, it is also difficult to observe all possible particles.

On the other hand, scoring sediment from the bottom of the bottle offers the advantages of speed and simplicity, since no preparation of sediment discs is necessary. In the routine examination of unhomogenized bottled milk, where emphasis is usually placed on the flavor quality of the milk, the observation for possible sediment on the bottom of the bottle is desirable, but it should be remembered that this method only furnishes an indication of the presence or absence of particles that are too large to be "rafted" upward into the cream layer.

In the sediment disc method, the sediment (or extraneous matter) is concentrated and firmly fixed on a white cotton or lintine disc, where it may be studied more carefully and "filed" for later reference. The sediment discs are prepared by filtering one pint of tempered milk through a round, white cotton pad of 0.4 in. (1.0 cm) diameter filtering area. The sediment discs are protected and stored for later reexamina-

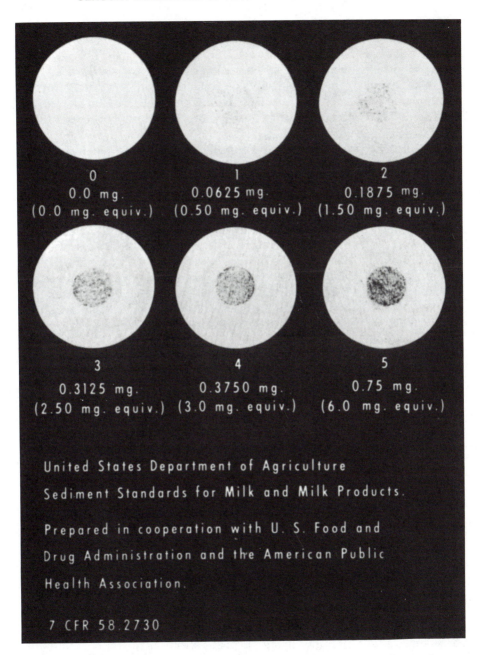

Fig. 5.9. Standard discs that represent known weights of sediment for a given volume of tempered milk sample (one pint).

tion by placing them on a cardboard receptacle (covered with cellophane) or placing them in a clean, covered petri dish.

For the occasional testing of raw milk from cans, the off-the-bottom method is used, which employs a sediment tester especially designed for this purpose. One pint of milk is collected from the bottom of an undisturbed can of milk, and the sediment is collected on a 1.25 in. (3.18 cm) disc. One-pint samples are more frequently collected from bulk tanks for sediment testing, after the milk has been well-agitated. The sediment tester for milk from bulk tanks is fitted with a 0.4 in. (1.0 cm) diameter orifice, so that the sediment is concentrated in a smaller cross-section. Pasteurized milk may be sampled for sediment only after thorough mixing in the original container.

Each disc may then be compared to a standard chart or photograph that reflects the appropriate sediment ratings. To score "perfect" on sediment, there should not even be a trace of foreign particles on the disc, or any discoloring of the disc, except that due to the natural pigments of milk. Deductions are made in accordance with the amount, kind, and size of foreign particles present, as well as for any smudgy appearance. If the milk were not strained or filtered on the farm, the amount of sediment on the disc would readily indicate the general cleanliness and care taken in production. However, if the milk were strained or filtered, the amount of sediment merely indicates the efficiency of that process or the amount of sediment subsequently accumulated.

Sediment standards for milk have been developed by the U.S. Department of Agriculture, and are published in the Code of Federal Regulations, Title 7, Parts 58.2728 through 58.2731. Standard discs containing known weights of sediment are shown in Figs. 5.9 and 5.10. Discs prepared from milk samples are evaluated by comparing them to these standard discs.

Consumers want and insist that milk be free of foreign matter, which is certainly a reasonable expectation. The critical factors that determine the entry of foreign or extraneous matter into milk are: (1) the sanitation and care during the milking process; (2) the efficiency of milk straining or filtering on the farm; (3) the efficiency of clarification at the plant; (4) the cleanliness of equipment and containers; and (5) avoidance of milk contamination whenever it is exposed to the atmosphere.

The recommended requirements of the U.S. Department of Agriculture for milk for manufacturing purposes specifies the highest amount of sediment permissible in raw milk. However, manufacturing grade raw milk may be graded using a scoring system which distinguishes finer details of sediment quality than simply a designation of "accept-

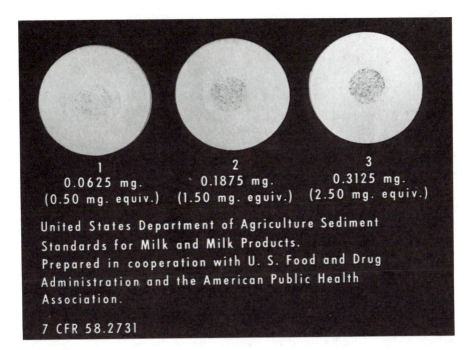

1
0.0625 mg.
(0.50 mg. equiv.)

2
0.1875 mg.
(1.50 mg. equiv.)

3
0.3125 mg.
(2.50 mg. equiv.)

United States Department of Agriculture Sediment
Standards for Milk and Milk Products.
Prepared in cooperation with U. S. Food and Drug
Administration and the American Public Health
Association.

7 CFR 58.2731

Fig. 5.10. A smaller grouping of sediment discs that demonstrate various weights of extraneous material per pint of milk.

able" or "not acceptable." A 5-point scale may be employed. When a pint sample is found to contain more sediment than the equivalent of a 2.5 mg/standard 0.4 inch disc, a score of "zero" is commonly given. No visible sediment would be required for assigning a score of "5." The intervening scores may be assigned as follows: 0.5 mg/disc—4; more than 0.5 mg but less than 2.5 mg/disc—3 or 2; 2.5 mg/disc—1.

The presence of any sediment in the finished product is serious, since the consumer may be quick in registering a complaint. Thus, anything over a trace of sediment may cause the product to be unmarketable and should receive a score of "zero." Obviously, products in containers, ready for the consumer, should be scored differently than raw milk. While 0.5 mg of sediment/pint may be "acceptable" for raw milk, this much sediment is excessive (and should receive a score of "zero") if found in any finished product. One possible scoring system for finished products is: No sediment—3, more than "no sediment" but less than 0.02 mg/disc—2, 0.025 mg/disc—1, and over 0.025 mg/disc—0 (see Table 5.4).

Table 5.4. A Suggested Scoring Guide for
Sediment in Processed Fluid Milk.

Amount of Sediment[a]	Score
0	3
<0.02 mg	2
<0.02 mg–0.025 mg	1
>0.025	0

[a] Stirred sample. Disks with so little sediment do not
reproduce clearly enough to be illustrated (see 7
CFR 58.2726).

Evaluating Bacterial Content

The examination of milk for bacterial content is a laboratory procedure
which can be performed by a qualified technician who may have no
experience in milk judging. The maximum permissible bacterial counts
for raw Grade A and pasteurized market milks are specified in the 1978
Grade A Pasteurized Milk Ordinance. For pasteurized milk, the upper
limit of the P.M.O. is 20,000/ml and is not to exceed 10 coliforms/ml.
Thus, a sample that has a standard bacterial plate count (SPC) of more
than 20,000/ml or a coliform count (performed by standard methods)
of more than 10/ml should receive a score of "zero" for bacteria. As
emphasized earlier, a report of the actual bacterial count is usually
more meaningful than a bacterial score for most quality control purposes.

The bacterial count of milk potentially reveals the general conditions
of sanitation and temperature control under which the milk was produced, handled, and held. High quality milk should be relatively low
in bacteria content, but milk with low bacterial counts may not always
necessarily exhibit satisfactory flavor characteristics. If off-flavors in
milk are the result of bacterial growth, the bacterial count is usually
in the millions per ml. However, serious off-flavors may also be found
in milk that is low in bacteria, since numerous milk off-flavors are not
due to bacterial activity. Frequently, there is no correlation between
milk bacterial count and milk flavor quality, unless there is sufficient
growth and development of microorganisms in the milk to form reaction end products such as lactic acid and/or volatile compounds from
proteolysis or lipolysis. However, in such instances, the physical appearance of milk may be changed. A significant consequence of this
(for quality determination) is that many off-flavors produced by bacteria in raw milk usually persist in the pasteurized milk, even though few
of the bacteria are likely to survive the heat treatment of pasteurization.

When evaluating market milk and other milk products for competitive purposes, the scoring system should be based on both the total

bacteria and coliform counts. A suggested scoring guide for total bacterial and coliform counts of milk is shown in Table 5.5. A sample may receive a score for bacterial content ranging from "0 to 5," based on the outcome of either the total bacterial count or the coliform count (or both counts). Typically the score is determined for each separately, and the lower of the two scores is the score assigned to the sample. For example, a sample with 13,000 bacteria/ml and 1 coliform/ml would receive a score of "2" on the basis of the bacterial count and a "4" on the basis of the coliform count. The lower score of "2" would be assigned to the sample.

Evaluating Milk Flavor

Desired Milk Properties. Typically, the flavor of whole milk should be pleasantly sweet and possess neither a foretaste nor an aftertaste other than that imparted by the natural richness, due to the milkfat and other milk solids. The evaluator should not assume or expect that a sample of good (high quality) flavor milk will have a "taste," per se. Judges should remember that when milk clearly exhibits a so-called "taste," there is usually something "wrong" with the flavor of that milk sample. Milk of excellent quality should seem pleasantly sweet and leave only a clean, pleasing sensation after the sample has been expectorated or swallowed. The mixed sample should also be perfectly homogeneous (i.e. exhibit no buttery particles or graininess). When the closure of the unshaken bottle is removed, there should be no evidence of adhering cream, foam, or butter granules, and the milk should not show a cream plug.

Placing Samples into Flavor Groups. With appropriate training, the evaluator should be able to classify the flavor quality of milk samples into categories of excellent (10), good (7-9), fair (4-6), poor (1-3) and unacceptable (0). The next step for a milk judge might be to rate the samples within the group into which it falls; that is, whether the flavor quality (relative merits) is such as to place it as average, high, or low in that group. Since each group has a range of numerical scores, it

Table 5.5. A Suggested Scoring Guide for Bacteria in Milk.

Standard Plate Count C.F.U./ml	Coliforms/ml	Score[a]
>20,000	>10	0
>16,000-20,000	10	1
>12,000-16,000	7-9	2
>8,000-12,000	4-6	3
>3,000-8,000	1-3	4
≤3,000	0	5

[a] The score for each of the criteria is determined separately; the lower of the two scores is assigned to the given sample.

should become relatively easy to place a numerical flavor score or grouping on the respective samples. Further assistance has been provided by various professional groups which have developed scoring guides that suggest scores for milk that possess varied intensities of specific defects. Two scoring guides for milk, one developed by the Committee on Evaluation of Dairy Products of the American Dairy Science Association, and the other by the authors, are illustrated in Tables 5.6 and 5.1, respectively. To use a guide, the evaluator should be somewhat proficient in the identification of various flavor defects of milk.

Undesirable Flavors. Milk is generally considered to have a flavor defect if it manifests an odor, a foretaste or an aftertaste, or does not leave the mouth in a clean, sweet, pleasant condition following tasting. Some samples may simultaneously have more than one flavor defect. In this case, the assigned flavor score usually corresponds to the most serious defect of the several noted. The scoring guide in Table 5.6 lists the most frequently encountered off-flavors of milk. Whenever a flavor defect is encountered that differs from those listed on the scoring guide (which happens occasionally), it should be described in the most de-

Table 5.6. The A.D.S.A. Scoring Guide for Off-Flavors of Milk and Cream.

Flavor Criticisms[a]	Intensity of Defect		
	Slight	Definite	Pronounced
Acid	3	1	0[b]
Astringent	8	7	6
Barny	5	3	1
Bitter	5	3	1
Cooked	9	8	6
Cowy	6	4	1
Feed	9	8	5
Fermented/Fruity	5	3	1
Flat	9	8	7
Foreign	5	3	1
Garlic/Onion	5	3	1
Lacks Freshness	8	7	6
Malty	5	3	1
Metallic	5	3	1
Oxidized			
Light-induced	6	4	1
Metal-induced	5	3	1
Rancid	4	1	0
Salty	8	6	4
Unclean	3	1	0

Source: American Dairy Science Association, 1987.
[a] "No criticisms" is assigned a score of "10." Normal range is 1–10 for a salable product.
[b] An assigned score of zero ("0") is indicative of an unsalable product.

scriptive or associative term(s) possible and entered on a blank line of the score card. In such a case, the assignment of a numerical score may be difficult, particularly when such a defect may be encountered for the first time. Evaluators must draw upon their experience and sound judgment in assessing the degree of seriousness of uncommon defects.

The description, taste, and smell sensations, and cause(s) of the different off-flavors of milk, follow in alphabetical order (as noted in Fig. 5.11 and Appendix X).

Acid. The "acid" or "sour" off-flavor of milk is detected by both the sense of smell and the sense of taste. When *S. lactis*, *S. cremoris*, or other acid-producing organisms grow in milk, and convert the lactose (milk sugar) into lactic acid and other by-products, a distinguishable, characteristic odor is emitted by the formed end products. Most milk

CONTEST MILK SCORE CARD

DATE: _____ ADSA   CONTESTANT NO. _____

		SAMPLE NO.								TOTAL GRADES
	CRITICISMS	1	2	3	4	5	6	7	8	
FLAVOR 10	CONTESTANT SCORE ➡									
	GRADE SCORE									
	CRITICISM									
NO CRITICISM = 10	ACID									
	ASTRINGENT									
	BARNY									
	BITTER									
	COOKED									
	COWY									
	FEED									
	FERMENTED/FRUITY									
	FLAT									
NORMAL RANGE = 1-10	FOREIGN									
	GARLIC/ONION									
	LACKS FRESHNESS									
	MALTY									
	OXIDIZED Light Induced									
	OXIDIZED Metal Induced									
	RANCID									
	SALTY									
	UNCLEAN									
SEDIMENT	ALLOWED PERFECT IN CONTEST	X	X	X	X	X	X	X	X	
PACKAGE	ALLOWED PERFECT IN CONTEST	X	X	X	X	X	X	X	X	
BACTERIA	ALLOWED PERFECT IN CONTEST	X	X	X	X	X	X	X	X	
TEMPERA-TURE	ALLOWED PERFECT IN CONTEST	X	X	X	X	X	X	X	X	
TOTAL	TOTAL SCORE OF EACH SAMPLE ➡									
	TOTAL GRADE PER SAMPLE									

FINAL GRADE _____
RANK _____

Fig. 5.11. The ADSA contest milk score card.

judges can readily detect this odor, despite the fact that up to this point, sufficient acid may not have as yet been produced to be detected by the sense of taste. As the fermentation progresses, the acid taste becomes more pronounced and the odor may become less offensive. Acid or sour milk imparts to the tip of the tongue a peeling or tingling effect. An acid taste tends to leave both the tongue and the mouth with a general feeling of "cleanliness" or an enhanced ability to taste.

Astringent. This sensory defect, "astringent." is not very common in beverage milk. Astringency is best noted by a peculiar mouthfeel after having rolled a sample of the milk about the mouth and expectorating it. In astringency, the tongue and linings of the mouth tend to feel shriveled, almost puckered. Some milk judges that have a relatively high threshold value for rancid taste may possibly perceive this astringent feel at the base and/or back of the tongue when they are tasting slightly rancid milk. Hence, experiencing an astringency sensation may serve as a hint to such judges to observe more carefully for possible rancidity.

Barny. In the past, a number of off-flavors were grouped together under the general heading of "barny." The distinction between "smothered," "cowy," barny, and unclean off-flavors was thought to be one of intensity rather than a difference in perceived sensory characteristics. Although the flavor sensations are similar, they should still be differentiated on the basis of what probably caused them. Currently, the term "smothered" is seldom used.

Barny basically represents a transmitted milk off-flavor. If cows are stabled and milked in a poorly ventilated, foul-smelling environment, a pronounced barny odor is quite likely to occur (transmitted) in milk. The cow may also transmit this foul odor to milk after inhaling off-odor producing compounds, along with air. The off-flavor is detected by sniffing and/or tasting. A characteristic unpleasant aftertaste is most noticeable immediately after sample expectoration. This off-flavor is suggestive of the odor of a poorly maintained barn and leaves a persistent, unclean aftertaste. Fortunately, this off-flavor is becoming encountered less often due to improved ventilation of barns, loafing sheds, and milking parlors and advances in dairy production management.

Bitter. A pure, unassociated "bitter" off-flavor can be detected by taste only. Compared to the other three basic tastes, the reaction time for bitterness is relatively slow, hence, the evaluator must guard against premature judgment. Bitterness is best detected at the base of the tongue (back of the mouth), and this taste sensation tends to persist for a relatively long time. Although a bitter off-flavor may be encountered as a singular defect in milk, it may also be associated with other defects. In some cases, an associated astringency may be noted. Some evaluators find bitterness a distinctive feature of the rancid off-flavor, which will be discussed in subsequent paragraphs. A foreign off- flavor may also exhibit a bitter note, if the foreign substance which entered the milk has a bitter taste. Two common causes of bitterness are specific weeds (consumed as part of the roughage by cows) and certain microorganisms, especially some psychrotropic bacteria.

Cooked. Although "cooked" is the only designation which commonly appears on milk score cards, this term actually represents a number of possible heat-induced sensations of milk and milk products. Upon storage, the heat-induced flavor of pasteurized milk tends to change both in intensity and character. Immediately after processing, the flavor may be quite intense, but after the elapse of 24 hours there is usually a marked reduction in its intensity. Thus, with respect to the cooked flavor, milk flavor may tend to improve during storage, or at least change in characteristics. This is not the case with highly heated products that have acquired a "caramelized" off-flavor. This flavor defect is produced by a different mechanism of chemical interaction of milk components; the caramel-like note frequently intensifies and becomes increasingly more objectionable with increased storage.

Gould and Sommer (1939) demonstrated that the cooked flavor of milk appeared abruptly within a very narrow limit at a temperature of 76°C to 78°C (168.8°F to 172.4°F). Below this processing temperature, heated milk did not appear to develop the cooked flavor. The flavor note that remains in moderately heated milk after refrigerated storage, particularly when higher processing temperatures are used, is generally described as "heated." This distinguishes it from the more aromatic sensation suggestive of sulfides, which is more typical of the cooked flavor (Patton and Day 1956).

In the report of the A.D.S.A Committee on Flavor Nomenclature and Reference Standards, the authors (Shipe *et al.* 1978) recognized four kinds of heat-induced flavors: (1) cooked or sulfurous; (2) heated or rich; (3) caramelized; and (4) scorched. The variety of heat-induced flavor that is encountered depends on a combination of the heating time and the attained temperature, the length of refrigerated storage time for pasteurized milk, and the amount of "product burn-on" in the heat exchanger.

Both the heated and cooked flavors are easily identified. Taste reaction time is relatively quick, and the taste sensation that remains after sample expectoration is usually considered to be pleasant. Cooked flavor may especially be noted by the sense of smell. As the sampling container is brought to the lips and in close proximity of the nose, the characteristic volatility of the cooked note should provide the judge with a hint of what particular flavor is present in the milk.

The presence of "moderately heated" flavors in milk is not particularly objectionable to consumers (or judges), but a pronounced degree of "cooked" flavor is frowned upon. Of particular note, when a heated flavor occurs in milk or cream products, an accompanying oxidized off-flavor is seldom, if ever, present. Thus, in ice cream or butter, a cooked or heated flavor is often recognized as "the flavor of assurance" for the improved keeping quality of milk products, insofar as possible auto-oxidation of milk lipids is involved. Fortunately, natural antioxidants are formed in milk by the heating process. Additional merits of a cooked flavor in milk and cream are that it: (1) serves to help mask more objectionable feed off-flavors, and (2) may provide improved richness and/or mouthfeel sensations in the product.

Cowy (Acetone). Usually, a "cowy" flavor defect implies a distinct cow's breath-like odor and a persistent unpleasant, medicinal, or chemical aftertaste. Though rarely encountered in bulk tank milk, this off-flavor seems to be associated with the presence of acetone-bodies in milk. Hence, milk from individual cows that have acetonemia or ketosis may exhibit this type of off-flavor. Other possible causes of a cowy off-flavor have also been suggested, but additional research would be beneficial in providing a clearer understanding of this defect.

Feed. Some feeds, especially high volume roughages, impart aromatic taints to milk if fed to cows within a critical time frame before milking. The 0.5–3 hour time period is the most critical (Hedrick 1955). This is especially true of succulent feeds, silage, some commodities, brewery wastes, and some of the hays (see Table 5.7). A "feed" off-flavor is characteristic in that it is aromatic, sometimes pleasant (i.e., alfalfa), and can usually be readily detected by the sense of smell. A characteristic note (and mild aftertaste) of "cleanliness" is associated with most feed off-flavors, when the milk sample is expectorated. This distinguishes the feed off-flavor from cowy, barny, or unclean off-flavors. Feed off-flavors usually "disappear" rather quickly and thus leave the mouth in a clean state of condition. By contrast, cowy, barny, or unclean off-flavors tend to persist with an accompanying unpleasant, somewhat "dirty," aftertaste. Beginner judges may experience some difficulty in distinguishing between a slight barny and a feed off-flavor of moderate to definite intensity.

Obviously, the characteristic odor/taste of feed off-flavors varies with the type of feed consumed by lactating animals. The odor of a given raw milk supply is generally characteristic of a particular feed. In some U.S. dairy regions, a severe feed defect is often observed early

Table 5.7. Feed Flavors Transmitted to Milk in Relation to the Quantity of Roughage and the Length of Interval Prior to Milking.

No.	Feed	Amount of Feed (lbs)	Interval Before Milking (hrs)	Flavor of Resulting Milk
1	Alfalfa hay	2–6	2	Objectionable feed
2	Alfalfa hay	2–6	4	Occasional feed
3	Alfalfa hay	2–6	5	No criticism
4	Alfalfa silage	5	1	Definite feed
5	Alfalfa silage	15–25	11	No criticism
6	Clover hay	6	2	Pronounced feed
7	Clover hay	15–20	11	No criticism
8	Clover silage	5	1	Definite feed
9	Clover silage	15–20	11	No criticism
10	Green corn	25	1	Slight feed
11	Green corn	25	11	No criticism
12	Dry beet pulp	7	1	Slight feed
13	Oat hay	12	2	No criticism

Adapted from Hedrick (1955).

in the spring when the all-dry winter ration is terminated, and changed to one that includes fresh green pasture. Also, severe feed off-flavors are likely to occur when there is a sudden change to a new, more odorous form of roughage, such as from alfalfa hay to corn or grass silage.

To minimize the occurrence of objectionable feed off-flavors, milk producers must be aware of the need to avoid the feeding of highly aromatic roughages in the 0.5 to 3 hours just prior to milking. This is an important production management task if milk of *good* flavor quality is to be produced.

Fermented/Fruity. Certain microorganisms produce aromatic fermentation end products that seriously taint milk; this off-flavor is variously described as "fermented" or "fruity" (Morgan 1976). The off-flavor is quickly and easily detected by its odor, which may resemble that of sauerkraut or vinegar (fermented), or pineapple, apples, or other fruits (fruity). This is considered a rather serious defect; it is often found after extended storage of bulk raw milk, as well as in older pasteurized milk. This off-flavor is commonly caused by the growth of psychrotrophic bacteria, especially certain *Pseudomonas* sp. (e.g., *P. fragi*).

Flat. Since "flat" as a flavor defect is not associated with an odor, the sense of smell furnishes absolutely no indication of its possible presence. However, when such a milk is tasted, flatness is apparent soon after the sample reaches the tongue, partly as the result of a marked change in perceived mouthfeel. This flavor defect can be simulated by adding water to a sample of milk and noting the alteration of mouthfeel of the mixture. A flat flavor should not be confused with a "lack of richness" sensation in milk. The latter usually exhibits a sweetness, whereas the former does not. For some evaluators, a slightly oxidized off-flavor may suggest a flat taste on initial tasting.

Foreign. As the name would seem to imply, a "foreign" off-flavor is not commonly developed in or associated with milk; in fact, it is most atypical. In some instances, a foreign off-flavor in milk may be detected by the sense of smell; in other cases, it may not be readily noted until the sample is tasted. The sensory characteristics of this off-flavor differ with the causative agent(s). Foreign off-flavors in milk may be caused by the improper use of various chemicals such as detergents, disinfectants, and sanitizers; exposure to fumes from the combustion of gasoline or kerosene; contamination from insecticides; drenching cows with treatment chemicals; or from treatment of the udder with ointments or medications. Dairy producers must exercise utmost caution in handling various farm chemicals and medications, if milk adulteration is to be avoided.

Garlic/Onion (Weedy). "Garlic" and "onion" off-flavors in milk are recognized by the characteristic pungent odor and a somewhat persistent aftertaste (if tasted). This most objectionable flavor defect may be expected in the spring through fall seasons in those regions where pastures or hay crops become infested with these weeds. In addition to garlic and onion, there are many other weeds that can taint milk when they are consumed by cows; especially if consumed a short time before milking. The character and intensity of weed off-flavors depends on the kind of weed and the time elapsed between cow consumption and milking. Frequently, a weed off-flavor is accompanied by a bitter aftertaste.

Milk judges should familiarize themselves with any potential or unique weed problems in their locality. Evaluators and field department personnel should learn the characteristics of each weed off-flavor (when found in milk) and then be able to suggest a feeding routine to dairy producers that will either minimize or eliminate these flavor defects. The flavor score assigned to milk with a weedy off-flavor depends on the intensity and whether it is caused by a common or an obnoxious weed.

Lacks Freshness (Stale). This mild to moderate flavor defect lacks specific characteristics to make description or identification easy. As the designation "lacks freshness" or "stale" suggests, milk with this off-flavor yields a taste reaction that indicates a loss of those fine, pleasing taste qualities typically noted in excellent or high quality milk. Difficulty may be encountered in attempting to find something specifically wrong with the flavor, yet the astute milk judge senses a certain inherent shortcoming in the milk sample. In some cases, a perceived slight "chalky" taste, perhaps reminiscent of some reconstituted nonfat dry milks, is one way to describe this off-flavor. Such a milk is not as pleasantly sweet and refreshing or as free of an aftertaste as is typically desired in milk. The lacks freshness defect in milk can be a "forerunner" of either oxidized or rancid off-flavors, or off-flavors caused by psychrotrophic bacteria.

Malty. A "malty" off-flavor in milk is usually of either definite or pronounced intensity, and is quite suggestive of malt. Variations of the off-flavor may be encountered; one variation may suggest a "Grape Nuts®"-like flavor. The malty off-flavor is generally caused by the growth of *S. lactis* var. *maltigenes* bacteria in the milk as the result of temperature abuse ($\geq$ 18.2°C [$\geq$ 65°F]) for 2-3 hours (Morgan 1976). This off-flavor can be detected by either smelling or tasting the milk. The bacterial population of such milk will generally be in the millions per ml. Hence, this off-flavor is frequently a forerunner of acid or sour milk. It is not uncommon to simultaneously perceive the malty aroma and the acid taste (or odor).

Oxidized (Metal-Induced). The "oxidized" off-flavor results from lipid oxidation, which is commonly induced by the catalytic action of certain metals. Metallic, oily, cappy, cardboardy, stale, tallowy, painty, and fishy are terms that have been used to describe qualitative differences of the generic "oxidized" off-flavor. The oxidized off-flavor is characterized: (1) by a "quick" taste reaction when the sample is placed into the mouth; (2) by its resemblance to some of the off-flavors mentioned above (see Table 5.8); and (3) by its relatively short adaptation time. When intense, the defect can be detected by smelling; oxidized products are especially perceptible when tasted. This off-flavor is moderately persistent after the sample has been expectorated. A puckery mouthfeel characterizes the oxidized off-flavor, especially when the intensity is relatively high. Unhomogenized or cream-line milk is substantially more susceptible to the development of this off-flavor than homogenized milk, for reasons that are not clearly understood.

Fortunately, the "pure" metallic off-flavor of milk is only encountered occasionally. Its presence may be noted by a definite, peculiar mouthfeel, somewhat like that when a piece of metal foil or rusty metal is placed into the mouth. Both the reaction and adaptation times are quite short. Frequently with the metallic off-flavor, an initial flatness is suggested. The metallic off-flavor is generally associated with the early stages of metal-induced oxidation (cardboardy or papery).

Oxidized (Light-Induced). This off-flavor has been variously described as burnt, burnt protein, burnt feathers, cabbagey, and as medicinal or chemical-like by different authorities. Other names by which

Table 5.8. A Comparison of Hydrolytic Rancidity (Lipolytic), Oxidative Rancidity (Oxidized), and Light-Activated Off-Flavors in Milk.

Factors	Lipolytic (Rancid)	Oxidized (Auto-Oxidation)	Light-Activated
"Substrate(s)" or component(s) involved	Tri- or diglycerides of milk fat	Unsaturated fatty acids (i.e., phospholipids)	Protein (methionine) possibly
End-products of reaction	Short-chain free fatty acids, salts of free fatty acids (soaps)	Short chain volatile aldehydes, ketones	Methional (aldehyde)
Sensory characteristics exhibited	Soapy, bitter, "sour," "blue cheese"-like aroma, vomit	Papery, cardboardy, metallic, painty, fishy	"Burnt" or chemical odor/taste—may eventually become similar to oxidized defect
Chemical mechanism(s)	Hydrolysis of the ester linkage of a short-chain fatty acid	Peroxide radical formation on adjacent carbon-atom of a double bond	"Oxidation" of an amino acid, with the participation of riboflavin
Causes or "triggers" of reaction	Physical abuse of milk ruptures fat globule membrane, then lipase attacks "unprotected" triglycerides. Mixing raw and homogenized milk	Oxygen incorporation; Divalent cations (i.e., Cu^{++}, Fe^{++}, Mn^{++}); Lack of antioxidants; Low bacteria counts; High grain concentrates in ration	Exposure to sunlight or fluorescent light
Measurement of defect	Sensory; Acid degree value (ADV)	Sensory; TBA test[a]; Peroxide value	Sensory
Other features	Foaming of raw milk; Freezing of milk; Extreme temperature changes; Late lactation milk; Enzymatic	High heat treatments minimize occurrence, also homogenization; Nonenzymatic	Protective packaging and eliminate exposure to light; Nonenzymatic

[a] TBA—Thiobarbituric acid, colorimetric determination of malonaldehyde.

this off-flavor is known are light-activated, sunlight flavor, or sunshine flavor. When milk is exposed to sunlight or fluorescent light (i.e., light of wavelengths below 620μ), two different off-flavors may develop. Light catalyzes a lipid oxidation, as well as a protein (amino acid) degradation, both of which are involved in the development of the light-induced flavor defect. The latter reaction requires the presence of the vitamin riboflavin (which is naturally abundant in milk).

As pointed out earlier, homogenized milk is more susceptible to the development of the light-induced off-flavor, but it is more resistant to the generic oxidized off-flavor, which results from lipid oxidation. The opposite is true for cream-line (unhomogenized) milk. Thus, the efficiency of homogenization may determine the proportion of each of these two off-flavors which result from a given exposure to light. Another interesting complication is that the off-flavor due to lipid oxidation becomes more intense during storage of the milk, while the intensity of the light-induced form of oxidized off-flavor diminishes upon storage. Interestingly, a given milk sample may therefore exhibit different off-flavor characteristics when evaluated on successive days.

The light-induced type of oxidized off-flavor may be detected by smell; its odor is quite different from that of the metal-induced, oxidized off-flavor. Difficulties in differentiating between the two off-flavors are largely due to the fact that the light-activated form of off-flavor is not commonly found to be free of lipid oxidation components. Hence, true oxidized and light-induced off-flavors tend to overlap each other. This complicates our efforts at detection, or at least detection with full confidence of which form of oxidation a given milk sample may be guilty of possessing. See Table 5.8 for a summary of the similar and dissimilar characteristics of light-induced and the generic oxidized off-flavor.

Rancid. Some evaluators find "rancid" milk samples extremely unpleasant; by contrast, other persons may find no particular fault or objectionable characteristics in rancid milk. Some people appear to be insensitive or have a relatively high threshold for the taste and odor of free fatty acids and their salts. Some of these individuals may, with guidance and practice, learn to recognize the defect, but may still not find it objectionable (refer to Table 5.8).

There are several characteristics of the rancid off-flavor, as it is perceived, that may be noted in succession. The characteristic odor of rancid milk is derived from the unpleasant volatile fatty acids that are formed as the result of fat hydrolysis. Immediately after placing the rancid sample in the mouth, the flavor may not be too revealing, but a growing awareness of the defect should commence as the sample is manipulated toward the back of the mouth. The perceived sensation should now suggest rancidity—a soapy, bitter, and possibly unclean aftertaste. At this stage, highly sensitive evaluators may find this flavor experience somewhat nauseating or revolting. When the sample is expectorated, the soapiness and bitterness (or rancidity) tend to fade only gradually, and an astringency or "roughness" of the interior mouth surface may occur. Most notably, the rancid aftertaste is persistent and unpleasant.

Salty. The "salty" taste of milk is perceived rather quickly upon placing the sample into the mouth. The sense of smell is valueless in detecting this off-taste, as there is no odor related to salty milk unless the off-flavor is in association with another defect. Saltiness (like acidity) lends a cleansing feeling to the mouth. This off-taste is commonly associated with milk from individual cows that are in the most advanced stages of lactation or with milk from cows that have clinical stages of mastitis. A salty taste is most infrequently encountered in commingled milk supplies or market milk.

Unclean (Psychrotrophic). Some forms of this off-flavor are becoming less common in raw milk supplies due to the general improvement in farm sanitation and more effective temperature control of milk. In either raw or pasteurized milk, this off-flavor may develop by the action of certain psychrotrophic bacteria, particularly when the storage temperature is too high ($\geq 7.2\,^\circ C$ or $\geq 45\,^\circ F$) or milk is stored too long. The end products of bacterial growth that are responsible for this highly objectionable off-flavor may be produced either: (1) directly by the bacteria when they grow in the milk, or (2) indirectly when they grow on improperly cleaned equipment surfaces from which they are transferred into the milk. Spoilage by psychrotrophic bacteria has been the subject of many studies (e.g., Bodyfelt 1974, 1980a, 1980b; Bradley 1983; Cousin 1978; Hankin and Anderson 1969, 1972; Hankin *et al.* 1977; Mikolajcik and Simon 1978; and Moseley 1975).

The presence of an unclean off-flavor in milk may generally be readily noted by its somewhat offensive odor and a failure of the mouth to clean-up after tasting and expectorating the sample. This objectionable off-flavor sometimes suggests extreme staleness, mustiness, a putrid or spoiled ("dirty socks") odor, or foul stable air.

TRACING THE CAUSES OF MILK OFF-FLAVORS: A GUIDE

The examination of innumerable milk samples for off-flavors has disclosed that certain understandings and techniques are helpful in diagnosing the causes or factors contributing to the formation of milk flavor defects. The causes of most milk flavor defects can be classified in one of several ways. Recognizing the more distinguishing characteristics of each possible defect should help the fieldman, plant superintendent, or quality control person to trace the given off-flavor to its source; from here, hopefully, the cause may be eliminated, or at least minimized.

Distinguishing Characteristics of the General Causes of Off-flavors. Different groupings or classifications of the causes of milk off-flavors have been suggested, but the following classification, modified from those offered by Hammer (1938) appears to be the most comprehensive:

Bacterial growth	Processing and handling of milk
Feed or weed	Chemical changes
Absorption (direct and indirect)	(enzymatic and catalytic)
Chemical composition of milk	Addition of foreign material

Each of these groups of off-flavor causes has some unique or distinguishing characteristics, which aid in the eventual identification of the flavor defect. From this point, hopefully, the source(s) or the "trigger(s)" for the flavor problem can be pinpointed and remedial action taken to eliminate, or at least minimize, the impact of the given flavor defect. The general distinguishing characteristics of the above grouping of milk off-flavors are summarized in Table 5.9.

Troubleshooting Causes of Off-flavors. To eliminate or minimize the occurrence of a milk flavor defect, its cause or source must first be identified. To find the possible cause, the milk judge should attempt to review the sensory problem by seeking answers to a number of questions, such as those enumerated in Table 5.10.

Although any of the flavor defects discussed may be encountered by the fluid milk industry, the most frequent consumer complaints relate to the keeping quality of milk and cream. Unfortunately, psychrotrophic bacteria are common postpasteurization contaminants that can easily produce objectionable spoilage off-flavors such as the fruity, unclean, rancid, and bitter off-flavors. With the increased usage of plastic milk containers, the light-induced form of the oxidized off-flavor has also become more and more prevalent in recent years.

The Seasonal Occurrence of Flavor Defects. An awareness and knowledge of the general occurrence of certain milk flavor defects at different months of the year may be helpful in determining the cause. These seasonal differences in milk flavor hinge on the availability of different feeds and on the stage of lactation (Kim *et al.* 1979; Stadhoud-

Table 5.9. Distinguishing Characteristics of Milk Off-Flavors by Category.

Cause of Off-flavor(s)	Distinguishing Characteristics of Off-flavor(s)
Bacterial Growth Typically, $3.0–5.0 \times 10^6$ C.F.U./ml.	High bacterial count in raw milk. The standard plate count of pasteurized milk will be high if the bacterial growth occurred after pasteurization.
Feed or Weed	Bacteria count low; usually off-flavor is present when milk is drawn; commonly more intense in evening milk; occurs when cows have had access to offending feed shortly before milking; odor pronounced (except bitterweed).
Direct Absorption:	Encountered infrequently; occurs after long exposure of thc milk to an odoriferous atmosphere; odor not present when milk is first drawn or handled. Some types of milk containers are pervious to highly odoriferous substances.
Indirect Absorption: (from cow breathing foul air)	Bacteria count usually low; odor of milk suggests "uncleanliness"; odor present when milk is first harvested from the cow. Milk may smell "barny."
Chemical Composition of Milk	Flavor defect is noticeable when the milk is first drawn; milk may be distinctly salty or cowy; inherent to individual animal, rarely noted in mixed milk; defect more likely from an animal in advanced stage of lactation, with an infected udder, or a disease condition.
Processing and Handling of Milk	Pasteurized "heated" or "cooked" flavor. A sulfur-like odor detectable immediately after processing; flavor tends to disappear with increased storage time.
Chemical Changes	Off-flavor not present when milk is first drawn; develops readily at low temperatures—below 4.4°C (40°F); bacteria count usually low. *Three types:* a. Rancidity—In raw milk; bitter, soapy off-flavor; defect more intense in cream than in milk, and more intense in butter than in cream. b. Oxidized—Occurs most often in raw and unhomogenized pasteurized milk; cardboardy; metallic; tallowy; odor similar to wet cardboard. c. Light-induced—In pasteurized milk exposed to light; odor suggests "burnt" protein.
Addition of Foreign Material to Milk	Defect present in either raw or pasteurized milk; rarely increases in intensity during storage; taints varied; may resemble brine, medicine, paint, insecticides, or any other chemical substance with which the milk may have been contaminated.

Table 5.10. A List of Questions to Facilitate the Troubleshooting of Sensory Problems Related to Milk.

1. What does the off-taste of the milk in question resemble?
2. Can customer complaints be categorized as: (1) occasional, or (2) general?
3. Is the defect limited to the raw milk, or does it occur following pasteurization?
4. Does the defect occur sporadically, or has it persisted over an extended period of time?
5. Is the defect present immediately after the milk is drawn from the cow(s)?
6. If the defect is not present when the milk is first drawn, how long does it take to develop a definite intensity?
7. What is the bacteria count of the milk?
8. Does the defect occur in commingled milk or only in the milk from individual cows or individual herds (producers)?
9. What kind and amount of roughage is fed to the cows?
10. How much time elapses between the time of feeding the roughage and the milking time?
11. Has the milk come in direct contact with any copper or rusty equipment? Do farm water supplies or grain rations include elevated levels of copper, iron, or manganese levels?
12. How long has the milk been held in refrigerated storage?
13. What is the storage temperature history of the milk?
14. In what type and/or size of containers does the defect develop?
15. Do various microbiological test results or keeping-quality tests reveal any potential problems?
16. Can line-sample tests (microbiological results) pinpoint the source of the problem?

ers 1972; Tracy *et al.* 1933). Also, dry lot feeding (with either none or minimal pasture or green feeds) has become quite prevalent with U.S. dairy producers. Flavor defects of milk from drylot fed cows may occur at anytime. Increasingly, the stage of lactation also has become less of a factor, as cows are bred to freshen the year around in order to maintain production quotas throughout the calender year. The off-flavors closely associated with drylot feeding are the oxidized, rancid, and feed (silage) off-flavors. Late lactation tends to promote the rancid and salty off-flavors of milk. The evaluator should be alert to the possible occurrence of any flavor defect, regardless of the season.

The Flavor of Milk from Individual Cows. Milk from individual cows tends to differ in flavor and in its susceptibility to the development of certain off-flavors, especially the oxidized and rancid off-flavors. Theoretically, a relatively high proportion of cows within a herd, whose milk is susceptible to the oxidized or rancid off-flavor, could cause a whole shipment of milk to develop these off-flavors. Usually, however, there is an adequate dilution with normal milk, so that no apparent problem may be encountered due to the shortcomings of one or several cows. On rare occasions, the plant fieldman may elect to trace the possible source of a given flavor problem to individual cows. However, with large dairy herds, this can be a formidable task, Unfortunately, little research has been conducted on heredity factors and their possible effects on milk flavor.

CHOCOLATE MILK

Of the flavored milk products (including lowfat milk and skim milk), chocolate milk is by far the most popular one in the U.S. Dairy product judges are frequently asked to evaluate these products, although it must be remembered that chocolate character and intensity, color, and viscosity are a matter of consumer preference in a given market. Since it would be presumptuous for the judge to tell consumers what to like and dislike, product evaluation should allow for a wide range of differences in sensory properties which merit a "no criticism" judgment. On the other hand, actual milk off-flavors and other apparent or obvious sensory defects should be noted. Chocolate flavoring tends to mask (cover up) some of the off-flavors that might be present in milk, but any serious ones may be detected. Sour (high-acid) chocolate milk, for instance, is perceived as extremely unpleasant by most consumers of this product.

The examination of the container and closure of chocolate milk products should be conducted similarly to the approach used for judging milk. These packaging items are subject to the same defects and are given a corresponding evaluation. In evaluating the other qualities of chocolate milk, however, an entirely different set of standards is usually employed. Emphasis is placed on the appearance, color, viscosity, flavor, and freedom from cocoa sedimentation.

Appearance. Chocolate milk should show a uniformity of appearance throughout. The defects in the appearance of chocolate milk, with which the judge should be familiar, are: (1) stratification; (2) mottled or curdy; and (3) the presence of air bubbles. These defects should be recognized easily, but when they are present to a slight degree, they may often be overlooked in a casual examination of the product.

Color. Chocolate milks may vary widely in their color, but the product should probably not be criticized in this respect if the color ranges from a light to a reddish-brown color, such as ordinarily associated with certain cocoas or chocolate. The intensity of color should neither be so light nor so dark as to lack visual appeal. Possible defects of the color of chocolate milk are: (1) unnatural; (2) too light; (3) too dark; and (4) lack of uniformity.

Viscosity. Wide differences in opinion exist as to the most desired viscosity for chocolate milk. Some persons believe that chocolate milk should have the same viscosity as normal milk. Other people prefer a thick, more viscous product. When a small percentage of product stabilizer is added, elevated heat treatment is used, and/or the product is homogenized, the chocolate milk will be more viscous than regular milk. Development of a viscosity so thick that the chocolate milk pours like syrup is not desirable, nor is a body which creates a "slick" sensation when placed into the mouth. Acceptance of a slightly increased viscosity to inhibit creaming is typical, but a heavy, viscous product should probably be criticized by the evaluator(s).

Flavor. Chocolate milk should have a chocolate flavor similar to that of fresh, high-quality chocolate candy. The sweetness should be of medium intensity, so the appetite will not be quickly satiated. Different varieties and manufacturing processes of cocoas and chocolate liquors may be used in the preparation of the syrup or flavoring material for use in chocolate milk. Various attempts may be made to enhance or fortify the chocolate flavor by the addition of one or more of the following adjuncts: malt, salt, vanilla, cinnamon, nutmeg, or other spices; consequently, a variety of flavor notes may be observed. Furthermore, the type of sweetener used may impart a nonchocolate flavor; molasses or excessive corn syrup are examples. Flavor defects of chocolate milk which may be encountered are: (1) unnatural; (2) too sweet; (3) lacks sweetness; (4) syrup flavor; (5) lacks chocolate; and (6) harsh (or coarse) chocolate. It should be born in mind by the evaluator of any chocolate-flavored products that different consumers prefer different types and levels of sweetener and chocolate.

Sedimentation. The "settling out" or precipitation of chocolate and cocoa solids in chocolate milk is very common. While not particularly objectionable, it does have the disadvantage of contributing toward an unfavorable appearance. In aggravated cases, the dark chocolate can form a distinct layer (or strata) under a light "white-livered" upper layer. Furthermore, the consumer is then obliged to agitate the milk vigorously to make the product homogeneous.

In judging chocolate milk for cocoa sedimentation, the evaluator should raise the bottle slightly above the level of the eyes. Next, the judge should note the amount of sedimentation, the quality or fineness of cocoa sediment and the ease or resistance with which it remixes with the milk. Homogenized chocolate milk generally shows more tendency toward sedimentation than the same product that has not been homogenized. Sedimentation of chocolate milk in paper containers may be ascertained to an adequate degree of accuracy by first carefully decanting the liquid, and then observing the inside bottom of the container.

A more quantitative way to measure sedimentation is to pour the agitated contents of a carton of fresh product into a transparent graduated cylinder; then store this test sample in a refrigerator for the shelf-life period of the product. Observations can be made at appropriate intervals, and the extent of cocoa sedimentation quantitated if desired.

OTHER UNCULTURED FLUID DAIRY PRODUCTS

Included in this category are skim milk, lowfat milk, half-and-half, light cream, light whipping cream, and heavy cream. Federal Standards of Identity for these products permit the addition of specific optional ingredients, including characterizing flavors. Many possible products, therefore, are included within this group. As emphasized in the previous discussion on chocolate milk, flavored products can be evaluated for quality, but appropriate allowances must be made for differences in consumer preference. The sensory properties of various unflavored milk products may be assessed by applying the milk score card and scoring guide, with a few modifications.

Additional evaluation categories may be desirable for some of these products, particularly in the case of those that have certain functional properties. A logical test for whipping cream is a determination of its whipping properties, since even the best-flavored whipping cream is of little value to the consumer if it will not whip. Certainly, the coffee "whitening power" and freedom from "feathering" in coffee cream (half-and-half) or light cream are important functional properties.

Obviously, cream and skim milk typically taste different from each other, as well as different from whole milk, but this fact is of little consequence in the evaluation for quality. The judge must memorize or "bear in mind" the normal or typical flavor, and criticize the product only when flavor defects are present. Generally, many of the same off-flavors may be found in skim milk, lowfat milk, whole milk, and the various creams. They may appear to have different characteristics, but much of that is due to the different flavor background. Flavor-producing chemical compounds that are fat soluble are more concentrated in cream than in skim milk. Since the concentration of an odorant may influence both the intensity and qualitative characteristics of the odor, one may expect to perceive the same off-flavor somewhat differently in skim milk than in cream. Similar reasoning would also apply to aromatic compounds that exhibit greater water solubility. This helps explain flavor perception differences in low and high fat products. In any case, most of the defects in low or high fat products will be readily recognized by an evaluator familar with these off-flavors in milk.

Skim Milk

The Code of Federal Regulations description of skim milk was given earlier in this chapter. The product can vary in fat content from less than 0.1% to just under 0.5%. Milk-solids-not-fat (MSNF) may range from 8.25% to 10% or slightly more. Both flavor and mouthfeel characteristics may be affected by the differences in composition within the ranges for fat and MSNF. In the protein-fortified product, the flavor quality of the source of concentrated milk solids can be a significant factor in determining the sensory characteristics of the finished product.

An assumed form of storage flavor commonly encountered in skim milk is variously described as stale, lacks freshness, chalky, or wet paper. The factors responsible for this off-flavor are not known. Skim milk is the test medium of choice for the sensory examination of preparations of vitamin concentrate used in fortifying milk. If a defective vitamin concentrate is likely to impart an off-flavor, skim milk is a more sensitive detection medium than higher fat milks.

A hay-like off-flavor was first reported by Weckel and Chicoye (1954) in lowfat milk fortified with vitamin A. Fluid milk processors continue to occasionally experience puzzling off-flavors in vitamin-fortified milk, apparently caused by the auto-oxidation of vegetable oil carriers for the vitamin concentrates. The most common descriptors used by evaluators (when this off-flavor is noted) is hay-like, or a peculiar stale note. Lowfat milk and skim milk seem to be more vulnerable than homogenized milk to this off-flavor, which may be imparted by sporadic "off batches" of vitamin concentrate.

Lowfat Milk

The CFR definition for lowfat milk was given earlier in this chapter. Since the milkfat content may vary from 0.5% to 2%, the sensory properties of lowfat milk may be similar to skim milk at one extreme, or approach the properties of milk at the upper end of the fat range. The label declaration must clearly specify the actual milkfat content to the closest 0.1%.

Half-and-Half

This product is basically defined in the CFR as that food that consists of a mixture of milk and cream which contains milkfat specifically limited to the range of 10.5% to 18%. It is either pasteurized or ultrapasteurized, and is practically always homogenized. Optional ingredients may include "safe and suitable" emulsifiers, stabilizers, nutritive sweeteners, and "characterizing flavoring" ingredients (with or without coloring), which could include fruit, fruit juice, and/or natural or artificial food flavoring. The majority of half-and-half on the market is pasteurized, homogenized, and unflavored. The principal uses of this product are as coffee cream and as a cereal or fruit topping.

The sensory qualities of half-and-half should be evaluated with the same approach used for milk; the evaluator should be alert for the same defects. Factors which may impact on quality, but which are not typically listed on the milk score card are: appearance (possible cream or oil separation or a cream plug); viscosity (appropriate for the product of a given composition); acidity (there should be no developed acidity); and feathering (or other developed defects when added to coffee).

The viscosity of half-and-half may be measured instrumentally, by the use of one of several commercially available viscosimeters. Since viscosity is substantially influenced by sample temperature, all measurements must be made at a standardized temperature. The logical temperature to use is $4.4\,^{\circ}C-10\,^{\circ}C$ ($40\,^{\circ}F-50\,^{\circ}F$) since this is the typical temperature range at which the consumer will subsequently use the product and observe the viscosity. Both the instrument and the sample should be tempered to the preset standard temperature for conducting the viscosity measurement.

There are three possible defects which may be noted when half-and-half is added to hot coffee: feathering, oiling-off, and off-color (in coffee). Of these, feathering is probably the most commonly encountered and the most objectionable.

Feathering. Feathering is evident is several ways, depending upon the intensity of the defect. Such a product may initially appear immiscible in coffee, wherein the cream may rise in flocculent masses to the surface, and thus reflect a lack of homogeneity. Frequently, this defect appears as a light, evenly serrated scum on the coffee surface, after the coffee and half-and-half mixture has become quiescent. Occasionally, this defect may be so extensive that most of the added cream rises in mass to the coffee surface immediately after the half-and-half has been poured into it; wherein it may appear like distinct chunks of sour cream. When the homogenization pressure is excessive, the half-and-half may be more susceptible to feathering under certain conditions, particularly when the water used for coffee making has a high calcium content. Actually, with half-and-half of normal composition, the susceptibility to feathering is not unduly affected by homogenization, even at high pressures. However, if the milkfat content is high, the effect of homogenization (and higher homogenization pressures) becomes more apparent. The susceptibility of light cream (to be dis-cussed next) to feathering is considerably enhanced by higher homoge-nization pressures. Additionally, half-and-half suffering from elevated titratable acidity ($\geq 0.12\%$ as lactic acid) may be more susceptible to feathering. The presence of this developed acidity will be reflected as an acid or slightly sour off-flavor in the product. Unfortunately, regard-less of the cause of cream feathering in coffee, the consumer usually believes that the cream is sour; hence, this can represent a rather seri-ous defect of half-and-half.

Oiling Off and Off-color. These defects are more apt to occur with light cream than with half-and-half, particularly a cream that tends to have an "oily" body. Freezing of the cream product or improper homogenization contribute to these difficulties. Droplets of butter oil may be noted on the coffee surface, and instead of developing a light brown color, the coffee appears slate gray. Also, on occasion, a cream plug, partial churning, and/or coalescence of fat globules may be ob-served in the product before its addition to hot coffee. When such de-stabilized cream is added to the hot beverage, oiling-off (and a possible off-color) is most likely to occur.

Light Cream

Light cream is basically described in the CFR as a cream which con-tains not less than 18%, but less than 30%, milkfat. With respect to processing and optional ingredients, the definition of light cream does not differ from that of half-and-half. In some isolated localities, this product may still have some significance for individual milk proces-sors, but in most U.S. markets the demand for light cream no longer exists and the product is not available. Imitation "cream" toppings (or "coffee whiteners") and half-and-half have essentially replaced light cream in consumer food service markets. All of the potential defects enumerated for half-and-half also apply to light cream. In fact, light cream is generally even more susceptible to these developed quality shortcomings. The body and viscosity of light cream is somewhat more difficult to control than that of half-and-half; thus, this merits more detailed discussion.

The body of light cream should be smooth, uniform, and reasonably viscous, given the higher percentage of milkfat. When poured into hot coffee, the cream should be readily miscible, and exhibit neither "feath-ering" nor "oiling-off." It should impart a pleasant color to the coffee. Some body defects are readily apparent to the eye, while others may require physical examination of the cream and/or tests that employ the use of hot coffee. The more common body defects of table cream that are readily apparent by direct visual examination are listed in the following paragraphs.

Cream Plug. A cream plug may be exhibited by: (1) a lack of uniformity in the cream, particularly at the surface; (2) a layer of frothy and sometimes heavy cream that adheres to the bottle closure; (3) butter particles on the surface of the cream; or (4) a distinct, heavy, leathery plug that obstructs the flow of cream from the container. A cream plug should not be confused with "ropy cream," which is a bacterial spoilage defect of somewhat similar appearance. Cream displaying a definite cream plug often has a distinctly thin body throughout the remainder of the product. When such cream is poured into hot coffee, droplets of milkfat are generally noted on the surface. This defect varies widely in its intensity. The various intensities of the cream plug defect, listed in increasing order of relative seriousness of the defect and degree of being objectionable are: a slightly soft, foamy plug; a distinct, soft plug; a buttery plug; and a leathery plug.

Oiling-off. Oily cream is inclined to be seasonal; it is observed more frequently when cows have just been placed on pasture or green grass. In reality, this defect is closely associated with the cream plug defect; in the aggravated state of oiling-off, a cream plug invariably forms. Cream that has this defect generally appears shiny and usually has a thin body. The presence of a distinct skim milk layer is commonly found with oily cream.

Separation of a Skim Milk Layer. Separation of a skim milk layer is more common on the lower fat-content cream products. It results from the rising of fat particles (creaming-off). The defect is best described as a bluish, watery-like layer, that is from one-sixteenth to one-half inch in depth, at the bottom of the container. Its presence in cream connotes to the customer a dilution of the product with skim milk.

Two qualities must be considered in observing the serum or skim milk layer of cream; namely, the depth of the layer and its distinctness. The latter quality seems to be the more serious of the two. A relatively obscure, deep skim milk layer is probably less objectionable than a distinct, shallow layer that displays a pronounced line of demarcation.

Certain associations with a skim milk layer may be noted in cream. Usually, cream with this defect does not exhibit a thin body, but instead manifests a relatively viscous body, considering the amount of fat present. Sometimes an old, stale, or oxidized off-flavor may be noted and associated with a cream displaying this body defect. The skim milk layer in light cream becomes more distinct upon extended storage.

Thin Body. Thin body is a quite common body defect of light cream.

It is evidenced by a tendency to drip as it is slowly poured from the container, and/or a tendency to definitely "splash" (similar to milk) as the product is poured onto a flat surface, from a distance of six inches or more. Thin body may sometimes be associated with the cream plug defect, but it will rarely be associated with the separation of a skim milk layer. While this defect may be objectionable on the basis that it suggests to the cream customer a low milkfat percentage in the cream, it is not as serious as some other body defects.

Defects such as a cream plug, oily cream, and the separation of a skim milk layer can also occur in light cream that is packaged in paper. However, these conditions cannot be observed by examining the unopened container. The cream itself must be examined; sometimes after decanting the product into a glass container (such as a graduated cylinder) and storing for a time period sufficient to "restore" the defect. If cream in paperboard cartons has a thin body, this defect may sometimes be detected by shaking the container and carefully noting the apparent "difference in sound."

Whipping Cream

The Code of Federal Regulations recognizes two products in this category, *light whipping cream* and *heavy cream*. Except for milkfat content, the definitions for these products do not differ from those of light cream and half-and-half. Light whipping cream must not have less than 30%, but less than 36%, milkfat. The fat content of heavy cream must not be less than 36%.

Whipping cream constitutes only a modest volume of the annual total production of Grade A milk and cream products in the U.S. In fact, many processors no longer produce it because low demand results in excessively long storage times, which can lead to substantial losses due to spoilage. This serious spoilage problem is perhaps best addressed by specialized plants that produce an ultrapasteurized version of whipping cream and in turn solicit the same milk processors to serve as product distributors. Without question, much of the U.S. sales for whipping cream has been lost to imitations and substitutes, which come in many forms—powders, frozen, frozen prewhipped, and in pressurized containers.

In general, a highly desired whipping cream possesses a clean, sweet, nutty flavor, a relatively heavy body (which is uniform throughout), and a smooth texture. The flavor, bacterial count, sediment, container, and closure defects may be the same as those encountered in milk, half-and-half, and light cream. The most critical quality criterion is the whipping test (see Appendix VII). When performed under standardized conditions, it should provide data on the required time to produce the desired stiffness and appearance of whip; whether or not the desired stiffness and dry, velvety appearance is achievable; the final overrun; the stability of the whipped cream; and the mouthfeel properties of the whipped cream.

Fat Content of Whipping Cream. As long as the percentage of fat in whipping cream conforms to the legal milkfat standard, the product cannot be faulted, despite the possibility of higher percentages of milkfat in other samples. Most research workers concur that the percentage of milkfat in whipping cream should be between 30% and 35%. Such a cream should be expected to respond to whipping, and to subsequently yield a reasonably stiff, stable, whipped cream of normal overrun (approximately 100–200%).

Body Defects of Whipping Cream. Whipping cream is subject to the same general body defects as light cream, but to different degrees of intensity. The viscosity of whipping cream, although higher than light cream, may sometimes be too low, given the higher percentage of milkfat present; cream plug defects may be accentuated; serum separation may be reduced to a minimum; and the feathering and oiling-off problems (of the lighter creams) may be of little or no consequence.

Whipped Cream from Pressurized Containers

A specially formulated whipped cream, dispensed from pressurized containers, is commonly used by the general retail market. Product formulation, type of gas, and the design of the container and valve, are under proprietary control. Upon release of the gas, a saturated, pressurized cream is formed and removed through a special valve. The cream explodes instantly into a relatively stable, sometimes almost frothy, product similar to traditional whipped cream. The increase in volume is proportional to the pressure at which the cream is saturated before being released; the volume is independent of the milkfat content. The quality criteria for this product are the same as those for whipped cream prepared by traditional methods, namely: flavor, stiffness, dryness, stability (as exhibited by resistance to air cell collapse and drainage or leakage), and the overrun.

Eggnog

Part 131.170 of Title 21 of the CFR describes eggnog as the food containing one or more of a set of listed dairy ingredients (cream, milk, skim milk, or partially skimmed milk), one or more of the optional ingredients that provide egg yolks (liquid, frozen, or dried egg yolks or whole eggs), and one or more of listed nutritive carbohydrate sweeteners (sugar, invert sugar, brown sugar, high-fructose corn syrup, and others). Other optional ingredients for eggnog include: certain other milk-derived products, such as nonfat dry milk, whey, lactose, etc.; salt; flavoring ingredients; color additives (except those that impart a color simulating egg yolk or milkfat); and approved stabilizers. All ingredients used must be considered safe and suitable. Eggnog must contain not less than 6% milkfat and not less than 8.25% MSNF. The egg yolk solids content of eggnog must not be less than 1% by weight of the finished food. The product must be pasteurized or ultrapasteurized and may be homogenized.

Important components of the sensory quality of eggnog are flavor, body (consistency), and product appearance (Feet et al. 1963; Hedrick et al. 1962). As in other flavored milk or cream products, consumer preference plays an important part, but typical milk-related off-flavors can be a quality problem. Since milk and its derivatives make up the major portion of eggnog, the evaluator should be alert to any off-flavor or flavor deterioration that may occur during processing and/or storage. The potential off-flavor concerns of eggnog probably more closely resemble those of ice cream than of milk or cream (see Chapter 6 for details).

There seem to be differing views as to the most desired viscosity of eggnog, but industry authorities generally agree that the body should be smooth, thicker than milk, and uniform throughout. The color should be characteristic of eggs and cream, and if particles of sweet spices have been incorporated into the product, they should be uniformly distributed.

REFERENCES AND BIBLIOGRAPHY

American Dairy Science Association. 1987. Committee on Evaluation of Dairy Products, Score Card Sub-Committee. Champaign, IL.

Bodyfelt, F. W., Tesdal, A. E., Briody, G., and Rackleff, E. R. 1985. *HTST Pasteurizer Operation Manual* 4th Ed. OSU Bookstores, Inc. Corvallis, OR. 66pp.

Bodyfelt, F. W. 1974. Temperature control monitoring II: A method for fluid milk processing plants. *J. Dairy Sci.* 57:592.

Bodyfelt, F. W. and Davidson, W. D. 1975. Temperature control I: A procedure for profiling temperatures of dairy products in stores. *J. Milk and Food Technol.* 38:734.

Bodyfelt, F. W. and Davidson, W. D. 1976. Temperature control means longer shelf-life. *Dairy and Ice Cr. Field.* 158(6):34.

Bodyfelt, F. W., Morgan, M. E., Scanlan, R. A., and Bills, D. D. 1976. A critical study of the multiuse polyethylene plastic milk container system. *J. Milk Food Technol.* 39:481.

Bodyfelt, F. W. 1980a. Is it time for a temperature audit? *Dairy Record.* 81(9):96.

Bodyfelt, F. W. 1980b. Heat resistant psychrotrophs affect quality of fluid milk. *Dairy Record. 81*(3):96.

Bodyfelt, F. W. 1983. Quality the consumer can taste: A primer on quality assurance procedures that produce excellent milk flavor. *Dairy Record. 84*(11):170.

Bradley, R., Jr. 1983. How to minimize off-flavors in milk. *Dairy Record. 84*(2):93.

Code of Federal Regulations. 1985. *Title 21—Food and Drugs, Part 131—Milk and Cream.* U.S. Government Printing Office. Washington, DC.

Cousin, M. A. 1978. Psychrotrophs in relation to keeping quality of milk products. *J. Food Protect. 41*:830.

Doan, F. J. 1933. Critical preheating temperatures for inhibiting rancidity in homogenized milk. *Milk Dealer 23*(2):40.

Dorner, W. and Widmer, A. 1931. Development of rancidity in milk during homogenization. *LeLait 11*:545.

Feet, O., Hedrick, T. I., Dawson, L. E., and Larzelere, H. E. 1963. Factors influencing acceptability of eggnog. *Mich. Agr. Expt. Sta. Quart. Bul. 46*:293.

Gasaway, J. M. and Lindsay, R. C. 1979. Flavor evaluation of milk held in returnable Lexan® polycarbonate resin containers after commercial usage. *J. Dairy Sci. 62*:888.

Gould, I. A. and Sommer, H. H. 1939. Effect of heat on milk with special reference to the cooked flavor. *Mich. Agr. Exp. Sta. Tech. Bul.* 164.

Gould, I. A. 1940. Control of flavor in milk heated to high temperature. *Milk Dealer 29*(8):70.

Hall, C. W. and Trout, G. M. 1967. *Milk Pasteurization.* AVI Publishing Co. Westport, CT. 234 pp.

Halloran, C. P. and Trout, G. M. 1932. The effect of viscolization on some of the physical properties of milk. *Abs. 27th Ann. Mtg. Amer. Dairy Sci. Ass'n.* 17.

Hammer, B. W. 1938. *Dairy Bacteriology.* 2nd Ed. John Wiley and Sons. New York.

Hankin, L. and Anderson, E. O. 1969. Correlations between flavor score, flavor criticism, standard plate count and oxidase content on pasteurized milks. *J. Milk Food Technol. 32*:49.

Hankin, L. and Stephans, G. R. 1972. What tests usefully predict keeping quality of perishable foods? *J. Milk Food Technol. 35*:574.

Hankin, L., Dillman, W. F., and Stephens, G. R. 1977. Keeping quality of pasteurized milk for retail sale related to code date, storage temperature and microbial counts. *J. Food Protect. 40*:842.

Harper, W. J. and Hall, C. W. 1976. *Dairy Technology and Engineering.* AVI Publishing Co. Westport, CT. 631 pp.

Hedrick, T. I., Dawson, L. E., and Feet, O. 1962. A proposed eggnog score card. *Mich. Agr. Expt. Sta. Quart. Bul. 45*:293.

Hedrick, R. R. 1955. Feed flavor transmission to milk. Montana State University, Bozeman, MT. Mimeographed material.

Henderson, J. L. 1971. *The Fluid Milk Industry.* AVI Publishing Co. Westport, CT.

Herreid, E. O. and Tobias, J. 1959. Ultra-high temperature short-time experimental studies on fluid milk products. Experimental equipment. *J. Dairy Sci. 42*:1486.

Hood, E. G. and White, A. H. 1934. Homogenization of market milk. Canad. Dept. Agr. Dairy and Cold Storage. Br. Mimeo 25. Ottawa, Canada.

Kim, H. S., Gilliland, S. E., and Von Gunten, R. L. 1979. Chemical test for detecting wheat pasture flavor in cow's milk. *J. Dairy Sci. 63*:368.

Landsberg, J. D., Bodyfelt, F. W., and Morgan, M. E. 1977. Retention of chemical contaminants by glass, polyethylene, and polycarbonate multi-use milk containers. *J. Food Protect. 40*:772.

Mikolajcik, E. M. and Simon, N. T. 1978. Heat-resistant psychrotrophic bacteria in raw milk and their growth at 7°C. *J. Food Protect. 41*:93.

Morgan, M. E. 1976. The chemistry of some microbiologically induced flavor defects in milk and dairy foods. *Biotechnol. and Bioeng. 18*:953.

Moseley, W. 1975. Improving and maintaining shelf-life of dairy products. *Dairy and Ice Cr. Field 158*:44.

Patton, S., Forss, D. A., and Day, E. A. 1956. Methyl sulfide and the flavor of milk. *J. Dairy Sci. 30*:1469.

Pillay, V. T., Myhr, A. N., and Gray, J. I. 1980. Lipolysis in milk. I. Determinaton of free fatty acids and threshold values for lipolyzed flavor detection. *J. Dairy Sci. 63*:1213.

Schroder, M. J. A. 1982. Effect of oxygen on the keeping quality of milk. I. Oxidized flavour development and oxygen uptake in milk in relation to oxygen availability. *J. Dairy Res. 49*:407.

Shipe, W. F., Bassette, R., Deane, D. D., Dunkley, W. L., Hammond, E. G., Harper, W. J., Kleyn, D. H., Morgan, M. E., Nelson, J. H., and Scanlan, R. A. 1978. Off-flavors in milk: Nomenclature, standards, and bibliography. *J. Dairy Sci. 61*:855.

Shipe, W. F., Senyk, G. F., and Fountain, K. B. 1980. A modified copper soap solvent extraction method for measuring the free fatty acids in milk. *J. Dairy Sci. 63*:193.

Sliwkowski, M. X. and Swaisgood, H. E. 1980. Characteristics of immobilized sulfhydryl oxidase reactors used for treatment of UHT milk. *J. Dairy Sci. 63(Suppl.1)*:60.

Stadhouders, J. 1972. Technological aspects of the quality of raw milk. *Neth. Milk Dairy J. 26*:80.

Strobel, D. R., Bryan, W. G., and Babcock, C. J. 1953. Flavors of milk. A review of literature. *U.S. Department of Agriculture Bulletin.* 91pp.

Swaisgood, H. E. 1980. Sulphydryl oxidases properties and applications. *Enzyme Microb. Technol. 2*:265.

Thomas, E. L., Nelson, A. J., and Olson, J. C., Jr. 1955. Hydrolytic rancidity in milk— A simplified method for estimating extent of its development. *Amer. Milk Rev. 17*:50.

Tracy, P. H., Ramsey, R. J., and Ruehe, H. A. 1933. Certain biological factors related to tallowiness in milk and cream. *Ill. Agr. Exp. Sta. Bul.* 352.

Trout, G. M. 1940. Watery appearance of frozen homogenized milk. *Mich. Agr. Exp. Sta. Quart. Bul. 23(1)*:10.

Trout, G. M. 1941. The freezing and thawing of milk homogenized at various pressures. *J. Dairy Sci. 24*:277.

Trout, G. M. 1945. Tracing the off-flavors in milk: A guide. *Mich. Agr. Exp. Sta. Quart. Bul. 27(3)*:266.

Trout, G. M. 1950. *Homogenized Milk*. Michigan State Univ. Press. East Lansing, MI. 233pp.

U.S. Department of Agriculture. 1975. Judging and scoring milk and cheese. *Farmers' Bulletin* No. 2259. 16pp.

U.S. Department of Health and Human Services. 1978. Grade A Pasteurized Milk Ordinance. U.S. Public Health Service/Food and Drug Administration. Washington, DC.

Weckel, K. G. and Chicoye, E. 1954. Factors responsible for the development of hay-like flavor in vitamin A fortified lowfat milk. *J. Dairy Sci. 37*:1346.

6

Sensory Evaluation of Ice Cream and Related Products

Ice cream enjoyed only limited success until the advent of mechanical refrigeration in 1878; after this technical breakthrough, the U.S. ice cream industry underwent rapid growth. The impetus for industry growth was the development and implementation of the "iceless cabinet," which provided a route for direct-to-the-consumer sales through retail food stores.

Ice cream and related products are members of the "frozen desserts family" and are defined in Part 135 of the Code of Federal Regulations. Each product category may differ in the type of flavoring, the composition in terms of dairy ingredients and other food solids, and the extent of product overrun. Table 6.1 summarizes the compositional differences of the major classes of frozen dairy desserts. The optional milk ingredients which these frozen dairy desserts may contain are listed in Table 6.2.

Within the restrictions imposed by the Federal Standards of Identity (Table 6.1), ice cream is basically defined as that food produced as a result of freezing, while stirring, a pasteurized mix which consists of one or more of the dairy ingredients listed in Table 6.2 and other non-milk-derived ingredients (that are safe and suitable). The latter serve as nutritive carbohydrate sweeteners, stabilizers, emulsifiers, flavorings, and coloring agents.

The Sweeteners. The sweeteners commonly used in ice cream are sucrose (cane or beet sugar), dextrose (corn sugar), and various corn syrups. Honey, when used, imparts both sweetness and a characteristic flavor. Corn syrup is produced by converting starch into a mixture of simpler sugars including dextrose, maltose, maltotriose, maltotetraose, and dextrins (in ascending order of molecular weights). Members of the mixture with lower molecular weights exhibit greater sweetness, while the higher molecular-weight members have the ability to bind water more effectively. The Dextrose Equivalent (DE) designation of a corn syrup provides an indication of the distribution of starch conversion sugars present. High DE values imply a high degree of con-

Table 6.1. Federal Standards of Identity for the Composition of Frozen Dairy Desserts.

Product	Weight (lb/gal)	Total food solids (lb/gal)	Total milk solids[a] (%)	Milkfat (%)	Whey solids[b] (%)	Egg yolk solids (%)
Ice cream	≥4.5	≥1.6	≥20	≥10	≤2.5	<1.4
Bulky-flavored ice cream[j]	≥4.5	≥1.6	≥16	≥8	≤2.0	i
Frozen custard[c]	≥4.5	≥1.6	≥20	≥10	≤2.5	≥1.4
Bulky-flavored frozen custard[j]	≥4.5	≥1.6	≥16	≥8	≤2	≥1.12
Mellorine	≥4.5	≥1.6	e,g	f	g	d
Ice milk	≥4.5	≥1.3	≥11	≥2	≤2.25	d
Ice milk	≥4.5	≥1.3	≥11	≤7	≤1.0	d
Bulky-flavored ice milk[h,j]	≥4.5	≥1.3	h	h	b	d
Sherbet	≥6.0	e	2 to 5	1 to 2	0 to 4	d
Water ices	≥6.0	e	0	0	0	0

Adapted from Code of Federal Regulations Title 21, Part 135.

[a] Caseinates may not be used to satisfy any part of the total milk solids requirement. Increases in milk fat may be offset with corresponding decreases in nonfat milk solids, but the latter must be at least 6% in frozen custard and ice cream and 4% in ice milk. Corresponding adjustments may be made in bulky-flavored products.

[b] Solids from concentrated, dried, and modified whey used singly or in combination may not exceed 25% of the total nonfat milk solids content.

[c] Also designated French Ice Cream or French Custard Ice Cream.

[d] Permitted.

[e] No standard.

[f] Milkfat replaced by a minimum of 6% vegetable or animal fat.

[g] At least 2.7% milk-derived protein having a protein efficiency ratio (PER) not less than that of whole milk protein, 108% of casein.

[h] Composition is determined by calculation based on actual quantity of the bulky flavoring used. However, the milkfat content and the nonfat milk solids content must never be lower than 2% and 7%, respectively. (Total milk solids must be not less than 9%.)

[i] Less than 1.4% egg yolk solids by weight of the food exclusive of the weight of any bulky-flavor ingredients.

[j] Adjustment in composition in bulky-flavored frozen desserts is determined by calculation based on the actual quantity of bulky flavor used. However, the analysis must never be lower than the minima given in the table.

Table 6.2. List of Optional Dairy Ingredients Approved for Use in Frozen Dairy Desserts, Federal Standards[a].

Cream:	fresh, dried, plastic
Butter and butteroil	
Milk:	fresh, concentrated, evaporated, sweetened condensed, superheated condensed, dried
Skim milk (nonfat milk):	fresh, concentrated, evaporated, dried, condensed, sweetened condensed, superheated condensed, concentrated and partially delactosed, concentrated or dried after modifying by treatment with calcium hydroxide or disodium phosphate
Cheese whey:[b]	concentrated, dried, modified
Casein:[c]	precipitated with gums
Caseinate:[c]	salt of ammonium, calcium, potassium, or sodium
Buttermilk:	fresh, condensed, or dried; from churning of sweet cream

Adapted from Code of Federal Regulations Title 21, Part 135.
[a] The Federal Standards of Identity provide quality standards for certain of the above ingredients.
[b] If recognized by FDA as safe (GRAS) for this type of food.
[c] Not considered to be milk solids (does not satisfy milk solids requirements).

version into dextrose, the simplest sugar made from starch, and maltose. Other available corn syrups are designated as high maltose and high fructose; the latter is produced by an additional processing step which converts dextrose into fructose. Fructose provides the most sweetness for a given amount of added sweetener.

For the purpose of evaluating the sensory properties of ice cream, knowledge of certain facts about sweeteners may be important. On an equal weight basis, sugars vary in their ability to impart sweetness. In descending order of sweetening power, the sugars commonly found in ice cream rank as follows: fructose, sucrose, dextrose, maltose, and lactose. In an aqueous solution, such as found in ice cream, approximately 2 parts of 42 DE corn syrup, 3 parts of lactose, or 1 part of high fructose syrup are required to impart the equivalent sweetness of 1 part of sucrose. The generally accepted sweetness level for vanilla ice cream is a 13 to 15% sucrose equivalent (equal to 13%–15% sucrose in the mix).

The relative hardness of ice cream at any given temperature depends on what proportion of water is frozen at that temperature, which in turn largely depends on the freezing point of the ice cream mix (Arbuckle 1986; Bodyfelt 1983; Tobias 1981b, 1982c). As the freezing point of the mix is reduced, more water will remain unfrozen at a given temperature and the final ice cream will seem softer. The monosaccharides fructose and dextrose equally lower the freezing point of a solution (or a mix), and concomitantly reduce the freezing point to a greater extent than the disaccharides sucrose, maltose, and lactose. The higher molecular-weight sugars that are present in corn syrup depress the freezing point to a lesser extent than do disaccharides, when compared on an equal weight basis.

One useful definition of "bound water" is water that does not freeze; thus, bound water behaves as though it were a solid. A high proportion of bound water in ice cream, or other frozen dairy desserts, serves to reduce the amount of water to be frozen; hence, improving the body and texture of the product. Each of the various sugars used in ice cream bind water to a different extent. The higher sugars and dextrins in corn syrup are the most effective binders of water. The low DE corn syrups (e.g., 36 DE and 42 DE) lack sweetening power compared to the higher DE corn syrups, but the low DE sweeteners have greater water binding and "bodying" properties in ice cream and ice milk.

Liquid sugars of poor quality or corn syrups can be sources of off-flavors in frozen dairy desserts, especially in vanilla-flavored products. Dark syrups in which nonenzymatic browning (Maillard reaction or carmelization) has taken place may impart a stale, caramelized flavor. Certainly more serious is the fermentation of liquid sugars or corn syrups, which generally makes them unusable in ice cream. When evaluating ice cream, one should be alert to the possible flavor shortcomings which can stem from sweeteners.

Stabilizers and Emulsifiers. Although most commercial ice creams contain stabilizers and emulsifiers in small concentrations, some manufacturers exclude these body- and texture-modifying agents from the formulation of certain brands, especially those products categorized and promoted as "premium quality" or "all natural" (Bodyfelt 1983; Tobias 1981b; 1982c; 1983). The primary function of stabilizers in ice cream and related products is to bind water, which in turn promotes small ice crystal formation and helps keep ice crystals from growing in size when storage temperatures fluctuate or become too high (referred to as "heat shock"). Stabilizers are usually proprietary blends of gums such as guar, locust, carrageenan, alginates, and carboxymethyl cellulose (CMC), to name a few. Depending on the type and concentration of gums in the mixture, and the milkfat and solids content of the ice cream, stabilizers are used at levels from 0.15% to 0.5%.

Commonly used emulsifiers include lecithin, mono- and diglycerides of fatty acids, Polysorbate 80 (polyoxyethylene (20) sorbitan monooleate), and Polysorbate 65 (polyoxyethylene (20) sorbitan tristearate). Depending on which specific emulsifier(s) is used, the concentration may vary from 0.03% to 0.2%. Emulsifiers contribute to the formation of small, uniformly dispersed air cells; protect against texture deterioration due to heat shock; and provide a semblance of a "richness" sensation. The latter function is probably achieved as the result of partial agglomeration of fat globules around the air cell walls. Over-emulsification may result in fat churning, a greasy-like mouth coating, and/or an "emulsifier" taste. At times, even lower levels of emulsifiers may be perceived as an aftertaste, particularly when this or other ingredients are old, oxidized, or have deteriorated in some other way.

Flavoring Agents. Space does not permit the listing of all the possible flavorings used in ice cream. As a general principle, there is no point in comparing one flavor type against another, as the choice is generally a matter of personal preference. The evaluator should be aware that flavorings range from natural to artificial; as a general rule, the natural source may be preferred from several viewpoints. However, the use of natural flavoring is not always a guarantee of high quality. For example, some sources of fresh or frozen strawberries (as well as certain other berries or fruits) may be deficient (lacking) in flavor intensity, though used at the recommended level. Other possible problems with berries or fruits may involve: (1) the utilization of the wrong, or a less satisfactory, variety; (2) improper stage of ripeness at harvest; (3) physical damage prior to preservation; (4) excessive and/or improper storage prior to preservation; (5) high and fluctuating temperatures in frozen storage; and/or (6) an inadequate quantity of fruit incorporated into the product.

The most popular flavor of ice cream in the U.S. is vanilla, which accounts for nearly one-half of all ice cream sales. Since vanilla is a most delicate flavoring, it will not "cover-up" or mask any off-flavors which may reside in the background. Off-flavors in the mix are more difficult to detect in the presence of stronger flavorings, such as chocolate. To manufacture a vanilla ice cream with an ideal flavor requires that (1) the dairy products, sweeteners, and all other ingredients be free of flavor defects; (2) the mix be correctly processed; and (3) the vanilla flavoring be of the highest quality. The perceived flavor should not only exhibit the desired intensity, but should also blend pleasingly with the background or the complementary flavor provided by the mix. While vanilla ice cream provides a rigid test for overall sensory and quality control, these general manufacturing requirements also apply to other flavors. A common axiom in the manufacture of dairy products is that "the quality of the finished product can be no better than the quality of the ingredients."

THE ICE CREAM SCORE CARD

The ice cream score card illustrated in Fig. 6.1 represents a modification of the card developed by the American Dairy Science Association

ICE CREAM SCORE CARD

PRODUCT: _____   DATE: _____

FLAVOR: _____

CRITICISM		SCORE	SAMPLE NO. 1	2	3	4	5	6	7	8	9	10
FLAVOR 10		▶										
	FLAVORING SYSTEM											
NO	LACKS FINE FLAVOR											
CRITICISM	LACKS FLAVORING											
= 10	TOO HIGH FLAVOR											
	UNNATURAL FLAVOR											
	SWEETENERS											
UNSALABLE	LACKS SWEETNESS											
= 0	TOO SWEET											
	SYRUP FLAVOR											
	PROCESSING											
NORMAL	COOKED											
RANGE	DAIRY INGREDIENTS											
= 1-10	ACID											
	SALTY											
	LACKS FRESHNESS											
	OLD INGREDIENT											
	OXIDIZED											
	METALLIC											
	RANCID											
	WHEY											
	OTHERS											
	STORAGE (ABSORBED)											
	STABILIZER/EMULSIFIER											
	NEUTRALIZER											
	FOREIGN											
BODY & TEXTURE 5		▶										
NO	COARSE/ICY											
CRITICISM	CRUMBLY											
= 10	FLUFFY											
UNSALABLE	GUMMY											
= 0	SANDY											
NORMAL	SOGGY											
RANGE	WEAK											
= 1-5												
COLOR, APPEARANCE												
& PACKAGE 5		▶										
NO	DULL COLOR											
CRITICISM	NON-UNIFORM COLOR											
= 5	TOO HIGH COLOR											
UNSALABLE	TOO PALE COLOR											
= 0	UNNATURAL COLOR											
NORMAL	DAMAGED CONTAINER											
RANGE	DEFECTIVE SEAL											
= 1-5	ILL-SHAPED CONTAINER											
	SOILED CONTAINER (DIRT)											
	SOILED CONTAINER (PRODUCT)											
	UNDER FILLED											
	OVER FILLED											
MELTING QUALITY 3		▶										
NO CRITICISM	CURDY											
= 3	DOES NOT MELT											
UNSALABLE	FLAKY											
= 0	FOAMY											
NORMAL RANGE	WATERY											
= 1-3	WHEYED OFF											
BACTERIAL CONTENT 2		▶										
	STANDARD PLATE COUNT											
	COLIFORM COUNT											
TOTAL 25	TOTAL SCORE											
	OF EACH SAMPLE											
	TOTAL SOLIDS (%)											
	FAT CONTENT (%)											
	NET WEIGHT (LBS/GAL)											
	OVERRUN (%)											

SIGNATURE(S) OF EVALUATORS: _____ _____ _____

Fig. 6.1.   A modified version of the ADSA ice cream score card.

(ADSA) for use in the Annual Collegiate Dairy Products Evaluation Contest. With this proposed scoring format, those product samples (representative of a lot) that receive a "zero" in any one or more quality categories should generally be regarded as unsalable products.

The rating for bacteria content must be performed in the laboratory, where equipment, laboratory technique, and additional time are required. In many situations, the results of the standard plate count and coliform count may not be available at the time the product is evaluated, in which case the "full score" may be allowed with a notation that the data were not available. As in milk evaluation, actual microbial counts are more meaningful than point scores. Coliform counts of either 50 or 500,000/ml require a score of "zero," but obviously the latter would reflect a more inferior product.

Suggested scoring guides which accompany the score card are presented in Table 6.3 and Tables 6.5 through 6.8. Scoring guides are useful in training new evaluators and in promoting standardization of judgments among different evaluators. Further modifications of the score card will be suggested later in this chapter when other frozen products are discussed. The so-called "official" score card (as of 1987) for ice cream (which is approved, revised as deemed appropriate, and published through the ADSA Committee on Evaluation of Dairy Products) is shown in Fig. 6.2; the associated "official" scoring guide is duplicated in Table 6.4. A version of the ADSA score card that has been adapted to electronic assessment of contestant performance is shown in Appendix XIII.

TECHNIQUES OF ICE CREAM SCORING

Tempering the Samples. The technique of judging ice cream is markedly different in many respects from the judging of other dairy products. Since ice cream is a frozen product, it must be evaluated, in part, in that condition in order to ascertain the typical or desired body and texture characteristics. Consequently, arrangements must be made to store (temper) the samples at a uniformly low temperature so that the ice cream retains its appropriate physical properties; yet, the temperature maintained must not be so low that the ice cream is intensely cold and unnecessarily hard. When ice cream is too cold, the recovery of the sense of taste from temporary anesthesia, due to extreme cold, requires a longer period than is expedient for satisfactory and efficient work. Furthermore, evaluators will have greater difficulty in determining the actual body and texture properties if the ice cream is too firm.

Generally, temperatures between $-18\,°C$ to $-15\,°C$ ($0\,°F$ to $5\,°F$) are

Table 6.3. A Suggested Scoring Guide for the Flavor of Vanilla Ice Cream (Suggested Flavor Scores for Designated Defect Intensities).

Flavor Defect[a]	Intensity of Defect				
	Slight[b]	Moderate	Definite	Strong	Pronounced[c]
Flavoring System					
Lacks fine flavor	9	8	7	—	—[d]
Lacks flavoring	9	8	7	6	—
Too high flavor	9	8	7	6	—
Unnatural flavor	8	7	6	5	0–4[e]
Sweeteners					
Lacks sweetness	9	8	7	—	—
Too sweet	9	8	7	—	—
Syrup flavor	9	8	7	6	0–5
Processing					
Cooked	9	8	7	6	0–5
Dairy Ingredients					
Acid	3	2	1	0	0
Salty	9	8	7	6	0–5
Lacks freshness	8	7	6	—	—
Old ingredient	6	5	4	3	0–2
Oxidized	5	4	3	2	0–1
Metallic	6	5	4	3	0–2
Rancid	4	3	2	1	0
Whey	7	6	5	4	0–3
Others					
Storage (absorbed)	7	6	5	4	0–3
Stabilizer/Emulsifier	7	6	5	4	0–3
Neutralizer	3	2	1	0	0
Foreign[f]	4	3	2	1	0

[a] "No criticism" is assigned a score of "10." Normal range is 1 to 10 for a salable product.
[b] Highest assignable score for defects of slight intensity.
[c] Highest assignable score for defect of pronounced intensity. However, a sample may be assigned a score of "0" (zero) (unsalable product).
[d] A dash (—) indicates that the defect is unlikely to occur at this intensity level.
[e] The assignable score range at the pronounced level of intensity. A numerical score of "0" (zero) is indicative of an unsalable product.
[f] Due to the variety of possible foreign off-flavors, suggesting a fixed scoring range is not appropriate. Some foreign flavor defects warrant a zero ("0") score even when their intensity is slight (e.g., gasoline, pesticides, lubricating oil).

173

CONTEST ICE CREAM SCORE CARD

DATE: _____ A.D.S.A.   CONTESTANT NO. _____

	CRITICISMS	SAMPLE NO.								TOTAL GRADES
		1	2	3	4	5	6	7	8	
FLAVOR 10	CONTESTANT SCORE ➡									
	GRADE SCORE									
	CRITICISM									
NO CRITICISM 10	ACID (SOUR)									
	COOKED									
	LACKS FLAVORING									
	TOO HIGH FLAVOR									
	UNNATURAL FLAVOR									
	LACKS FINE FLAVOR									
	LACKS FRESHNESS									
	METALLIC									
	OLD INGREDIENT									
NORMAL RANGE 1-10	OXIDIZED									
	RANCID									
	SALTY									
	STORAGE									
	LACKS SWEETNESS									
	SYRUP FLAVOR									
	TOO SWEET									
	WHEY									
BODY AND TEXTURE 5	CONTESTANT SCORE ➡									
	GRADE SCORE									
	CRITICISM									
NO CRITI-CISM 5	COARSE/ICY									
	CRUMBLY									
	FLUFFY									
	GUMMY									
NORMAL RANGE 1-5	SANDY									
	SOGGY									
	WEAK									
COLOR	ALLOWED PERFECT IN CONTEST ➡	X	X	X	X	X	X	X	X	
MELTING QUALITY	ALLOWED PERFECT IN CONTEST	X	X	X	X	X	X	X	X	
BACTERIA	ALLOWED PERFECT IN CONTEST	X	X	X	X	X	X	X	X	
TOTAL	TOTAL SCORE IN EACH SAMPLE ➡									
	TOTAL GRADE PER SAMPLE									

FINAL GRADE _____
RANK _____

Source: American Dairy Science Association (1987).

Fig. 6.2. The ADSA contest ice cream score card.

satisfactory for tempering ice cream prior to judging. This can be best achieved by transfering the ice cream samples from the hardening room to a dispensing cabinet at least several hours prior to judging, or preferably tempered overnight. This insures that the ice cream tempers uniformly. Exposing ice cream to room temperatures for tempering purposes is most unsatisfactory, since the ice cream rapidly melts along the outer edges while the center remains too firm for dipping. Ice cream in small packages (pints or smaller) may be tempered satisfactorily for evaluation by grouping them on a table and covering the containers with several clean, heavy towels. As a sample is needed for examination, it can be removed, evaluated, and then discarded. If sat-

Table 6.4.   The A.D.S.A. Scoring Guide for Sensory Defects of Ice Cream (Suggested Flavor and Body and Texture Score for Designated Defect Intensities).

	Intensity of Defect		
Criticisms[a]	Slight	Definite	Pronounced
Flavor			
Acid (sour)	4	2	0[b]
Cooked	9	7	5
Flavoring:			
Lacks flavoring	9	8	7
Too high	9	8	7
Unnatural	8	6	4
Lacks fine flavor	9	8	7
Lacks freshness	8	7	6
Metallic	6	4	2
Old ingredient	6	4	2
Oxidized	6	4	1
Rancid	4	2	0
Salty	8	7	5
Storage	7	6	4
Sweetener:			
Lacks	9	8	7
Too high	9	8	7
Syrup flavor	9	7	5
Whey	7	6	4
Body and Texture[c]			
Coarse/Icy	4	2	1
Crumbly (brittle, friable)	4	3	2
Fluffy (foamy)	3	2	1
Gummy (pasty, sticky)	4	2	1
Sandy	2	1	0
Soggy (heavy, pudding-like)	4	3	2
Weak (watery)	4	2	1

Source: American Dairy Science Association (1987).
[a] "No criticism" is assigned a score of "10." Normal range is 1 to 10 for a salable product.
[b] An assigned score of zero ("0") is indicative of an unsalable product.
[c] "No criticism" is assigned a score of "5". Normal range is 1 to 5 for a salable product.

isfactory evaluation is to be performed, the importance of proper tempering of ice cream and related products cannot be minimized.

Some freezer cabinets are not satisfactory for product tempering, as they do not maintain a uniform temperature throughout the unit. Temperatures should be measured at different locations throughout the cabinet to help insure uniform tempering of samples. Overfilling a tempering cabinet can cause some samples to be warmer than others, since crowded conditions inhibit the movement of air. Placement of all samples, if possible, at the same height within the cabinet (with air space between containers) usually helps insure uniform tempering.

Conditions for Best Work. Convenience is an important adjunct to efficient evaluation. The samples, therefore, should be arranged so that they are easily accessible without causing too much inconvenience in securing portions for sensory examination. This involves providing ample spacing of the samples to minimize or eliminate possible congestion when a number of people are conducting the product evaluation. Placing an especially designed "dolly" under the ice cream case so that the cabinet may be moved and/or arranged at will has been found to be a convenient form of mobility in the laboratory or evaluation setting. Thus, the ice cream is readily accessible, conveniently located, and, hopefully, properly tempered. The temperature of the room should be comfortably warm. Attempting to judge ice cream in a chilly room usually results in hurried work and hasty, questionable judgments; in fact, it is better that the room be too warm than too cold.

Sampling. When ice cream (or another frozen dairy dessert) is properly tempered, sample portions may be easily secured for completing all aspects of the sensory evaluation. Generally, a regular ice cream dipper, scoop or spade, rather than a spoon, is preferred for obtaining samples (Fig. 6.3).

Exercising certain precautions is deemed advisable for the sampling process. If the product surface has been exposed, then any dried surface layer (to a depth of approximately 1/4 in. [0.8 cm]) should be removed before securing the sample for evaluation. If a meltdown test is conducted, the test sample need not be large, but its volume must be uniform across all lots of ice cream being compared. For the meltdown examination, a No. 30 scoopful of ice cream placed on a clean, numbered petri dish is quite satisfactory. The petri dish should be set in a convenient place (but away from heat sources) where melting qualities may be observed from time to time during the overall evaluation process. Small samples for tasting may be removed from the product package by either a metal or plastic scoop (dipper) when desired. Individual, 6-, 8-, or 10-in. (15.2-, 20.3-, or 25.4 cm) paper plates have been used satisfactorily for holding the individual samples during the course of tasting. One or more samples may be placed on the same plate for study and comparison. Care must be exercised that portions of several samples are not intermixed.

The manipulation and conveying of sample portions to the mouth for tasting may be done by means of a clean plastic, bright metal, compressed paper, fiber, or wooden spoon. Some judges prefer metal or plastic spoons to all others for judging ice cream. Spoons should be easy to clean between samples. It is important that spoons not impart

Fig. 6.3. Several types of scoops and spades used for dipping ice cream samples.

any atypical or foreign off-flavors to the product. Plastic, compressed paper, fiber, and wooden spoons are all generally satisfactory, providing an adequate supply is available so that heavily used or worn spoons may be discarded at will. Single-service plastic spoons are most commonly used. In using wooden spoons, precautions must be taken to guard against a slightly "woody" taste.

Intermittent or unrestricted dipping of "used" spoons into the container of ice cream should absolutely not be tolerated for reasons of personal hygiene. Having placed a reasonable sized portion (a small scoopful) of ice cream onto an individual plate for sensory study, the evaluator can then taste from this "individual" sample as often as needed. The evaluator is free to secure additional samples from any product container (with the appropriate dipper) when needed, in order to complete the process of product evaluation.

Sequence of Observations

Since the physical condition of ice cream changes so rapidly when exposed to ordinary temperatures, the evaluator must be alert and constantly observing during the "time-restrictive" sampling and evaluation process, in order not to overlook any possible sensory defects associated with a given product sample, particularly body and texture features. An orderly sequence of observations has been found to be most effective in evaluating ice cream for sensory characteristics. The steps are listed in the following paragraphs.

Examine the Container. Note the type and condition of the container, the presence or absence of a liner and cover on bulk containers, and any package defects that may be present.

Note the Color of the Ice Cream. Observe the color of the ice cream, its intensity and uniformity, and whether the hue is natural and typical of the given flavor of ice cream being judged.

Sample the Ice Cream. During the course of dipping the sample, carefully note the way the product cuts and the feel of the dipper as its cutting edge passes through the frozen mass. Note particularly whether the ice cream tends to curl up or roll in serrated layers behind the dipper, thus indicating excessive gumminess or stickiness. The "feel" of dipping (i.e., the resistance offered), the evenness of cutting, the presence of spiny ice particles, and whether the ice cream is heavy or light and fluffy should be especially noted. The way the sample responds in the dipping process often gives a fairly accurate impression of its body and texture characteristics.

Begin Judging. After a sample portion has been secured, the examination for further body and texture characteristics and for flavor should begin immediately. As a general rule, little conception of the flavor may be gained by smelling the sample. Until the ice cream is melted within the mouth, the sample portion is so cold that for all practical purposes the odoriferous substances remain practically nonvolatile, and therefore little or no aroma may be detected. When the sample is liquefied and warmed to near body temperature, detection of the flavor characteristics is not particularly difficult. This is best accomplished by placing a small teaspoonful or bite of frozen product directly into the mouth, quickly manipulating the sample between the teeth and palate, and simultaneously noting the taste and odor sensations.

Since the body and texture characteristics of a frozen product are to be determined, the sample placed into the mouth should initially be in the natural frozen state. Immediately after placing a portion into the mouth, roll the sample between the incisors and bring them together very gently, noting (relatively) how far apart the teeth may be held by the ice crystals and for how long. The evaluator should note also whether any grittiness is apparent between the teeth. A small portion between the incisors may reveal the presence of minute traces of a gritty or sandy texture (lactose crystallization). By pressing a small portion of the frozen ice cream against the roof of the mouth, thus melting the sample quickly, the relative degrees of smoothness, coarseness, coldness, the presence or absence of sandiness, and the relative size of ice crystals may be determined. Certain body characteristics of the ice cream may become apparent by the resistance to mastication that the product offers in the mouth.

Expect Delayed Taste Reaction. When ice cream is first placed into the mouth, its low temperature temporarily numbs the sense of taste. The sensation of cold is usually predominant. Until the sensory nerve centers recover from the temporary anesthesia, a flavor sensation is usually not experienced. The duration of this temporary impairment of taste is dependent upon the size of the sample, its temperature, and its heat conductivity. In order not to needlessly impair the sense of taste, an evaluator should use as small or modest a sample as possible to accommodate evaluation of body and texture.

Sense the Flavor. While manipulating the sample about the mouth to ascertain some of its body and texture characteristics, the evaluator should be aware that: (1) the physical properties of the ice cream are constantly changing; (2) the period of temporary taste anesthesia (from coldness) is of fairly short duration; and (3) a hint of the flavor will soon manifest itself as an initial taste sensation. The judge should be alert and prepared to detect this sensation, whether it is prompt or otherwise.

The first perceived sensory reaction will probably be one of the fundamental tastes (if present), and in the order of salty, sweet, sour and/or bitter. As the sample is warmed in the mouth, the volatile, flavor-contributing substance(s) will soon evoke a perceived aroma (smell). Since sweetness is practically always perceived prior to detection of volatile, odor-contributing substances, the characteristics of the sweetener should be noted at once. Ice cream may be perceived as pleasantly sweet, intensely sweet, lacking in sweetness, or "syrupy"; the latter denotes a departure from a simple, basic sweet taste.

By the time the quality and quantity of sweetness is assessed, other flavor notes will likely have registered with the taster; including possible off-flavors that may be traceable to the dairy ingredients. The judge should note particularly whether the flavor is harsh (coarse) or delicate, mild or pronounced; whether the flavor seems creamy, pleasantly rich, or possesses a pronounced, objectionable, unnatural taste; and whether the mouth readily "cleans up" after the sample has been expectorated. These are but a few of the numerous characteristics which should be observed and noted in the process of evaluating ice cream flavor.

After the sample has been held in the mouth for sufficient time to nearly attain body temperature, and the flavor characteristics noted, it should be expectorated. Occasionally, a sample may be swallowed, but this is the exception rather than the rule. When the sensory evaluation is in progress, the judge's focus should be on tasting and observing, not on satisfying one's sense of hunger. Unfortunately, in ice cream scoring, the keenness of flavor perception may soon be lost or destroyed. Some experienced judges may actually consume a small amount of ice cream just before judging begins in order to adjust their palates and mental processes to this product. But once judging is underway, absolutely all samples are expectorated after completing the flavor evaluation task.

Note the Melting Qualities. By the time the flavor attributes have been determined, the samples previously set aside for the observation of melting properties should have softened sufficiently to yield an impression of those characteristics. The judge should observe whether each ice cream sample has retained its form and approximate size, even though some free liquid may have leaked (oozed) out, and whether the melted liquid appears homogenous and creamy, curdled, foamy, or watery (wheyed-off).

Record the Results. Once all of the sensory observations have been completed, the judge should record the sensory observations on a score card and assign the appropriate numerical values. If the ice cream judge is to make efficient use of limited time and be reasonably accurate in one's observations, a certain routine or technique similar to that just described should be followed.

REQUIREMENTS OF HIGH-QUALITY VANILLA ICE CREAM

There are specific criteria for sensory quality that apply to each flavor of ice cream. However, since so many flavors of ice cream (and other related products) are produced in the U.S., only a select few will be discussed in depth. Vanilla ice cream is a logical candidate for in-depth coverage due to consumer popularity and to its vulnerability to off-flavors.

Color and Package

Color. The color of vanilla ice cream or ice milk should be attractive, uniform, pleasing, and typical of the flavor stated on the label. Colorants may or may not be added to dairy frozen desserts. As long as the shade of color reasonably resembles the natural color (carotene pigment) of cream, and is neither too pale not too vivid, color criticisms are generally resisted for vanilla-flavored products. Ice cream flavors other than vanilla should also exhibit a color that is in harmony with the stated flavor on the package. The possible color defects of vanilla ice cream may be listed as follows:

Gray, dull	Too pale, chalky, lacking
Not uniform	Unnatural
Too high, vivid	

Table 6.5 is a suggested guide for scoring the color, the appearance, and the package of vanilla ice cream; however, with minor revisions it can be adapted for all ice cream flavors.

Gray, Dull. Though infrequently encountered any more, a gray, dull color is easily recognized by its "dead," soiled-white, and unattractive appearance. Such ice cream suggests uncleanliness in manufacture and, therefore, it is one of the more serious and objectionable color defects. If the gray color is caused by the use of flavoring with ground vanilla beans, which may be apparent by the presence of small pepper-like particles of the ground bean, the color should not be criticized. Ice cream that displays ground particles of vanilla bean is in demand and is actually preferred in some locales of the U.S.

Not Uniform. Lack of color uniformity in vanilla ice cream is comparatively uncommon but may be easily recognized when it occurs. Although the most appealing color for vanilla ice cream may be a moderate creamy shade of white, certain portions may be darker or lighter than others. Particularly, this may be true of the top or bottom surface or portions next to the side of the container where some desiccation may have occurred. This defect is often associated with age (extended product storage).

If the color uniformity defect is restricted to the surface layer (which is usually discarded when taking samples), it is not considered serious. At times, streaks or waves of different color may be encountered throughout the mass of a vanilla ice cream. This can be caused by varying overruns attained from multi-barrel freezers or may derive from different freezers that have a common discharge. Sometimes, a nonuniform color may originate from successive changes in the flavor source (and associated color) throughout the freezing and packaging process.

Too High, Vivid. A high color level is often objectionable because it appears unattractive and often connotes an "artificial" impression. Although individual preferences for color vary, evaluators have a general tendency to downgrade products that have an obvious, excessive

Table 6.5. A Suggested Scoring Guide for Color, Appearance, and Package of Vanilla Ice Cream.

Defect[a]	Intensity of Defect				
	Slight[b]	Moderate	Definite	Strong	Pronounced[c]
Dull color	4	3	2	1	—[d]
Nonuniform color	4	3	2	—[d]	—[d]
Too high color	4	3	2	—[d]	—[d]
Too pale color	4	3	2	—[d]	—[d]
Unnatural color	4	3	2	1	0
Soiled container	3	2	1	0	0
Product on container	4	3	2	1	—[d]
Underfill/Overfill	4	3	2	1	0
Damaged container	3	2	1	0	0
Defective seal	2	1	0	0	0
Ill-shaped containers	4	3	2	1	0

[a] "No criticism" is assigned a score of "5." Normal range is 1 to 5 for a salable product. An assigned score of "0" (zero) is indicative of an unsalable product.
[b] Highest assignable score for defect of slight intensity.
[c] Highest assignable score for defect of pronounced intensity.
[d] A dash (—) indicates that the defect is unlikely to occur at this intensity level.

intensity of color. Such a product conveys the idea of cheapness, imitation, poor workmanship, or a general lack of understanding and care on the part of the manufacturer.

Too Pale, Chalky, Lacking. A pale, chalky, or snow-like color is the opposite of too high in color. This defect is not particularly serious, although a lighter-colored product may not have as much eye appeal as a creamy shade of white color. However, uncolored ice cream, especially vanilla, should not necessarily be criticized for lack of color. For special markets, ice cream without any form of added color is a must; many products meet that marketing objective and it does not seem logical to penalize the color in those circumstances.

Unnatural. Unnatural color of ice cream should be recognized at a glance; the product appearance is not "in keeping" with the impression conveyed by cream (or milkfat). An unnatural color may be any shade of yellow, orange or tan; colors that do not correspond to the true color characteristics of milkfat. Some more common off-shades of color in vanilla ice cream include: lemon-yellows, light greenish-yellows, orange-yellows, and occasionally reddish-yellows or tannish-browns. Where the use of food colors is permitted, some manufacturers may select a particular one or combination of colorants which makes their vanilla ice cream(s) appear unique or distinctive. While the selected color may accomplish this purpose, it may nevertheless be faulted by some ice cream judges. Unnatural color may also arise from the use of "rerun," remelted ice cream or commingling of successive freezer runs of product (that have contrasting colors).

The criticism for unnatural color is a broad designation. As a general rule, this descriptor of appearance is applied to the various deficiencies or shortcomings in the hue of natural cream color. "Unnatural" color might also describe an ice cream whose color is gray, dull, high, vivid, pale, chalky, or nonuniform. Application of the most descriptive terminology possible helps in pinpointing the source of the problem within manufacturing operations.

Generally, the several color defects of vanilla ice cream do not occur at the "serious" level. Since different types of lighting will significantly affect color characteristics as viewed by human subjects, the type of light employed during examinations should certainly be standardized. Several so-called "all-natural" products have appeared in the U.S. marketplace, which absolutely have no added color to any of the flavors of ice cream. Many consumers seem to prefer products that comply with the claim, "no color added." However, in turn, many ice cream judges tend to severely criticize such aforementioned products (other than vanilla) for their appearance; the most common descriptor involved is "unnatural color."

Package. The ideal frozen dessert package or container should be clean, undamaged, full, neat, attractive (pleasant eye appeal), and protective of the product. Multiuse containers (if used) should be free of dents, rust, paint, battered edges, or rough, irregular surfaces. In general, ice cream packages should reflect neatness and cleanliness throughout, giving the consumer the impression that by use of a clean, well-formed container, the manufacturer is definitely interested in supplying a high-quality product. Some more common package defects which may be encountered are: A slack-filled container, bulging container, improperly sealed container, ill-shaped retail packages or product adhering to the outside of the container, ink smears, lack of a parchment liner on the top of bulk containers, and a container that is soiled, rusty, or damaged (the last two defects pertain to refillable containers).

These packaging defects, when they occur, are generally so obvious that additional descriptors or discussion hardly seems necessary. Encountering a high proportion of defectively packaged products from a production run is most unlikely, but such a problem might occur in the absence of adequate supervision. Just a few defective packages or containers present a problem of some magnitude, because consumers will simply not select and purchase damaged units of products from the retail ice cream cabinet. Thus, evaluators must keep in mind an appropriate perspective that defective containers generally render a product unsalable.

Melting Quality

High-quality ice cream should show little resistance toward melting when a dish is exposed to room temperature for at least 10 to 15 minutes. During the melting phase, the mix should flow from the center of the portion (initially placed in the dish), as rapidly as it melts. The melted product should be expected to form a smooth, uniform and homogeneous liquid in the dish.

The melting quality may be observed by placing a scoopful of the sample on a dish and noting its meltdown response from time to time, as the other sensory qualities are being examined. Although fiber dishes may be used, petri dishes seem to permit more accurate observation of the melted ice cream; the contrast between the product and the dish background is greater. In setting out the samples and examining them for meltdown, some precautions are necessary, namely:

Select a uniformly heated, well-lit area for placing and observing the samples (as close to 20°C (70°F) as possible).

Set the sample out for meltdown at the beginning of the judging (if feasible).

Absolutely, avoid dipping some of the samples with a warm dipper and others with a cold dipper.

Be sure that the sizes of the reasonably small samples used for the meltdown test are uniform in volume (use the same scoop or spoon for each sample).

Always use a flat-bottom dish (not a cup), so the melted ice cream is free to spread out.

Once melting has started, do not disturb the samples by tilting or swirling the containers.

Observe the melting quality at various stages of melting (Fig. 6.4) and score on the basis of the scheme suggested in Table 6.6.

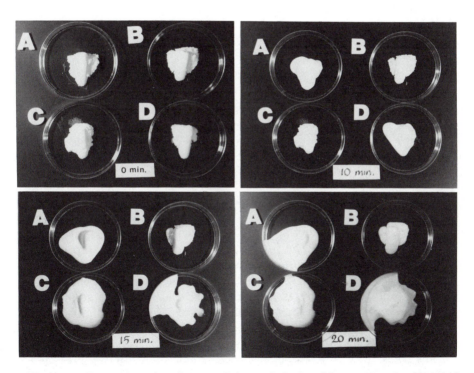

Fig. 6.4. Examples of various meltdown defects of ice cream as observed after elapse of 0, 10, 15, and 20 minutes: A—The "ideal" melting characteristics; B—Does not melt; C—Curdy meltdown (nonhomogenous); D—Wheyed off (watery separation).

Table 6.6. A Suggested Scoring Guide for the Melting Quality of Ice Cream.

	Intensity of Defect		
Defect[a]	Slight[b]	Definite	Pronounced[c]
Does not melt	3	2	1
Flaky	3	2	1
Foamy	3	2	1
Curdy	3	2	1
Wheying off	3	2	1
Watery	3	2	1

[a] "No criticism" is assigned a score of "3." Normal range is 1–3 for a salable product.
[b] Highest assignable score for defect of slight intensity.
[c] Highest assignable score for defect of pronounced intensity.

The defects of melting quality frequently observed in ice cream judging are as follows:

Does not melt, delayed melting
Flaky, lacks uniformity
Foamy, frothy, large air bubbles
Curdy
Wheying-off (syneresis)
Watery, low melting resistance.

Does Not Melt, Delayed Melting. This defect is easily recognized since the ice cream either retains (or tends to retain) its original shape after it has been exposed to ambient temperature for a period in excess of 10 to 15 minutes. This defect is related to the use of an excess of certain stabilizers and emulsifiers, high overrun, the age of the ice cream, and several processing and product composition interactions which promote formation of a highly stable gel (even when the temperature is above the freezing point). This defect is considered quite objectionable, as it conveys the impression of poor workmanship or that excessive amounts of product thickeners were used (Fig. 6.4).

Flaky, Lacks Uniformity. This defect may be noted when the sample is about half-melted, but it is more noticeable when the sample has completely melted. Flakiness is shown by a feathery, light-colored scum formation on the surface. Sometimes it resembles a fragment of crust. Usually, no indication of wheying-off (water separation) accompanies the defect. The defect is not very common; furthermore, it is not particularly objectionable. However, it is not in keeping with an impression of the highest quality, since the product is not uniform or homogeneous in appearance (Fig. 6.4).

Foamy, Frothy, Large Air Bubbles. A foamy meltdown is usually only noted when the sample is completely melted. Ice cream which exhibits many small, fine bubbles upon melting is not commonly criticized, but a sample that demonstrates a mass of large bubbles, 1/8 to 3/16 in. (0.3–0.5 cm) in diameter, is criticized. The meltdown should be uniform and attractive; this is not the case when large air bubbles or excessive foam occur. The consumer may associate the presence of foam with excessive overrun, even though this defect may not be associated with high overrun, but more often (or rather) with some of the constituents used in the mix.

Curdy. A meltdown with a curd-like appearance lacks product uniformity and is, for the most part, unattractive. The melted ice cream appears flaky; it separates from the mass in small distinct pieces rather than leaving the impression of a creamy fluid. The surface layer may exhibit formation of dry, irregular curd particles. To the layman, this defect suggests souring of the milk or cream, although the cause is usually another matter. Any conditions which lead to the destabilization of proteins are potential causes of this defect in frozen dairy desserts. A combination of factors may be responsible, including: (1) high acidity; (2) the salt balance (related to calcium and magnesium salts); (3) age of the ice cream; (4) certain adverse processing conditions (involving temperature, time, and method of heating, homogenization pressure and temperature, and rate of freezing and hardening); and (5) the type and concentration of stabilizers and emulsifiers.

Wheying-off (Syneresis). Wheying-off may usually be noted by the appearance of a bluish fluid leaking from the melting ice cream at the initiation of the meltdown test. If the sample is disturbed during melting or the observation is delayed, it may be difficult to see this condition. Whey separation may be noted in some ice cream and ice milk mixes even before they are frozen. This is a common complaint of operators of soft-serve freezers who buy their mix from a wholesale manufacturer. These mixes tend to be stored longer and are subjected to more abuse than those mixes which are made and frozen within the same plant. Factors contributing to the difficulty include the salt balance of milk ingredients, the mix composition (a product with a high-protein-in-water concentration can be expected to be less stable than one with a lower concentration), certain adverse processing conditions, and the extent of mix abuse (excessive agitation, air incorporation, and "heat shock").

Watery, Low Melting Resistance. This defect is not particularly objectionable, but it is not consistent with the characteristics of the highest quality ice cream. As the terms suggest, the sample melts quickly and the resultant meltdown has a thin, watery consistency. This defect is commonly associated with low solids in the mix and may often be associated with a coarse, weak bodied ice cream or ice milk.

Curdiness and delayed melting are the two most common meltdown defects; they may occur simultaneously. Whey separation may be observed frequently, since protein destabilization is a common problem (Fig. 6.4).

Body and Texture

Unfortunately, the terms "body" and "texture" are often used indiscriminately and loosely; adding to the confusion may be the combined use of the two terms, either in reference to one or to the other term. As it relates to ice cream, *body* is best defined as the property or quality of the ice cream as a whole. *Texture* refers to the parts or structure of ice cream which make up the whole.

Both the body and texture of ice cream may be partially determined by applying the senses of touch and sight when the evaluator observes the product's appearance on dipping. *The desired body in ice cream is that which is firm, has substance (has some resistance), responds rapidly to dipping, and is not unduly cold when placed into the mouth.* Firmness, resistance, and coldness are strongly influenced by the product's temperature. As emphasized earlier, proper tempering of the samples to $-18\,^{\circ}C$ to $-15\,^{\circ}C$ ($0\,^{\circ}F$ to $5\,^{\circ}F$) is essential, particularly for properly assessing the body of samples.

The desired texture of ice cream is that which is fine, smooth, velvety, and carries the perception of creaminess and homogeneity throughout. Small ice crystals and small air cells are required for portraying good product texture. If the product is too cold when evaluated, the texture may appear worse than it actually is. Just the opposite is true when the product is too warm. An experienced evaluator of ice cream will have learned to partially compensate for a less than optimum tempering effort on the samples, but will still definitely prefer to observe body and texture characteristics when the product is properly tempered. This assures a competent, conscientious ice cream judge that more relevant and objective assessments of the body and texture are being achieved. Suggested scoring guides for the body and texture of ice cream are given in Tables 6.4 and 6.7.

Due to differences in composition, all frozen dairy desserts do not behave in the same manner, even at a given temperature. The various *body* defects which may be encountered in ice cream are termed or classified as follows:

Table 6.7. A Suggested Scoring Guide for Body and Texture of Ice Cream.

Defect[a]	Intensity of Defect				
	Slight[b]	Moderate	Definite	Strong	Pronounced[c]
Crumbly	4	3	2	1	0[d]
Gummy	4	3	2	1	—[e]
Heavy	4	3	2	—	—
Shrunken	3	2	1	0	0
Weak	4	3	2	1	0
Buttery	3	2	1	0	0
Coarse	4	3	2	1	0
Flaky	4	3	2	1	—
Fluffy	3	2	1	0	0
Sandy	2	1	0	0	0

[a] "No cricitism" is assigned a score of "5." Normal range is 1–5 for a salable product.
[b] Highest assignable score for defect of slight intensity.
[c] Highest assignable score for defect of pronounced intensity.
[d] When a product is determined to be unsalable because of the severity of a given body and texture defect, a "0" (zero) numerical score is assigned for body and texture.
[e] A dash (—) indicates that the defect is unlikely to occur at this intensity level.

Crumbly, brittle, friable   Heavy, doughy, pudding-like
Gummy, pasty, sticky, elastic   Weak, watery
Shrunken

The *texture* defects of ice cream which may be noted are:

Buttery, greasy, churned   Fluffy, foamy, spongy
Coarse, grainy, icy, ice   Lumpy, gelatin lumps
    pellets, spiny   Sandy, gritty
Flaky, snowy

Description of Body Defects

Crumbly, Brittle, Friable. A brittle, crumbly, and friable body is evident by a tendency of the ice cream to fall apart when dipped. The product appears to be dry, open, and sometimes as friable as freshly fallen snow. The particles seem to lack the needed property to stick together or be retained as a common mass (Fig. 6.5). When such a sample is dipped, many loose particles are likely to be noted on the remaining ice cream or the dipping implement. The defect may be provoked

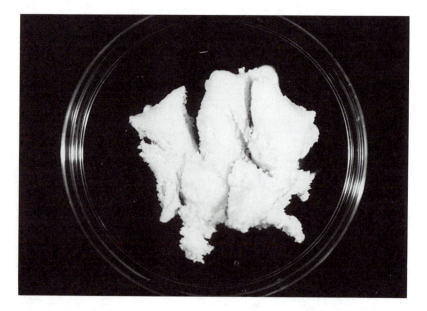

Fig. 6.5. An example of the crumbly or friable body defect in ice cream.

by the use of certain gums, inadequate stabilization, too high an over-run, and/or low total solids in the mix.

Gummy, Pasty, Sticky, Elastic. A gummy or sticky body is the exact opposite of a crumbly body. Such ice cream seems pasty, putty-like, and, under certain conditions of temperature and manipulation with a spoon, it somewhat resembles taffy (Fig. 6.6). The ice cream hangs together, so much so that it has a marked tendency to "curl" just be-hind the scoop as it is pulled across the surface, which leaves coarse, deep, irregular waves. Frequently, there is a correlation between a gummy body and a high resistance to melting; gummy ice cream often resists melting. If melting does occur, the mass often tends to retain its original shape.

The gummy body defect is associated with an excessive use of stabi-lizers, certain corn syrup sweeteners, or both. One should recognize that all ice cream is sticky to some extent, due to the concentration of carbohydrates in the product. Ice cream should only be severely criti-cized when the stickiness is so severe that it is obviously pasty and would probably be difficult to dip or scoop. As an important economic consideration, gummy (or sticky) ice cream fails to yield as many scoops per unit volume as typical bodied products.

Shrunken. A shrunken ice cream manifests itself by the product

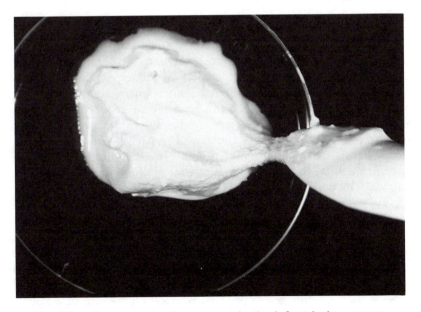

Fig. 6.6. An example of a gummy body defect in ice cream.

mass being withdrawn from the sides of the container. This defect is readily obvious when the package is first opened for examination. This defect may be associated with low solids, high overrun, and/or unfavorable storage conditions; however, any ice cream may shrink. Since "heat shocking" may be one of the contributing causes, the judge should be alert to correlate, if possible, this defect with a coarse, icy texture. All the reasons or causes of shrinkage are not clear to technologists; occurrences of the problem are often quite unpredictable. Product shrinkage may suddenly be encountered where none existed before, even when no changes were made in the product's composition or manufacturing procedures. A basic predisposition to shrinkage is apparently imparted to frozen dairy desserts by certain milk components, especially proteins. Certain environmental conditions, such as season of the year, stage of lactation, feed, etc., may unfavorably affect the normal formation of strong air-cell walls (which contain proteins) in the frozen mix. Other associated factors seem to merely aggravate the conditions that predispose ice cream to shrinkage (see Fig. 6.7).

Heavy, Doughy, Pudding-like. A heavy, resistant body is best described by the terms heavy, doughy, or pudding-like. The descriptor, soggy, has also been used in association with this defect, although perhaps inappropriately. This defect can readily be noted when the product is dipped. Portions of an ice cream with this criticism, when placed in the mouth, seem colder than those free of the defect. Apparently, this is due to a greater heat conductivity of heavy-bodied products. This defect is associated with a high solids content of the mix, possibly too much stabilizer, and/or a low overrun. Through product formulation, individual ice cream manufacturers can control the "degree of bite resistance" in the body of their ice cream. Some processors may purposely strive for an extremely heavy body in order to achieve product uniqueness. Many consumers seem to prefer a product with a great deal of bite resistance. The ice cream judge should be aware of the wide range of consumer preferences and only criticize a heavy body as a defect when it is obviously "out-of-line." In fact, more and more consumers are willing to pay a premium price for high-solids, low-overrun ice cream. The body of such products is generally quite resistant, firm, or heavy.

Weak, Watery. A weak, watery body is usually associated with a low melting resistance and a thin, milky, low viscosity meltdown. A weak-bodied ice cream conveys the impression of having a low proportion of food solids, when a sample is placed into the mouth. The mouthfeel of the sample may more likely resemble ice milk than ice cream. Such an ice cream may be easily compressed by slight pressure of a spoon or scoop. This defect may also be associated with coarse texture; low

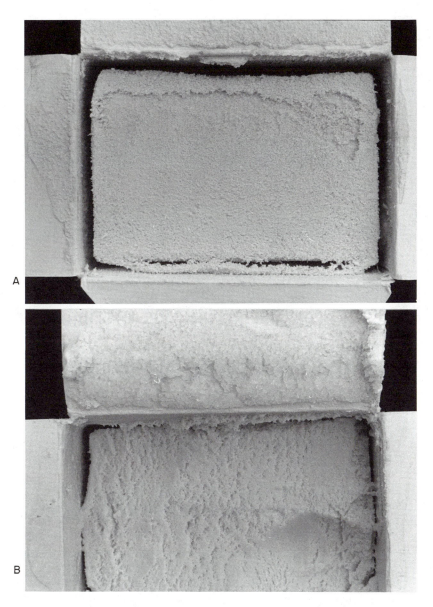

Fig. 6.7. Two examples of the shrinkage defect in ice cream. Note the formation of large ice crystals on the upper lid flap of sample B.

Fig. 6.8. Sensory evaluation of ice cream by contestants in a collegiate competition.

solids and high overrun also contribute to causing a weak bodied ice cream.

Description of Texture Defects

Buttery, Greasy, Churned. This defect may be noted by the presence of actual butter particles in the mouth after the ice cream has melted, or by a distinct greasy coating of the mouth surface after expectoration. Common causes of a greasy mouthfeel are inadequate homogenization, a relatively high milkfat content and overemulsification of the product. In soft-serve frozen dairy desserts, churning may be due to deemulsification of milkfat during prolonged agitation in the soft-serve freezer. High fat mixes are also more susceptible; incomplete homogenization and overemulsification aggravate this problem.

Coarse, Grainy, Icy, Ice Pellets, Spiny. This defect ranks as the most commonly encountered texture defect in frozen dairy desserts. Such a product may be characterized by its structural make-up of comparatively large particles, a feeling of unusual coldness within the mouth, a simultaneous lack of a smooth, velvety character and a frequently associated rough visual effect. When a sample of a coarse/icy product (most common descriptors) is placed between the upper and lower incisors, a temporary resistance is exhibited before the incisors are finally permitted to come together. This form of a slight, temporary resistance should not be mistaken for another form of bite resistance provoked by another texture defect, sandiness. The resistance of coarse/iciness is quite temporary, almost instantaneous, while that of sandiness is of longer duration.

A coarse texture is due to comparatively large particles of frozen water; each ice crystal is sufficiently large that the coarseness is obvious. When extreme conditions of coarse, grainy texture are noted, the product is criticized as being icy or spiny. Ice cream samples with a pronounced icy texture may be readily noted during the dipping process from the "feel" of the scoop or spade as it strikes or breaks the tiny icicles or spines.

A coarse, icy texture may be manifested by either the presence of localized, layer-like, ice crystals, or by grainy ice particles distributed throughout the product. The layer-like crystals are frequently found along the sides of the container where melting and subsequent re-freezing may have occurred. Both kinds of ice crystals are objectionable, since the product lacks the smooth, homogenous, and velvety texture which is typically deemed most desirable for high-quality ice cream.

Ice crystals can be felt between the teeth and/or with the tongue. As continuous melting of ice cream occurs in the mouth, larger ice particles are momentarily left behind; and they register a distinct sensation of coldness.

Ice crystals have a natural tendency to increase in size with increased storage time; the larger crystals selectively become larger at the expense of the small ice crystals, which disappear. Thus, ice cream frequently gets progressively more coarse with time in storage. Much of the technology of ice cream formulation, freezing, and storage is designed to produce small ice crystals and delay their growth during storage or distribution. Since, almost invariably, ice cream will be exposed to some "heat shock" (temperature fluctuations and storage at higher than ideal temperatures), specific steps are advisedly taken to provide protection against fluctuations in storage temperature (Bodyfelt 1976; Lucas 1941; Tobias 1982c; Tobias and Muck 1981). Effective stabilizers and emulsifiers, microcrystalline cellulose, and corn syrups (particularly the low conversion types) are commonly used as "protective" agents. Close control of production, inventories, and rotation of product to help ensure that the oldest product is used first, are important measures to help keep storage time minimal.

Among the many possible causes of coarse-textured ice cream are the following:

Faulty formulation
Inadequate protection against heat shock
Ineffective or improper stabilization and/or emulsification
Inadequate hydration of dry mix constituents
Incomplete protein hydration
Inadequate homogenization
Insufficient aging of the mix
Too high a product temperature out of the freezer
Extended interval between freezing, packaging, and/or transfer to
     the hardening system
Slow hardening
Too high a hardening temperature
Fluctuating storage temperatures
Extended storage and distribution times.

Some of these production problems are mechanical, such as dull freezer blades, which prevent the ice cream mix from freezing properly, while other product quality shortcomings are traceable to inadequate management and supervision. Sensory evaluation helps to identify the nature of product defects and pinpoint deficiencies of production and distribution.

Flaky, Snowy. A flaky, snowy-textured ice cream manifests itself by a tendency to fall apart when dipped. In this respect it has the same characteristics as that noted in a crumbly body. The condition seems to be associated with low solids, low stabilizer, and/or high overrun in the product.

Fluffy, Foamy, Spongy. A fluffy texture may be noted by the presence of large air cells and a general "openness" throughout the product. Such an ice cream tends to compress substantially upon dipping or applied pressure with a flat object. This defect is closely associated with a high overrun. A fluffy ice cream usually melts slowly in the dish, yielding a relatively small proportion of liquid which is often foamy and spongy.

Sandy, Gritty. A sandy texture is certainly one of the most objectionable texture defects encountered in frozen dairy desserts, but it is also one of the easiest to detect. Such a texture conveys to the tongue and palate a definite lack of smoothness and an associated distinct form of grittiness. When the sample melts, there remains in the mouth fine, hard, uniform particles which suggest fine sand, which are crystals of lactose.

The presence of these sand-like particles can be noted in several ways: (1) by pressing a thin layer of the suspect ice cream against the roof of the mouth with the tongue to secure quick melting; (2) by bringing the teeth together slowly on a portion of it; or (3) by pressing a small quantity of the product between the thumb and forefinger. Sandy texture should not be confused with the coarse, grainy texture which results from the presence of comparatively large ice crystals. The lactose crystals dissolve markedly more slowly than ice crystals; therefore, they may be noted even after the ice cream has fully melted.

A high percentage of serum solids, high total food solids, product age, and "heat shock" are factors that seem to provoke the development of this defect. When sandiness occurs, the judge should be alert to the likely presence of other defects which are commonly associated with frozen dairy desserts stored under unfavorable conditions (coarse/icy and shrinkage).

Flavor

High-quality vanilla ice cream should be pleasantly sweet, suggest a creamy background sensation, exhibit a delicate "bouquet" of vanilla flavor, and leave a most pleasant, but brief, rich aftertaste. The flavor intensity of the vanilla, the sweetener and the various dairy ingredients should not be so pronounced that, when first tasted, one component of the overall flavor seems to predominate over the others. All of the ingredients should blend to yield a pleasant, balanced flavor.

The flavor evaluation of ice cream offers some difficulties unlike those encountered in the scoring of butter, cheese, and milk. In comparison to most other dairy products, ice cream is intensely sweet. This is the first obstacle confronted by the ice cream judge. The sweetness is often so pronounced to inexperienced judges that they frequently find it difficult to identify other flavor notes which may be (or should be) present. A second obstacle to the successful evaluation of ice cream flavor is simply tastebud fatigue due to the combined effect of sweetness and coldness on the organs of taste. A third obstacle for the ice cream judge is the mouth-coating effect of milkfat. Some of the tastebud sites may be partially coated or blocked by milkfat, and hence lessen the ease of taste perception.

Usually, inexperienced evaluators look forward to the judging of ice cream with considerable enthusiasm. After tasting a few samples, however, this enthusiasm probably begins to wane. The appetite is satisfied and novice judges may have to force themselves to continue judging a set of samples that have started "to taste alike." Fortunately, experienced judges score ice cream with about the same ease as they evaluate other products. Some evaluators initially condition their mouths by tasting several samples, in order to adapt to the sweetness and coldness before actually placing flavor judgments on any of them. Frequent rinsing of the mouth with water between ice cream samples is apparently helpful for some evaluators, but this is primarily an individual matter as to whether it is a beneficial technique.

When evaluating ice cream for flavor, tasting is usually performed from a scooped sample on a plate or in a cup. Taste-sampling directly from the original container is not generally advised; there is a potential risk of personal sanitation (hygiene) problems, as well as irreversible temperature abuse of ice cream samples. The technique of securing several samples on the same plate for facilitating taste comparisons is appropriate, but caution is advised, since melted samples may intermix and thus lead to confusion. A preferred practice of the authors and most ice cream judges is to evaluate one sample at a time. In this approach, the judge compares the flavor, body, and texture with a fixed, mental standard of the so-called "ideal" product, rather than with that of another sample.

Due to the severe coldness of ice cream and ice milk, some off-flavors may not be sufficiently volatile to be immediately detectable or recognizable. As pointed out earlier, the body and texture of the ice cream must be determined on the ice cream at the typical serving temperature, but any off-flavors present will become more apparent as the sample warms up. Warming occurs within the mouth as well as on the sample plate. After first assessing the body and texture of the sample, the evaluator may taste a warmer sample portion for at least one of several phases of the flavor judgment that should be completed. This approach may be somewhat complicated by the fact that the "flavor balance" may change with temperature and, hence, cause some of the flavor notes to dominate others at the higher temperatures, but not at the lower ones. Thus, the best observations of the actual "flavor balance" should be undertaken at normal consumption temperatures for frozen desserts. This approach is especially important when a number of samples must be evaluated in succession. The evaluator must try to maintain accuracy and objectivity, and in the process, avoid both mental and physical fatigue as well as taste, touch, and odor adaptation. When the human senses are continuously exposed to a given stimulant, sensory perception diminishes because of the phenomenon of adaptation (discussed in Chapter 2).

Due to the numerous ingredients which may be used in ice cream manufacture, one may expect a wide variety of flavors and potential off-flavors. In general, all frozen dairy desserts are susceptible to the

Table 6.8. A Suggested Scoring Guide for the Bacterial Content of Ice Cream.

Standard Plate Count C.F.U./g	Coliform/g	Assigned Score[a]	Remarks
< 3,000	<1	2	No Criticism
3,000–Regulatory Limit	1–10	1	
Above Regulatory Limit	>10	0	

[a] The score for each of the criteria is determined separately; the lower of the two scores is assigned to the given sample.

development of most of the off-flavors encountered in other dairy foods. The flavoring systems used for ice cream, ice milk, and sherbet may be obtained from several sources and each one is manufactured by different processes. Consequently, the given source of flavoring itself may contribute to a surprising variety of flavors or flavor notes. Additionally, ice cream possesses varying degrees and qualities of sweetness. The major flavor defects of ice cream and ice milk may be classified according to their origin, as summarized in Tables 6.3 and 6.9.

Knowledge of the possible source of off-flavors is quite useful when troubleshooting, pinpointing, and correcting difficulties with sensory quality. While the aforementioned tables cover most of the anticipated problems, there is always the chance for the highly unusual or extraordinary to happen. For instance, the eggs may be oxidized, the cream may have an intense absorbed or medicinal off-flavor, or the liquid sugar or corn syrup may be fermented. Occasionally, when production and quality control personnel least expect it, an off-flavor may be encountered which defies description.

As an aid to problem solving, a description of some of the more common flavor defects of frozen dairy desserts is presented as a review for the prospective ice cream judge. Table 6.9 classifies some of the possible flavor defects of ice cream according to their cause or origin.

OFF-FLAVORS FROM THE INGREDIENTS USED

The Flavoring System

Typically, the first perceived flavor or off-flavor in a frozen dairy dessert is one associated with the flavoring material used. Due to the volatility of flavor substances, it tends to "register early" with the olfac-

Table 6.9. Classification of Ice Cream Flavor Defects According to Their Cause or Origin.

I. Off-flavors Due to the Ingredients Used:
   A. *The Flavoring System:*
     1. Lacks (deficient)
     2. Lacks fine flavor (harsh, lacks balance)
     3. Too high (excessive)
     4. Unnatural (atypical)
   B. *Sweeteners:*
     1. Lacks sweetness
     2. Too sweet
     3. Syrup flavor (malty, Karo-like)
   C. *Dairy Products:*
     1. Acid (sour)
     2. Cooked (rich, nutty, eggy)
     3. Lacks freshness (stale)
     4. Old ingredient
     5. Oxidized (cardboardy, metallic)
     6. Rancid (lipolytic)
     7. Salty
     8. Whey (graham cracker-like)
   D. *Other Ingredients:*
     1. Eggs (eggy)
     2. Stabilizer/Emulsifier
     3. Nonmilk food solids
II. Off-flavors Due to Chemical Changes (in the Mix or Product):
     1. Lacks freshness (stale, old)
     2. Rancid (lipolytic)
     3. Oxidized (cardboardy, metallic)
     4. Storage
III. Off-flavors Due to Mix Processing:
     1. Cooked (rich, nutty, eggy)
     2. Caramelized/Scorched
IV. Off-flavors Due to Microbial Growth in the Mix:
     1. Acid (sour)
     2. Psychrotrophic (fruity/fermented, cheesy, musty, unclean)
V. Off-flavors Due to Other Causes:
     1. Foreign contaminants
     2. Neutralizer

tory center. In evaluating ice cream, the judge should particularly note the kind, the quantity, and the relative quality of the flavoring used in the product. If the ice cream is vanilla, for instance, the judge should constantly keep in mind the desired delicate "bouquet" (aroma note) that is so highly prized and sought in a high-quality vanilla ice cream. The judge should not deviate from an established mental standard or predetermined "flavor profile" of the "ideal" vanilla ice cream. Both the pure vanilla (if used) or the vanilla/vanillin blend, and the amount used, should blend with the other ingredients to provide a pleasing, refreshing, and appetizing product. The judge should be eager for a second (and a third) bite of the ice cream—if it is one of high quality. Four flavor defects related to the product flavoring system may be experienced, which are described in the following paragraphs.

Lacks Flavor (Deficient). An ice cream with this defect is often criticized as flat, bland, or deficient in the amount of added flavoring. Even though the ice cream may be pleasantly sweet and free from any dairy ingredient off-flavor, it seems to lack the characteristic delicate "bouquet" of excellent vanilla; the desired intensity is missing. The obvious cause of this defect is failure to use sufficient quantities of flavoring. However, there are instances when certain ingredients mask (or hide) the vanilla flavor, thus invoking the "lacks flavor" criticism, even though the added quantity of flavoring seemed adequate to the manufacture. Also, a defective source of flavoring could contribute to this flavor defect.

Lacks Fine Flavor (Harsh, Coarse). This criticism is generally used to describe an ice cream which is basically "good" or "very good," but for some less-than-clear reason, it just barely seems to fall short of being "perfect" or "ideal." In some instances, such an ice cream may simply lack an overall "flavor balance" (blend), but otherwise the product appears to be free of any hint of detectable flavor shortcomings. In other instances, the sensory dimensions of a pure (real) vanilla or a vanilla/vanillin blend may be determined by close sensory examination to be slightly less than expected or desired. Experienced ice cream judges are able to recognize the desirable, delicate, balanced flavor notes of a high-quality flavor. The novice judge should remember that "lacks fine flavor" is not readily described in more definitive or specific terms. Thus, this descriptor should practically be considered a "last resort" in describing a minor flavor defect related to the flavoring system.

Too High Flavor (Excessive). This flavor condition, when it occurs, is best recognized when the sample is first placed into the mouth. The intensity of the flavoring seems so striking or sharp, that the desired, pleasant flavor blend is not achieved due to the harsh tones imparted by the flavoring level observed in the product. Ice cream that is too highly or excessively flavored is not severely criticized as a rule, especially if the quality of the flavoring used is high.

Unnatural Flavoring (Atypical). Frequently, the manifestation of "unnatural flavoring" in ice cream may convey the sensation of being too high in flavoring. The impression of unnatural flavoring may be of several types and intensities, depending upon the kinds and proportions of constituents used in preparation of the extract, emulsion, or flavor concentrate. For example, synthetic or imitation vanilla, which is often used to fortify vanilla extracts, may tend to produce a "quick," sharp, piercing, or burning sensation on the sides and base of the tongue. Generally speaking, the unnatural flavor criticism is observed more frequently in ice creams that are labeled "vanilla flavored" or "artificially flavored vanilla," than in products labeled "vanilla" or "real vanilla." Details of ice cream classification and associated labeling requirements (as a function of vanilla or vanilla flavoring category added to the product) are summarized in Table 6.10. To minimize bias in ice cream judging, it is most important that the sensory observations be conducted without the evaluators examining the product labels before completion of the task.

Another form of unnatural flavor may occur due to the addition (usually unintentional) of extracts other than vanilla to the ice cream mix; the imparted flavors may be suggestive of spices, coconut, marshmallows, custard, candy, nuts, lemon, cherry, maple, "buttery" or "smoky." Numerous other unnatural flavors are possible in frozen dairy desserts, depending on the circumstances of manufacture. If one of the aforementioned or another atypical flavor note is perceived in vanilla ice cream, the appropriate recourse is to criticize the sample for "unnatural flavor." This flavor problem also frequently arises through the accidental intermixing of two or more product flavors when ice cream freezing machines are converted from one flavor to another. In fact, this is probably the most common cause of this type of unnatural (or atypical) off-flavor in U.S. commercial ice cream. This is unfortunate, since numerous consumers (through surveys) have indicated that they were the recipient of a "surprise flavor;" a "flavor" they didn't bargain for at the time of purchase.

Sweetener Defects

Lacks Sweetness. An ice cream that lacks sweetness is readily noted upon tasting; the product simply manifests a distinct flat or bland taste. The desired or anticipated blend of flavor is missing. An adequate amount of sweetener is required to bring out the full flavor "bloom" in a given flavor, whether it is vanilla, fruit, or chocolate ice cream. Since preferences for the desired level of sweetness vary among individuals, the product is not severely criticized for lacking sweetness, within reasonable limits, if that is its only flavor defect encountered. However, a severe deficiency in sweetener solids may give rise to readily evident defects in body and texture or mouthfeel.

Too Sweet. An ice cream that is observed to be excessively sweet tends to exhibit a candy-like taste sensation; this defect is readily noted upon the first stages of tasting. Too much sugar (or other form of sweetener) tends to interfere with the overall desirable blend of flavor(s). Another unfortunate characteristic of a given ice cream that is perceived as being too sweet is a general lack of refreshing property.

Syrup Flavor (Malty or "Karo®"-like). A desired property of sweeteners (in ice cream as well as other food systems) is that they impart the basic sweet taste, and simultaneously be free of other flavor notes. Some flavor technologists have coined the term "clean sweet" for su-

Table 6.10. Labeling Requirements for Various Categories of Vanilla Ice Cream According to the Flavor Source.

| Ice Cream Type or Category | Flavor Declaration | | Flavor Requirements | |
	Characterizing Flavor Declaration	Subsidiary Flavor Declaration	Sources	Quantity
Category 1	*Vanilla*	None	Vanilla beans, extract, or powder; *no artificial flavor* permitted	Sufficient to impart characterizing flavor
Category 2	*Vanilla* flavored	"Vanilla and artificial vanilla flavor" or "Artificial flavor added" or "Artificial Vanilla flavor added"	Vanilla beans, extract, or powder plus artificial vanilla; i.e., twofold or fourfold vanilla-vanillin extract (or powder)	Vanilla beans, extract, or powder in combination with vanillin, not to exceed 1 oz per "unit of vanilla constituent" as defined in vanilla standards. Concentrations may be used where ratio of "vanilla constituent" and vanillin remains 1 to 1[a]
Category 3	Artificially flavored *Vanilla* or artificial *Vanilla*	None	Artificial vanilla, with or without vanilla beans, extract, or powder.	If the amount of vanillin used is >1.0 oz per "unit of vanilla constituent," the product must be labeled in accordance with this category. Product may be flavored exclusively or in part with other artificial vanillas, e.g., ethyl vanillin.

Source: Adapted from Code of Federal Regulations 1985. Title 21, Part 135.
[a] For example, if 1 gal of vanilla extract contains extractives from 26.7 oz beans, a maximum of 2 oz vanillin may be used. One (1.0) unit "vanilla constituent" = total extractable flavor components of 13.35 oz of vanilla beans with a moisture content ≤25%, or a proportionally greater amount of vanilla beans if >25% H_2O.

crose. In the past, the more complex flavor imparted by some sweeteners was termed "unnatural sweetness." This sweetener off-flavor is still commonly encountered in certain forms of corn syrups and corn syrup solids; hence "syrup flavor" is the common descriptor for this characteristic defect. When honey is used as a sweetener, the resulting sweetness may be criticized as syrupy unless the ice cream is intended to be honey-flavored. More frequently encountered descriptions for syrup flavor might be: malty, "Karo®"-like, caramel-like, molasses-like, or like low levels of burnt sugar. Certain forms or sources of corn syrup solids, corn syrup, and some liquid sugar blends (with excessive levels of corn syrup), when used in ice cream in high proportion to sucrose, may convey a slight to distinct malty or caramel-like off-flavor. Too often, a syrupy off-flavor may mask or otherwise interfere with the release of the given flavoring (especially delicate flavors like vanilla). Additionally, a syrup off-flavor tends to be enhanced by the cooked flavor note of the mix. Simultaneously, a gummy or sticky body can often be associated with an ice cream or ice milk that has also been criticized for "syrup flavor."

Dairy Products as a Source of Defects

Acid (Sour). An acid or sour off-flavor in frozen dairy desserts may be distinguished from other off-flavors by a sudden, tingly, taste sensation (on the tip or top of the tongue), plus an associated "clean and refreshing" mouthfeel. Since this off-flavor results from uncontrolled bacterial activity at elevated temperature, other bacterial off-flavors may also be present. In such case, the flavor defect(s) may be more appropriately described as a combination acid (sour) and psychrotrophic bacteria-caused off-flavor (unclean, fruity, or putrid). The acidity (and/or psychrotrophic defect) may have developed in one or more of the dairy ingredients used, or the mix may have been stored at a favorable growth temperature for lactic-acid-forming or other types of bacteria. In any severe temperature abuse situation, the bacterial count would ordinarily be expected to exceed established regulatory limits. A serious processing and product handling error or disregard for quality control is evident when an acid taste is so intense that the evaluator is inclined to think of the sample as a sour product. Such a product should never reach the marketplace; the consumer would often be offended by the presence of this unusual off-flavor in a sweetened product such as ice cream.

Cooked. The "cooked" flavor of ice cream is commonly experienced. It is also referred to as "rich," "eggy," "custard," scalded milk-, condensed milk-, or caramel-like. These mix flavors, although they may differ slightly in some respects, actually have much in common. A cooked milk or cream "background flavor" is the characteristic flavor note of this group of heated flavor sensations. Depending on its intensity, this flavor sensation is usually somewhat delayed in terms of the initial perception, but then it tends to persist after the sample has been expectorated. A highly cooked or heated flavor of the product may tend to "mask" or modify the vanilla flavoring. The resulting flavor sensation may be rather pleasant, although it would usually be perceived differently than a pure vanilla flavor.

Cooked (or rich) flavor is not considered a serious defect in ice cream, unless it is so intense as to be perceived as caramel, scorched, or burnt. In fact, some manufacturers intentionally strive for a slight to moderate degree of cooked (rich, nutty, custard-like) flavor in vanilla ice cream. They believe, as do the authors, that a slight to modest cooked flavor note helps convey a fuller, smoother, richer flavor in the product.

Quite commonly, the dairy ingredients incorporated into ice cream will have already been pasteurized, but federal and state regulations require that the assembled or final ice cream mix must also be pasteurized. Second, or subsequent, heat treatment is likely to produce some degree of cooked flavor in the mix. As indicated earlier, this is not typically objectionable in ice cream; in fact it may be quite desirable or preferred in many instances.

An excessive cooked off-flavor usually results from using ingredients which have received such severe heat treatment that a scorched or burnt effect is attained. Mix pasteurization, under some adverse conditions, may also develop a cooked off-flavor. Even though pasteurization standards require heating at a minimum of 79.4°C (175°F) for 25 or more consecutive seconds, some manufacturers may opt to heat to near the boiling point or above. Some mixes may be ultrapasteurized or commercially sterilized and aseptically packaged. Again, it should be emphasized that a moderate cooked flavor is not particularly objectionable. However, an obvious scorched or burnt off-flavor is to be avoided.

Lacks Freshness (Stale). The descriptor, "lacks freshness," or "stale," refers to a moderate off-flavor of ice cream and related frozen desserts. This flavor defect is generally assumed to result from either a general flavor deterioration of the mix during storage, or from the use of one or more marginal-quality dairy ingredients in mix formulation. For instance, some old milk or old cream, or stale milk powder (nonfat milk solids), may have been incorporated as an ingredient. If the off-flavor imparted by the "marginal" ingredients were quite intense, then "old ingredient" would probably be the most appropriate criticism. However, if the other milk components and/or mix ingredients dilute the adverse sensory aspects of the dairy ingredient(s) in question, a lacks freshness (or stale) descriptor is more applicable. Occasionally, relatively small quantities of cream or milk used as mix ingredients may manifest an old ingredient, oxidized, rancid, or unclean defect. But, unfortunately, this situation was "missed" or overlooked by production and quality control personnel. Subsequently, dilution of the "offending" dairy ingredient(s) (by higher volume "quality" ingredients) results in an overall deterioration of flavor quality, which is commonly described as stale or lacks freshness.

When ice cream and ice milk lack freshness, there may or may not be a slight aftertaste. However, if the aftertaste is strong or persistent, the judge should look for or consider more serious defects such as old ingredient, storage, oxidized, or rancid.

Old Ingredient. Nearly all dairy ingredients used in ice cream are subject to flavor deterioration with age (extended storage). Poor sanitation in milk handling and processing and subsequent bacterial action may produce psychrotrophic off-flavors or an "old milk" or "old cream" flavor. Through chemical reactions, milk and whey powders may become stale and caramelized in storage. Caseinates may acquire a stale and glue-like off-flavor; syrups may ferment. With storage, various deteriorative processes may occur in stabilizers, emulsifiers, and flavoring agents. The same descriptor, "old ingredient," is used to describe a relatively large number of possible flavor defects. The cause of the problem should be pinpointed by checking all possible ingredients, through sensory examination, for their potential for adversely affecting the delicate flavor of the product.

To some evaluators, old ingredient and oxidized off-flavors may resemble each other to some extent. With increased age (storage), the judge can expect that some auto-oxidation may have occurred, along with other possible deteriorative changes. In many instances, the old ingredient defect will not be noted immediately after the sample is placed into the mouth; but, usually an ice cream with this defect will exhibit a persistent aftertaste. Typically, the aftertaste will not be pleasant; the tastebuds will fail to "clean-up."

Oxidized (Cardboardy, Metallic). In dairy products, the oxidized off-flavor may vary so widely in character and intensity that several terms or descriptors are used to distinguish between the various stages. In ice cream (or ice milk), this off-flavor may be encountered to such a slight intensity that the product flavor seems flat or "missing." A further development of this off-flavor may be described more accurately as astringent, metallic, or puckery (with an associated mouthfeel of shrinking of the mucous membranes). Other, more moderate intensities of the off-flavor might be described progressively as oxidized, papery, or cardboardy. In the most intense stages of the oxidation of milk products, oily, tallowy, painty, or fishy are common descriptors. The oxidized off-flavor is usually noted soon after the sample is placed into the mouth; if intense, it may persist long after the sample has been expectorated. Depending on the intensity, such an ice cream may not be entirely repulsive to the evaluator (or consumer). However, an oxidized defect definitely conveys the idea that the product is not made from high-quality ingredients, is not refreshing, or may be stale or old. Generally, the evaluator or consumer is not very eager for a second bite of such a product. Hence, when an oxidized off-flavor occurs in frozen dairy desserts, repeat sales for the product (or brand) are not as likely to occur.

Some evaluators think of a metallic off-flavor as a distinctly separate defect, even though this off-flavor is commonly considered another stage or degree of the generic oxidized off-flavor. Since stainless steel has replaced monel or "white metal" in milk handling and processing equipment, the metallic defect has substantially decreased as a problem. Historically, the conditions associated with the occurrence of a metallic off-flavor were: equipment made of copper or copper alloys, improperly tinned equipment, rusty milk cans and utensils, and/or storage of milk products in nonstainless steel containers or vessels. The metallic off-flavor is characterized as having a peculiarly rough, astringent, puckery mouthfeel. As indicated previously, the metallic defect is often considered one of several stages in the series of off-flavors due to lipid oxidation. The light-induced form of the oxidized off-flavor (protein oxidation) is much less likely to occur in ice cream than the metal-induced form of oxidation. Occasionally, a light-activated defect might be encountered in frozen desserts packaged in containers that employ the transparent, "see-through" lid, but it is usually highly localized on the top surface and only after direct exposure to light.

Rancid. Fortunately, a rancid off-flavor is infrequently observed in ice cream (Tobias 1983a). A specific, delayed, reaction time of perception is characteristic of rancidity, and it has an attendant persistent repulsiveness. However, the sweeteners and flavoring may tend to mask any potential rancidity to the extent that unless the defect is quite pronounced, this off-flavor may not be recognized for what it actually is. If rancidity were to occur in ice cream, the peculiar blend of flavors and off-flavors would typically terminate as an unclean or unpleasant aftertaste, which is characteristic of the rancid defect. Rancidity is severely criticized, since it indicates either utilization of mishandled dairy ingredients or serious processing errors that led to mixing raw milk or cream with homogenized milk ingredients.

Salty. Occasionally, a salty off-taste may be encountered in frozen dairy desserts. This taste may be readily detected, since the reaction time is relatively short; hence, it is a quickly perceived taste. A salty taste could be due to added salt, the use of salted butter as a milkfat source, or it may be associated with use of a high percentage of concentrated whey, whey solids, or milk-solids-not-fat (MSNF) in the formulation. High displacement rates of MSNF with whey solids (i.e., in excess of 20–25% replacement) seems to occasionally lead to a slight salty off-taste in ice cream or ice milk. Other sensory defects may accompany the higher usage rates of some sources of dry whey (see the following discussion on the whey off-flavor). To most evaluators, a salty taste in frozen dairy desserts seems distinctly "out-of-place" for this form of product; hence, it is usually criticized in line with the level of intensity and the specific flavor involved.

Whey ("Graham Cracker"-like). Federal Standards of Identity limit the maximum concentration of whey solids in ice cream to 25% of the MSNF (for products engaged in interstate commerce). While the quantity of whey used in the mix is certainly a factor in the possible transmission of a whey off-flavor, an even more important aspect is the whey quality. The quality of whey solids should be carefully determined; especially important is a close scrutiny of the flavor characteristics (freshness and freedom from stale, old ingredient, or oxidized-like off-flavors. Freedom from off-colors, caking (free-flowing), or lumping is also critical for dry whey. Preferably, the level of whey solids used in ice cream or ice milk should be below the flavor detection threshold for the "whey flavor." However, even lower levels of whey (15–17% displacement of MSNF) may be detected by sensory test when it is of poor quality.

A whey off-flavor in frozen dairy desserts is probably best described as being "graham cracker"-like; or similar to stale condensed milk (Bodyfelt *et al.* 1979), with an associated slight taste of salt. Extremely old or poor quality whey solids may reflect oxidized, cheesy, rancid, and/or unclean defects, and subsequently transmit these off-flavors to the ice cream. An unpleasant aftertaste may prevail, due to the amount and/or quality of whey solids used in the mix. Sometimes ice cream and related products that exhibit a whey off-flavor may simultaneously display slight off-colors (reddish-orange), as well as a friable, crumbly body and/or a gritty texture.

Other Ingredients

Eggs (Eggy). Part 135 of the CFR permits the use of egg solids, but regular ice cream must contain less than 1.4% egg yolk solids by weight, exclusive of the weight of any bulky flavoring ingredients used. When the content of egg yolk solids (by weight) is 1.4% or more, the product must be labeled "frozen custard," "french vanilla," or "french custard" ice cream. Although not widely used in contemporary ice cream, eggs have, or have had, definite functional roles in ice cream; namely stabilization and emulsification.

Egg yolks, whether in liquid, dry, or frozen form, do not necessarily impart an off-flavor to ice cream, but they may impart a characteristic "eggy" flavor note. This derived flavor is typical for egg yolks. However, off-flavored egg solids have the capacity, similar to off-flavored milk solids, to introduce certain unwanted off-flavors. Deteriorated, poor quality whole eggs or egg yolks readily impart a flavor defect to ice cream. A characteristic "egg flavor," imparted by high-quality egg solids, is not that easy to distinguish, since this flavor note resembles the cooked (custard or nutty) sensation, although an eggy flavor is usually more persistent. When used at low levels in ice cream (less than 1.4%), high-quality egg solids are usually compatible with the desired flavor blend. Since egg yolks have good emulsifying properties, some ice creams are formulated to contain them as a supplement to, or a substitute for, stabilizers and/or emulsifiers.

Stabilizer/Emulsifier. These off-flavors are due to the incorporation of poor quality, deteriorated, or excessive amounts of stabilizers and/or emulsifiers. Ice milk may be more susceptible since it generally contains higher concentrations of these body and texture modifying agents than does ice cream. Substances used as emulsifiers are somewhat prone to imparting an off-flavor generally described as "stabilizer-like" or "emulsifier-like." Occasionally, some of the mono- and diglycerides and other emulsifiers in proprietary blends of stabilizer-emulsifiers may exhibit some degree of lipid auto-oxidation. Hence, this form of stabilizer/emulsifier off-flavor may be confused with the generic oxidized flavor defect. Certain soft-serve (ice milk) and ice cream novelty products are more likely to manifest a slight to moderate intensity of emulsifier off-flavor than conventional ice cream. The novelty products and ice milk rely on higher concentrations of polysorbates, mono- and diglycerides, or lecithin, to provide "drier," firmer products when drawn from the freezer; hence, they are more prone to this off-flavor than ice cream.

Nonmilk Food Solids. On a rare occasion, other approved food solids (other than dairy-derived, sweeteners, flavoring agents, and stabilizers/emulsifiers) may be incorporated into frozen dairy desserts for a special flavor effect, body and texture, or appearance function. Cookies, cake, and cheesecake are several examples that come to mind. It is conceivable that certain off-flavors could be imparted to ice cream from such sources, especially if used in relatively large quantities. Examples of materials cited here, however, should not be encountered in vanilla ice cream.

Off-flavors Due to Chemical Changes in the Mix or Product

Lacks Freshness (Stale, Old) or Oxidized (Cardboardy, Metallic). These off-flavors may develop due to chemical changes that can readily occur as the result of adverse conditions of producing, storing, transporting, processing, and distributing such perishable milk products as ice cream and ice milk mixes and finished products. Processes of staling, "aging," auto-oxidation of milk lipids, hydrolytic rancidity, and microbially-induced deterioration of milk proteins and milkfat represent a set of complex chemical and enzymatic activities that "takes its toll" on flavor stability of frozen dairy products and their mixes. The specifics of the possible off-flavors that can develop from these chemical changes have been described earlier in this chapter, but one new category that should be addressed is the so-called "storage" off-flavor.

Storage. The "storage" off-flavor generally refers to an off-flavor that may develop either in the mix or frozen ice cream (or ice milk) during the storage period. When ice cream is stored for an extended period of time, the flavor loses its initial luster, even though no specific defects seem to stand out. In one instance, the product may simply lack the sensation of freshness. In another case, absorption of odors from the environment can cause the product to acquire a "storage" off-flavor, a form of absorbed flavor defect. Smoke, ammonia, and various chemical odors are but a few examples of absorbed substances that may be responsible. Serious storage flavor defects have been known to develop when odor, absorption, and chemical change or deterioration in storage occurred simultaneously. The storage off-flavor is commonly considered more serious or objectionable than the "lacks freshness" (stale) defect in ice cream.

Off-flavors Due to Mix Processing

Cooked (Rich, Nutty, Eggy) and Caramelized/Scorched. These heat-induced off-flavors that might occur in ice cream were disussed earlier under the heading of "cooked," within the section of this chapter on the role of dairy products imparted off-flavors.

Off-flavor Due to Microbial Growth in the Mix

Acid (Sour), Fruity-Fermented, Cheesy, Musty, and Unclean (Psychrotrophic). Each of these microbially-induced off-flavors is likely to occur as the result of varied degrees of temperature abuse in the handling of milk and cream ingredients and/or excessive storage temperatures of perishable mixes (i.e., higher than 4.4°C [40°F]). For descriptions of each defect enumerated above, the reader is directed to the discussion of microbial off-flavors of milk and cream (Chapter 5).

Off-flavors Due to Other Causes

Foreign. As a rule, a foreign off-flavor may be easily detected, but the exact substance or specific contaminant is often difficult to positively identify. This flavor defect is definitely atypical (foreign) for dairy products or the ingredients ordinarily associated with good ice cream. Detergents, sanitizers, paint, gasoline, pesticides, and other chemicals are some of the possible serious offenders. Unfortunately, chemical substances may not only impart off-flavors, but may also be nauseating or toxic. Obviously, any products found to contain this defect must be severely downgraded and not marketed for human consumption.

Neutralizer. Although neutralization of lactic acid is not currently an accepted step in ice cream manufacture, the judge should be familiar with the flavor defects which may result from such a practice. When neutralizer is used to reduce the acidity of milk ingredients or the mix, the end products formed by the chemical reaction of neutralization are left as residual compounds in the frozen product, where they may become apparent upon tasting. This off-flavor is recognized by a peculiar alkaline off-flavor (reminiscent of sodium bicarbonate (baking soda) or milk of magnesia). Sometimes, a slight bitter taste can be associated with neutralizer off-flavors, though this bitter note is usually rather mild. The taste reaction time for a neutralizer off-flavor is somewhat delayed, but the peculiar taste persists for some time after the sample has been expectorated. Any frozen dairy desserts exhibiting a neutralizer off-flavor are usually severely criticized by ice cream judges.

In this age, the use of neutralizers in ice cream manufacture, or any type of dairy product, should certainly be discouraged, if not altogether eliminated. In those instances where a neutralizing agent might be used, the ice cream manufacturer is also likely to experience the development of other associated serious off-flavors (besides the neutralizer defect); namely, lacks freshness or stale, old ingredient, storage, and/or spoilage (psychrotroph) bacteria-related off-flavors.

OTHER FROZEN PRODUCTS

Ice Milk. As can be noted from Table 6.1, this product differs from ice cream principally in the milkfat content. Although ice milk is offered in a variety of flavors, vanilla is the most popular. For evaluating the sensory properties of ice milk, the ice cream score card and scoring guide (Fig. 6.1 and Table 6.3) are appropriate for all sensory quality parameters. Due to the lower milkfat content, ice milk would be expected to lack the typical richness, mouthfeel characteristics, and the overall flavor blend that most ice cream possesses. Also, the body and texture, as expected, can differ considerably from ice cream, due to the lower total solids content of ice milk. However, in spite of these inherent problems, many manufacturers have mastered the required technology and art for producing ice milk of excellent flavor, body, and texture. In fact, the sensory properties of many samples of ice milk may be practically free of criticism, even though they might be evaluated on the same general criteria as ice cream.

Mellorine. Despite the different language in the Federal Standards of Identity, except for the source and type of fat, this product generally resembles either ice milk (usually) or ice cream in composition. The ice cream score card and guide are generally applicable for conducting sensory evaluation, but certain additional defects that may be derived from vegetable or animal fats may be encountered and recorded as appropriate on the score card. Flavor defects of main concern in mellorine are the possibilities of oxidation, rancidity, the presence of a distinctive off-flavor derived from the specific fat source, and a lack of flavor or "blandness" (which can be attributed to the fat). The relative hardness and melting properties of the fatty acids that constitute the fat can influence the body and mouthfeel of frozen mellorine.

Frozen Custard. Basically, this product is identical to ice cream except for the addition of egg yolk solids at a concentration of at least 1.4% by weight. Based on this requirement, frozen custard should not be criticized for having an egg solids flavor, unless a characteristic "poor egg solids" off-flavor is sensed (due to use of poor quality egg ingredients). A greater tolerance for a "cooked" or "eggy" flavor should be extended in evaluating those products labeled "frozen custard," "french custard," or "french vanilla" ice cream.

Frozen Bulky-Flavored Products. Due to the relatively small quantity of required flavoring (and a minimum dilution effect), ice cream composition remains essentially unchanged when it is flavored with vanilla or other extracts. However, some flavorings such as chocolate, fruits, bakery products, candy, and nuts are often added in relatively high proportions; hence, the applied term of "bulky flavors." Bulky flavors may be added to either ice cream, ice milk, or frozen custard. Federal standards allow for alteration of the product composition by bulky flavors, as indicated in Table 6.1. Numerous bulky flavoring ingredients are used in ice cream; a few will be discussed to illustrate the applicable principles when sensory qualities are assessed by sensory methods.

In ascertaining the quality of bulky-flavored frozen desserts (actually any flavor), the evaluator should be alert to the possible occurrence of any of the defects which may be manifested in vanilla ice cream. Some of the milder off-flavors of ice cream may be masked or partially masked by some flavorings, but not by others. However, the judge should bear in mind that even a masked off-flavor may modify the overall perception of some flavorings in an undesirable way. A smooth, creamy texture is usually desired regardless of the type of flavorings used, but different or altered characteristics of body and texture should be recognized as the norm with some flavors of ice cream. Generally speaking, the higher the quantity of bulky flavorings incorporated into ice cream (or ice milk), the greater the tendency for development of a coarse or icy texture, and possibly a weaker body. This is primarily due to the dilution of solids, added moisture from some sources of bulky flavorings and/or higher overrun. When the added flavoring does not incorporate air, the ice cream portion may be excessively whipped to maintain minimum weight (e.g., 4.5 lbs/gal).

Chocolate

The principal forms of chocolate flavoring for frozen dairy desserts are cocoa, chocolate liquor, or a combination of the two. Chocolate liquor contains all of the usable portion of the cocoa bean, including about 50% cocoa butter. Cocoas are made by removing varying amounts of cocoa butter from the liquor. However, the flavor character of cocoa or chocolate liquor from different sources can vary significantly. These flavor variations may be due to the source of the cocoa beans, climatic conditions during growth, fermentation conditions, whether Dutch processed (treated with alkali) or naturally processed, and the roasting conditions. Aside from flavor variations, the resulting cocoa may be light, dark, or reddish-colored. Although the bulk of the characteristic flavor of chocolate is retained in the cocoa, some delicate, unique aroma constituents may be lost into the cocoa butter. Thus, the fat content of the given cocoa and the selected proportion of chocolate liquor to cocoa used in flavoring the ice cream will influence the flavor balance of the chocolate.

Chocolate ice cream often employs a substance to modify or enhance the chocolate flavor; vanilla is most frequently used, but on occasion coffee, cinnamon or salt may be added. The intent of the selected flavor modifier may be to mellow the chocolate sensation, diminish a certain harshness note, or simply to enhance or "bring out" chocolate flavor. However, the flavor modifier or enhancer should not be so intense as to actually predominate over the chocolate flavor of the ice cream.

The sweetness level of chocolate ice cream requires full consideration. Cocoa and chocolate liquor are quite bitter and thus they demand a higher sweetness level in ice cream than vanilla or most other flavors. As an illustration, the sweetness level of vanilla ice cream is commonly between 13% and 16%, expressed as sucrose, while that of chocolate ice cream may be 17% to 18% (expressed as sucrose).

Obviously, there are distinct variations in consumer preferences for the type and intensity of chocolate flavor in ice cream. Individual preferences may span the intensity range from "just a hint of chocolate" to an overwhelming "double chocolate," from a light to a very dark color, and from a mellow-sweet to a bitter, harsh chocolate. In evaluating the flavor of chocolate ice cream, the judge's personal preference should not prejudice the rating, insofar as possible. The overriding requirements for a regular or conventional chocolate ice cream are that the true-chocolate flavor be readily recognizable in a supposed "blindfold test," that the cocoa and/or chocolate liquor that is used be of high quality, that no off-flavors be present, and that any nonchocolate flavor notes "contribute, but not predominate."

Although some additional definitions of flavor terms and some new descriptors may need to be added, the ice cream score card and scoring guide in Figs. 6.1 and 6.2 can be applied to chocolate ice cream. Modifications are suggested in the following paragraphs.

Lacks Fine Flavor/Harsh, Coarse. These terms describe a lack of proper or desired chocolate flavor blend; an otherwise unidentifiable flavor defect of the chocolate; a flavor that is somewhat lacking in the desired delicate volatile components of chocolate; or a product which merely seems not to project a "perfect," "ideal," or highly desirable flavor.

Lacks Sweetness/Bitter. This flavor defect of chocolate ice cream is self-explanatory. Adjustment of the sweetener level (upwards) usually eliminates the defect in subsequent lots of the product.

Unnatural Flavor/Lacks Chocolate Character. These terms describe an artificial flavor; a chocolate flavor which is not readily recognizable as chocolate per se; or a flavor in which the nonchocolate components predominate. Selection of another source of chocolate flavoring is suggested.

Other Quality Factors of Chocolate Ice Cream. The body characteristics of chocolate ice cream are influenced by the relative proportions of cocoa and chocolate liquor used, as well as by the sugar content of the mix. Approximately 1.67 lb (0.74 kg) of chocolate liquor is required to impart the equivalent flavor intensity of 1 lb (0.45 kg) of cocoa; hence, ice cream has a higher total solids content when chocolate liquor is used exclusively or there is a high proportion of chocolate liquor to cocoa. But even when cocoa is used exclusively as the source of chocolate flavoring, the solids content of the mix is increased, and in either case, additional sugar (solids) is usually required and incorporated. The general effect of a product with a higher solids content is a chewier, more resistant body. The descriptors listed on a conventional ice cream score card to describe body and texture defects are generally applicable to chocolate ice cream.

The various color defects listed on the regular (vanilla) score card also apply to chocolate ice cream, except that a gray off-color would not be expected to occur in chocolate. Departures from the desired range of chocolate color may be variously described as dull, not uniform, too high (too dark), too pale (too light), or unnatural (atypical).

When evaluating the meltdown characteristics, the package, or bacterial content, the same criteria apply equally to vanilla and chocolate ice creams.

Chocolate ice cream is also made and/or packaged in combination with other flavors. Several examples are chocolate-almond, chocolate-marshmallow, chocolate mint, and other chocolate-based products sold under proprietary names; this is by no means an all-inclusive list.

Fruit Ice Creams

The flavor of berries and fruits (strawberries, peaches, etc.) may be imparted to frozen dairy desserts by fresh, frozen, or processed fruits, natural extracts (that sometimes contain other natural flavors), imitation flavors, or various combinations of these. The flavor character, body, and texture and the appearance of the finished product are influenced by the type of flavoring used.

Generally, the flavor of the given ice cream should be reminiscent of sweetened fresh fruit and cream (e.g., strawberries-and-cream or peaches-and-cream). To overcome the problem of seasonality, availability, and perishability of fresh fruit, frozen fruit is quite commonly used. The choice of the particular variety or lot of frozen fruit should be based on quality and its suitability for ice cream. For example, a considerably softer, riper, and more flavorful peach is required for ice cream than for pie baking. Processed fruit may often exhibit a cooked, "fruit preserves" type of flavor which may not be objectionable in itself, but it is unlike the typical or more preferred flavor of fresh fruit. Processed preparations of some fruits may be used alone, quite successfully, in combination with other forms of flavorings, or as a part of a more complex flavoring system. Processed cherries and some types of processed berries produce popular ice cream flavorings, and processed pineapple has been successfully used in combination with other flavors (especially for sherbet).

The sweetness level of fruit ice creams tends to be slightly higher than that of vanilla; the sweetener should blend smoothly into the overall flavor sensation in a well-made ice cream. There are two basic reasons for the incorporation of more sugar into fruit ice creams. The first is to compensate for the tartness of the fruit and optimize the intensity of the fruit flavor. Actually, the sweetness level of ice cream (from the mix) may already be sufficiently high to accomplish that for some fruits, hence the second reason becomes more important for quality considerations of the product. Secondly, sugar is generally required in the fruit preparation to reduce the freezing point of the fruit particles to prevent them from being ice-hard when the ice cream is consumed. Frozen fruits typically contain about 20% added sugar (1 part of sugar to 4 parts of fruit).

A few flavor terms on the regular ice cream score card must be redefined in order to apply this scoring tool to fruit-flavored ice cream. The suggested changes are enumerated in the following paragraphs.

Lacks Fine Flavor. This describes the lack of a highly desirable flavor blend; an otherwise unidentifiable flavor defect of the fruit and/or fruit flavoring; a flavor which lacks the full impact of fruit at the peak of its flavor development; or a flavor which just seems to fall short of being "perfect" or "ideal."

Cooked/Processed. The terms "cooked" or "processed" describe a moderate off-flavor produced by heat treatment of the mix and/or an off-flavor that resulted from heat processing of the fruit.

Unnatural Flavor/Lacks Specific Fruit Character. These terms attempt to describe an artificial-like or atypical fruit off-flavor; a flavor sensation in which the specific fruit is not readily recognizable; or a flavor note in which other fruit or nonfruit components seem to predominate.

Lacks Freshness/Stale Fruit. This set of flavor defect descriptors is generally self-explanatory.

Body and Texture of Fruit Ice Cream. Since fruit preparations may be used in rather high concentration in ice cream (15–24%), there is considerable dilution of the mix, which, unless it is compensated for in some manner, can lead to a coarse texture and a decidedly weaker body. For fruit ice creams, one slight modification, listed following, seems appropriate for the body and texture segment of the ice cream score card.

Coarse/Icy/Icy Fruit. The descriptor used to describe the relative coldness and size of ice crystals in frozen dairy desserts is "expanded" to encompass potential problems that may arise from fruit particles added to the product.

Other Quality Factors of Fruit Ice Cream. Both the color and appearance of fruit ice cream should be closely evaluated for esthetic appeal. As with other flavors of ice cream, the color may be dull, not uniform, too deep, too light, or unnatural (atypical). The appearance also should be checked for any of the following possible defects (where applicable):

Fruit particles too small

Fruit particles too large

Too few fruit particles

Too many fruit particles

Poor distribution of fruit

Atypical color of fruit particles.

Nut Ice Cream

Pecans, walnuts, almonds, peanuts, macadamia nuts, hazelnuts (filberts), and pistachio nuts are among the most popular nuts added to ice cream in the U.S. Generally, ice cream is flavored with either an appropriate background flavor for the nuts (butter pecan, chocolate almond, etc.), or with a concentrate of the same basic nut flavor (e.g., pistachio, black walnut). The degree and the method of roasting the nuts (light or heavy roast; dry or butter roasted) provide interesting variables that manifest themselves in the sensory properties of the ice cream. The initial quality and freshness of the nuts must be good; no deterioration should occur as a result of storage. Since some types of nuts contain a high proportion of unsaturated oil, they can be highly susceptible to auto-oxidation. Some nuts (walnuts and hazelnuts) are also prone to the development of hydrolytic rancidity due to the presence of lipolytic enzymes.

The size of nuts in ice cream may range from intact, whole nuts to small, broken, or sliced pieces. Except in special cases, medium- to larger-sized pieces are generally favored. In any case, the nuts should retain their firmness, crispness, and freshness in the frozen product.

Vanilla (or chocolate) ice cream score cards are generally applicable to nut-flavored ice creams. The following revisions of flavor descriptors are suggested for the flavor of nut ice creams.

Lacks Fine Flavor. This term describes a general lack of the desired flavor blend; an otherwise unidentifiable, slight flavor defect of the nuts or background flavor; or a flavor that simply does not quite attain the "ideal" or anticipated flavor.

Unnatural Flavor. An artificial or atypical background flavor for the particular nut is described by the term "unnatural" off-flavor.

Salty/Excessively Salty Nuts. These self-explanatory descriptors cover the instances of excessive incorporation of salt on the nuts or in the ice cream.

Oxidized/Oxidized Nuts/Rancid Nuts. Within nondairy segments of the food industry, a generic "oxidized" off-flavor is often referred to as a "rancid" off-flavor. However, walnuts and hazelnuts may also exhibit an actual rancid (lipolyzed) off-flavor due to the lipase content of these nuts, if they have not been roasted.

For assessing the body and texture of nut ice creams, one additional criticism is suggested below.

Nut Meats Lack Crispness. This term is generally self-explanatory; the nut pieces absorb moisture and become somewhat waterlogged or soft in consistency.

Other Quality Factors of Nut Ice Cream. Both color and appearance are important criteria in measuring the sensory qualities of nut ice cream. Appearance is primarily influenced by the size and uniform distribution of the nut meats, which help determine the eye appeal of the product. In addition to obvious color defects, the following defects of appearance are possible in nut ice creams:

Nut particles too small   Poor distribution of nut meats
Too few nut particles   Atypical color of nut meats
Too many nut particles   Inclusion of nutshell fragments.

Candy Ice Cream

Chocolate chip and mint candy are probably the most popular representatives of this group of products, though many others are produced by U.S. ice cream manufacturers. The background flavor may be vanilla, chocolate, or another flavor that is compatible with the candy (e.g., mint chocolate chip). As with fruit and nut ice creams, the evaluator should be somewhat familiar with the quality criteria of the added materials. General quality requirements for candy flavored ice creams are: (1) a pleasing flavor blend; (2) crispness of the candy components; (3) attractive color and appearance (size and shape); (4) good distribution of candy pieces throughout product; and (5) minimal or no color migration through the ice cream. Some ice cream manufacturers have reported some success with minimizing the occurrence of overly softened candy pieces and color migration by freezing the candy before addition to the frozen product. The suggested sensory descriptors of defects for fruit and nut ice creams also apply to candy ice cream. The judge should try to note whether a given defect seems to pertain to the background flavor or to the candy itself. The various flavor defect definitions for chocolate ice cream also apply to the flavor of any added chocolate chips or pieces.

Variegated Ice Cream

A variegated ice cream should basically emulate an ice cream sundae, although the flavored syrup, sauce, or puree is dispersed throughout the product. Chocolate, fudge, marshmallow, butterscotch, peanut butter, strawberry, and raspberry are just a few of the flavors that are variegated or marbled. The flavoring (or slurry) syrup is usually pumped directly into the ice cream as it emerges from the ice cream freezer; the variegating substance is intended to form a definite pattern within the product. Although some indication of the regularity or uniformity of the variegation pattern is obtained in the course of normal sampling of the ice cream, a more objective visual impression can usually be realized by examining both exposed surfaces, after cutting through the center of the container. Sometimes, several cross-sectional cuts may have to be made to properly assess the distribution or the "pattern" of the variegating material with the frozen product. Typically, the ribbon of syrup should be of medium thickness, and the pattern should essentially reach into all segments of the container.

Other quality criteria include the flavor and consistency of the variegating syrups used in the ice cream. In general, the flavor should be readily identifiable, be free of off-flavors, and produce a pleasing blend with the background or the "other" flavor(s) of the product. The syrup should not "settle out" or mix with the ice cream, but simultaneously, it should not be overly hard, crusty, or icy.

The following modified definitions of flavor defects are suggested for better application in evaluating variegated ice creams.

Lacks Fine Flavor. A lack of the desired flavor blend; an otherwise unidentifiable flavor defect of the variegating syrup or background; or a flavor which just falls short of being "perfect" is implied by this descriptor.

Lacks Flavor/Variegating Syrup Lacks Flavor. Self-explanatory.

Unnatural Flavor. "Unnatural" describes an artificial or atypical off-flavor in the background flavor and/or in the variegating syrup.

Other Quality Factors in Variegated Ice Cream. The body and texture of variegated ice cream should be similar to that of its unvariegated counterpart. However, less "heat shock" resistance is a typical property of variegated ice creams; consequently, it can be expected that frequently the body will be weaker and the texture more coarse than plain or regular ice creams. Another reason for a weak, coarse body in variegated ice creams is in the "overrun gradient" between the variegating syrups and the ice cream. The variegating syrups are usually quite heavy; at the time of freezing, air is incorporated only into the mix portion. If product is drawn at the same weight/unit as that of the product without variegating syrup, the ice cream mix portion obviously has to be much lighter. The same problem may be encountered in other bulky-flavored ice creams in which no overrun is formed within the more dense or solid flavoring material.

A specific defect is likely to occur within the body and texture of variegated ice cream, which is explicated below.

Variegating Syrup Too Hard, Icy, or Chewy. Due to the difference in physical and chemical properties, especially the "overrun gradient" between the variegating syrup and ice cream, a certain crustiness, chewiness, or iciness can occur in variegated ice cream. A correct composition of the variegating syrup should help guard against this defect. Under color and appearance, the following possible criticisms for variegated ice creams are likely to occur:

Poor pattern of distribution	Syrup settled out (precipitated)
Too thick a ribbon	Syrup mixed with ice cream
Too-thin a ribbon	Unnatural or atypical color (of the ice cream or the variegating syrup).

Frozen Novelties

A group of products referred to as frozen novelties may be made of ice cream, ice milk, mellorine, sherbet, ice, frozen yogurt, pudding, or combinations of several of these. They may be in many forms, such as bars (with or without a stick), coated or uncoated, "sandwiches," prepackaged cones, and other numerous forms. Although they should be evaluated by the processor in ongoing quality assurance procedures, novelties are seldom, if ever, judged competitively. The flavor, body, and texture of these types of products should be evaluated just as critically as their packaged counterparts, but there are some unique, potential problem areas that should be identified (Tobias 1980b). A listing of some of the more common quality problems of various types of frozen novelties that require special attention include:

Incomplete coverage with coatings	Cracked coating
Coating too far down the stick	Slipped coating
Incorrect volumes	Overrun too high
Coating too thick	Overrun too low
Coating too thin	Defective flavor
	Defective texture

Damaged wrappers	"Soggy" wafers or cones
Sticking wrappers	(lack crispness)
Broken sticks	High coliform count
Sugar "bleeding" from bars	Brine contamination.

Due to their relatively small size, frozen novelties are markedly susceptible to the irreversible, damaging effects of temperature fluctuations. "Heat shock" is probably the most serious problem, but unfortunately, once the product enters the distribution system, there is limited control of frozen storage temperatures.

Sherbet

According to the Federal Standards (21 CRF 135.140), there are two types of sherbet, fruit and nonfruit sherbet. Compositional standards shown in Table 6.1 are the same for both types of sherbets, except that the titratable acidity of fruit sherbet must not be less than 0.35%, calculated as lactic acid. Sugar solids, the major constituent of sherbets, may exceed 30% by weight in concentration. Some of the nonfruit characterizing ingredients include ground spice; coffee or tea; chocolate or cocoa, including syrups; confections; distilled alcoholic beverages, including liqueurs or wine (in an amount not to exceed that required for flavoring); and any natural or artificial food flavoring (except characteristic fruit or fruit-like flavor). Some of the more common fruits used in fruit sherbets are orange, pineapple, raspberry, strawberry, lemon, lime, and cranberry. Of the two types of sherbets, fruit sherbets are by far the more popular in the U.S.

Though poor quality dairy ingredients may infrequently cause an off-flavor in sherbets, the mandatory low concentration of milk solids (less than 5%) somewhat reduces this likelihood. In fruit sherbet, the quality is usually determined by the overall flavor blend of sweetness, tartness, fruit flavor intensity, and by how closely the given fruit flavoring emulates the true fruit flavor (at its peak of quality). In nonfruit sherbet, quality differs with each specific flavoring; therefore, only a vague, general statement pertaining to the desired flavor can be made. In nonfruit sherbets, the flavoring and the sherbet base (mix) should be free of perceptible defects, and the frozen product should have a pleasing flavor blend.

The ice cream score card may be applied as a tool to evaluate the flavor of sherbets, if the evaluator takes into account the following additional criticisms and revisions of definitions.

Defective Flavoring/Peel Flavor. Defective flavoring may be any off-flavor due to a manufacturing error, an oversight, or due to quality deterioration of the flavoring materials during shipment or storage. A "peel" off-flavor is commonly encountered in citrus fruits and is suggestive of the concentration of essential oil of citrus, which is found in the peel.

Unnatural Flavor. This describes an artificial flavor, a flavor that is lacking in true fruit character, or an off-flavor which is not recognizable as the flavor stated on the product's label.

Lacks Tartness or Excessive Tartness. Self-explanatory.

Other Quality Factors of Sherbet. The texture of sherbets can be nearly as smooth as that of ice cream. The body of sherbet may range from weak to resistant, although a heavy or even slightly gummy body need not be considered defective. Probably the most common defects of sherbet body and texture are severe coarseness and crumbliness. Inadequate stabilization, "heat shock," high overrun, low solids content, and prolonged storage are usually responsible for the development of a coarse and icy texture. Inadequate stabilization may also be responsible for crumbliness. This defect seems to be more frequently encountered in orange-flavored sherbet, presumably due to some unexplained property of one or more orange oil constituents. Addition of an emulsifier to the mix is helpful in correcting or limiting the severity of the problem.

The sugars commonly used in sherbets are sucrose, corn syrups, and, to a lesser extent, dextrose (corn sugar). The body of sherbet may be *hard* or *soft*, depending on whether too little or too much sugar was used in the formulation. Several other sherbet defects, common in yesteryear, may still be encountered occasionally. *"Surface crustation"* may occur, particularly when the product surface is exposed to air. Effective stabilization and partial replacement (25–50%) of sucrose with corn syrup are good precautionary steps. *"Ice separation"* may occur in the continuous freezer by the action to centrifugal force. Ice builds up on the freezer wall and eventually breaks away and "lands" in the product. Increasing the viscosity of the unfrozen portion of the mix by proper stabilization helps control this problem. *"Separation, drainage, or bleeding"* of the unfrozen syrup within the sherbet may also be a problem of inadequate stabilization and/or holding the sherbet at too high a temperature.

The ice cream score card is satisfactory for evaluating the body and texture of sherbets with the following minor modification.

Heavy/Hard. The formulation and lower overruns ($\leq 60\%$) of sherbet generally leads to a heavier or harder product at the typical serving temperature. Sherbets that may be formulated with lower levels of sweetener may not depress the freezing point adequately; hence, a greater likelihood of a heavy/harder product at or near the serving temperature.

Both the color and appearance should be evaluated in sherbets, particularly in multiflavored products (e.g., rainbow sherbet) in which the distribution pattern of the different flavored products is a quality criterion, and in products to which fruit particles or confectionery were added. Suggested descriptors for these possible color defects of sherbet are:

Defective pattern   Poor distribution of added
Too little added material   material
                                      Poor appearance of added
                                         material.

Water Ices

The federal standards describe water ices as a food which is prepared from the same ingredients as sherbets, except that no milkfat, milk-derived ingredients, nor egg ingredients (other than possibly egg whites) are used. As indicated in Table 6.1, the minimum weight (Federal Standard) for water ices is 6 lbs/gal. Sensory evaluation procedures for water ices differ little from those used for sherbet.

Frozen Yogurt

In some respects, frozen yogurt resembles ice cream, ice milk, and sherbet. This product is available in packaged, novelty, or soft-serve form and in a variety of flavors, most commonly fruit flavors (Bodyfelt 1978). The general criteria used in the sensory evaluation of frozen yogurts are comparable to those used for sherbets or ice milk. "Chalkiness" may sometimes be observed in the mouthfeel of frozen yogurt; this is quite possibly the result of dehydration of proteins by the combined action of heat and acidity. The levels of sweetness and acidity (sweetness/acidity balance) in association with the given flavor are important considerations for frozen yogurt quality.

Soft-Serve Frozen Desserts

These products (usually ice milk or frozen yogurt) are commonly dispensed from a special freezer for immediate consumption by the consumer. Since the serving temperature is about $-7.2\,°C$ (19°F), the hardening step is omitted, which eliminates the "damaging effects" of slow freezing and subsequent temperature fluctuations. As a result, soft-serve should generally exhibit creamy, smooth mouthfeel properties, as well as provide excellent "flavor release."

Generally, the same requirements apply to the flavor of soft-serve as to the corresponding hard frozen product (ice milk or frozen yogurt). Most of the body and texture criteria also apply, except that the desired or optimum characteristics should be partially redefined. The body should be fairly resistant and firm (to retain shape on a cone), but obviously not as firm as that of hardened products which are stored and consumed at much lower temperatures ($-13°C$ [$8°F$]). The desirable characteristics of soft-serve (Tobias 1969) may be summarized as follows:

A desirable flavor blend and absence of off-flavors.
Smooth texture: small ice crystals; no lactose crystals; no butter granules; and no excessive coldness.
Dry appearance; a pleasing color.
Some modest resistance to melting.
A reasonably firm, resistant body.
A neatly shaped serving portion; shape maintained for a reasonable time before consumption.

When sensory problems are encountered with soft-serve frozen desserts, they may be traced to mix ingredients, mix composition, mix processing, age of mix, mix handling, mechanical and sanitary condition of the freezer, freezer operation procedures, and numerous other factors. For instance, on "slow business" days the product remains in the freezer under intermittent agitation for an extended time. The effect on quality may be: a progressively wetter, weaker-bodied product (even though the temperature may be unaffected or even decreased); problems with overrun (weight of serving); fat separation (due to churning); and lactose crystallization (sandiness). A well-formulated mix, along with good mechanical condition of the freezer and a properly operated freezing machine, can minimize most of these problems.

Most of the soft-serve on the market is ice milk (by federal definition), but ice cream, sherbet, water ices, and especially frozen yogurt are also available in many localities. Although vanilla is the predominant flavor (along with a number of "sundae" options); chocolate, fruit, or berry flavors and other flavor options are offered by more and more retail stores.

Direct-draw Shakes

This product, similar in composition to ice milk, emulates the traditional milk shake (Holsinger *et al.* 1987). Depending on composition and whether a "thick" or "thin" shake is desired, the product is drawn from the freezer in the temperature range of $-3.3\,°C$ to $-1.1\,°C$ ($26\,°F$ to $30\,°F$). The mix may be flavored prior to freezing, or a flavoring syrup may be added to the frozen shake and dispersed in a spindle type mixer.

The finished product should possess a pleasing blend of flavor (chocolate is the most popular flavor) and be free of off-flavors. Opinions may vary as to the desired body and texture that appeals to the widest group of consumers. A thick, smooth-textured shake that draws through a straw is probably the choice of a majority of consumers. Product overrun is still another factor that affects coldness and mouthfeel. A product with a high overrun yields comparably less liquid as it melts in the mouth. A desirable range appears to be 40–60% overrun for direct-draw shakes.

Just as with soft-serve, the sensory characteristics of shakes are also traceable to either the mix, the freezer, or to the procedures of the freezer operator (Tobias 1969). The resolution of a particular sensory defect may be as simple as resetting a freezer control knob or as complex as reformulating the mix.

REFERENCES AND BIBLIOGRAPHY

American Dairy Science Association. 1987. Committee on Evaluation of Dairy Products. Score card and scoring guide for ice cream. Champaign, IL.

Arbuckle, W. S. 1986. *Ice Cream.* 4th Ed. AVI Publishing Co. Westport, CT. 482 pp.

Bodyfelt, F. W. 1973. Ice cream quality: Consumer preference vs. 'the experts'. *Amer. Dairy Rev. 35*(9):24.

Bodyfelt, F. W. 1976. The role of temperature control for assurance of high-quality ice cream. *Amer. Dairy Rev. 38*(4):18L,N.

Bodyfelt, F. W. 1977. Is real vanilla the best flavoring approach? Nat. Ice Cr. Retail. Assn. Production Tips. Jan. 3 pp.

Bodyfelt, F. W. 1978. Has frozen yogurt joined the frozen dairy dessert club? Yearbook Nat. Ice Cr. Retail. Assn. Muncie, IN. p. 10.

Bodyfelt, F. W. 1979. Ice cream quality—who should be the judge? Nat. Ice Cr. Retail. Assn. Production Tips. Mar. 2 pp.

Bodyfelt, F. W. 1983a. Ice cream: What really determines the quality? *Dairy and Food Sanit. 3*(4):85.

Bodyfelt, F. W. 1983b. Quality assurance for ice cream manufacture. *Dairy and Food Sanit. 3*(5):164.

Bodyfelt, F. W., Andrews, M. V., and Morgan, M. E. 1979. Flavors associated with the use of cheddar cheese whey powder in ice cream mix. *J. Dairy Sci. 621:*51.

Bodyfelt, F. W., McGill, L. A., Morgan, M. E., Bills, D. D., and Scanlan, R. A. 1973. Flavor preferences for strawberry ice cream: Commercial products versus special formulation. *J. Dairy Sci. 56:*626.

Code of Federal Regulations. 1985. Title 21—Food and Drugs, Part 135—Ice Cream and Frozen Desserts. U.S. Government Printing Office. Washington, DC.

Holsinger, V. H., Smith, P. W., Talley, F. B., Edmundson, L. F., and Tobias, J. 1987.

Preparation and evaluation of chocolate-flavored shakes of reduced sweetener content. *J. Dairy Sci. 70*:1159.

Hyde, K. A. and Rothwell, T. 1973. *Ice Cream.* Churchill Livingston, Edinburgh. U.K.

Keeney, P. G. and Kroger, M. 1974. Frozen dairy products. In: *Fundamentals of Dairy Chemistry.* Webb, B. H., Johnson, A. H., and Alford, J. A., Eds. AVI Publishing, Westport, Conn. p. 873.

Lucas, P. S. 1941. Common defects of ice cream, their cause and control; a review. *J. Dairy Sci. 24*:339.

Sommer, H. H. 1938. *Theory and Practice of Ice Cream Making.* Published by the author. Madison, WI. 639 pp.

Tobias, J. 1969. Defects of soft-serve and direct draw milk shakes. In: *A Success Manual by the Mixmaker,* 148–154. Published by CAH Enterprises. Whitehaven, TN.

Tobias, J. 1980a. Ice cream in the 20th century: A compendium of industry growth. *Dairy Record 81*(8):96.

Tobias, J. 1980b. Quality, production rate spell success in novelties. *Dairy Record 81*(10):85.

Tobias, J. 1981a. Is super-premium ice cream always better? *Dairy Record 82*(8):122.

Tobias, J. 1981b. Unraveling the mysteries of ice cream body. *Dairy Record 82*(10):64.

Tobias, J. 1982a. Evaluating the sensory qualities of your ice cream. *Dairy Record 83*(2):48.

Tobias, J. 1982b. HTST pasteurized mix, how it affects the sensory properties of ice cream. *Dairy Record 83*(11):109.

Tobias, J. 1982c. Frozen dairy products. In: *Handbook of Processing and Utilization in Agriculture.* Vol. 1. Animal Products. I. H. Wolff Ed. CRC Press. Boca Raton, FL. p. 315.

Tobias, J. 1983a. Controlling rancidity in ice cream. *Dairy Record 84*(4):112.

Tobias, J. 1983b. Ice cream and frozen products. Some "natural" concerns. *Dairy Record 84*(2):74.

Tobias, J. and Muck, G. A. 1981. Ice cream and frozen desserts. *J. Dairy Sci. 64*:1077.

Turnbow, G. D., Tracy, P. H., and Raffetto, L. A. 1947. *The Ice Cream Industry.* John Wiley and Sons. New York. 654 pp.

7

Sensory Evaluation of Cultured Milk Products

Long ago, it was observed that milk generally soured soon after it was harvested. It was also noted that sour milk does not readily undergo undesirable proteolysis and other unwanted physical and chemical changes. Hence, milk was usually handled in a manner to insure souring and thus preserve it for several days or longer. Each tribe or ethnic group developed their own method of handling or treating milk; consequently, the final products differed. This helps explain why a variety of cultured (fermented) milk and cream products have originated, each known and referred to by a unique name. The common denominator was that each product required the presence or addition of lactic acid producing bacteria to accomplish the preservation. Additionally, some of these products underwent an alcoholic fermentation.

In many countries, fermented milk foods are distinctly favored over fresh, fluid milk. This frequent preference for "sour milk" is based on a combination of public safety, preferred flavor and texture, and purported therapeutic effects. Where inadequate facilities for transport, storage, refrigeration, pasteurization and/or distribution of milk exist around the world, many health authorities prefer that milk turns "sour" in the earliest stages of handling. In this approach, the presence of high populations of harmless lactic acid bacteria and their metabolic end products discourage the growth of food spoilage, food poisoning, and disease-producing bacteria. In many countries, nutritionists and pediatricians definitely prefer certain fermented milk products over fresh milk as a weaning food for infants. In other locales, fermented milk foods are blended with cereals and other food ingredients to provide a nutritionally balanced food for the populace.

For those countries where few or none of the above described conditions or health philosophies exist, the acceptance of cultured milk products relates more to "slimming diets," cost considerations, adaptation of ethnic foods, recent food trends, and new technologies of food processing and distribution. In numerous countries, fresh fluid milk is the dominant product of commerce, but certain cultured milk foods enjoy increasing attention, modification and modest popularity. Modern cultured products are produced by inoculating adequately pasteurized milk or cream bases with appropriate lactic acid-producing bacteria.

Cultured (fermented) Grade A milk products, which include cottage cheese, sour cream, sour half-and-half, buttermilk, kefir, and yogurt, play an important role in the U.S. dairy industry. These and other forms of cultured dairy products have made quite important contributions to civilization. They have enabled some populations to survive periods of famine. Nutritionally, cultured dairy foods provide nutrients vital to good health, which make them desirable staples in the human diet. The lower pH and extended shelf-life of cultured milk products lends itself to realistic production in many developing countries.

THE ROLE OF SENSORY EVALUATION

Sensory evaluation of cultured milk products in the U.S. has not progressed to the same extent as the art and science of sensory discrimination for milk and many other manufactured milk products. This was probably due to the more limited popularity and a restricted technical focus for most cultured dairy foods, up until the last several decades. There are no commercial market or U.S.D.A. grades for any of the cultured milk products, and the standards or guidelines for sensory quality characteristics are not as clearly defined as for other dairy products. There also seem to be specific geographical differences in consumer preferences for flavor intensity, body and texture characteristics, and/or color and appearance features of many cultured dairy foods. An example of this is the apparent preference of West Coast consumers for a less acid, blander flavor of creamed cottage cheese. This contrasts with a Midwestern consumer preference for a more acidic, aromatic (diacetyl) flavor in cottage cheese. There are also apparent regional preferences for the relative size and degree of firmness for the curd particles of cottage cheese (Kosikowski 1977). For yogurt of fruit-on-bottom types, there are two distinct styles; a conventional sundae style (an uncolored milk base) and a Western sundae style (a colored, flavored, and commonly sweetened milk base).

Unfortunately, systematic or routine sensory evaluation of cultured milk products has received less attention than most other traditional U.S. milk products. Research workers in university, government, and industry laboratories have examined lactic cultures and cultured milk products critically, and have defined some desired qualities for cultured products. When practiced, sensory evaluation of commercial fermented milk and cream products has frequently involved more of a comparison of the products of current manufacture with those made previously. This procedure has some merits, provided a definite set of sensory attributes has been established, but this approach may tend to result in a progressively lower-quality product. This may help explain why U.S. per capita consumption of many cultured milk products is static (Tamime and Robinson 1985). Desirable standards for some of the more common sensory features of various fermented dairy products will be discussed in this chapter.

Cultured Dairy Product Score Cards

A common score card applicable to all types of cultured milk products is not feasible, since the products differ widely in their sensory characteristics. However, a common score card may be satisfactory for the sensory evaluation of lactic cultures (starters), cultured buttermilk, cultured half-and-half, and cultured sour cream. Following the pattern of other dairy product score cards (which have been approved by the Committee for Evaluation of Dairy Products of the American Dairy Science Association), the authors have proposed a score card (illustrated in Fig. 7.1) for the sensory evaluation of cultured milk products.

This score card not only recognizes the importance of flavor, body, and texture in a good cultured dairy food, but also provides for consideration of such other quality criteria as the acidity level, appearance and color, container, and closure. The relative importance of various quality attributes is illustrated in the scoring guide shown in Table 7.1. In using this score card for the sensory evaluation of fermented milk foods, procedures similar to those in scoring milk or cream should be followed.

CULTURES AND CULTURED PRODUCTS: AN OVERVIEW

Lactic cultures contain viable, lactose-fermenting bacteria; each culture may contain single or mixed bacterial strains (different species or genera). The controlled growth of the selected bacterial strains should yield a smooth, glossy coagulum.

Traditionally, the first culture obtained after inoculation of the growth media with the selected organisms was known as a "mother" culture. Mother cultures, carefully transferred (re-inoculated) on a routine basis, were used to inoculate either whole milk, lowfat milk, or skim milk to produce commercial cultured buttermilk, kefir, and yogurt, or to inoculate a sweet cream base to produce cultured sour cream or half-and-half. Current lactic culture technology relies on the use of

CULTURED DAIRY PRODUCT SCORE CARD

DATE: _____   PRODUCT:

PLACE: _____   Buttermilk _____   Kefir _____
                                        Sour Cream _____   Other _____

	CRITICISMS		SAMPLE NO.							
			1	2	3	4	5	6	7	8
FLAVOR 10	SCORE ➡										
	ASTRINGENT										
	BITTER										
NO CRITICISM 10	CHALKY										
	CHEESY										
	COARSE (Harsh)										
	COOKED										
	FERMENTED										
	FOREIGN										
	GREEN (Acetaldehyde)										
	HIGH ACID (Sour)										
	LACKS ACID (Flat)										
NORMAL RANGE	LACKS CULTURE FLAVOR										
1-10	LACKS FRESHNESS										
	METALLIC/OXIDIZED										
	RANCID										
	SALTY (Too high)										
	SAUERKRAUT-LIKE										
	STABILIZER/EMULSIFIER										
	UNCLEAN										
	VINEGAR-LIKE										
	YEASTY										
BODY AND TEXTURE 5	SCORE ➡										
	CURDY										
NO CRITICISM 5	GASSY										
	GRAINY/GRITTY										
	LUMPY										
NORMAL RANGE	TOO FIRM (Over-stabilized)										
1-5	TOO THIN (Weak)										
APPEARANCE 5	SCORE ➡										
NO CRITICISM 5	CHURNED FAT										
	DULL (Lacks gloss)										
	LACKS UNIFORMITY										
NORMAL RANGE	UNNATURAL COLOR										
1-5	WHEYED-OFF (Syneresis)										
PRODUCT ACIDITY 2	SCORE ➡										
	% TITRATABLE ACIDITY										
	pH										
CONTAINER AND CLOSURE 3	SCORE ➡										
	SHORT-FILL										
	OVER-FILL										
	SOILED										
	DUSTY										
TOTAL SCORE 25	SCORE PER SAMPLE ➡										

EVALUATORS: _____   _____   _____

Fig. 7.1. A suggested score card for the sensory evaluation of cultured milk products.

Table 7.1. A Suggested Scoring Guide for the Sensory Defects of Cultured Milk Products with Assigned Scores for Designated Defect Intensities.

	Intensity of Defect		
	Slight[b]	Definite	Pronounced[c]
Flavor Defects[a]			
Astringent	7	5	3
Bitter	8	5	2
Chalky	8	5	2
Coarse (harsh)	8	6	4
Cooked	9	8	6
Fermented (vinegary)	7	5	2
Foreign[d]	6	3	0[e]
Green (acetaldehyde)	8	7	6
High acid (sour)	9	8	7
Lacks acid (flat)	9	8	7
Lacks freshness	8	7	6
Metallic/oxidized	6	4	2
Rancid	4	2	0
Salty (too high)	9	8	6
Sauerkraut-like	7	6	5
Stabilizer/emulsifier	8	7	5
Unclean	4	2	0
Yeasty	5	3	0
Body and Texture Defects[f]			
Curdy	4	3	2
Gassy	4	3	2
Grainy/gritty	4	3	2
Lumpy	4	3	2
Too firm (overstabilized)	4	3	2
Too thin (weak)	4	3	2
Wheyed-off (syneresis)	4	3	2
Appearance[f]			
Churned fat	4	3	2
Dull (lacks gloss)	4	3	2
Lacks uniformity	4	3	2
Surface growth	1	0	0
Unnatural color	4	3	2
Wheyed-off (syneresis)	4	3	2
Product Acidity[g]			
pH			
% Titratable acidity			
Container and Closure[h]	2	1	0

[a] "No criticism" for flavor is assigned a score of "10." Normal range is 1 to 10 for a salable product.
[b] Highest assignable score for defect of slight intensity.
[c] Highest assignable score for defect of pronounced intensity. A score of "0" (zero) may be assigned if the defect renders the product unsalable.
[d] Due to the variety of possible foreign off-flavors, suggesting a fixed scoring guide is not appropriate. Some foreign flavor defects warrant a "0" (zero) score even when their intensity is slight (e.g., gasoline, pesticides, lubricating oil).
[e] An assigned score of zero ("0") indicates an unsalable product.
[f] "No criticism" for body and texture and appearance categories is assigned a score of "5." Normal range for either category is 1 to 5 for a salable product.
[g] "No criticism" for product acidity is assigned a score of "2"; penalty point deductions for pH or % T.A. would have to be devised for each cultured product evaluated by this scoring system. Normal range is 1 to 2 for a salable product.
[h] "No criticism" for container and closure is assigned a score of "3"; penalty point deductions would have to be devised for any assessed defects or criticisms. Normal range is 1 to 3 for a salable product.

concentrated cultures which either have been lyophilized (freeze dried) or frozen at extremely low temperatures.

Selected lactic cultures that form diacetyl (buttery aroma) can also be used to inoculate cream to produce culture-flavored butter, which is common practice in Europe (Scandinavian or Danish style) and many other countries. Extensive use is made of various lactic cultures in the cheese industry to form the required lactic acid (and other fermentation end products) in most cheesemaking processes.

Superior-quality dairy products require carefully selected and properly maintained lactic cultures. Consequently, careful attention must be directed to the physical evaluation and the functional roles of lactic cultures.

Prerequisites for the satisfactory sensory evaluation of cultures involve an awareness of desired qualities in a culture, as it pertains to flavor, body, acidity, and general appearance, as well as an understanding of the techniques for culture evaluation.

Flavor Aspects of Cultures

Cultures for products such as buttermilk, sour cream, creamed cottage cheese, and cultured cream butter should impart a pleasing "bouquet flavor" which results from the overall blend of a delicate, diacetyl (buttery-like) odor and a distinctly clean, acid taste (Bodyfelt 1981a; Connolly et al. 1984). Any suggestion of an atypical or foreign odor and/or taste should not be tolerated. Once the aroma and taste characteristics of a good culture or cultured milk product are "fixed in the mind" of the evaluator, they are not easily forgotten. The evaluator should always be alert to the possible occurrence of one or more of several off-flavors associated with cultures used in the production of milk products, which are listed below.

Bitter. A "bitter" off-taste is generally detected near the end of the tasting period; this defect is often more pronounced after the sample has been expectorated. Bitter cultures may or may not have a normal amount of aroma; the aroma itself may suggest uncleanliness. Although certain strains of lactic cultures may naturally develop modest degrees of bitterness, contaminant microorganisms should be suspected in the instances of definite to pronounced bitterness.

Cheesy. A "cheesy" off-flavor is quite uncommon and is more often associated with products which have been stored for some time. The flavor characteristics of cheesy are: (1) a lack of typical culture flavor; (2) a definite proteolytic flavor note; and sometimes (3) a slightly bitter aftertaste. Contamination of the lactic culture with a proteolytic microorganism is highly suspect in most instances of this defect.

Lacks Desired Aroma or Flavor (Coarse). Frequently, cultured products lack the delicate appeal, flavor balance, or bouquet sought in such foods. This defect is often associated with a high titratable acidity, as a result of overripening, and it may lack the appropriate volatile compounds which constitute a desirable aroma. Also, coarseness may simply be due to selection of inappropriate culture strain(s) or incorrect propagation methods.

Lacks Flavor or Aroma (Flat). The "lacks flavor" or "flat" defect is easily identified, since it is characterized by an absence of desired aroma, and often associated with a sharp (high) acid taste in lieu of odor. Samples that are considered flat have little or no flavor appeal. Usually, the lack of flavor intensity is apparent upon noting a distinct absence of characteristic aroma within the product container and/or soon after the culture is introduced to the mouth. This defect could be due to the wrong selection of strain(s) and/or improper culture handling or milk base preparation. Also, it is possible that various microbial inhibitors could limit or prevent propagation of the aroma-producing strains within the culture; i.e., bacteriophages, antibiotics, or quaternary ammonium compounds.

"Green" Flavor (Acetaldehyde). This characteristic off-flavor, due to formation of relatively high concentrations of acetaldehyde by the culture, may remind one of the flavor of "green apples" or "plain" yogurt. To some tasters, this off-flavor may suggest high acid or a somewhat astringent character. This aromatic defect is readily noted upon either "whiffing" or tasting the sample. Causes of the green off-flavor may simply be selection of the wrong culture, incubation at incorrect temperatures, and/or overincubation. There is a tendency for *Streptococcus lactis subsp. diacetylactis* strains to dominate other lactic strains in mixed culture and produce excess amounts of acetaldehyde.

High Acid. A sharp, "high acid" taste in cultured products is most common. A high lactic acid level is readily detected by the distinct tingle or slight painful sensation on the tongue when the product is first placed into the mouth. This off-flavor also may be accompanied by either a decided lack of pleasing aroma or occasionally an excess of aroma (overripening). Control factors for this defect include selection of the proper culture(s) for aroma production, addition of citrate to milk (precursor for diacetyl), use of an appropriate inoculation rate, and closer control of incubation temperatures and times.

Metallic/Oxidized. Fortunately, the "metallic" or oxidized off-flavor in cultured foods occurs infrequently. The first impression may be that the product flavor is flat. However, as the sample is held in the mouth, a sort of puckery, astringent, papery, or cardboard-like flavor defect may appear. If the metallic (oxidized) defect is intense, this off-flavor sensation tends to remain after the sample has been expectorated. Use of poor-quality equipment for handling milk for culture production is a common cause of this serious defect. Certain non-stainless steel metals used for handling or storing milk, or the presence of rust or copper in equipment or water supplies, can catalyze changes in milkfat that provoke auto-oxidation of lipids.

Yeasty. A "yeasty" off-flavor is usually characterized by an aroma somewhat suggestive of alcohol or acetic acid, or an earthy aroma; it may be associated with a quick, sharp-acid taste. This flavor defect is generally caused by unwanted yeast contamination through faulty equipment sanitation.

Miscellaneous Off-flavors. Various flavor defects other than those just described may be infrequently detected in cultured products. Possible off-flavors such as cooked, feed, foreign, rancid, and unclean may be traceable to the milk from which the product was made and may not be completely obscured by the flavors resulting from the growth of the culture organisms. Contamination by undesirable microorganisms may also result in the production of several varied off-flavors (e.g., sauerkraut-like, cheesy, rancid, unclean) which may not be specifically identified, but may be recognized as uncharacteristic of a good cultured product.

Body of Cultures

Before being shaken, the body of a good, properly cultured product should appear firm or solid and generally be uniform in appearance. It should only show a few beads of whey exuded from the surface. The culture should break away cleanly from the side of the container upon tilting, and reveal an intact, "liver-like" body that resembles a soft custard. However, the body should not be so firm that it fails to break down to a cream-like consistency upon agitation. The shaken or agitated product should remain somewhat mounded when a teaspoonful is poured onto a smooth surface. The mixed sample should appear smooth, somewhat resembling rich cream; no curd particles or lumps should appear when it is spread in a thin layer on a glass surface or diluted with water. Some of the more common body defects of cultured milk products are described in the following paragraphs.

Curdy. A "curdy" body tends to lack uniformity, smoothness, or homogeneity. The curd particles may be sufficiently large to be readily observed upon pouring, or so small in size that close examination is necessary to see the "feathery curds." Agitation or movement during the coagulum-forming process may be responsible; sometimes culture contaminants may cause curdiness.

Gassy. A "gassy" product is denoted by excessive gas bubbles (CO_2), or by streaks in the coagulum due to the rise of gas bubbles to the surface. If accompanied by whey separation, a gassy sample will likely whey-off at the bottom or at the center of the container. Certain strains of lactic cultures have a tendency to produce excessive amounts of CO_2; the use of lower incubation temperatures or reduced incubation times may prevent or minimize gassy cultures (or products). It should be remembered that a modest degree of carbonation is generally desirable in cultured buttermilk.

Lumpy. A "lumpy" body is often an aggravated case of curdy consistency; the particle size is larger in the lumpy defect. Sometimes, the overall body of the product may be smooth, but there may be lumps of firm curd interspersed throughout the container contents. The causes of lumpiness in lactic cultures are the same as those listed for the curdy defect (discussed above).

Ropy. A so-called "ropy" culture tends to stretch or "string-out" when poured. Sometimes the defect is so pronounced that the product "strings-out" like a thin syrup or mucous substance. The defect is generally due to a capsular layer of polysaccharides (slime formation) around the cells of certain bacterial strains. In some cultured products, a slightly ropy body may actually be considered desirable (e.g., certain yogurts).

Thin (Weak) Body. A "thin" or "weak" body may be observed by tilting the unagitated sample to an angle of $45°$, whereupon the culture will often break and flow without sufficient viscosity. The agitated curd of such a culture will seem to pour too readily; the culture almost seems watery. This defect is often accompanied by low culture acidity ($\leq 0.75\%$ titratable acidity (T.A.)). Frequent causes of weak body are inadequate heat treatment of the product base, low solids, and/or inadequate culture activity.

Too Firm (Excess Viscosity). When a culture seems too viscous or heavy, it resists pouring and flow even after agitating the container contents. "Excessive viscosity" may be due to the extent of proteolysis or lactic acid formation by the culture and/or too high a milk solids level in the product base.

Wheyed-Off (Syneresis). This defect is manifest by a shrunken curd or coagulum and the presence of liberated or "free whey" in areas around the side and on the surface of the container. There are a number of causes (e.g., inadequate heat treatment of milk/cream, low viscosity, and the strain of culture used), but movement or agitation during incubation is probably the most frequent cause. Production of excessive acid can also cause "syneresis" (free whey).

Acidity of Cultures

The inclusion of acidity on the cultured product score card as a separate quality criterion, as well as a component of flavor, seems somewhat redundant. For example, if a product was definitely "high acid," it would likely be criticized for having a sharp, acid flavor and be "scored down" accordingly; in addition, the culture could also be penalized for high T.A. or low pH. Despite this situation, the inclusion of acidity as an item on the score card is advised, since an appropriate acidity level (or pH) is important in developing the desirable aroma and flavor balance in cultured milk products.

Although the T.A. of many cultured milk products generally falls between 0.70 and 0.85% as lactic acid (pH 4.3–4.6), the final acidity reading is dependent upon several factors. Titratable acidity is also a function of the MSNF content, but pH is not markedly affected by MSNF level. There is not a given acidity which is always the most desirable; some cultured products must be ripened more than others to develop the desired sensory properties. For a given product, a moderate variation in acidity (or pH) from day to day is tolerated. A close relationship exists between developed acidity and desired intensity of acid flavor. Generally, if a product is underripened, it will tend to lack flavor; if overripened, the product's flavor usually will not be delicate or properly balanced; instead, it will tend to be coarse or harsh. The flavor defects of lactic cultures and cultured milk foods that can be related to the amount of lactic acid are listed in the following paragraphs.

High Acidity. This defect is quite common in lactic cultures. "High acidity" is readily recognized by a sharp, acid taste sensation on the tip and sides of the tongue; it can be confirmed by a relatively high titration value (in excess of 0.90% as lactic acid), or by pH readings below the norm for the given product. Control of inoculation rates, the incubation time and temperature, and/or the rate of product cooling (following incubation) are critical for minimizing problems of high acidity in cultured products. Personnel should be aware that there are differences in the activity of various lactic cultures, hence, inoculation rates and incubation temperatures and periods may require adjustments to control the developed levels of acidity.

Low Acidity. This defect is less common than high acidity. It is commonly exhibited by a titratable acidity (≤0.70%) below the acidity level that is necessary for good flavor and aroma development. Low-acid products usually have a flat or insipid flavor. The "low acidity" defect is possibly due to insufficient incubation time, low propagation temperature, low inoculation rate, slow growing cultures, or the presence of microbial inhibitors.

CULTURED MILK PRODUCTS

A cultured product of high quality tends to have a glossy to a semi-glossy, velvety, and uniform (homogenous) appearance. In reflected light, the surface of the product should appear to have a definite sheen. Samples that exhibit a chalky, dull, flat, lifeless, or watery appearance are generally undesirable. Cultured products made from whole milk will usually show a slightly more yellow or cream color than those produced from skim milk.

Techniques for Evaluating Cultured Products

The evaluation of cultured milk products should be undertaken in a systematic manner. The first step of product evaluation should be examination of the unshaken or unagitated product. The evaluator should notice the relative smoothness and solidity of the coagulum and its freedom from streaks, gas bubbles, and whey separation. After the aforementioned characteristics are observed, the judge should gently tilt the sample and note whether the product remains intact or breaks easily into smaller segments (the more desirable response).

After careful examination of the physical state of the coagulum, the evaluator should swirl, shake, or agitate the product vigorously to render it "smooth and creamy-like." Next, some of the sample should be poured into a glass beaker or other receptacle for tasting and observation; the judge should especially note how the sample pours. The evaluator should smell the poured sample at once (before the delicate aroma escapes); it is advisable to rotate the contents of the beaker again and note the odor characteristics. A delicate, pleasant, buttery aroma (due to diacetyl and certain volatile fatty acids) should be perceived in high-quality buttermilk, sour cream, cultured half-and-half, cultured butter, and cultured dressing for cottage cheese.

After carefully noting the aroma of the product, the judge should place the approximate equivalent of a teaspoonful of the product into the mouth for tasting. The first flavor impression noted when the sample is brought into contact with the sensory organs is important. The evaluator should note how long the sensory impression remains; i.e., whether it seems to persist throughout the overall time period of flavor perception. Finally, the judge should expectorate the sample, and note any possible aftertaste. Following a brief rest of approximately 30 seconds, the judge should retaste the sample. The subsequent tasting attempt is not an effort to reproduce the initial taste sensation, but rather it is an exercise to confirm whether the second taste impression agrees with or confirms the first perception of the sample. Astute judges should try to concentrate their thought processes and compare the various sensory qualities observed in the sample(s) against the sensory properties of a high-quality sample (used as a reference) and/or against a mental image of the so-called "ideal product."

Some Precautions in Evaluating Cultured Products

To facilitate the evaluation of cultured products with some degree of procedural uniformity, certain precautions need to be observed. For reliable evaluation, temperatures of samples presented for flavor assessment should be reasonably uniform from day to day. A designated refrigerated temperature (7.2°C (45°F)) should be selected as the "routine temperature" for product evaluation. Most consumers will consume or utilize the product within the 4.4°–10°C (40°–50°F) range. When a sample is poured into a container, it should be evaluated immediately, since product contact with air (oxygen) for only a brief period of time may alter the perceived flavor. To enhance an evaluator's sensitivity for recognition of varied characteristics or qualities of lactic cultures, it is considered helpful to compare several different strains of lactic culture that may be available or in day to day use by the processor.

Definitions of Cultured Products

The definitions, composition, optional ingredients, and other requirements, including nomenclature, of acidified (acidity developed by addition of approved, food grade acids) and cultured milks and cream are addressed in the Code of Federal Regulations, Title 21, Part 131. (1987). A listing of the paragraphs pertaining to specific products is provided below:

Product	CFR, Title 21, 1987 paragraph:
Acidified skim milk	131.144
Cultured skim milk	131.146
Acidified lowfat milk	131.136
Cultured lowfat milk	131.138
Acidified milk	131.111

Product	CFR, Title 21, 1987 paragraph:
Cultured milk	131.112
Acidified sour half-and-half	131.187
Sour half-and-half	131.185
Acidified sour cream	131.162
Sour cream	131.160
Nonfat yogurt	131.206
Lowfat yogurt	131.203
Yogurt	131.200
Lowfat cottage cheese	133.131
Cottage cheese	133.128
Dry curd cottage cheese	133.129

Readers should consult this source for the legal background on product characteristics and nomenclature for this class of products. For example, the definitions spell out when a product can be designated as "cultured buttermilk" or "cultured nonfat buttermilk"; or when the product should be labeled "kefir cultured milk" or "acidophilus cultured milk."

Since the sensory properties within certain groups of cultured products are quite similar, it would be redundant to discuss each of the defined products separately. The acidified products have been introduced largely to provide an alternative to culturing and to emulate the cultured products. Thus acidified products are generally evaluated for the same sensory properties as their cultured counterparts. Also, some of the nomenclature is relatively new and in some cases represents a departure from the traditional (familiar) designations. For these reasons, the sensory properties will be discussed under familiar headings and in a "generic" manner as they apply to logical groupings of cultured products.

CULTURED BUTTERMILK

In the U.S., cultured buttermilk replaced naturally soured buttermilk several decades ago. The latter product form, which was obtained as a by-product of butter churning, generally lacked uniformity in acidity, appearance, consistency, and flavor. This by-product of butter manufacture was also most susceptible to development of the generic oxidized off-flavor, due to partitioning of phospholipids (and their associated unsaturated fatty acids) into buttermilk after the fat emulsion was broken by the churning process.

Cultured buttermilk is the product resulting from the growth of especially selected culture(s) (*Streptococcus cremoris, S. lactis,* and/or *S. lactis* subsp. *diacetylactis* with *Leuconostoc* sp.) in heat-treated skim milk or lowfat milk or whole milk. Bulgarian-style buttermilk is usually made from whole milk and generally employs *Lactobacilli* sp. and/ or *Streptococcus thermophilus,* in addition to a mesophilic lactic culture. Once the desired ripeness or acidity level of the buttermilk is reached, the coagulum is broken, cooled, and packaged. In many respects, the sensory qualities of commercial buttermilk should be comparable to those of lactic cultures (which have been previously described). However, various trade demands and regional preferences by consumers or institutional users have resulted in the establishment of certain desired or specified properties for cultured buttermilk, which are somewhat different or unique in certain locales or marketing areas.

The importance of evaluating the aroma, taste, and body profile of cultured buttermilk as a routine production procedure cannot be overemphasized. The sensory qualities of cultured buttermilk are dependent upon so many factors that the flavor attributes must not be taken for granted. Interestingly, the quality attributes noted in cultured buttermilk from a given processor (or brand) often become a customer's criterion for the "quality image" developed for other dairy products manufactured and distributed by the dairy processor in question (Keenan *et al.* 1968; Lindsay 1965).

Flavor

In general, the same delicate, pleasant, diacetyl (buttery) and volatile acid flavor notes sought in a lactic culture are the flavor characteristics sought in a cultured buttermilk of high quality. This particular flavor may be enhanced or enriched by the presence of milkfat, added as a small amount of cream or supplied by the lowfat, whole, or partially-skimmed milk from which it was made. High acidity and a lack of (full) culture aroma seem to be the two most common flavor defects of cultured buttermilk. Other possible off-flavor defects of cultured buttermilk include: astringent (chalky), coarse (harsh), cooked (heated), foreign, excess diacetyl, fruity/fermented, green (acetaldehyde), lacks acidity, lacks freshness (stale), oxidized/metallic, rancid, sauerkraut-like, stabilizer, unclean, and yeasty.

Body and Texture

Due to regional market expectations, consumer demand varies as to the preferred body characteristics for cultured buttermilk. Some buttermilk customers definitely prefer a heavy, viscous body; others insist upon a decidedly thinner body. Consequently, no single or definite standard for the body (viscosity level) of cultured buttermilk has been established that would be aggreeable to all product users. However, a medium-bodied, semiviscous cultured buttermilk, that pours similarly to a thin gravy, seems to be a product style most frequently in demand by consumers and food service users.

Although the body and texture defects found in commercial cultured buttermilks are quite similar to those of lactic cultures, some of the quality shortcomings require further explanation in light of current efforts to manufacture products of high quality on a uniform basis. Some of the more pronounced body and texture defects of cultured buttermilk are as listed in the following paragraphs.

Curdy Texture. A "curdy-textured" buttermilk appears to have a rough, coarse, nonhomogenous body; it is particularly noticeable as the buttermilk recedes from the glass surface or is poured slowly over the lip of the container. The individual curd particles may be quite small, sometimes as small as a pinhead and some particles may be as large as a wheat grain. By diluting a small sample of the buttermilk in a transparent glass container with a high proportion of cold water, the individual curd particles will settle to the bottom where they may be readily observed upon decanting the liquid. Curdy buttermilk is often associated with a thin, weak body and/or wheying-off (syneresis). Causes of curdy-textured buttermilk may be: (1) low milk solids in the product base; (2) movement or agitation of the coagulum during the incubation period; or (3) selection of the wrong culture for producing buttermilk.

Heavy Body. Generally, a "heavy-bodied" or high-viscosity cultured buttermilk is not preferred by product customers or food service users. Such buttermilk not only pours from the container with extreme difficulty, but also may be somewhat difficult to drink. Sometimes, an overly viscous condition is so pronounced that the product appears and behaves like whipped cream. This defect may be associated with: (1) certain lactic cultures that impart heavy body; (2) too high a level of milk solids in the buttermilk base, (3) excessive heat treatment of the product base; (4) entrapped air in the buttermilk; and/or (5) excessive product stabilization.

Thin Body. A "thin-bodied," low-viscosity buttermilk is easily recognized when poured. It breaks, drips, and splashes similar to water. Frequently, products displaying this body defect also exhibit a dull, lusterless appearance. Furthermore, this body defect may be associated with a flat taste and lack of aroma; or with a weak body that is quite subject to wheying-off. Possible causes of a thin body in cultured buttermilk include: (1) low milk solids levels; (2) insufficient heat-treatment (or contrastingly, too severe heat-treatment) of the product base; (3) culture with impaired activity; or (4) low proteolytic activity of certain starter culture strains.

Wheyed-Off (Syneresis). This troublesome defect manifests itself by the presence of free or liberated whey, usually at or near the surface. Sometimes syneresis may be found at the bottom or middle of the container. The location of syneresis within a product is expected to be a function of the density differential between the whey and the coagulum. Should "wheying-off" occur other than at the surface of buttermilk, the defect could be associated with the presence of entrapped gas as the result of an abnormal fermentation. A culture contaminant might be suspected. Wheying-off at the mid-level or bottom of whole milk buttermilks may be due to the presence of trapped milkfat within the curd, which markedly increases curd buoyancy. In this instance, the wheying-off may or may not be associated with the presence of gas. Occasionally, whey separation may occur as the result of insufficient acid development before the coagulum was broken and cooled. One of the best deterrents to syneresis in cultured milk products is proper heat treatment of the milk base.

General Appearance and Color

The general appearance and color of cultured buttermilk should be pleasing, attractive, and uniform, whether the product is made from skim milk or from whole milk. The desired color is a luster-white or a white tinged with yellow, due to the carotene pigment of milkfat. Pronounced or off-shaded colors are discriminated against in most markets. By placing a glass of an "off-color" buttermilk alongside a glass of whole milk, the color may be compared and judged more reliably.

Acidity

The perceived and/or measured acidity of cultured buttermilk often shows a wide range. "High acidity" may result from: (1) a failure to cool the vat of buttermilk promptly when it is sufficiently ripened; (2) too slow or inadequate cooling; and (3) extended storage or delayed distribution, especially if temperatures exceed 13°C (55°F). Sometimes, buttermilk of low acidity may be partially accounted for by dilution of the ripened product with either sweet skim milk or whole milk, which may be added to reduce viscosity or adjust milkfat content. Generally, a low acid buttermilk may have less than 0.70% T.A. Cultured buttermilk that has a T.A. above 0.90% is often criticized for a "high acid" taste. Generally, it is considered more serious if the product lacks the desired culture aroma (i.e., diacetyl and other volatiles), than if the acidity level is slightly too high or too low.

In some markets, butter granules, flakes, or specks may be incorporated into cultured buttermilk. This is done primarily to emulate the appearance of "escaped" butter granules, as was the case for the original buttermilks which were drawn directly from churns. These "butter particles" (if added to cultured buttermilks) should be relatively uniform in size and have a pleasant, light yellow-orange color. Unfortunately, the addition of certain simulated "butter particles" to cultured milk sometimes serves as a source of unwanted microbial contamination (e.g., coliform bacteria); this often leads to the development of unpleasant, unclean off-flavors.

Container and Closure

If the container and closure for cultured buttermilk are evaluated, the procedure is conducted exactly as is done for commercial fluid milk. The defects encountered are generally similar for both product types. The defect "short-" or "slack-fill," seems to be encountered more frequently with cultured buttermilk than with fluid milk products; this is probably due to the heavier viscosity and/or escape of CO_2 from the fermented milk.

Sensory Evaluation Procedures
for Cultured Buttermilk

For training purposes, it is advisable to have several samples of buttermilk available for comparative evaluation. Samples may be obtained by setting aside several packages from successive batches of buttermilk, thereby providing a number of samples for sensory evaluation, representative of various stages of freshness and maturity. Additionally, products of competitive processors should be purchased from retail stores. To mitigate against bias, product evaluation should proceed (if possible) without brand identification being known to the judge(s). An adequate portion of the sample (at least 250 ml) should be transferred to another container (equipped with an adequate closure to retain volatiles). Containers must be properly coded for subsequent identification. The coding of containers and product transference should definitely be done by someone other than the person(s) performing the sensory evaluation. By obscuring the brand identity of all samples, a more objective sensory analysis of buttermilk samples can be undertaken.

FLAKE OR GRANULATED BUTTERMILK

A style of cultured buttermilk that incorporates butter granules, flakes, or specks is in demand in some localities. The "butter particles" may be present as a result of: (1) the addition of melted butter; (2) the addition of separately-churned butter granules; (3) churning all or a portion of the creamed buttermilk; (4) circulating the entire lot of creamed buttermilk through a specially-designed pump; or (5) the addition of commercial preparations that impart the appearance of butter granules. This product has some characteristics that differ somewhat from those of conventional cultured buttermilk (described earlier).

Flavor

The flavor of flake (granule) buttermilk should be clean, rich, and have a distinct diacetyl aroma. Provided that the small quantities of added butter or cream are of high quality and the equipment and manufacturing process is sanitary, the presence of butter granules should help enhance the buttery flavor of the product. Many of the same flavor defects of regular cultured buttermilk are also found in flake buttermilk, along with certain flavor defects due to its unique composition, which are listed in the following paragraphs.

Lack of Buttery Flavor. Flake or granulated buttermilk is discriminated against if it does not have at least some trace of a buttery flavor (diacetyl aroma). The presence of the granules should suggest a high, full buttery flavor. If this flavor is not present in sufficient intensity, the product is considered to be flat or lacking flavor.

Poor Blend of Flavor. This defect occurs when the serum (buttermilk base) appears to yield one characteristic flavor—and the butter granules impart another flavor. This defect may be associated with larger-than-desired-size butter granules, or the source of the butter flakes having imparted an off-flavor.

Body and Texture

Typically, the body of flake buttermilk should be slightly heavier or more viscous than that of regular cultured buttermilk. The viscosity should be sufficient to stabilize or suspend the butter granules evenly throughout the product. Two defects are sometimes noted in the body and texture of flake buttermilk, and are listed below.

Butter Granule Plug. If the body of flake buttermilk is weak and the granules are relatively large and sticky, the combination may result in the formation of a butter granule plug (large mass). This is generally objectionable, due to loss of product functionality and an undesirable appearance.

Uneven Body. When the upper portion of product in the container exhibits extensive viscosity, and the lower portion definitely seems to lack viscosity, the defect is referred to as "uneven body." This defect usually stems from an uneven distribution of butter granules or flakes within the product, or improper culture propagation procedures.

General Appearance and Color

Two quality features are desired in the general appearance and color of flake buttermilk, namely, uniform distribution of butter granules and a pleasant, golden-yellow color of the granules against a white background. Defects encountered in the general appearance of flake buttermilk are listed in the following paragraphs.

Lack of Color Contrast. Sometimes the granules do not contain sufficient color. When the product is observed from a short distance (approximately 0.5 m (approx. 20 in.)), the flake buttermilk may merely resemble plain cultured milk. While this appearance defect may not seem as serious as unnaturally- or highly-colored granules, a bleached color certainly merits criticism. Butter granules should exhibit an appealing, medium intensity creamy-yellow color. To make this color as uniform as possible, the fat within the added granules (flakes) should be color-adjusted with annatto coloring (as necessary in those seasons of the year when the cream-yellow shade does not occur naturally).

Lack of Uniformity. Occasionally, highly-colored butter granules may not be evenly dispersed, or may be so uneven in size that the buttermilk appears spotty or lacks a pleasing uniformity of appearance.

Unnatural Color of Granules. The medium-yellow color of the butter granules should contrast with the white background of the buttermilk. To accomplish this effect, the added granules may be more intensely colored than typical butter. Sometimes the added coloring may not be thoroughly blended with the butter granules, and thus cause a reddish-yellow appearance of the granules, which is generally considered objectionable.

OTHER CULTURED MILK PRODUCTS

In addition to the cultured milk products previously discussed, other fermented milk foods that have market significance will be considered. Bulgarian buttermilk, sweet acidophilus milk, kefir, kumiss, taete, and ymer have sensory characteristics which are quite specific for each product (Foster *et al.* 1957; Kemp 1984; Kosikowski 1977; Robinson 1985). Although all of these cultured milks do not currently have commercial significance, they each have certain expected sensory features.

Bulgarian Buttermilk

This product is the result of inoculating a culture of *L. bulgaricus,* sometimes in association with *Streptococcus thermophilus* and mesophilic lactics, into highly-heated or sterile lowfat or whole milk and incubating the product base at a relatively high temperature (32.2°C to 43.3°C (90°F–110°F)). The final product is comparatively high in lactic acid, ranging from 1.5 to 3.0%. Typically, the body of Bulgarian buttermilk is heavy, thick, viscous, and may be somewhat gelatinous in character, accompanied by a glossy, velvety appearance. Bulgarian buttermilk commonly has a distinctly sharp, puckery, acid taste (resembling unsweetened rhubarb juice), or a "green" or plain yogurt-like flavor. The aroma is sharp, acidy, and somewhat "green"; it does not resemble the more delicate, pleasing aroma which is typical of regular cultured buttermilk. Bulgarian buttermilk may be subject to certain flavor defects due to the higher heat-treatment of the product base (cooked or heated) and the higher growth temperature of the culture microorganisms. Too high acid and/or a bitter taste may occasionally occur; this may give the product an unpleasant flavor and a distinctive puckery (mouthfeel) sensation.

Some *L. bulgaricus* cultures have a tendency to develop a ropy or stringy body. So-called "ropy" cultures, due to their ability to form a polysaccharide layer (or mucous capsule) around the cells, usually develop a characteristic and desirable ropy appearance of Bulgarian buttermilk.

Acidophilus Milk

High temperature pasteurized or sterilized whole milk (or lowfat milk or skim milk) that has been fermented by *Lactobacillus acidophilus* is retailed under the product name, "acidophilus milk." The optimum growth temperature for *L. acidophilus* is between 35°C and 37.8°C (95°F–100°F). This organism grows slowly by comparison to other lactics, even at the optimum temperature. This may permit any viable contaminant organisms, or spores in the product milk base to outgrow the culture, and result in an off-flavor. Due to the required high heat-treatment ($\geq 100°C$, ≥ 10 psi steam pressure and ≥ 10 min) for the milk base, acidophilus milk usually has a distinctive, light brown, caramel-like color. This is not ordinarily considered objectionable, since it merely indicates proper preparation of the product base for producing regular acidolphilus milk.

The flavor of regular acidophilus milk is quite distinctive, if not outright objectionable to a majority of people. It is a blend of a caramel-like flavor (which results from high temperature heat-treatment), plus a sharp, harsh-acid taste that results from the development of 1.2–1.6% lactic acid and certain other volatile compounds.

The body of regular acidolphilus milk should be similar to that of cultured buttermilk. Wheying-off (whey separation) is a common body defect encountered in this product.

"Sweet acidolphilus milk" is a more recently reintroduced product (1972) that has essentially displaced the harsh-flavored acidolphilus milk in the U.S. marketplace. Cell concentrates of *L. acidophilus* are added to lowfat milk to provide minimum viable cell counts of 1–2 million/ml. This product does not perceptibly differ in flavor from lowfat milk, yet contains significant numbers of this lactic microorganism, which has purported health and/or therapeutic benefits for both humans and livestock production (Duggan *et al.* 1959; Sandine *et al.* 1971; Speck 1972, 1976; Gilliland 1979, 1986; Klaenhammer 1982).

CULTURED SOUR CREAM
AND HALF-AND-HALF

Cultured sour cream is manufactured with a process similar to cultured buttermilk, hence, production problems and sensory defects are similar for each of these products. Cultured sour cream is generally a relatively heavy, viscous product with a glossy sheen that should possess a delicate, lactic acid taste and a balanced, pleasant, buttery-like (diacetyl) aroma (Bodyfelt 1981a; Connolly *et al.* 1984). Common applications of cultured sour cream include use as a garnish, for dressings, dips, toppings, in ethnic food dishes, and as a cooking and baking adjunct for flavor enhancement.

Typical manufacture of cultured sour cream includes standardization of a cream and milk base to 18–20% milkfat and 25–28% total milk solids content. The cream base is either vat pasteurized at approximately 82°C (180°F) for 30 minutes or 85°C (185°F) for 25–60 seconds by the HTST method. Next, the product base is homogenized (single-stage only) at 2000–3000 psi at temperatures ranging from 40°C–85°C (102°F–185°F). The sour cream base is then cooled to approximately 21°C–30°C (70°F–86°F) and inoculated with 0.5–2.0% of an aroma-producing lactic culture (*S. cremoris* and/or *S. lactis* with *Leuconostoc* sp.). The incubation is generally conducted at 21°C–30°C (70°F–86°F) until the titratable acidity reaches 0.70% or slightly higher, before cooling to 15°C (60°F) to terminate the fermentation process.

Inasmuch as a heavy, viscous body is desired for cultured sour cream, agitation of the ripened product should be minimized in spite of the need to rapidly cool it before packaging. The most successful approach for initiating cooling and achieving simultaneous gentle agitation is attained by introduction of ice water or coolant into the jacket of the vat or processor. The agitator should be rotated for only several revolutions, at intermittant intervals, throughout the time required to cool the product below 18°C (65°F). Obviously, it is advantageous to use the lowest possible incubation temperature (for maintaining lactic culture activity) to accommodate a more rapid and efficient cooling of the sour cream and simultaneously optimize product viscosity.

Sensory Properties

Cultured sour cream of high quality possesses a distinct, clean lactic acid taste, while simultaneously exhibiting a well-balanced, delicate aroma of diacetyl and other volatile compounds. The proper balance of acidic taste and diacetyl aroma makes cultured sour cream of high quality a pleasing accompaniment to many other foods.

Generally, the body and texture of cultured sour cream should be reasonably firm, smooth, homogenous, and free of whey separation (in an undisturbed container). The product should have a glossy appearance (sheen). Excessive addition of stabilizers and/or emulsifiers may impart excessively heavy body characteristics and thus detract from the delicate, desirable flavor of cultured sour cream. Frequently, a masking affect or an interference with "flavor release" occurs when combined stabilizer/emulsifier levels exceed a product usage level of 0.30–0.40% (w/w).

Sensory Defects. The categories of sensory defects for cultured sour cream and half-and-half include flavor, body and texture, and color/appearance. The cultured dairy product score card illustrated in Fig. 7.1 is sufficiently versatile, in the opinion of the authors, to be used for evaluating cultured sour creams, as well as for cultured buttermilk and kefir. Only minor modifications or adjustments for terminology of certain defects may be necessary when this score card is used for the various products just mentioned.

Flavor. The same possible off-flavors described earlier for buttermilk may also occur in cultured sour cream (or cultured half-and-half). Due to the substantially higher milkfat level in cultured creams, oxidized, metallic, and/or rancid defects might be expected to occur at a higher frequency than for buttermilk. Also, stabilizer- or emulsifier-induced off-flavors (foreign) may occur in sour cream with some frequency. The stale, oily, cardboardy, or oxidized-like aftertaste of certain stabilizer and emulsifier ingredients used in sour cream manufacture are often associated with a firm, stiff, gel-like, or overly viscous body, in the experience of the authors.

Body and Texture. Attaining the ideal body and texture in sour cream is frequently difficult, due to the wide range of firmness and viscosity demanded by various customers. Some food service establishments and segments of the fast-food industry, (e.g., baked potato shops and Mexican food restaurants), frequently demand stiff, heavy-bodied sour cream for their use. Meanwhile, other restaurants, food service, and household users may prefer a less viscous product. Interestingly, the less firm-bodied products tend to exhibit more cultured aroma and substantially higher "flavor release." Some dairy processors may find it necessary or advisable to make two distinct kinds of sour cream to satisfy both extremes of customer preference: (1) a firm-bodied product, and (2) a weaker-bodied product. However, the observation of the authors is that the more aromatic, more delightfully flavored sour creams (or cultured half-and-halfs) are generally not overstabilized. Apparently, when cultured sour cream is stabilized/emulsified heavily, there is a definite interference with perception of the desired, delicate flavor (especially the "buttery" aroma).

Some of the more common body and texture defects (see Fig. 7.2) of cultured sour cream (as well as cultured half-and-half) are listed in the following paragraphs.

Curdy. A "curdy" texture in sour cream detracts from the smooth, homogenous mouthfeel sought in high-quality sour cream. Possible causes of this defect may be: (1) unwanted agitation or movement of the weak coagulum in the late stages of incubation; (2) incomplete mixing of the culture inoculum; and/or (3) incomplete hydration of all of the dry ingredients added to the liquid components of the product base.

Gassy. Occasionally, small bubbles or pockets of carbon dioxide may form in sour cream; this defect is termed "gassy." This detracts from the desired smooth, homogenous consistency of quality sour cream. Depending on the source of the CO_2-producing organisms, associated off-flavors or alterations of the typical flavor may occur. Gassy (CO_2) lactic cultures may produce an objectionable effervescent (carbonation) effect in sour cream products. Undesirable microbial contaminants in sour cream may produce unclean, earthy, or yeasty off-flavors, in addition to gassiness.

Grainy. "Grainy" textured sour cream is usually best detected by mouthfeel. Mealiness or small, hard particles in sour cream are best detected by pressing the top of the tongue against the roof of the

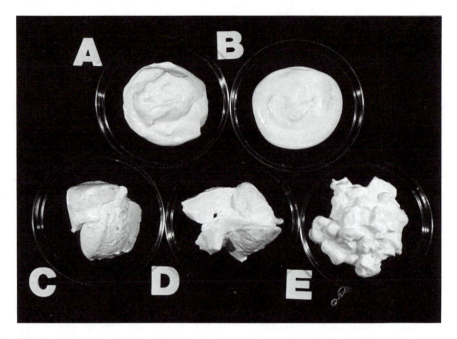

Fig. 7.2. Examples of some of the more common body defects of sour cream: A—An "ideal" consistency; B—Weak or too thin; C—Too firm or rigid; D—Gel like; E—Lumpy (nonhomogenous).

mouth and noting any possible grainy mouthfeel. On close visual examination of the product, the evaluator may also observe small but distinct particles. This defect is often due to incompletely dissolved dry ingredients within the product base (i.e., nonfat dry milk and/or stabilizers or emulsifiers).

Too Firm (Overstabilized). This body defect seems to occur all too often in sour-cream-type products. U.S. manufacturers seem to have a tendency to overstabilize the product base, rather than rely on processing techniques for developing controlled viscosity (e.g., variation of the homogenization and pasteurization parameters). As an illustration, some sour creams cannot be stirred with a brittle, light-weight plastic spoon, without the handle completely breaking off when it is pulled through the sample. Another practical way to determine excessive viscosity and/or gel-like properties in sour cream is to vertically insert a spoon or small spatula near the outer edge of the container (quart or smaller size). With the first attempt to laterally move the spoon or spatula, the entire contents (coagulum) of the container may "spin" or move as a solid object, rather than "giving way" when a stirring action is applied. Quite often, an overstabilized or too-firm sour cream appears dull; it seems to lack the desired gloss or sheen.

Too Thin (Weak). This product defect is the opposite of too firm; it lacks viscosity and the desired degree of moderate firmness for user functionality. The sour cream may be liquid-like or appear as if it would be difficult to spoon, apply to a baked potato, or would be inadequate served as a dip or salad dressing base. Modified processing techniques or incorporation of a modest amount ($\leq 0.35\%$) of stabilizer/emulsifier are two possible methods of correcting this defect.

Wheyed-Off (Syneresis). In the development of "syneresis," or "wheying-off," liquid whey separates from the coagulum. This tends to make the consumer or food service customer suspicious of the raw materials or overall product quality. An inability of the whey proteins to adequately bind water is the general cause of this defect. Utilization of higher pasteurization temperatures, a longer holding time, and/or more careful selection of stabilizing agents should correct most wheying-off problems.

Appearance/Color. High-quality sour cream generally possesses a pleasing, light-straw color, a distinct sheen (or gloss), is homogenous, uniform in appearance, and has an apparent moderate viscosity. Some of the more common appearance defects are listed below.

Dull. When sour cream lacks the desired gloss, sheen, or silky appearance, it may appear dull, nonhomogeneous, and unattractive. There may be an associated stiffness due to excessive use of stabilizer/emulsifier in the sour cream base.

Lacks Uniformity. A distinct lack of a homogenous appearance in sour cream is often referred to as a "lack of uniformity." Careless production methods may lead to this defect, especially improper conditions of blending ingredients and/or inadequate homogenization.

Unnatural Color. Seldom is sour cream considered to be unnatural in color. But occasionally, the cream source may be low in carotene content; consequently the sour cream may appear to be atypically light in color (snowy-white).

YOGURT

One of the oldest forms of fermented milk is yogurt, a cultured product which is known by at least thirteen different names throughout the world. Since at least 5000 B.C., yogurt has been an important food staple of people in the Middle East, especially in those countries bordering the east Mediterranean coast. Yogurt is a quickly-curdled, decidedly acid milk-based product, with little or no alcohol content. Yogurt results from the associative growth of *Lactobacillus bulgaricus* and *Streptococcus thermophilus* in warm milk (29°C–45°C (85°F–114°F)).

Typically, yogurt is characterized by a smooth, viscous gel, with a characteristic taste of sharp acid and a "green-like" or "green apple" flavor. Some yogurts exhibit a heavy consistency that closely resembles custard or milk pudding; by contrast, other yogurts are purposely soft-bodied and essentially "drinkable" (Bodyfelt 1976; Connolly *et al.* 1984; Duthie *et al.* 1977; Kosikowski 1977; Ryan *et al.* 1984).

Commercial production of yogurt increased most rapidly in Europe early in the twentieth century after Dr. Eli Metchnikoff published on his advocacy of regularly consuming cultured milks (especially yogurt) for the "prolongation of life." Apparently, the earliest successful introduction of commercial yogurt into the U.S. occurred in 1939 in New York City.

Traditionally, yogurt was made from either whole milk or skim milk that had been boiled for a considerable time to evaporate part of the water and thus concentrate the solids. Currently, a similar effect is achieved either by: (1) removing one-fourth to one-third of the water in a vacuum pan; (2) adding four to five percent nonfat dry milk to lowfat or whole milk; or (3) blending appropriate quantities of condensed milk with whole milk or skim milk (Kosikowski 1977; Tamime and Robinson 1985).

Yogurt Defined

The name "yogurt" is but one of many terms that peoples of the world have used to describe this particular form of cultured milk product. For example, in India the name is "dadhi"; in Egyptian or Arab cookbooks it is called "leben"; "madzoon" in Armenia; or "kesselo-mleko" in Bulgaria. The Turks and West Europeans have used the term "yoghurt" or "yogurt" for many decades.

The Code of Federal Regulations (1985) definition of yogurt is as follows:

§131.200 Yogurt.—

(a) *Description.*—Yogurt is the food produced by culturing one or more of the optional dairy ingredients specified in paragraph (c) of this section with a characterizing bacterial culture that contains the lactic acid-producing bacteria, *Lactobacillus bulgaricus* and *Streptococcus thermophilus*. One or more of the other optional ingredients specified in paragraph (d) of this section may also be added. When one or more of the ingredients specified in paragraph (d)(1) of this section are used, they shall be included in the culturing process. All ingredients used are safe and suitable. Yogurt, before the addition of bulky flavors, contains not less than 3.25% milkfat and not less than 8.25% milk solids not fat, and has a titratable acidity of not less than 0.9%, expressed as lactic acid. The food may be homogenized and shall be pasteurized or ultrapasteurized prior to the addition of the bacterial culture and bulky flavoring material. To extend the shelf-life of the food, yogurt may be heat treated after culturing is completed, to destroy viable microorganisms. . . .

(c) *Optional Dairy Ingredients.*—Cream, milk, partially skimmed milk, or skim milk, used alone or in combination.

(d) *Other Optional Ingredients.*—(1) Concentrated skim milk, nonfat dry milk, buttermilk, whey, lactose, lactalbumins, lactoglobulins, or whey modified by partial or complete removal of lactose and/or minerals, to increase the nonfat solids content of the food: *Provided,* That the ratio of protein to total nonfat solids of the food, and the protein efficiency ratio of all protein present shall not be decreased as a result of adding such ingredients. . . .

(2) Nutritive carbohydrate sweeteners. Sugar (sucrose), beet or cane; invert sugar (in paste or syrup form); brown sugar, refiner's syrup; molasses (other than blackstrap); high fructose corn syrup; fructose; fructose syrup; maltors, maltose syrup, dried maltose syrup; malt extract, dried malt extract; malt syrup, dried malt syrup; honey; maple sugar; or any of the sweeteners listed in Part 168 of this chapter, except table syrup.

(3) Flavoring ingredients.

(4) Color additives.

(5) Stabilizers.

Characteristics of Plain Yogurt

The lactic acid bacteria that constitute a yogurt culture are the thermophilic organisms *S. thermophilus* and *L. bulgaricus*. Cells of these two genera are generally present in approximately a 1:1 ratio in a typical yogurt culture. Relatively high populations of viable culture bacteria (unless heat-treated as a final processing step) are expected to remain in the product until it is consumed. Frequently, selected strains of *Lactobacillus acidolphilus* or *Bifidobacterium* sp. may be added to the yogurt culture for dietary adjunct purposes (Gilliland 1979, 1986; Tamime and Robinson 1985).

The typical acetaldehyde ("green" or "green-apple") flavor of "plain yogurt" is achieved through a symbiotic bacterial relationship influenced by such factors as: (1) temperature of incubation; (2) amount of inoculum; (3) period of incubation; (4) source of the culture (strains employed); (5) heat treatment of yogurt milk base; and (6) pH of the finished product. The flavor of plain (unflavored) yogurt is somewhat unique and unlike that encountered in any other type of fermented milk. Several volatile flavor components of plain "yogurt flavor" include: acetaldehyde, acetic acid, diacetyl, and several volatile fatty acids. Acetaldehyde is recognized as the most characteristic volatile; this 2-carbon aldehyde is noted for its "green" or "green-apple" character.

Styles of Yogurt

Sundae-style. The second most popular type of yogurt in the U.S. is the "sundae-style," wherein the fruit or other flavoring material is initially placed in the bottom of the container. This product is also referred to as fruit-on-the-bottom (F.O.B.) -style yogurt. The amount of fruit or flavoring varies from 1.25 to 2.0 oz (37–59 g) per 8 oz (0.23 kg) cup; proportionally lesser amounts of flavoring are added to 6 oz (0.17 kg) cups. Apparently, in an effort to attain greater eye appeal, yogurt processors on the West Coast developed a so-called "Western sundae-style" yogurt in the 1960s. This product form uses a colored, sweetened syrup within the upper yogurt portion of the container. This syrup may be flavored with extract and added to the yogurt base in an effort to attain enhanced flavor and improve the color and/or appearance of the yogurt.

Swiss-style. The most popular category of flavored yogurt in the U.S. is "Swiss-style" (also referred to as either "French-style," "preblended" or "prestirred" yogurt. In this product form, the flavoring and/or fruit is blended uniformly throughout the yogurt base. By definition, this category includes all nonfruit flavored yogurts, such as coffee and vanilla, as well as those yogurts that include preblended fruits (with or without added flavors). Uniform preblending of the flavoring is usually accomplished by: (1) comminuting the fruit or berries to the most desirable particle size, and (2) incorporation of selected stabilizing agents. Close acidity control of the fruit is helpful; to minimize effect on tactile quality, the pH of the flavoring should be maintained in the range of pH 3.5–4.5. The various categories of flavored yogurt are summarized in Table 7.2.

Other Yogurt Styles. Advanced processing and packaging equipment has prompted the introduction of variegated fruit yogurts. Another yogurt style has been facilitated by special cup filling, which facilitates a "sundae-on-top" impression for the product. Artificially-sweetened and noncaloric-sweetened fruits, in place of fruit preserves or other fruit products that are processed with sugar, offer lower calo-

Table 7.2.   Characteristics of the Various Styles of Flavored Yogurts in the U.S.

Yogurt Style (Type)	Characteristics
1. *Swiss-Style*[a] (French-, Prestirred, or Preblended)	Precultured yogurt base and fruit or berry flavoring (15–25%) blended prior to packaging
2. *Sundae-Style*[b] (Fruit-on-bottom)	Flavoring (15–25%) added to the container, yogurt base added to top of flavoring.
a. Eastern-type	No coloring agent, flavoring, or sweetener added to yogurt base (milk base is white).
b. Western-type	Coloring agent, flavor extract, or concentrate and/or sweetener added to yogurt base (milk base is color of given flavor).
c. Fruit-on-top	Yogurt cups filled in a manner so that flavoring material is on top portion of container.
3. *Extract Flavored* (or Concentrates)	Flavor extracts and/or concentrates are sole source of flavor, plus sweetener(s) (i.e., coffee, chocolate, lemon, etc.).
4. *Frozen Product Forms*	
a. Soft-serve	Served as cones, dish, or sundaes.
b. Hard frozen	Pint and quart size.
c. Novelties	On-a-stick, coated bars, "push-ups."
d. Yogurt pies	In "pie" crusts.
5. Miscellaneous types	A variant of the sundae-style Western type: A firm, flavored yogurt with additional flavoring cascading over its exterior when emptied up-side down.

[a] Commonly available in either 4, 6, or 8 oz individual portions, or quarts (bulk).
[b] Commonly available in either 4, 6, or 8 oz individual portions.

rie yogurt products for consumers. Other forms of yogurt include hard-frozen, soft-serve, various novelties, yogurt bars, and yogurt stick-confections. A recently introduced "premium" yogurt is sweetened with fruit juice concentrates only; no sugar or corn syrup sweeteners have been added to the fruit flavorings. Other recent introductions for yogurt-based products include the addition of cereal grains and liqueurs to the yogurt base for special flavor effects and variety. Soft-serve yogurt sundaes and "yogurt salads" represent additional marketing concepts for yogurt.

Plain Yogurt. The original yogurt marketed in the U.S. was the so-called "plain" or "natural-style." This style of yogurt has no added flavoring material or fruit; it is seldom, if ever, sweetened. Plain yogurt is characterized generally as having a tart or sour, somewhat green apple-like flavor, due to the levels of lactic acid (0.9–1.2%) and acetaldehyde (5–40 ppm) present. Acetaldehyde is readily produced by both *L. bulgaricus* and *S. thermophilus,* and is the chemical compound most responsible for providing characteristic "yogurt flavor." In flavored yogurts, the acetaldehyde flavor note appears subdued. One reason may be that added sugar, in the 4–16% concentration range, tends to limit production of acetaldehyde and lactic acid by yogurt microorganisms. However, it is primarily the masking effect of fruit, berries, and other flavorings that restricts the perception of acetaldehyde in flavored yogurts, according to Bills *et al.* (1972).

Some authorities describe the viscosity or body of the ideal yogurt as having the consistency of a light custard or a light sour cream. It is difficult to indicate the preferred degree of viscosity/firmness/softness for U.S. commercial yogurt in subjective terms, let alone in objective terms. Perhaps the most objective observation for yogurt body (consistency) is to denote the manner in which a moderately stirred (moderate stirring or agitation is achieved by five (5) complete "sweeps" (rotations) of an adequate sized spoon or spatula for the given carton size) yogurt sample "mounds" or "heaps up" in a spoon. The eye level, visual impression attained from a side profile of the yogurt on a spoon is a somewhat objective approach. When the contents of the spoon tend to spill, or flow from the lip of the spoon and/or simultaneously appear flat (level or concave) within the bowl of the spoon, then the yogurt body should be considered "weak" or "thin." By contrast, when a spoonful of the yogurt tends to markedly "mound up" (i.e., an obvious high, convex, rigid, structure, or exhibit a distinct or sharp profile when viewed at eye level), it is considered "too firm." The more ideal or preferred yogurt body is one that has a modest convex configuration (i.e., a moderate rounded, heaping effect) when a spoonful is viewed horizontally (from the side) at eye level. Figure 7.3 illustrates examples of the desirable, weak, too firm and gel-like defects of yogurt body.

Defects of Plain Yogurt

Defects in plain yogurt may be controlled by careful selection of milk components, stabilizers (if used), cultures, and correct formulation. Possible flavor defects include high or low acid, excessive or low acetaldehyde, bitter, cooked, and rancid. (Refer to pages 271–275 for further details.)

Color and appearance considerations for plain yogurt are rather simple and straightforward, compared to the complexities of flavored yogurt. Generally, the appearance of plain yogurt should convey a smooth, homogenous, moderately firm gel or custard-like body and texture and a uniform off-white color. The more common color and appearance defects of plain yogurt are reviewed here. (See also discussion beginning on page 265).

Frey Whey. There should be none or a minimum of visible whey sepa-

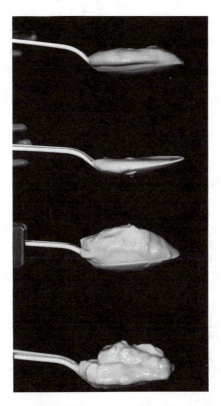

Fig. 7.3. Use of the side profile approach to illustrate the more common body defects of yogurt: A—An "ideal" body or consistency; B—Weak or too thin; C—Too firm or rigid; D—Gel-like.

ration (or "free whey") on the yogurt surface. Frequently, a watery exudate (whey separation) may appear at the outer edge of the yogurt container (i.e., at the interface between the yogurt and the container wall). Whey separation, unless excessive, is primarily a concern of aesthetics or eye appeal, and usually does not detract from the overall flavor characteristics of yogurt. The initial impact is that of markedly detracting from product appearance. Usually, any separated whey may be reincorporated easily into the yogurt through modest stirring.

Gel-like. This condition may be considered as both an appearance and a body and texture defect. The term "gel-like" is used to describe the appearance of excessive product firmness, or a severe gelatin (liver-like) consistency. This defect often stems from excessive use of stabilizing agents in yogurt. Yogurt that exhibits the gel-like defect tends to resist "breaking down"; frequently, the product appears rough and

nonhomogenous following a moderate amount of stirring. Furthermore, there often may be an associated slick, rigid, semisolid mouthfeel
when a sample portion of gel-like yogurt is placed into the mouth. This
defect occurs with much higher frequency in flavored yogurts than in
the plain style.

Shrunken. Occasionally in yogurt, the gel or coagulum tends to
shrink in size within the container (or pull away from the carton side
wall); this leaves the impression of reduced or "shrunken" contents.
Quite often, free whey will fill the void that results from this "shrinking" of the coagulum. Unfortunately, the combined defects of
shrunken and free whey in yogurt tend to reduce the impression of
product quality by the discriminating consumer.

Surface Growth. Probably the most serious defect of yogurt appearance is "surface growth." This defect consists of visible colonies of
yeast and/or mold growth on the top surface of the yogurt. In addition
to being aesthetically displeasing, such microbial outgrowth generally
suggests inadequate processing and packaging sanitation, or an overall lack of satisfactory housekeeping within the processing environment. Furthermore, an earthy or musty off-flavor may stem from this
unwanted microbial growth. Products exhibiting this defect should not
be salable and should not be consumed.

Body and texture defects that may be encountered with some frequency in plain yogurt include: graininess, ropiness, too firm, and
weak. They are discussed in the following paragraphs. (See discussion
beginning on page 268).

Grainy. In the instance of "graininess," the product lacks the desired smoothness and uniformity of appearance. Small particles of a
grit or grain size may actually be visible; graininess is quite often detectable by mouthfeel. Graininess probably arises from undissolved
milk or other food solids added to the product base. Complete hydration of all added solids, such as milk powder and stabilizers, must be
achieved in the blending process to prevent the grainy defect in yogurt.

Ropy. "Ropy" is an interesting body characteristic wherein the yogurt distinctly stretches or "strings out" when a spoon is placed into
the product and is slowly lifted about 5 to 8 cm (2–3 in.) above the top
surface. Instead of the yogurt cutting- or breaking-off distinctly, the
product manifests a marked stringiness similar to mucous-like materials. This characteristic (or defect in the view of many observers) is
usually due to the inclusion of certain strains of bacteria in the yogurt
culture that produce a polysaccharide capsular layer (slime-like)
around the cell. This property of certain yogurt cultures may serve to
minimize the need for stabilizers to achieve the desired consistency.

Occasionally, certain stabilizing and emulsifying agents may cause a slight ropy-like character in yogurt.

Too Firm. When the body of plain yogurt is considered "too firm," it conveys the impression of being too rigid or resistant to mastication when placed in the mouth. Also, a too firm body is often apparent by visually examining a side profile of a spoonful of product. Firm or rigid edges can be noted, rather than a more preferred "soft-rounding" impression of a spoonful of product.

Weak. A "weak" body defect is the exact opposite of too firm; the product consistency conveys the distinct impression that it would probably be easier to consume the product as a beverage than to "spoon" it. Viewed from side profile, the product may appear practically level in the spoon, or it may spill over the lip of the spoon. For a drinkable-style yogurt, a weak body is a prerequisite.

Procedures for Yogurt Evaluation

Since yogurt and flavored yogurt are comparatively new products within the U.S. dairy foods industry (relative to other manufactured dairy products), product quality standards for sensory evaluation are not yet as uniform, or widely accepted, as for butter, cheese, cottage cheese, ice cream or milk, and cream. Flavored yogurt (strawberry Swiss-style) was evaluated for the first time in the 1977 National Collegiate Dairy Products Evaluation Contest in Denver. Prior to this time there was no accepted score card or scoring guide for industry use in assessing the sensory qualities of yogurt.

The Yogurt Score Card. A score card for Swiss-style flavored yogurt was developed and adopted by the Committee on Evaluation of Dairy Products of the American Dairy Science Association in 1977. This yogurt score card was cooperatively designed through the suggestions and efforts of ingredient suppliers, commercial yogurt manufacturers, and university teaching and research personnel, following the initiation of a course project on the design of a yogurt score card by food science students at Oregon State University coordinated by the first author (Bodyfelt 1976). Following several minor modifications during the first eight years of use, the score card for Swiss-style yogurt currently appears as shown in Fig. 7.4. The scoring guide for Swiss-style yogurt is shown in Table 7.3. Another yogurt score card was proposed by Duthie *et al.* (1977).

Sample Preparation. As a rule, consumer size units (individual servings) of plain or flavored yogurt are selected for sensory evaluation (6 oz [0.17 kg] or 8 oz [0.23 kg] sizes). Sensory evaluation of yogurt is

CONTEST SWISS STYLE YOGURT SCORE CARD

FLAVOR:_____   CONTESTANT NO. _____
DATE: _____ A.D.S.A.

			SAMPLE NO.								TOTAL
			1	2	3	4	5	6	7	8	GRADES
FLAVOR 10	CRITICISMS CONTESTANT SCORE ➤											
	SCORE											
	GRADE _____											
	CRITICISM											
NO CRITICISM 10	ACETALDEHYDE (Coarse)											
	BITTER											
	COOKED											
	FOREIGN											
	HIGH ACID											
	LACKS FINE FLAVOR											
	LACKS FLAVORING											
	LACKS FRESHNESS											
	LACKS SWEETNESS											
	LOW ACID											
	OLD INGREDIENT											
NORMAL RANGE 1-10	OXIDIZED											
	RANCID											
	TOO HIGH FLAVORING											
	TOO SWEET											
	UNNATURAL FLAVORING											
	UNCLEAN											
BODY AND TEXTURE 5	CONTESTANT SCORE ➤											
	SCORE											
	GRADE _____											
	CRITICISM											
NO CRITICISM 5	GEL-LIKE											
	GRAINY											
	ROPY											
	TOO FIRM											
NORMAL RANGE 1-5	WEAK											
APPEARANCE 5	CONTESTANT SCORE ➤											
	SCORE											
	GRADE _____											
	CRITICISM											
NO CRITICISM 5	ATYPICAL COLOR											
	COLOR LEACHING											
	EXCESS FRUIT											
	FREE WHEY											
NORMAL RANGE 1-5	LACKS FRUIT											
	LUMPY											
	SHRUNKEN											
	SURFACE GROWTH											
TOTAL ___	TOTAL SCORE OF EACH SAMPLE ➤											
	TOTAL GRADE PER SAMPLE											

FINAL GRADE
RANK

Source: American Dairy Science Association (1987)

Fig. 7.4.   The ADSA contest swiss-style yogurt score card.

usually conducted without tempering the samples after they have been obtained from a refrigerated source. Conducting flavor evaluation within the temperature range of 1.7°C to 10.0°C (35°F to 50°F) helps insure portrayal of the typical body characteristics of yogurt. It is generally advisable to remove or cover the identity or brand name of selected yogurt samples. This can be achieved by having a disinterested or noninvolved person place the yogurt samples to be examined inside

Table 7.3. The ADSA Scoring Guide for the Sensory Defects of Swiss-Style Yogurt. Suggested Flavor, Body and Texture, and Color and Appearance Scores for Designated Defect Intensities.

	Intensity of Defect		
	Slight[b]	Definite	Pronounced[c]
Flavor Criticisms[a]			
Acetaldehyde (green)	9	7	5
Acid (too high)	9	7	5
Acid (too low)	9	8	6
Bitter	9	7	5
Cooked	9	8	6
Foreign	5	3	0[d]
Lacks fine flavor	9	8	7
Lacks flavoring	9	8	7
Lacks freshness	8	7	6
Lacks sweetness	9	8	7
Old ingredient	7	5	3
Oxidized/metallic	6	4	1
Rancid	4	2	0
Too high flavoring	9	8	7
Too sweet	9	8	7
Unclean	6	4	1
Unnatural flavoring	8	6	4
Body and Texture Criticisms[e]			
Gel-like	4	3	2
Grainy/gritty	4	3	2
Ropy	3	2	1
Too firm	4	3	2
Weak/too thin	4	3	2
Appearance Criticisms[f]			
Atypical color	4	3	2
Color leaching	4	3	2
Excess fruit	4	3	2
Lacks fruit	4	3	2
Lumpy	4	3	2
Shrunken	4	3	2
Surface growth	2	1	0
Wheyed-off (syneresis)	4	3	2

Source: American Dairy Science Association (1987)
[a] "No criticism" for flavor is assigned a score of "10." Normal range is 1 to 10 for a salable product.
[b] Highest assignable score for defect of slight intensity.
[c] Highest assignable score for defect of pronounced intensity.
[d] An assigned score of "0" (zero) is indicative of an unsalable product.
[e] "No criticism" for body and texture is assigned a score of "5." Normal range is 1 to 5 for a salable product.
[f] "No criticism" for appearance is assigned a score of "5." Normal range is 1 to 5 for a salable product.

blank or "discontinued" containers of similar shape and size. Another method of obliterating carton identity is to apply a combination of aluminum foil and masking tape around the container sidewall. Samples should be numbered in sequence or otherwise marked with a sample code on the container sidewall, followed by removal of lids just prior to presentation of samples to the evaluator(s). Figure 7.5 illustrates several methods for coding and presenting yogurt samples in an unbiassed manner.

Yogurt Sampling. Prior to stirring or spooning the sample, it is common practice to closely observe the top, undisturbed surface of the yo-

Fig. 7.5. Several methods for obscuring the brand or label identity and coding commercial yogurt samples prior to sensory evaluation: A—Samples inserted into "neutral" containers; B—Use of either masking tape or aluminum foil.

gurt (within the container) for the possible presence of any of several color and appearance defects (i.e., surface growth, shrunkenness, and wheying-off). Next, it is customary to briefly stir or agitate the yogurt by folding the container contents over itself a minimum number of times (no more than 5 or 6 strokes) in an attempt to achieve a near-homogenous blend. The objectives of this minimum stirring are to: (1) sufficiently break down the yogurt structure to separate it from the container bottom and sidewall, and (2) obtain a homogenous sample for subsequent placement on a plate. Understirring or excessive agitation at this point can markedly alter the observed consistency (body and texture characteristics) of the yogurt. Plastic or metal spoons should be used.

Sequence of Observations. Following a specific sequence of observations is most helpful for sensory evaluation of yogurt when several samples are available for comparative discrimination. These observations should include the four steps listed in the following paragraphs.

1. *Color and Appearance.* The judge should observe the color and overall appearance of each yogurt sample, and while doing so refer to the score card as a check list of possible appearance defects. For certain styles, types, or flavors of yogurt, unique or specific defects may occur; these shortcomings of appearance should simply be noted and written-in on the provided blank lines of the score card. With reference to product color, the evaluator should note the color intensity and uniformity and whether the hue is natural or typical of the style and/or flavor of the given yogurt. Various body and texture characteristics or possible flavor defects of the yogurt sample may sometimes by apparent by closely examining: (1) the representative sample portion spooned onto a plate, as well as (2) noting the top surface of the product within the original container.

2. *Sample the Yogurt.* During the process of obtaining spoonfuls of the product from the original container, the evaluator should carefully note the manner in which the product "fills" the spoon. When examined at eye level, the side profile of the yogurt in the spoon will give one of several possible impressions. (1) May form a moderate but definitive extent of product mound (the ideal); (2) may appear to have a quite rigid or strong structure for the topmost edge of the product mound (too firm and/or gel-like); or (3) may be nearly level with the edges of the utensil, with little or no visible product mound (weak). Figure 7.3 illustrates some of the more common body and texture defects of yogurt (i.e., weak, too firm, and gel-like).

3. *Initiate Sensory Evaluation.* After a spoonful of yogurt has been obtained and the initial color, appearance, body, and texture observations have been made, the sample may be examined for additional

body, texture, and flavor characteristics. Quite possibly, an impression of the flavor characteristics may be gained by noting the aroma of the sample. With experience, some flavor defects or sensory characteristics may be ascertained by initially concentrating on the presence or lack of certain aroma notes.

4. *Sense the Flavor.* While manipulating the sample in the mouth to ascertain various body and texture characteristics, the evaluator should take special note of the initial, mid-point, and delayed taste and aroma sensations. The relative levels of acidity and/or sweetness, and the nature of the "yogurt flavor" are often noted promptly; the characteristics of the flavoring system (if a flavored yogurt) are generally noted at mid-point in the flavor perception process. The volatile, flavor-contributing substances usually require a short warming or tempering period within the mouth to "register" with the evaluator. However, product sweetness or acidity are usually noted prior to the volatile components of the flavoring system, unless the latter is excessively out-of-balance. The evaluator should notice particularly whether the acidity and sweetness levels of the yogurt are either: (1) pleasant and typical; (2) too intense; (3) lacking; (4) unnatural; or (5) slightly out-of-balance (lacks fine flavor).

By the time the intensity of acidity and sweetness has been assessed, other flavor notes will begin to be noted by the evaluator. At this point, the evaluator should try to determine whether the overall product flavor is: (1) coarse (harsh or severe acetaldehyde (green-like)); (2) too delicate or mild (lacks flavoring); (3) pronounced (too high in flavoring); or (4) "just right" (good flavor balance). The judge should try to answer the following questions: Is the flavor typical of the stated flavor on the container, or does the flavor tend to be unnatural or atypical? Does the yogurt seem refreshing or is there a failure of the mouth to quickly "clean up" after the sample is expectorated? These are basic sensory impressions that should serve to establish a flavor character for the yogurt sample.

The yogurt sample should be held in the mouth until it is warmed to approximately body temperature, and simultaneously masticated and manipulated by the tongue and between the teeth until the mouthfeel and flavor characteristics have been appropriately noted. Following this, the sample should be expectorated. Some evaluators find it helpful to consume several small bites of yogurt just before the actual scoring of samples commences, in order to "adjust their taste buds" or "acclimate" themselves to yogurt acidity and flavors.

The authors suggest the following evaluation technique for novice yogurt judges, a so-called "4-step analysis of flavor components" approach. The new yogurt judge is advised to taste a given yogurt sample as many as four times, but each time the evaluator should concentrate on just one segment of the overall flavor characteristics. The typical order of sequence for the four perceived elements of flavor in flavored yogurt are: (1) acidity level—ask whether it is too high, too low, atypical (citric or malic acid rather than lactic), or (preferably) 'just right' for the given type or flavor of yogurt; (2) sweetness level—ask whether it is too high, too low, or 'just right'; (3) the nature and balance of the flavoring system—ask whether it is too high, too low, lacks fine flavor (balance), or unnatural for the given flavor; and (4) the lack or presence of an objectionable aftertaste—in yogurt, most of the perceived sensations of aftertaste are the result of poor-quality dairy ingredients; i.e., staleness or lacks freshness, old ingredient, oxidized/metallic, rancid, unclean off-flavors and/or atypical fermentation (contaminants, yeasts, poor cultures, etc.).

The authors assume that the one sensory characteristic most sought by consumers of flavored yogurt is distinct refreshingness. A refreshing flavor in yogurt implies: no unpleasant aftertaste(s), a pleasant level of acidity, a pleasing flavor balance or bouquet (of the given or stated flavor), and a barely perceptible sensation of product sweetness, if perceived at all.

REQUIREMENTS OF HIGH-QUALITY FLAVORED YOGURT

Evaluation of Color and Appearance

The color of flavored yogurt should be uniform, attractive, pleasing, and typical of the flavor represented. The appearance of yogurt encompasses all visible aspects that are important from the standpoint of visual appeal. Not all flavored yogurts require added color, some flavoring systems have adequate levels of pigments in the fruit, berries, or other flavoring material to provide the desired color level. In order to meet certain market demands, to comply with labeling requirements, or to minimize the addition of ingredients (food additives), many yogurt manufacturers refrain from incorporating any form of food coloring materials.

Assuming that coloring material is added to yogurt (or even if it is not), the preferred shade of color for a given yogurt flavor is generally the same hue of color as possessed by the fruit, berries, or other flavor source. Color and appearance defects of Swiss-style (preblended) yogurt may be listed as follows:

Atypical color

Color leaching (color bleeding)

Excess fruit

Lacks fruit

Lumpy (nonhomogenous)

Free whey

Shrunken (coagulum)

Surface growth (yeast or mold)

Atypical Color. Basically, the hue and color intensity of flavored yogurt should be pleasing to the eye and suggestive or typical of the given or represented flavor. Preferably, there should be no color dullness, unpleasant grey shades, off-whites, or drab colors. An overcolored product that appears vivid, overly bright, or actually suggests a different flavor (other than the one stated on the label) is usually considered objectionable. An uncolored, pale, or too-lightly colored product is generally preferred over a product that appears overcolored or incorrectly colored for the given flavor.

The use of recommended levels of good-quality fruit or berry flavorings and/or coloring materials which have appropriate pH stability usually serves to help minimize the occurrence of atypical color. Proper storage of the above ingredients (according to the suppliers' recommendations) also helps minimize oxidation/reduction effects on coloring agents and pigments.

Color Leaching. "Color leaching" is a frequently encountered defect in some berry-flavored yogurts; it gives the impression that a portion of the color has been drained or "leached" away from berry pieces and has migrated as a "pigment trail" throughout the yogurt. Another possible impression is that the berry pigment(s) have separated from pieces of the flavoring material and "sought a new location." The defect was originally called "color bleeding," but the term "color leaching" was later designated as being more descriptive of this defect. More careful selection of properly processed berry flavoring materials for Swiss-style yogurt is one solution of this problem of product appearance; sometimes, fruit pieces of larger than optimum size may provoke this defect. Color leaching can also be caused by excessive acidity (pH ≤ 3.8) in the final product. This suggests that acid stable colors within the typical pH range of yogurt (pH 3.8–4.3) need to be employed. Another cause of this appearance defect may be incomplete blending of the yogurt base with the berry flavoring material before commencement of the filling operation.

Excess Fruit. The defect "excess fruit" is encountered seldomly in commercial yogurt for obvious reasons—primarily economic. However, production personnel may occasionally err in terms of blending proper ratios of flavoring(s) with yogurt base. This results in a product that "appears different from the norm"; it may be too liquid ("soupy") and it may tend to be overflavored and/or too sweet.

Lacks Fruit. The defect "lacks fruit" occurs quite frequently in U.S. yogurt, since some yogurt flavoring systems seem to utilize a minimum of puree or fruit pieces. As a result, there may be little or no visible "fruit show" in the final product, in spite of the fact that a reasonable flavor intensity may be present. Sufficient pieces of fruit should appear over a representative cross-section of the product to leave a natural-like or pleasing impression. A reasonable amount of visible fruit is expected in berry or fruit-flavored yogurt.

Free Whey. When syneresis, or the obvious expulsion of moisture (wheying off) from the gel occurs, it is commonly referred to as "free whey." This readily visible defect is frequently caused by contraction of the coagulum or gel structure, due to a low milk solids content. Free whey also may result from excess acid development or it may occur if the yogurt is agitated or moved at a critical time during incubation— when the pH of the yogurt is at or above the isoelectric point of casein (pH = 4.65). Free whey also may be caused by: (1) an insufficient milk pasteurization temperature and holding time; (2) poor product formulation; and/or (3) use of an inadequate stabilizer system. Another cause of free whey in yogurt is severe temperature change in storage or distribution (i.e., either extremely high temperature or subfreezing). Efforts to minimize the occurrence of free whey include: (1) good overall process control; (2) the accurate monitoring of pH; and (3) halting acid development within the pH range of 4.5 to 4.0 (depending on the type of yogurt produced). Careful selection and control of product stabilizer(s) are essential requirements. The various stabilizing agents that are employed should be compatible with both the milk ingredients and the flavoring system.

Shrunken (Coagulum). In yogurt, when the gel structure or coagulum appears to shrink in size or pull away from the vertical walls of the container, the resultant appearance phenomenon is referred to as being "shrunken." The yogurt mass within the container conveys the impression that a reduction in product volume has occurred. More often than not, free whey collects in the formed space between the shrunken (reduced) product mass and the container sidewall. The various production and quality control measures that help minimize the occurrence of shrunkenness are quite similar to the steps required to prevent syneresis (free whey).

Surface Growth. Surface growth refers to the visible yeast (or mold) colonies that occasionally develop on the surface of yogurt. The easiest way to observe surface growth is to carefully look for a slightly raised or convex spot (of 0.5–5.0 mm diameter), which is usually slightly less intense in color than the yogurt surface (except for plain or uncolored yogurts). Good housekeeping, dust control and the minimization of particulate matter in the air, and/or use of laminar airflow units (installed directly above container-filling equipment) serve to minimize the occurrence of yeast and mold colony growth on yogurt.

Evaluation of Body and Texture

There is a distinct range in the observed body and texture properties of North American yogurt. There is disagreement among manufacturers and authorities on yogurt quality, as to what constitutes the "ideal" or perfect consistency of commercial yogurt. Some consumers (and experts) seem to definitely prefer a heavy, gel-like structure; others obviously prefer a weaker or soft-bodied product.

After stirring the yogurt sample briefly (at most 5 or 6 rotations in order to obtain a representative portion), an ample spoonful of the product is removed from the container. The spoonful of yogurt should promptly be raised to eye level and viewed from a side profile. The visual impact that the spoonful of yogurt portrays is a good indicator of the presence or lack of certain body and texture defects. A modest degree of "rounded" mounding is the visual effect deemed most desirable for the viscosity or primary body characteristic in most (but not all) yogurts; by contrast, a soft-bodied yogurt should portray a flatter or nearly level profile when viewed from the side.

Gel-like. If a spoonful of yogurt (viewed at eye level) provides an impression of a relatively high, firm ridge, with sharp edges and a tendency to wiggle or "move" similar to a gelatin dessert ("Jello™"-like) when the spoon is gently shaken, the body is considered to be gel-like. The evaluator can confirm the occurrence of gel-like consistency by: (1) placing several spoonfuls of the product on a plate and observe for any presence of sharp-edged ridges or peaks and a higher degree of gloss, sheen, or wetness of the surface, and/or (2) placing a spoonful of the yogurt into the mouth and observing for a distinct gelatin-like or a "slick" mouthfeel. Upon initial mastication, the product tends to offer some resistance, break apart into pieces or small chunks, and may feel slightly slippery to tongue-touch. The product may not seem smooth and homogenous, as the evaluator typically expects.

The type and/or an excess quantity of gelatin (or other stabilizing agents) are generally responsible for the gel-like defect. Apparently, some yogurt manufacturers feel that wheying-off or possible breakdown of yogurt in distribution channels can be minimized by higher levels of stabilization, or that their customers actually prefer the gel-like characteristic in the yogurt. Unfortunately, some stabilizers may tend to mask or block release of the delicate flavor characteristics of yogurt.

Some fruit-on-the-bottom or sundae-style yogurts exhibit a heavy custard or moderate gel-like consistency. Some types of sundae-style yogurts are expected to stand-up or retain the shape of the container, when the cup is inverted on a plate and the contents dispensed en masse (see Fig. 7.6). Flavoring syrups (from the bottom of the con-

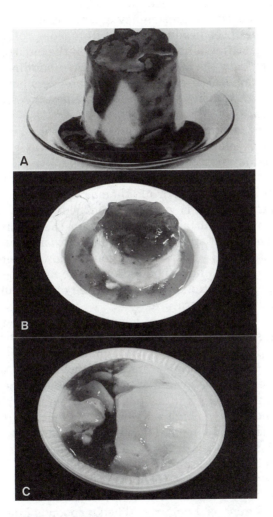

Fig. 7.6. Appearance or eye appeal of sundae-style flavored yogurt. The flavoring syrup should cascade over the vertical surface of a distinct yogurt mound (Sample A). Sample B represents a less satisfactory product appearance. A most unsatisfactory product appearance is noted in Sample C.

tainer) are expected to cascade down the external surface of the "yogurt mound." In this instance, the yogurt could be criticized if it did not exhibit a somewhat "gel-like," firm body.

Too Firm. When yogurt is "too firm" in body, it appears to be "overset" or overstabilized. If the yogurt is so firm that it appears like a heavy milk pudding, its mouthfeel (and appearance) may be objectionable to the consumer. In essence, the product may exhibit too heavy a body; it has exceeded the consistency of a moderate smooth, light custard. A spoonful of the product, viewed from side-profile, appears as a highly rounded, stiff mound. Upon placing in the mouth, a yogurt that is too firm may leave the impression of a heavy pudding, or be suggestive of starch (tapioca), if these types of stabilizers have been used in the formulation. Excessive stabilization is generally responsible for the too firm defect.

Weak. "Weak" yogurt generally lacks sufficient structural strength or appropriate firmness, or it may appear to be watery due to lack of food solids; it may not be readily "spoonable." In the extreme case, the impression might be that a weak-bodied product could be more readily consumed as a liquid drink, than as a semisolid product. When a spoonful of the product is viewed by the "side profile" procedure, an obvious flat or level surface appears within the spoon (some of the spoon's contents may spill over the edge). Possible causes of weak-bodied yogurt are: (1) understabilization; (2) low casein or low total milk solids content; (3) underincubation in the fermentation stage; (4) too low of a pasteurization temperature; and occasionally (5) too high of a pasteurization temperature.

Grainy. The "grainy" defect of yogurt is most frequently perceived by mouthfeel; sometimes it may be observed as small, barely perceptible grain-size particles scattered throughout the yogurt. Such a product generally lacks uniformity and smoothness, but the associated rough mouthfeel may be even more objectionable to the evaluator. In addition to the causes given on page 258 some possible reasons for yogurt graininess are: (1) high acid milk ingredient; (2) unstable casein; (3) too-rapid acid development; (4) too-high incubation temperature; (5) homogenization at too-high temperature; (6) an excessive amount of culture; (7) an incorrect stabilization system; and (8) a failure to properly blend the yogurt base with the fruit flavoring.

Ropy. The "ropy" characteristic (or defect) is infrequently experienced in U.S. commercial yogurts. When a spoon is immersed slightly into ropy yogurt and then slowly lifted about 5 to 8 cm (2–3 in.) above the yogurt surface, the yogurt exhibits a tendency to "stretch out" like taffy candy or glue between the two points of attachment. A ropy characteristic may develop on rare occasions due to the incorporation of certain types of gums or stabilizers into the yogurt base. Usually, ropiness in yogurt is caused by certain strains of *L. bulgaricus* used as part of the yogurt culture. These unique bacterial strains produce polysaccharides (as capsular material) that can result in a ropy consistency. In some countries, ropy strains of yogurt bacteria are purposely selected and used to develop this desired consistency (ropiness), without having to resort to various stabilizing agents in the product.

Evaluation of Flavor

In spite of the attention given to body, texture, color, and appearance, the flavor characteristics of yogurt are probably the most critical and important determinants of consumer acceptance. Generally, most consumers have a certain expectation of a reasonable level of acidity (or tartness) in yogurt, which is determined primarily by the amount of lactic acid formed. For most palatable yogurts in the U.S., the pH range is 3.8 to 4.4. In the case of fruit- or berry-flavored yogurts, there is an anticipation by the consumer of a pleasant blend (or balance) of the given flavor with the acidity of the yogurt and the added sweetener. A final, but critical dimension of "good yogurt flavor" is the absence or lack of any unpleasant aftertaste or off-flavor in the product. An overall flavor impression left with the evaluator of yogurt should be one of simply general refreshingness. In essence, the evaluator should basically wish or desire to consume (or taste) second, and subsequent spoonfuls, of the given product.

The various off-flavors of flavored yogurt can be summarized as follows:

High acid	Lacks sweetness
Low acid	Too sweet
Acetaldehyde (coarse, green)	Stabilizer flavor
Bitter	Lacks freshness
Cooked	Old ingredient
Lacks fine flavor	Unclean
Lacks flavor	Oxidized
Too high flavor	Rancid
Unnatural flavor	Storage

High Acid. "High acid," extreme tartness or sourness, is frequently encountered in both plain and flavored-types of U.S. commercial yogurts. Generally, it is due to excessive levels of lactic acid produced by lactic culture metabolism. However, each yogurt consumer seems to have a different "frame-of-reference" or range of tolerance (or intolerance) for yogurt tartness. The actual perception of high acid is often complicated by the tendency to overlap or confuse this specific taste defect with the "acetaldehyde" or "coarse" taste note (sometimes described as "green apple" or "plain yogurt" flavor). When yogurt pH drops below 3.8 (or exceeds a titratable acidity of 1.25%), a high acid flavor defect frequently ensues. An evaluator usually experiences a high acid note immediately if it is present. Generally it is perceived as a distinct, initial note of sharp tartness on the tip of the tongue; the distinct acid taste may readily dissipate or may sometimes persist, depending on other flavor notes that may be present. Frequent causes of high acid yogurt are: (1) an extended incubation period; (2) elevated incubation temperature; (3) "imbalance" of the bacterial genera in the yogurt culture; and (4) inadequate cooling to arrest culture activity at termination of the incubation period. Use of the criticism "sour" is indicative of not only too much lactic acid, but also the probable presence of other acids such as citric and/or malic acid(s) from the added fruit flavorings.

Low Acid. The "low acid" defect is less frequently encountered in flavored yogurt than is high acid. When this flavor defect occurs, the given yogurt seems to lack the typical full, refreshing flavor or usual level of modest tartness. Underincubation of the yogurt base, "imbalanced" yogurt cultures, or partial culture failure (due to inhibitors) are possible causes of low acid defect in yogurt. The product pH is most likely 4.5 or above when yogurt is criticized for being low in acidity. Occasionally the low acid defect may be due to: (1) insufficient heat treatment of the yogurt base (which may lead to less culture activity); (2) excessive sweetener level; and/or (3) an improperly balanced flavoring system for the product.

Acetaldehyde (Coarse, Green). Frequently stated synonyms for the "acetaldehyde" flavor defect in yogurt are "coarse," "green," or "green apple." In plain yogurt, acetaldehyde concentrations ranging from 5 to 40 ppm are typical. Acetaldehyde is primarily responsible for the characteristic or so-called "plain yogurt" flavor. Acetaldehyde is readily produced by *L. bulgaricus* at incubation temperatures in excess of 37.8°C (100°F). Although the optimum acetaldehyde note in plain or unflavored yogurt is definitely a matter of personal preference for both consumers and yogurt judges, the acetaldehyde intensity perceived in flavored yogurts apparently should be slight, if this flavor note is perceived at all. Typically, it seems that most yogurt evaluators (as well as a majority of U.S. consumers) prefer that the stated flavor of the given yogurt predominate, with the lactic acid taste and the acetaldehyde aroma serving as complementary or "fill-in" flavor notes. The refreshing component of yogurt flavor is probably due to the combined sensory perceptions of acidity, acetaldehyde (green), and sweetness, as well as the added flavoring material.

Researchers at Oregon State University (Bills *et al.* 1972) studied the effect of sucrose on the production of acetaldehyde and lactic acid by yogurt culture bacteria. The flavor threshold for acetaldehyde was found to be 11.7 ppm, determined in a system of milk plus 8% sucrose and strawberry flavoring. In this study, the masking effect imparted by the added flavoring was judged to have a greater impact on the perceived acetaldehyde note by panelists than the inhibition of yogurt culture by sucrose in strawberry yogurt. This seems to suggest that fruit or berry flavoring systems generally play a significant role in masking or "toning down" the plain yogurt (acetaldehyde) flavor note in many flavored yogurts.

Bitter. As in other dairy products, the "bitter" (quinine-like) defect of yogurt is usually perceived as a distinct aftertaste. Generally, there is a delayed perception (after the sample has been expectorated or swallowed) of this more objectionable yogurt defect. Typical causes of bitter yogurt include: (1) contaminated yogurt cultures; (2) poor quality milk ingredients (especially psychrotrophic bacterial activity); and occasionally (3) excessive use of certain stabilizing agents; or (4) extremely poor-quality fruit or flavoring.

Cooked. Due to the relatively high pasteurization temperature applied to yogurt bases, a "cooked" flavor note is sometimes encountered in commercial yogurts. This cooked note is generally perceived as a slightly nutty or sulfur-like note. It is not a particularly objectionable flavor, except when the intensity is most definite or pronounced (or possibly scorched). A cooked defect (when intense) may tend to detract from the intended refreshing characteristic of yogurt. Occasionally, this flavor criticism may be associated with either "lacks flavoring" or "lacks acid" flavor defects. Usually, when appropriate (or too high) levels of product flavoring occur, a cooked flavor note is not readily apparent, due to the masking effect of the flavoring system. More careful control of the pasteurization process and the addition of flavorings at a slightly higher concentration generally suffice to minimize the occurrence of "cooked" as a flavor criticism of yogurt.

Lacks Fine Flavor. The "lacks fine flavor" criticism in flavored yogurt generally refers to the perception by evaluator(s) of a product flavoring system that is slightly "out-of-balance," or it may be interpreted as being slightly "harsh." Yogurt samples critiqued for "lacks fine flavor" would most likely achieve a "perfect score" or "no criticism" flavor score with just the slightest change or modification of either: (1) the flavoring system; (2) the quantity of flavoring; and/or (3) the levels of lactic acid or acetaldehyde. Most often, the flavoring system is probably (though slightly) at fault; however, the product's flavor shortcoming(s) may occasionally be traceable to either the final product pH or the acetaldehyde concentration of the yogurt base.

Lacks Flavoring. This defect is simply the result of not using a sufficient amount of added flavoring to the yogurt base at the time of product blending. Improper mixing of fruit flavoring with the yogurt base is another cause of this defect and some flavoring bases simply lack sufficient flavor intensity. The occurrence of this defect is probably on the increase, since processors are depending on smaller and smaller quantities of concentrated fruit and berry preparations for the source of product flavoring; hence, it becomes more difficult to achieve uniform dispersement of the flavoring throughout the yogurt (in some types of blending systems). Inaccurate measurement or the purposeful addition of less-than-recommended amounts of flavoring (in an attempt to reduce ingredient costs), sometimes serve as a rationale for limiting the amounts of flavoring.

Too High Flavor. Miscalculations or a lack of control in blending flavor concentrates or essences into the yogurt are, most often, the cause of a "too high" or "too intense" product flavor. These developments tend to result in a harsh flavor note; in some cases "too high flavoring" may even suggest an artificial or unnatural flavor. In some instances of this defect, the yogurt may seem excessively aromatic; in essence, the anticipated, delicate flavor may simply seem out of balance or "out of character" for the stated flavor on the container.

Unnatural Flavor. In the occurrence of "unnatural flavoring," the flavoring character or "flavor balance" is simply not typical or representative of the stated product flavor. For example, the yogurt may be labeled "strawberry" flavor, but the flavor characteristics perceived by the evaluator may be more suggestive of raspberry (or boysenberry, prune-like, or some other fruit or berry type). This atypical off-flavor is most often due to: (1) certain inadequacies of the flavoring source; (2) some flavor enhancers; (3) WONF (with-other-natural-flavors) that are added; or (4) the particular combination of fruit and flavoring agents. Close and continuous monitoring of the selected flavoring system(s) is generally required by yogurt manufacturers to help insure that flavoring suppliers have not significantly altered yogurt flavorings from the original formulation. Occasionally, processing changes or lack of control in yogurt manufacture can lead to nonuniform lactic acid and acetaldehyde levels, which can substantially alter the flavor characteristics (as well as the body properties) of the yogurt.

Lacks Sweetness. One of the least encountered flavor defects of flavored yogurt is "lacks sweetness." Careless formulation and/or blending operations are probably responsible for this product shortcoming when it occurs. Generally, certain minimum levels of sweetener (approximately a 4 to 12% range of concentration) are required to enhance and/or balance most fruit and berry flavors in yogurt. The "lacks sweetness" defect may be one of the more confusing flavor defects of flavored yogurt, in terms of obtaining a consensus of opinion from both experienced evaluators and consumers. The interplay of acidity level, the acetaldeyde note, sweetness level and flavoring character, and/or intensity of flavored yogurt are not easily comprehended by many persons. Subtleties of flavor such as a slight "lacks sweetness" are perhaps best ignored or downplayed; however, more definite or pronounced observations for "lacks sweetness" (and other off-flavors) should be denoted if they occur and can be determined by sensory observation.

Too Sweet. Excessive levels of sweetener ("too sweet") in flavored yogurt is more frequently observed in U.S. yogurt than the "lacks sweetness" defect and certain other flavor defects. The evaluator should bear in mind that yogurt is basically a fermented milk product, hence, the distinct (but slight or moderate) taste of lactic acid or tartness is expected and desired. Meanwhile, the primary function of added sweetener in flavored yogurt is to balance the acid and enhance the fruit or berry flavor. Most experienced yogurt judges insist that a sweet taste sensation should be "barely evident" in the overall flavor profile of the yogurt. The sole function of sweetener in flavored yogurt is to help optimize or "bring out" the stated flavor.

Stabilizer Flavor. The descriptor "stabilizer flavor" is probably best interpreted as an "umbrella" term for a group of possible flavor defects that may occur in yogurt as the result of using excessive amounts of certain types of product stabilizers. The descriptors "foreign" or "chemical-like" do not adequately describe the sensory characteristics of this off-flavor when it occurs. When certain forms or combinations of stabilizing agents are used, characteristic (but objectionable) mouthfeel and associated body and texture properties may simultaneously occur. In those instances where excess amounts of modified starch (and/or tapioca) are incorporated as a stabilizer, the consequent texture and mouthfeel properties of yogurt may be variously described as "heavy," "gluey," or "pudding-like." Frequently, an associated flavor characteristic may be a delayed aftertaste that suggests staleness or "atypical" flavor. In addition to possible off-flavors, excessive use of gelatin (and/or pectin) in yogurt bases frequently conveys a slick, firm, and/or lumpy mouthfeel to the final product. The visual appearance of excessively stabilized yogurt is frequently gel-like, too firm, lumpy, and/or generally lacking in product uniformity.

Lacks Freshness. When yogurt exhibits a perceptible intensity of stale aftertaste, the flavor defect is usually due to use of less than the freshest possible or highest-quality dairy ingredients. This stale defect generally is called "lacks freshness." Typically, this moderately objectionable flavor defect is derived from the nonfat dry milk (or condensed skim milk) that is added to the yogurt base for the purpose of increasing the milk solids level in the yogurt base. Yogurt manufacturers can minimize this flavor defect by paying closer attention to the overall quality of the source of the nonfat milk solids added to the fresh milk ingredients.

The "lacks freshness" criticism is generally experienced as a delayed aftertaste; usually, it is perceived after the evaluator has taken account of the acidity, sweetness, and the yogurt's characterizing flavor. The "lacks freshness" defect tends to detract from the so-called "refreshing" attribute of yogurt. In essence, "lacks freshness" is usually an off-flavor derived from dairy ingredients, except for those infrequent occasions when stale berries or fruit may be the responsible factor.

Old Ingredient. There is logic in suggesting that "lacks freshness/old ingredient" be treated as a single defect of progressively increasing intensity. When the intensity of "stale/lacks freshness" exceeds a certain point or degree of objectionability, the defect is more appropriately termed "old ingredient." The after-taste associated with old ingredient is most perceptible or more intense; it definitely tends to linger after sample expectoration. The extent of this unpleasant aftertaste is such that it seems to predominate over the flavor sensations of acidity, sweetness, and/or the characterizing yogurt flavoring. Often, an old ingredient off-flavor (extremely stale and occasionally bitter) seems to completely mask the expected desirable taste and aroma notes of flavored yogurt. Old, stale, or overall poor-quality dairy ingredients, and occasionally, "deteriorated" stabilizing agents, are often responsible for this objectionable flavor defect.

Unclean. The next stage of a severe, unpleasant, aftertaste that can be noted in yogurt is simply and directly referred to as "unclean" (which refers to both flavor character and aftertaste). In no uncertain terms, the evaluator may note an intense "dirty" or unpleasant aftertaste that tends to linger and linger. The evaluator is usually most reluctant to try a second bite of such a product, due to the intensity and objectionable nature of the "unclean" flavor defect. Frequently, this sensory note is due to proteolysis of milk ingredients, stemming from resultant volatile end products, such as certain amines (e.g., putrescine or cadaverine). The authors have observed that some brands of nonfat flavored yogurts frequently manifest an unclean (and bitter) off-flavor; this is probably due to the unique ratio of protein to fat in this form of yogurt.

Oxidized. Fortunately, an "oxidized" off-flavor occurs most infrequently in U.S. commercial yogurts. Occasionally, some milk ingredients may have developed a typical oxidized off-flavor (cardboardy, metallic, or tallowy) prior to product formulation, blending and processing. Hence, the final product may exhibit an oxidized off-flavor, in spite of the significant flavor-masking imparted by flavoring, acidity, and/or sweetener. Careful and continuous monitoring of all milk ingredients is essential in preventing any occurrence of an oxidized off-flavor in yogurt.

Rancid. In the experience of the authors, more opportunity seems to exist for the occurrence of a "rancid" defect than an oxidized defect in yogurt. Apparently, the sequence of milk ingredient blending and pasteurizing steps for preparing the yogurt base occasionally lends itself to mixing homogenized milk with raw milk ingredients; this can lead to lipase activity within the yogurt base. Some yogurt processing operations can apparently experience some delays in final heat treatment; any ensuing lipolysis can lead to slight to moderate rancid off-flavors in flavored yogurts, especially in the milder product flavors. A rancid off-flavor, in yogurt as in other dairy products, is nearly always characterized by a delayed, slightly soapy, bitter aftertaste. Given the substantial masking effect of fruit flavorings, it may be difficult for an evaluator to consistently differentiate between the possible off-flavors of rancid and unclean in flavored yogurt. The dilemma for the evaluator may be the decision of "which is the more appropriate flavor criticism," since a bitter aftertaste is frequently a characteristic attribute of both of these sensory defects.

Storage. Yogurt stored for an extended period of time may develop a range of off-flavors due to enzymatic action and odor absorption. The problem is magnified when the product is contaminated with undesirable microorganisms. Other contributing factors are improper storage and temperature abuse. However, even the best yogurt stored under acceptable conditions has a defined, useful storage life, beyond which deterioration can be expected.

CREAMED COTTAGE CHEESE

Creamed cottage cheese is a soft, unripened cheese that is usually made by the acid coagulation of pasteurized skim milk by added lactic culture or acidulants, with or without the addition of minute quantities of milk-coagulating enzymes (as curd conditioners). The coagulum is cut into various-sized curd particles by special sets of knives, heated (cooked), and held for a sufficient time to facilitate firming of the curd and removal of the whey. Once the curd has developed the appropriate consistency (firmness or "meatiness"), the whey is drained. Then the curd is washed, creamed (usually), salted, and packaged.

Cottage cheese is consumed as a fresh product (within a maximum of three to four weeks). Consequently, the flavor attributes of this product depend on a combination of the sensory qualities of skim milk and cream dressing ingredients, as well as properties of the lactic cultures employed in the manufacturing process. The overall sanitation procedures and temperature control exercised in manufacture also play a key role in determining product shelf-life and sensory quality of this relatively perishable dairy product. Today, it is common practice among U.S. cottage cheese processors to incorporate especially-selected lactic cultures (S. lactis subsp. diacetylactis and/or Leuconostoc sp.) into the cream dressing to increase the "cultured aroma" (and coincidentally inhibit psychrotrophic bacteria). Hence, the addition of carefully selected lactic microorganisms to the dressing can simultaneously serve to significantly enhance flavor and increase the shelf-life of creamed cottage cheese by inhibiting psychrotrophs. Recently, the incorporation of "superpasteurized" cultures of Propionibacterium shermanii into the cream dressing (before addition to the curd) was shown to routinely extend the shelf-life of commercial cottage cheese up to four or five weeks (Sandine 1984).

Cottage Cheese Score Card

In 1957, a committee of the American Dairy Science Association (ADSA) developed a cottage cheese score card that included a scoring guide for flavor, body and texture, appearance, and color, as well as a glossary of terms to facilitate scoring this product. The current ADSA contest score card and the scoring guide for various sensory defects of creamed cottage cheese are presented in Fig. 7.7 and Table 7.4, respectively. A new collegiate contest score card for cottage cheese is shown in Appendix XIV.

Types of Cottage Cheese

Creamed cottage cheese is the general term used to designate the fresh, soft, uncured, high moisture (not over 80%) cheese made from pasteurized sweet skim milk, or from plain condensed skim milk, or from reconstituted nonfat dry milk (occasionally). Any observer will note several distinct types, forms, or styles of cottage cheese in North American supermarkets. Various terms, such as "schmierkase,"

CONTEST COTTAGE CHEESE SCORE CARD

DATE: _____ A.D.S.A.   CONTESTANT NO: _____

						SAMPLE NO.				
	CRITICISMS	1	2	3	4	5	6	7	8	
FLAVOR 10	CONTESTANT SCORE ➡									▔
	GRADE SCORE									
	CRITICISM									
	ACID (High)									
	BITTER									
NO CRITICISM	DIACETYL									
10	FEED									
	FERMENTED/FRUITY									
	FLAT									
	FOREIGN									
	HIGH SALT									
	LACKS FINE FLAVOR									
	LACKS FRESHNESS									
	MALTY									
NORMAL RANGE	METALLIC									
1-10	MUSTY									
	OXIDIZED									
	RANCID									
	UNCLEAN									
	YEASTY									
BODY AND										
TEXTURE 5	CONTESTANT SCORE ➡									▔
	GRADE SCORE									
	CRITICISM									
NO CRITICISM	FIRM/RUBBERY									
5	GELATINOUS									
	MEALY/GRAINY									
NORMAL RANGE	PASTY									
1-5	WEAK/SOFT									
APPEARANCE										
AND COLOR 5	CONTESTANT SCORE ➡									▔
	GRADE SCORE									
	CRITICISM									
NO CRITICISM	FREE CREAM									
5	FREE WHEY									
	LACKS CREAM									
NORMAL RANGE	MATTED									
1-5	SHATTERED CURD									
	SLIMY									
	SURFACE DISCOLORED									
	TRANSLUCENT									
	UNNATURAL COLOR									
PACKAGE	ALLOWED PERFECT IN CONTEST	X	X	X	X	X	X	X	X	
SCORE	TOTAL SCORE OF EACH SAMPLE ➡									
	TOTAL GRADE PER SAMPLE									

Source: American Dairy Science Association (1987)

FINAL GRADE _____
RANK [   ]

Fig. 7.7.   The ADSA contest score card for the sensory evaluation of cottage cheese.

Table 7.4. The ADSA Scoring Guide for the Sensory Defects of Creamed Cottage Cheese (Suggested Flavor, Body and Texture, and Color and Appearance Scores for Designated Defect Intensities).

	Intensity of defect		
	Slight[b]	Definite	Pronounced[c]
Flavor Criticisms[a]			
Acid (high)	9	7	5
Bitter	7	5	1
Diacetyl	9	8	7
Feed	9	7	5
Fermented/fruity	5	3	1
Flat	9	8	7
Foreign	7	4	0[d]
Garlic/onion	5	3	1
High salt	9	8	7
Lacks fine flavor	9	8	7
Lacks freshness	9	5	1
Malty	6	4	1
Metallic	5	3	1
Musty	5	3	1
Oxidized	5	3	1
Rancid	4	2	1
Unclean	6	3	1
Yeasty	4	2	1
Body and Texture[e]			
Firm/rubbery	4	3	2
Gelatinous	3	1	0
Mealy/grainy	4	2	—[f]
Pasty	3	1	—
Weak/soft	4	3	1
Appearance and Color[g]			
Free cream	4	3	2
Free whey	4	2	0
Lacks cream	4	3	1
Matted	4	3	1
Shattered curd	4	3	1
Slimy	2	0	0
Surface discolored	1	0	0
Translucent	3	1	0
Unnatural color	4	3	2

Source: American Dairy Science Association (1987).

[a] "No criticism" for flavor is assigned a score of "10." Normal range is 1 to 10 for a salable product.

[b] Highest assignable score for defect of slight intensity.

[c] Highest assignable score for defect of pronounced intensity.

[d] An assigned score of "0" (zero) is indicative of an unsalable product.

[e] "No criticism" for body and texture is assigned a score of "5." Normal range is 1 to 5 for a salable product.

[f] A dash (—) indicates that this level of intensity for the given defect is not likely to be observed.

[g] "No criticism" for appearance and color is assigned a score of "5." Normal range is 1 to 5 for a salable product.

"farmer-style," "country-style," "old-fashioned," "sweet curd," "small curd," "large curd," and "popcorn," are employed to describe the several products that result from variations in manufacture. Other product names that are sometimes used to designate certain product types of cottage cheese are "Dutch," "pot," "pressed," "baker's," and "hoop" cheese.

Creamed cottage cheese marketed in U.S. commercial channels may be classified according to the following methods of producing the curd or cream dressing:

1. *Producing the curd,* whether by:
   a. lactic acid development only (acid curd);
   b. lactic acid and milk-coagulating enzyme;
   c. addition of approved food acidulants (direct set or acidified).
2. *Breaking or cutting the coagulum* by:
   a. rigorous stirring (farmer-style, old-fashioned, pot);
   b. cutting with designed knife sets
      (1) small curd (¼–⅜ in.) (0.6–0.9 cm)
      (2) medium curd (½–⅝ in.) (1.3–1.6 cm)
      (3) large curd (⅝–¾ in.) (1.6–1.9 cm).
3. *Method of creaming* by:
   a. addition of cream dressing ($\geq 10\%$ milkfat) in an approximate ratio of 3 parts curd to 2 parts dressing ($\geq 4\%$ milkfat in final product);
   b. addition of a lower-milkfat creaming mixture (≥ 3–6% milkfat) of an appropriate amount to achieve either a 1% or 2% milkfat content in final product (lowfat cottage cheese);
   c. occasionally, addition of whipped cream or other high-fat cream dressings to the curd to achieve a special effect (usually marketed under a coined name for the product).
4. *Treatment of cream dressing* by:
   a. no addition of a lactic culture to the creaming mixture.
   b. direct addition of a lyophilized aroma-producing lactic culture (*S. lactis* subsp. *S. diacetylactis* and/or *Leuconostoc* sp.) to the creaming mixture.
   c. addition of a heat-treated milk culture of *P. shermanii* directly to the creaming mixture.

Examination of Creamed Cottage Cheese. Cottage cheese is examined for sensory properties in a manner similar to other dairy products—a combination of sight, mouthfeel, taste, and odor. Creamed cottage cheese should be examined en masse for the presence or lack of "free whey," nonabsorbed (free) cream dressing, and for curd appearance. If facilities and time are available, the equivalent of a large table-spoonful of creamed cottage cheese can be washed in a glass of cold water ($\leq 7.2°C$ [$\leq 45°F$]). The curd is allowed to settle and the milky water decanted. This process is usually repeated several times until practically dry curd is obtained. The washed curd is closely observed for the relative shape and size of the curd particles. Close examination of "washed" cottage cheese curd often reveals appearance defects which may have escaped identification otherwise (i.e., unwashed cheese).

REQUIREMENTS FOR HIGH QUALITY COTTAGE CHEESE

High quality creamed cottage cheese should have the following major sensory attributes:

1. *Flavor.* A clean, slightly acidic flavor with a slight "buttery" aroma (balanced culture flavor) and a creamy flavor note. There should be no particular aftertaste and only a sufficient salty taste to "bring out" the desired flavor.
2. *Body and Texture.* The body should have a "meat-like" (meaty) consistency, but not be overly firm, rubbery, or tough when it is first chewed or masticated (placed between the teeth). The product texture should be relatively smooth (silky, meaty) throughout. The evaluator should be able to feel (as well as see) distinct curd particles.
3. *Color and Appearance.* The curd particles should be reasonably uniform in both size and shape. The creamed cheese should exhibit a moderate gloss or sheen; the cream dressing should definitely cling or adhere to curd particles.

Evaluation of Package

Cottage cheese containers, whether plastic or paperboard, should be clean, dust-free, free of food substances on exterior surfaces, show no ink smears, and be neat and attractive. An opened package should reveal an appropriate volume of product for the given package size and withstand slight tilting without spilling; there should be no visible free whey. Protective closures should adequately cover the top lip of the carton. Most importantly, the product package must be nonabsorbent and contribute no off-flavor to the creamed cottage cheese.

Evaluation of Appearance and Color

The general appearance or visual impression of creamed cottage cheese should be attractive and pleasing. The curd particles should be separate and distinct, moderately uniform in both size and shape; the overall product should exhibit a glossy, creamy-white color. In creamed cottage cheese, the bulk of the cream should be absorbed by the curd particles. There should be a minimum of free or separated cream. The cream dressing should be reasonably viscous, relatively foam-free, and able to adhere to the curd particles. A limited amount of excess dressing should form a uniformly smooth coating on the curd particles and be void of separated water (free whey). Preferably, high-quality cottage cheese should exhibit little or no particle shattering (curd dust) and curd matting (lumps).

Most appearance and color defects of creamed cottage cheese should be obvious to the alert evaluator. Terminology for the various criticisms of appearance is specific and descriptive. The appearance defects of creamed cottage cheese are often related to deviations from recommended manufacturing procedures. Table 7.5 lists the common color and appearance defects of creamed cottage cheese, their possible cause and methods of control. (Also see Figs. 7.8 and 7.9.) The ADSA scoring guide for various sensory defects of creamed cottage cheese (including appearance and color) is presented in Table 7.4.

Table 7.5. Common Color and Appearance Defects of Creamed Cottage Cheese, Their Probable Causes and Remedial Measures.

Color/Appearance Defect	Probable Causes	Remedial Measures
Free cream	1. Excessive cooking which causes a firm, rubbery curd; this prevents dressing adsorption.	1. Reduce cooking temperature to avoid too firm a curd.
	2. Insufficient washing of curd (contact time).	2. Allow wash water to remain in contact with the curd for a longer time.
	3. Cutting pH of curd too high.	3. Cut curd at a pH of 4.65–4.70.
	4. Too-rapid temperature rise during cooking of curd (causes surface denaturation and loss of dressing permeability).	4. Exercise better control of curd cooking (i.e., do not cook too fast)

(continued)

Table 7.5. (*Continued*)

Color/Appearance Defect	Probable Causes	Remedial Measures
Free whey	1. Undercooking of curd retains an excess amount of whey.	1. Increase cooking temperature to help expel more whey.
	2. Insufficient washing of curd.	2. Increase curd washing time.
	3. Cutting pH of curd too high.	3. Cut curd at a pH of 4.65–4.70.
Lacks uniformity	1. Uneven cutting of coagulum.	1. Repair (replace) wires in the knives.
	2. Aggressive or abusive agitation during cooking.	2. Use proper cutting techniques.
		3. Train personnel in careful methods of cutting, agitating, and cooking curd.
Matted	1. Cutting pH of curd too high.	1. Cut curd at a pH of 4.65–4.70.
	2. Insufficient/inadequate agitation, especially during first hour of cooking.	2. Employ a standardized method of cooking and stirring.
	3. Curd cooked too rapidly.	3. Initiate cooking slowly and gradually, accelerate pace at mid-point of cooking.
	4. Missing wires in the knives.	4. Repair knives.
Shattered Curd	1. Excessive heat-treatment of skim milk.	1. Use minimum pasteurization conditions.
	2. Excessive acidity (pH too low) at cut	2. Cut curd at a pH of 4.65–4.70.
	3. Total solids content of skim milk too low.	3. Maintain total solids $\geq$ 8.75%.
	4. Overly severe agitation.	4. Stress gentle, "caring" form of agitation.
	5. Excessive amount of coagulator used.	5. Use the minimum amount of coagulator.
	6. Rough handling of curd during draining, pumping, creaming, and/or packaging.	6. Restrict curd handling to a minimum, if possible.

Source: Adapted from Connolly *et al.* 1984. Courtesy American Cultured Dairy Products Institute, Washington, D.C.

Evaluation of Body and Texture

Ideally, creamed cottage cheese should demonstrate a relatively firm but tender body and exhibit a silky-smooth and meaty-like texture. The curd particles should be relatively uniform in both size and configuration, for the given type of curd being considered. Understandably, size of curd particles and the relative degree of firmness of cottage cheese curd in the U.S. has not been fully standardized (at least from an objective standpoint). These particular characteristics are guided primarily by consumer preferences within a given market area of the country. However, with the advent of "popcorn" or sweet-curd type cheese, more emphasis has been placed on control of curd size and firmness than was the previous situation. Many manufacturers market two distinct types of cottage cheese; "small curd" and "large curd." Although large curd is usually firmer and tends to exhibit a more acid taste (due to more entrapped lactic acid), both product types are of comparable flavor character.

The most desirable body for cottage cheese is one that is apparently neither too firm nor too soft; it should have uniform consistency throughout the curd particle. The curd should be sufficiently firm to hold its general shape and maintain its individual identity against matting, yet simultaneously be soft enough to yield a silky, "tacky" smear when washed curd pieces are pressed lightly between the thumb and forefinger. Curd that is too firm tends to resist such pressing (i.e., there is a tendency for the curd to "spring back" or retain its original shape when the pressure is released). When a thoroughly washed curd particle is dropped onto the floor from waist level, an "appropriate bodied" curd particle will have a perceptible bounce (2.5–7.6 cm [1–3 in.]). A too-soft bodied curd will usually "splatter" and break apart when it strikes the floor; a too-firm (tough, rubbery) curd will generally bounce in excess of 7.6 cm (3 in.) when dropped from waist level.

The appropriate body and texture properties of cottage cheese are associated with consumer acceptance. In a laboratory, an evaluator can "wash" creamed cottage cheese with the aid of a fine-mesh sieve to void the dressing. This can serve to present a truer picture of curd uniformity (see Fig. 7.8). By tearing apart curd particles, the evaluator can more readily perceive the extent of "meatiness" and overall consistency of a cross-section of the curd (from the outer surface to the center).

Curd particles that are smooth, meaty, and tender tend to exhibit distinct striations of protein fiber when the particle is torn apart and closely examined. Such curd texture has been reported to exhibit good

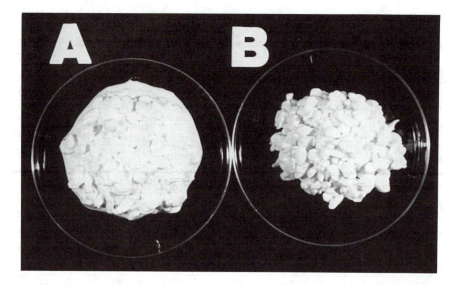

Fig. 7.8. A demonstration of the uniformity of curd size and curd shape by two different approaches: A—Careful visual examination of creamed curd; B—Visual examination of "washed" curd.

liquid capillarity, and thus this feature facilitates more complete absorption of added cream dressing.

Body and Texture Defects of Creamed Cottage Cheese

The more common body and texture defects of cottage cheese are:

Too firm, (rubbery, tough)	Pasty, (sticky, doughy)
Gelatinous	Weak/soft (mushy)
Mealy/grainy (gritty)	Overstabilized dressing (slick)

Brief descriptions of the characteristics which are indicative of the above-listed body and texture defects of creamed cottage cheese are given in the following paragraphs.

Too Firm (Rubbery, Tough). When the curd of overly "firm or rubbery" cottage cheese is pressed between the tongue and the roof of the mouth, a modest (but sometimes subtle) resistance to crushing or mastication can be noted by the careful observer. Further manipulation of the product in the mouth may suggest either a high solids level or low moisture content of the internal curd structure. Unless this firm-

Fig. 7.9.   Examples of some of the more common color and appearance defects of creamed cottage cheese: A—A reference sample for the "ideal" appearance; B—Shattered curd; C—Matted curd; D—Lacks cream; E—Free cream; F—Free whey.

ness is quite pronounced and/or associated with nonabsorption of cream dressing, this defect is not considered particularly serious. A tough, rubbery curd may be due to: (1) overuse of rennet or other milk coagulators; (2) too high a cooking temperature; (3) cooking too long; or (4) insufficient formed acid (pH ≥ 4.7) at the time of cutting the coagulum.

Gelatinous. "Gelatinous" cottage cheese tends to have a sticky or slightly "jelly-like" character. Occasionally, the product suffering this defect resembles tapioca pudding. Sometimes, this body defect has an accompanying bitter taste and a translucent curd appearance. A gelatinous defect is generally due to proliferation of psychrotrophic bacteria in the product and, hence, an indication of product spoilage; such a product is often unpalatable and, hence, unsalable.

Mealy/Grainy (Gritty). Unfortunately, this is a quite prevalent defect in U.S. cottage cheese. "Mealy/grainy" can be detected by briefly pressing (with the tongue), masticated curd against the roof of the mouth and carefully attempting to perceive the presence or absence of a gritty or corn meal-like sensation (just prior to expectorating or swallowing the sample). Another way of detecting curd graininess is to "wash" away the cream dressing, carefully knead the washed curd, and then smear it between the fingers. Instead of a silky, smooth smear, (which is characteristic of an "ideal" curd texture), the evaluator often will find a somewhat dry, rough, serrated curd mass instead. The uncreamed curd of "gritty" cottage cheese is similar to the curd formed in the manufacture of casein.

The mealy/grainy defect of cottage cheese may be caused by too-low moisture and/or overdeveloping the acid during coagulum and/or curd formation. To minimize this curd defect, more moisture can be incorporated by cooking the curd more gradually and by use of lower cooking temperatures. Curd cutting should only be undertaken when the coagulum reaches the isoelectric point of casein (pH 4.65–4.70). Mealiness/graininess may also be caused by: (1) nonuniform cutting of the curd; (2) uneven heating (cooking) of portions of the curd; (3) too-rapid cooking of the curd/whey mixture; (4) inadequate agitation during the cooking phase; and (5) allowing curd particles to contact extremely hot surfaces during cooking. The major techniques for controlling the extent of graininess/mealiness are cutting the coagulum at the proper pH (to avoid excess acidity) and maintaining sufficient, but gentle, agitation throughout the cooking stage of the process.

Pasty (Sticky, Doughy). The "pasty" defect in creamed cottage cheese is closely associated with soft, weak, high-moisture curd. En masse, pasty-bodied cheese resembles cereal dough, a flour-like paste, or glue. The curd particles have a tendency to mat or stick together in soft clumps. Authorities on cottage cheese quality simply regard the pasty defect as a possible extension or advanced degree of the weak/soft criticism (discussed next).

Weak/Soft (Mushy). This defect is characteristic of a high moisture cottage cheese of relatively low solids content. It is caused by careless manufacturing methods which favor substantial retention of whey (moisture) in the curd. Weak-bodied cottage cheese frequently may not meet the legal maximum of 80% moisture content. Following storage, a weak, soft-bodied cheese may often manifest a bitter taste, due to the entrapped whey (and associated peptides). According to a quality manual published by the American Cultured Dairy Products Institute (Connolly *et al.* 1984), probable causes of the weak/soft and/or pasty defects in creamed cottage cheese are:

1. Excessive heat treatment of cheese skim milk.
2. Excessive acidity (low pH) at time of cutting coagulum and during cooking.
3. Too-low cooking temperatures.

Overstabilized Dressing (Slick). In an attempt to "thicken" dressing, minimize free whey in the final product, and/or enhance adherence of the dressing to the curd, processors may occasionally overdevelop dressing viscosity through excessive use of nonfat dry milk, stabilizers, and/or emulsifiers. When this defect occurs, it is quite apparent; creamed cottage cheese may appear markedly dry and some individual curd particles may appear to be surrounded by a thick, pasty coating. Overstabilized dressing is not considered a serious defect unless it is so severe as to impart an off-flavor or unfavorable mouthfeel (slippery or slick) to the cottage cheese. Decreasing the quantity or changing the source of stabilizer can effectively eliminate the so-called "slick" or "overstabilized" defect in creamed cottage cheese.

The intensities of various body and texture defects are usually scored according to the guide for scoring creamed cottage cheese shown in Table 7.4. The various causes and methods for controlling body and texture defects of cottage cheese are summarized in Table 7.6.

Evaluation of Flavor

Creamed cottage cheese of high quality should have a fresh, pleasant, clean, delicate acid, mild diacetyl flavor, that imparts no aftertaste when the sample has been expectorated or swallowed. These flavor attributes are a "composite" of curd acidity, volatile compounds formed by the lactic culture fermentation, and/or from addition of aroma-producing microorganisms to the cream dressing. The composition of the cream dressing and the added salt also serve to greatly enhance the flavor of creamed cottage cheese. The amount of added salt should be just sufficient to enhance the pleasant flavor blend, but not produce a discernible salty taste.

Cream dressing should be added in such quantities that the curd can readily absorb it (within a reasonable time period) before marketing (≤ 24 hours). The evaluator should recognize two types of cream dressing: (1) one virtually devoid of aroma, but clean, sweet, and pleasantly acidic and (2) the other with an obvious diacetyl or culture aroma. Both types of flavor characteristics generally are considered equally appropriate in the discretion of experienced dairy product judges.

Cottage cheese curd (without cream) is referred to or labeled as "dry cottage cheese curd." Plain curd may be sold wholesale in bulk for later creaming, packaging, and retail distribution. Dry unsalted curd is also sold in retail packages for use in cooking, baking, salads and for use in special "low salt," "low fat," "low cholesterol," and/or "reduced calo-

Table 7.6. Common Body and Texture Defects of Creamed Cottage Cheese, Their Probable Causes and Remedial Measures.

Body and Texture Defects	Probable Causes	Remedial Measures
Firm/rubbery	1. Cutting pH of curd too high.	1. Cut curd at pH of 4.65–4.70.
	2. Excessive cooking time or temperature.	2. Carefully determine the optimum cooking endpoint.
Mealy/grainy	1. Cooking rate too rapid, especially during initial stages of cooking.	1. Slow, gradual cook temperature increments, accelerate rate at midpoint of cook.
	2. Excess acidity developed.	2. Cut curd at pH of 4.65–4.70.
	3. Inadequate vat agitation.	3. Controlled, steady agitation.
	4. Too much curd in direct contact with hot vat surfaces.	4. Minimize temperature gradient.
Pasty	An extreme case of weak/soft (see below).	
Slick	Excessive use of stabilizer in dressing.	Decrease amount of stabilizer in dressing.
Weak/soft	1. Excessive heat-treatment of skim milk.	1. Use minimum pasteurization conditions.
	2. Excessive acidity (low pH) at cut and during cook.	2. Cut curd at pH of 4.65–4.70.
	3. Inadequate cookout temperature.	3. Carefully determine optimum cook-out temperature.
	4. Overdressing the curd.	4. Calculate and blend curd and dressing at appropriate ratio.

Source: Adapted from Connolly *et al.* 1984. *Courtesy* American Cultured Dairy Products Institute, Washington, D.C.

rie" diets. Uncreamed cottage cheese is often evaluated by employing the same score card that is used for the creamed product. Much attention is given to the body and texture of dry curd, but one will not find it to possess a highly appealing flavor. Most likely, a distinctive flat, dull flavor will be obvious to most evaluators of dry cottage cheese curd. Most dry curd cottage cheese is virtually devoid of aroma, unless an especially selected diacetyl-producing culture was used for curd manufacture. The flavor of dry curd cottage cheese should be clean and pleasantly acidic, and show little persistence after the sample has been expectorated.

Flavor Defects of Creamed Cottage Cheese

As a rule, creamed cottage cheese is a highly perishable product, despite rigorous sanitation and product-handling precautions (Bodyfelt 1981b) that are usually practiced in manufacture. The specific flavor defects of creamed cottage cheese are:

High acid (sour)	Lacks freshness (stale, storage)
Bitter	Malty
Diacetyl (too high)	Oxidized (metallic)
Flat, lacks flavor	Musty
Foreign, chemical, medicinal	Rancid
Fruity/fermented	Salty (too high)
Lacks fine flavor (acetaldehyde,	Unclean
plain yogurt-like)	Yeasty (vinegar-like)

A brief description of the characteristic features of each off-flavor is helpful in trying to identify them; some flavor defects are distinctive and unique to cottage cheese.

High Acid (Sour). The terms "high acid" or "sour" basically designate various intensities of the same defect. They generally reflect an excess of lactic acid, a level of acidity beyond that which is considered desirable or highly acceptable to taste. However, this acid taste, it should be emphasized, is generally clean and sharp (no particular aftertaste). The so-called "sour" taste may be pronounced and may sometimes be associated with other bacterial defects, such as bitter or fruity/fermented.

The development of lactic acid by the culture inoculated into skim milk in making cottage cheese is essential for curd formation, unless the cheese milk is chemically acidified (direct set). Also, the formed lactic acid or added acidulant helps contribute to cheese flavor. However, if too much acid is developed in the course of curd formation or curd cooking, it usually results in a high-acid (sour) curd. A high acid curd tends to mask some of the more delicate, volatile, organic compounds responsible for the desirable flavor of cottage cheese. Insufficient washing(s) of the curd may result in too much whey retention and hence cause high acid flavor. Cottage cheese, such as just described, may sometimes merit the flavor criticism, "whey taint."

Bitter. A "bitter" off-taste in cottage cheese is characterized by its: (1) relatively slow reaction time and delayed perception; (2) detection at or near the back of the tongue; and (3) persistence after sample expectoration. Pronounced bitterness is not unlike the sensation imparted by quinine or caffeine. This defect is frequently encountered in older samples of cottage cheese or in cheese stored at favorable growth temperatures for psychrotrophic organisms (which are the principal causative agents). In the past, a bitter off-taste in cottage cheese may have resulted from the consumption of certain weeds by cows; however, bitter cottage cheese from this source would be extremely infrequent today.

Diacetyl (Too High). This flavor defect is noted by an overall lack of flavor balance which favors the distinct aroma of diacetyl, while masking other important flavor notes. It is often characterized by the presence of a harsh flavor and/or excess aroma, which seems "out-of-balance" for cottage cheese.

Flat (Lacks Flavor). A "flat" flavor in cottage cheese may be noted by an absence or lack of the characteristic flavor and aroma; identification is that simple and direct. A dry, unsalted, washed, "rennet curd" yields a distinctly flat taste, not unlike that of pure casein. A creamed cottage cheese may also tend to yield a flat taste and aroma during an early or intermediate stage of the development of an oxidized off-flavor. In this case, the initial "flatness" may lead to a delayed flavor perception that suggests a metallic off-flavor; the evaluator should be alert to this possible follow-up off-flavor. Even when pronounced, a flat flavor defect is not considered serious enough to classify the cottage cheese as a poor product (unless an associated and more objectionable off-flavor accompanies the flatness).

Foreign (Chemical, Medicinal). A "foreign" off-flavor, though only occasionally noted in creamed cottage cheese, distinguishes itself by being entirely unlike any off-flavor that might be anticipated in the product—it seems atypical or most unusual. Sometimes, the actual nature of the off-flavor betrays its identity. The persistent, atypical, or "out-of-place" off-flavor may suggest contamination by either cleaning compounds, chlorine, iodine, phenol, or various other chemical substances that may have accidentally gained entry to the product.

Fruity/Fermented. Surprisingly, a "fruity" or "fermented" defect may have a pleasant, aromatic quality, suggestive of pineapple, apples, bananas, or strawberries. A mere "whiff" of the just-opened package usually confirms the presence of this more serious defect. Tasting usually suffices to substantiate the already noted aroma and may also reveal an associated unpleasant, distinctive lingering aftertaste. The cottage cheese may be near its "sell-by" date and/or have been stored at elevated and favorable temperatures for psychrotrophic bacterial growth. The product may soon reach a point of unpalatability; complete spoilage is imminent.

Lacks Fine Flavor (Acetaldehyde, Plain Yogurt-like). When a given lactic culture which has been added to the cream dressing produces acetaldehyde as a principal volatile component, a "green-apple" or yogurt-like off-flavor often occurs in the final product. Such cottage cheese is said to "lack fine flavor," due to formation of substantial levels of acetaldehyde. Lacks fine flavor also suggests a note of the "coarseness" or "harshness" off-flavor in cottage cheese.

Lacks Freshness (Stale, Storage). These three off-flavors have been grouped together because they have much in common. The relative age of the product or ingredients seem to be the underlying factors for this group of flavor defects. A difference in defect intensity exists between "lacks freshness" and "stale." The latter is more obvious or intense, whereas the former defect tends to almost shield its true identity; it is simply a general lack of refreshingness in the product. Staleness may also be imparted by old ingredients (e.g., dry skim milk, cream, stabilizer).

Cottage cheese flavor is usually at its best or "peak" within one to five days after manufacture. When properly made and adequately refrigerated, cottage cheese should retain its "typical flavor" for a reasonable period of time (two to three weeks). Frequently during storage and distribution, even under adequate refrigeration ($\leq 4.4\,^{\circ}\mathrm{C}\,[\leq 40\,^{\circ}\mathrm{F}]$), cottage cheese progressively deteriorates in flavor quality. This is undoubtedly due to the simultaneous occurrence of microbiological and chemical changes. This resulting flavor deterioration can be referred to as "lacks freshness," since the cottage cheese seems to lack the refreshing flavor characteristics of a more recently made product. A storage off-flavor can develop in cottage cheese that is packaged and subsequently exposed to "volatiles" within the refrigerator or cold storage space. Hence, the "storage" off-flavor, if and when it does occur, is appropriately classified as an absorbed flavor defect.

Malty. A "malty" off-flavor defect in cottage cheese is rather specific or distinctive; maltiness tends to predominate over most any other flavor defect that may be present. This off-flavor, which resembles "Grape Nuts®" or malted milk, is quite easy to identify due to its uniqueness. It generally has a quick reaction time; the aftertaste is not prolonged. Since a malty off-flavor is the result of contamination by an outgrowth of *S. lactis* var. *maltigenes,* additional developed acidity (a sourness taste) may accompany a malty aroma defect.

Oxidized (Metallic). Fortunately, these two more serious off-flavors are infrequently encountered in cottage cheese. If they do occur, improper selection and/or handling of the cream for preparation of the curd dressing is usually indicated. "Metallic" has a slightly astringent, "rusty nail-like" taste, while "oxidized" is an off-flavor more reminiscent of wet cardboard or paper. Smelling the sample usually gives little indication of a metallic defect, but a weak off-odor may sometimes suggest the characteristic or "generic oxidized" off-flavor. Some research indicates that these two defects may be different intensities of the same basic defect (e.g., lipid auto-oxidation resulting from copper or iron contamination of susceptible milk or cream used for preparation of the cream dressing). This defect may be noted more frequently during late fall, winter, or early spring, when cows are more apt to be on dry feed for extended periods. Also, if a milk supply that is susceptible to milkfat auto-oxidation is used to produce cottage cheese curd, this potential off-flavor could conceivably be retained by the curd. An oxidized flavor defect will generally intensify during storage and may occasionally develop into a distinct, "tallowy" off-flavor. Any copper contamination, especially of the cream or milk used in preparing the dressing, can easily catalyze development of an oxidized off-flavor.

Musty. "Musty" cottage cheese exhibits an aroma that resembles that of a damp, poorly ventilated cellar. This serious, but seldom encountered, defect in cottage cheese is due to the outgrowth of various microbial contaminants, primarily molds, in cottage cheese. Cheese curd may sometimes become contaminated with certain psychrotrophic bacteria (*Psuedomonas taetrolens*) as the result of faulty plant sanitation. When this development is coupled with inadequate refrigeration and processing methods, the musty defect may occur; it usually intensifies as cottage cheese is held in storage. The product would soon become unpalatable, if such is not already the case.

Rancid. "Rancidity," in cottage cheese, as in milk, may be noted by an astringent, puckery feeling at the base of the tongue and throat, as well as an associated bitter aftertaste, following sample expectoration. The objectionable rancid off-flavor tends to persist as an unpleasant aftertaste for a considerable period of time.

If rancid milk or cream is used to manufacture cottage cheese curd and/or dressing, this serious off-flavor will usually carry over into the finished product. Since rancidity is due to the action of the enzyme lipase on milkfat, this flavor defect is more often derived from the added cream than from the curd. This defect may intensify as the cheese becomes older, particularly if the homogenized dressing was not adequately heat treated. Proper pasteurization of all milk products used in making cottage cheese prevents rancidity, providing the raw milk and cream supplies were free of this defect. However, cottage cheese processors should recognize that, because of its high acidity, cottage cheese may have a rancid flavor even if the off-flavor was below its detection threshold in the milk and/or cream from which the cottage cheese was made.

Salt (Too High). "High salt" manifests itself as an unwanted, sharp, piercing, biting taste sensation which detracts from the pleasant flavor of high-quality cottage cheese. Addition of the proper amount of salt (approximately 1% or less) enhances cottage cheese flavor; however, oversalting defeats the purpose of this product ingredient. Both the reaction and adaptation times of the tastebuds are of short duration for the salty taste sensation. The initial sensation encountered upon tasting high-salt cottage cheese is soon dissipated and relieved by an induced copious flow of saliva. Experienced evaluators of cottage cheese commonly recognize that 0.6 to 1.0% added salt is generally required to help enhance the flavor of cottage cheese. However, a distinct or obvious "salty taste" in creamed cottage cheese should not be consciously perceived by the product evaluator (Bodyfelt 1982, Wyatt 1983).

Unclean. The designation for this serious defect is self-explanatory. The off-flavor "unclean" cannot be easily expressed in other descriptor terms. Some judges have dared to use the term "dirty" to describe the unpleasant, objectionable, unclean-like off-flavor that sometimes proliferates as an undesirable aftertaste in cottage cheese that has commenced to spoil or exhibit microbial deterioration. This unpleasant flavor note, often accompanied by a distinct bitter off-taste, generally remains for some time after sample expectoration; product palatability is at stake.

Yeasty (Vinegar-like). "Yeasty" and "vinegar-like" defects in cottage cheese have a peculiar aromatic quality in addition to a possible associated high-acid note. While this defect may be caused by growth of yeasts, and tends to exhibit a yeasty or earthy off-odor, the often associated sharp, pungent taste may be suggestive of vinegar (possibly due to bacterial fermentation). Various microbial contaminants, including certain kinds of psychrotrophic bacteria, are generally responsible for this objectionable off-flavor. Usually, serious sanitation shortcomings in manufacture and/or packaging are at fault and in need of elimination to correct this serious off-flavor problem in cottage cheese. The shelf life of this relatively perishable product is significantly reduced by poor sanitation and lack of temperature control (Bodyfelt 1981b, 1981c; Morgan 1970a, 1970b).

KEFIR

Product Characteristics

One of the most recently introduced forms of fermented milk to the United States is kefir. This cultured milk product originated in the region of the Caucasus mountains many centuries ago. This effervescent, slightly alcoholic beverage is regularly produced in the USSR, south-western Asia, and Eastern Europe (Kosikowski 1977). The product, kefir, is also variously named, according to Kemp (1984). Some product names in various locales include: kephir, kiaphur, kefyr, képhir, kéfer, knapon, kepi, or kippi. The per-capita consumption of kefir in the USSR is approximately 4.5 kg (9.9 lb) per year (Davies and Law 1984).

The traditional method of producing kefir involves the addition and subsequent recovery of so-called kefir grains to especially prepared whole milk or skim milk from sheep, goats, or cows (depending on the locale of manufacture). Kefir grains are gelatinous, white to yellowish colored granules of irregular shape that range in size from wheat kernels to walnuts. The kefir granules often resemble the appearance of cauliflower flowerettes. After heat-treated milk (95°C [203°F]) is tempered to 21°C to 30°C (70°F–86°F), up to 5% by volume of kefir grains are added or suspended into milk within mesh netting. Upon becoming saturated with milk, the kefir granules swell and metabolize milk nutrients to initiate dual fermentations—lactic acid and ethanol. Some U.S. manufacturers have reportedly removed the alcohol-producing yeast from their kefir cultures.

It is widely acknowledged that the technology of kefir production is more complicated than the process of buttermilk or yogurt manufacture. Additional research activity seems to be needed to develop more practical methods of commercial scale manufacture of kefir. Several genera and species of bacteria and yeast exist in a symbiotic relationship within the kefir grains (and the final product).

Most of the complexity of kefir manufacture is probably due to the spectrum of microbial flora used to propagate this interesting beverage. The "seed" material, kefir grains, have been demonstrated by various investigators (according to Davies and Law 1984, Kemp 1984, Kosikowski 1977, and Robinson 1981) to consist of several genera of yeasts and lactic acid bacteria (LAB). The yeast have included: *Saccharomyces kefir, S. delbruekii, S. cerevisiae, S. exiguus, Candida (Torula) kefir,* and *C. psuedotropicalis.* Lactobacilli have included *L. (caucasicus) kefir, L. brevis, L. bulgaricus,* and *L. acidolphilus.* Lactic streptococci have also been isolated from kefir granules (i.e., *Leuconostoc cremoris* and *L. mesenteroides,* as well as *Streptococcus lactis, S. cremoris,* and *S. lactis* subsp. *diacetylactis*).

Quality Control of Kefir

The surfaces of kefir grains are frequently covered with a white mold, *Geotrichum candidum.* Furthermore, kefir granules may become contaminated with coliforms, bacilli, and/or micrococci in the handling, storage, and "washing" procedures. The aforementioned microbial contaminants can hasten product spoilage and associated off-flavors (e.g., cheesy, stale, unclean, etc.). The experience of several U.S. manufacturers has indicated that it is a challenge to consistently produce kefir that exhibits uniform sensory properties. Close control of the microbial inoculum, and rigorous attention to sanitation and the use of airtight packaging is necessary. Some improvement in the day-to-day uniformity of this product has been facilitated by the availability of freeze-dried kefir cultures from commercial suppliers.

Desired Sensory Characteristics of Kefir

In an effort to introduce kefir, some North American dairy processors have opted to market an atypical example of the traditional or original style of kefir (i.e., a hint of yeast-like flavor, plus variable amounts of CO_2 and ethanol). Kemp (1984) has described the sensory characteristics of high-quality kefir as follows:

> "It has a pH of about 4.0; a clean, pleasant acid taste without any bitterness (aftertaste); prickling and sparkling of CO_2; a slight taste (and aroma) of yeast; a smooth texture; altogether a very refreshing beverage."

Kosikowski (1977) has described the compositional properties of typical or high-quality kefir, in terms of major end products of fermentation: approximate pH of 4.4, 0.8% lactic acid, about 0.5 to 1.0% ethyl alcohol, and sufficient CO_2 content to make the beverage "foam and fizz like beer." Furthermore, Kosikowski (1977) emphasized that "Kefir's typical flavor is due mainly to an optimum ratio (3:1) of diacetyl (approximately 3 ppm) to acetaldehyde (approximately 1 ppm)." Apparently, the *Leuconostoc* sp. play a key role in synthesizing an important flavor compound in kefir—diacetyl (buttery aroma). Other authorities (Davies and Law 1984, Robinson 1981) indicate a range of 0.8 to 1.0% lactic acid and more moderate levels of ethanol (0.026 to 0.5%) in typical kefir.

In commercial distribution, glass bottles (sealed with crown caps) or foil-laminated paperboard cartons are required for the packaging of kefir, if the CO_2 and ethanol are to be retained in the final product at the time of consumption.

Flavored Kefir

It should be noted that all previous discussion on kefir has focused on plain or unflavored forms of the product. However, given the so-called "sweet tooth" or palate of the American consumer, the producer and marketer of kefir will undoubtedly find it necessary to simultaneously add fruit flavoring and sweetener to this beverage product to enhance consumer acceptance.

Currently, flavored kefir is available as a beverage in flavors such as strawberry, raspberry, boysenberry, cherry, peach, orange, and lemon. Similar to flavored yogurt, an expectation should be that flavored kefir definitely reflect the *true flavor* of the label statement on the package. In addition, it is assumed that flavored kefir should exhibit a slight to moderate level of clean acid taste and a slight hint of yeasty and alcohol notes. From a mouthfeel standpoint, high-quality flavored kefir should exhibit a marked refreshing effervescence (or "zip") due to the formed (and hopefully retained) carbon dioxide. Finally, high-quality kefir should manifest a distinctly viscous, but smooth and pourable viscosity (body).

REFERENCES AND BIBLIOGRAPHY

American Dairy Science Association. 1987. Committee on Evaluation of Dairy Products. Score card and guide for cottage cheese. Champaign, IL.

Bills, D. D., Yang, C. S., Morgan, M. E., and Bodyfelt, F. W. 1972. Effect of sucrose on the production of acetaldehyde and acids by yogurt culture bacteria. *J. Dairy Sci.* 55(11):1570.

Bodyfelt, F. W. 1976. Swiss-style yogurt score card and glossary of pertinent terms. Unpublished Communication. Oregon State Univ. Corvallis, OR.

Bodyfelt, F. W. 1981a. Cultured sour cream: Always good, always consistent. *Dairy Record.* 82(4):84.

Bodyfelt, F. W. 1981b. Temperature control monitoring for cottage cheese plants. *Dairy Record.* 82(1):65.

Bodyfelt, F. W. 1981c. Sensory and shelf-life characteristics of cottage cheese treated with sorbic acid. *Proc. Biennial Marschall Int'l Cheese Conf.* Madison, WI.

Bodyfelt, F. W. 1982. Processors need to put some pinch on salt. *Dairy Record* 83(4):83.

Code of Federal Regulations. 1985. Title 21. Part 131.200 Yogurt. U.S. Government Printing Office. Washington, D.C.

Code of Federal Regulations. 1987. Title 21. Part 131. U.S. Government Printing Office. Washington, D.C.

Connolly, E. J., White, C. H., Custer, E. W., and Vedamuthu, E. R. 1984. Cultured Dairy Foods Quality Improvement Manual. American Cultured Dairy Products Institute. Washington, D.C. 40 pp.

Davies, F. L. and Law, B. A. (Editors). 1984. *Advances in the Microbiology and Biochemistry of Cheese and Fermented Milk.* Elsevier Applied Sci. New York. 260 pp.

Duggan, D. E., Anderson, A. W., and Elliker, P. R. 1959. A frozen concentrate of *Lactobacillus acidophilus* for preparation of a palatable acidolphilus milk. *Food Technol.* 13(3):465.

Duthie, A. H., Nilson, K. M., Atherton, H. V., and Garrett, L. D. 1977. Proposed score card for yogurt. *Cultured Dairy Prod. J.* 12(3):10.

Foster, E. M., Nelson, F. E., Speck, M. L., Doetsch, R. N., and Olsen, J. C. 1957. *Dairy Microbiology*. Prentice-Hall, Inc. Englewood Cliffs, NJ. 492 pp.

Gilliland, S. E. 1979. Beneficial interrelationships between certain microorganisms and humans: Candidate microorganisms for use as dietary adjuncts. *J. Food Protect.* 42(2):164.

Gilliland, S. E. (Editor). 1986. *Bacterial Starter Cultures for Foods*. CRC Press, Inc. Cleveland, OH. 250 pp.

Keenan, T. W., Bodyfelt, F. W., and Lindsay, R. C. 1968. Quality of commercial buttermilks. *J. Dairy Sci.* 51(2):226.

Kemp, N. 1984. Kefir, the champagne of cultured dairy products. *Cult. Dairy Prod. J.* 19(3):29.

Klaenhammer, T. R. 1982. Microbiological considerations in selection and preparation of lactobacillus strains for use as dietary adjuncts. *J. Dairy Sci.* 65(7):1339.

Kosikowski, F. 1977. *Cheese and Fermented Milk Foods*. Edwards Brothers, Inc. Ann Arbor, MI. 709 pp.

Lindsay, R. C. 1965. Cultured dairy products. In: *Chemistry and Physiology of Flavors*, Schultz, H. W., Day, E. A., and Libbey, L. M. (Editors). AVI Pub. Co. Westport, CT. 552 pp.

Marshall, V. M. 1987. Fermented milks and their future trends. I. Microbiological aspects. *J. Dairy Res.* 54:559.

Morgan, M. E. 1970a. Microbial flavor defects in dairy products and methods for their simulation. I. Malty flavor. *J. Dairy Sci.* 53(3):270.

Morgan, M. E. 1970b. Microbial flavor defects in dairy products and methods for their simulation. II. Fruity flavor. *J. Dairy Sci.* 53(3):273.

Robinson, R. K. (Editor). 1981. *Dairy Microbiology*. Vol. 2. *The Microbiology of Milk Products*. Applied Science Publishers, Inc. Englewood, NJ.

Ryan, J. M., White, C. H., Gough, R. H., and Burns, A. C. 1984. Methodology for evaluating yogurt. *J. Dairy Sci.* 67:1369.

Sandine, W. E. 1984. Use of pasteurized milk cultures of *Propionibacterium shermanii* as a microbial inhibitor in cultured dairy foods. Unpublished Communication. Oregon State Univer., Corvallis, OR.

Sandine, W. E., Muralidhara, K. S., Elliker, P. R., and England, D. C. 1972. Lactic acid bacteria in food and health: A review with special reference to enteropathogenic *E. coli* as well as certain enteric diseases and their treatment with antibiotics and lactobacilli. *J. Milk Food Technol.* 35(12):691.

Speck, M. L. 1972. Control of food-borne pathogens by starter cultures. *J. Dairy Sci.* 45(9):1281.

Speck, M. L. 1976. Interactions among lactobacilli and man. *J. Dairy Sci.* 59(2):338.

Tamime, A. Y. and Robinson, R. K. 1985. *Yogurt Science and Technology*. Pergamon Press, New York. 431 pp.

Wyatt, C. J. 1983. Acceptability of reduced sodium in breads, cottage cheese and pickles. *J. Food Sci.* 48(4):1300.

8

Sensory Evaluation of Cheese

Of all the foods which mankind has created for eating pleasure, cheese is unique in many ways. No other group of foods possesses such variations in flavor, consistency, appearance or number of categories (Battistatti *et al.* 1984; Eckhof-Stork 1976; Kosikowski 1977; Scott 1981; Wilster 1980). Perhaps, no other form of food is more universally known and enjoyed around the world. Cheese provides a vast panorama of flavors and culinary experiences; it truly adds zest to eating or entertaining.

CHEESE DEFINED

In general, cheese is a dairy product made by coagulating either whole milk, part-skim (lowfat) milk, skim milk, or cream; removing much of the liquid portion while retaining the coagulum and entrapped milk solids. The retained milk solids may or may not undergo subsequent ripening. The liquid portion, known as "whey," consists primarily of water, lactose, and minor percentages of whey proteins, ash, and milkfat. The solid or semisolid portion remaining after the whey is removed is known as "curd." Milk solids in cheese curd made from whole milk are composed primarily of casein and fat, along with some ash, lactose, and whey proteins. Skim milk curd contains approximately the same milk components as curd manufactured from whole milk, except for a substantially less amount of fat.

The above definition of cheese encompasses hundreds of varieties of cheese. A USDA publication (Agricultural Research Service 1978) describes more than 400 cheese varieties and indexes more than 800 names for cheese; other publications list as many as 2,000 different names of cheese. Classification systems have been proposed for cheese on the basis of important composition variables such as moisture or fat, age, texture or general appearance, type of milk used, the ripening process, or the country of origin.

Definitions for different cheese types may be found in the Code of Federal Regulations, Title 21, Part 133 (1987). The specific paragraphs which address some of the cheese varieties covered in this chapter are given below. The reader is referred to 21 CFR 133 for definitions of many other cheese products.

Cheese Variety	CFR, Title 21, 1987 paragraph:
Cheddar	133.113
Colby	133.118
Swiss	133.195
Blue	133.106
Limburger	133.152
Soft ripened cheeses	133.182
Cream cheese	133.134
Mozzarella	133.155

ESSENTIAL STEPS OF CHEESEMAKING

The conversion of milk into finished cheese can generally be divided into several distinct steps. Numerous variations and subroutines within each of these general steps make possible the hundreds of cheese varieties. Five essential steps in cheesemaking are:

1. Preparing and inoculating the milk with lactic acid bacteria.
2. Curdling* the milk (forming a coagulum or gel).
3. Shrinking the gel (curd) and pressing the curd into forms.
4. Salting the curd or formed cheese.
5. Ripening or curing the cheese (optional).

CHEESE PROPERTIES

Moisture content and acidity are regarded as the two most important factors in the control of cheese properties (characteristics). Given a constant milkfat-to-casein ratio, the hardness of a given cheese is a function of moisture content. Generally, the firmer a cheese (due to low moisture), the slower the rate of ripening, the more selective the microflora, the milder the flavor, and the longer the product keeping-quality. On the basis of moisture content, cheese may be classified as: (1) very hard; (2) hard; (3) semihard (also known as semisoft) and (4) soft. The extent of protein hydrolysis, salt content, and the relative amounts of milkfat in cheese also help determine the extent of softness

---

*Coagulation is usually enzymatic, but acid coagulation is used for some cheese types.

or hardness. Cheeses may be: (1) unripened; (2) internally ripened by the action of bacteria, molds, and/or enzymes; or (3) externally ripened as the result of the surface growth of bacteria, yeasts, and/or molds.

Composition and Nutritive Value

In cheesemaking, marked changes in composition of the original cheese milk occur at two distinct stages: (1) during separation of curd from whey, and (2) during ripening. Whey removal (dipping or draining) concentrates, within the curd, many of the nutrients of milk. The degree of concentration depends largely on the type of cheese manufactured, the type of milk used (whole milk, part-skim milk, or skim milk), and on the method of coagulation. Nearly all water-insoluble and some water-soluble components are retained in the curd. In rennin coagulated cheese, this results in an approximate eight-to-ten fold concentration of protein, fat (if not made from skim milk), calcium, phosphorus, and vitamin A, compared to the amounts of these constituents found in milk. Most of the water-soluble components, including the water-dispersible whey proteins, are "lost" to the whey. As a result, lactose, whey proteins, and water-soluble salts are not appreciably retained by the curd, and thus, may be relatively lower in cheese than in the original milk. In cheese milk that is concentrated by membrane processing, most of the whey proteins are incorporated into the cheese curd.

The protein, fatty acids (both saturated and unsaturated), cholesterol, and lactose contents of common domestic cheeses are shown in Table 8.1. These values are based on a 28 g (1 oz) serving size. Table 8.2 summarizes the proximate analysis of common U.S. cheeses and percent contribution to the U.S. RDA for key nutrients.

Cheese (milk in concentrated form), contains the most important nutrients of milk. Most notable are: the nutritionally complete protein casein, calcium, phosphorus, and vitamin A. Cheese is considered to be one of nature's most versatile foods. Simultaneously, it is nutritious and readily digested.

DETERMINANTS OF CHEESE TYPES

As previously mentioned, there are many different kinds of cheese, depending upon the: (1) milkfat content (whole milk, part-skim milk, or skim milk); (2) method of milk coagulation (rennet or acid); (3) amount of whey retained within the curd; (4) ripened or unripened curd; (5) the ripening method; and (6) the source of milk (i.e., cow, goat, sheep, mare, water buffalo, llama, yak, or zebra).

Table 8.1. Protein, Total Lipids, Fatty Acids, Cholesterol, and Lactose Content of Selected Cheeses (amount of nutrient per 28 g (1 oz) edible portion).[a]

Cheese	Protein[b,c] g	Total Lipids[d] g	Fatty Acids[d] Saturated g	Fatty Acids[d] Unsaturated g	Cholesterol[e] mg	Lactose[c] g
Blue	6.03	8.29	5.35	2.55	21.06	N.D.
Brick	6.55	8.23	5.04	2.80	N.A.	N.D.
Camembert	5.47	7.31	4.59	2.38	20.17	N.D.
Cheddar	6.96	9.18	5.66	3.00	28.67	N.D.
Colby	6.67	8.62	5.43	2.74	26.57	N.D.
Cottage, creamed	3.49	1.12	0.73	0.34	3.89	0.17
Cottage, uncreamed	4.85[f]	0.11	0.06	0.03	1.88	0.13[f]
Cream	2.10	9.46	5.94	2.97	30.52	0.48
Edam	7.21	7.81	5.07	2.32	24.99	N.D.
Mozzarella, low-moisture part-skimmed	7.72	5.43	2.86	1.42	15.13	0.11
Neufchatel	3.25	6.78	4.31	2.16	N.A.	0.29
Parmesan	10.82[g]	7.42	4.70	2.32	20.49	N.A.
Provolone	7.27	7.28	4.62	2.27	19.28	N.D.
Ricotta skimmed	3.27	4.09	2.60	1.26	N.A.	0.41
Ricotta part-skimmed	3.11	2.41	1.45	0.76	N.A.	0.40
Swiss	8.14	7.73	4.93	2.46	24.05	N.D.
Pasteurized process American	5.56	8.09	5.04	2.66	N.A.	N.D.

From National Dairy Council. (1978).
[a] Calculated from data based on nutrients per 100 g edible portion.
[b] Total nitrogen × 6.36.
[c] *Source:* Feeley et al. 1975.
[d] *Source:* Posatim et al. 1974.
[e] *Source:* Lacroix et al. 1973.
[f] Average of values for long- and short-set types.
[g] *Source:* National Cheese Institute. 1982.
N.A.—not available.
N.D.—none detectable.

Table 8.2. **Proximate Analysis of Common Cheeses and Percent Contribution to U.S. RDA.**[a]

		Amount of Nutrient Per 28 g (1 oz) Edible Portion			% of U.S. Recommended Daily Allowances (U.S. RDA)[b]			
Cheese	Calories	Protein g	Carbohydrate g	Fat g	Protein	Vitamin A	Riboflavin	Calcium
Cheddar[c]	110	7	1	9	15	4	6	20
Swiss	100	8	0	8	15	4	4	25
Monterey	100	6	1	8	15	4	6	15
Mozzarella, low moisture part-skimmed								
<19% fat	80	7	1	5	15	2	4	20
>19% fat	90	7	1	6	15	2	4	20
Parmesan	110	10	1	7	20	2	4	30
Romano	100	9	1	7	20	4	4	25
Edam or Gouda	100	7	1	8	15	4	4	15
Blue	100	6	1	8	10	4	4	15
Provolone	90	7	1	7	15	4	4	15
Cream Cheese	100	2	1	10	4	2	2	2
Pasteurized Process American	110	6	1	9	10	6	6	15

[a] *Source*: National Cheese Institute. 1975.
[b] Contains less than 2% of the U.S. RDA of vitamin C, thiamin, niacin, and iron.
[c] Nutrition information also for washed curd, stirred curd, or Colby.

Milk coagulated by lactic acid development yields a curd that tends to have a rough, grainy texture. By contrast, rennin (chymosin) or microbial-derived milk-clotting enzymes tend to yield a curd that is more pasty and pliable. The state of the protein in the former is pure casein, while in the latter a complex of calcium paracaseinate and calcium phosphate is formed. A different type of curd (coagulum) is formed under these conditions. The relative softness or hardness of fresh cheese curd generally depends on the amount of whey retained in it, as well as the milkfat content.

CHEESE GRADING

A cheese judge is often called upon to evaluate one or more varieties or types of cheese. To be proficient, the evaluator should be knowledgeable in the sensory characteristics and the desirable and undesirable qualities of each cheese type under consideration. The processes and techniques involved in cheesemaking are so varied that different cheese varieties often have little resemblance to each other in sensory properties.

The relative amounts of various milk components, and the amount of whey retained in the curd, have much to do with the flavor and body characteristics of the finished cheese. Chemical changes that result from the controlled growth of various microorganisms and associated enzymatic activity during manufacturing and ripening processes help develop desired sensory characteristics in matured cheese. Hence, a combination of factors is responsible for yielding the many kinds of cheese.

Much can be learned about the quality of a given cheese by its appearance. By careful observation of the external appearance, and the internal body, texture, and color characteristics of a cheese, an experienced judge can often place a given cheese into a quality classification without actually tasting it.

Certain soft cheeses such as cream cheese or cottage cheese, which primarily derive their flavor from added lactic cultures and/or a cream dressing, are generally consumed while fresh. Hard or semihard cheese varieties are generally made from whole or part-skim milks coagulated by rennet or other milk-coagulating enzymes, and are usually ripened or aged before they are consumed. Cheese properties such as an intense aroma or piquant taste can also be a function of the bacteriological or enzymatic treatment of cheese milk before coagulation. The addition of proteolytic and lipolytic microbial cultures (usually mold) to curd before pressing can also determine cheese characteristics.

Methods of judging and grading cheese types are described later in this chapter. According to a USDA Handbook, cheese types are often classified (Agricultural Research Service 1978) (with cited examples) as follows:

Group A—*unripened* (45–75% H_2O)

1. Creamed cottage cheese and dry cottage cheese curd
2. Cream and Neufchatel cheese

Group B—*Ripened*

1. Soft cheese ($\geq$ 50% H_2O)
   Brie and Camembert
2. Semihard cheese (40–49% H_2O)
   Brick cheese
   Blue-veined cheese
3. Hard cheese (30–39% H_2O)
   Cheddar
   Swiss and Swiss types
4. Very hard ($\leq$ 30% H_2O)
   Romano and Parmesan
   Sap Sago

CHEDDAR CHEESE

Cheddar cheese is the most common type of cheese produced in the U.S. and is sometimes referred to as "American" cheese. It is a hard, ripened cheese made from either raw, flash-heated, or pasteurized whole milk to which about 0.5–1.0% lactic starter culture has been added. The curd formed by the addition of milk-coagulating enzyme(s) is firmed by heating (cooking) and stirring to about 38°C (100°F). The characteristic body of Cheddar cheese is developed by a process of matting the curd known as "cheddaring." The curd may be pressed in several different styles of hoops or in large barrel forms. More recently, granular forms of stirred curd, which have not undergone the traditional "Cheddaring" step are finding popularity as "barrel Cheddar." After removal from the forms, the cheese is generally vacuum wrapped in an air-impervious wrapper to protect it from mold growth and dessication during the ripening period.

The method of coagulating milk, cooking, cheddaring and salting the curd, and the ripening conditions impart definite flavor, body, and texture characteristics to the finished product. Some Cheddar cheese is sometimes referred to as "full cream cheese" because it is made from whole milk (~ 4.0% fat). However, most Cheddar cheese is manufactured from standardized milk, wherein the relative fat and casein proportions are adjusted, usually by adjusting the milkfat content of the cheese milk to approximately 3.8%. To produce 1 pound (0.45 kg) of Cheddar cheese requires approximately 10 pounds (4.54 kg) of whole milk (almost 5 quarts). Nearly one-half of the total solids of whole milk remain in the cheese curd, including about 75% of the milk protein. The milk fat content is about 31–35% of total weight.

Federal (CFR) Definition and Standards

Cheddar cheese is defined by the U.S. government as cheese made by the Cheddar process or by another procedure which produces a finished cheese having the same physical and chemical properties as that produced by the Cheddar process. This cheese is generally made from cow's milk, with or without the addition of coloring matter (usually annatto bean extract). Common salt (NaCl) may be added. Cheddar cheese must contain not more than 39% moisture and not less than 50% milkfat in the water-free substance. Such cheese must also comply with the regulations of the Federal Food, Drug, and Cosmetics Act (CFR Title 21, Part 133) (1985).

Degree of Ripening

Much pasteurized milk Cheddar cheese is marketed shortly after manufacture ($\leq$ 90 days), as a mild cheese or for use in producing processed cheese. Consequently, it may be found on the market in various stages of ripeness. The ripening or curing of Cheddar cheese to develop characteristic Cheddar cheese flavor is a slow, complex, bacteriological, chemical, and enzymatic process which requires months (sometimes years, for extra-sharp cheese flavor). For best results, cheese ripening requires carefully controlled temperature and humidity.

Unripened Cheddar cheese is often referred to as "fresh" or "green" cheese. Cheese at this stage is characterized as having a flat or weak flavor and a relatively tough, curdy, or corky body. Good-quality Cheddar cheese that has been properly cured for at least three months or longer, has a moderate, slightly nutty, "Cheddar" flavor, and is generally referred to as a "young" or "mild" cheese. At six to eight months of age, more of the distinct, aromatic Cheddar flavor should be evident;

Table 8.3. Cheddar Cheese Classifications Based on the Extent of Ripening

Classification	Aging Time
Mild	2 to 3 months
Medium or Mellow	4 to 7 months
Sharp or Aged	8 to 12 months
Extra-Sharp	Over 1 year

Note: These are typical aging times and may vary slightly between cheese manufacturers.

such cheese is considered as "semi-" or "medium-aged." Generally, a year or longer is required to develop the fully aromatic or robust Cheddar cheese flavor desired in an "aged," "sharp," or "matured" cheese. "Extra-sharp" Cheddar cheese is usually aged in excess of one and one-half to two years (Table 8.3).

Whether the flavor of Cheddar cheese is mild or pronounced does not depend exclusively on the aging process. The quality of the milk, the bacteriological and chemical control in manufacture, moisture, salt content, and the temperature of curing have much to do with the nature and intensity of flavor in the cured product. However, development of typical Cheddar cheese flavor is so dependent on the age factor that it is not generally advisable to evaluate cheese of various age categories within the same class. In educational cheese clinics, exhibits, and/or contests where Cheddar cheese is to compete for awards, the cheese should be entered into different age classes or categories. Young (mild) cheese (under four months old), semi-aged (medium) cheese (from four to eight months old), and aged (sharp) cheese (over eight (or twelve) months old) are logical age classifications.

Form or Style

As market demands are identified, Cheddar cheese may be made in several sizes, forms, or shapes, which are generally called styles. Usually, a judge will not be concerned with cheese style, except to remember that larger-sized cheeses are not as prone to drying out as smaller ones; this may slightly affect the texture and flavor of cured cheese.

The Cheddar cheese industry has developed a multiplicity of small sizes and shapes (Fig. 8.1), but it has also recently produced larger, more utilitarian sizes of cheese, as well. The rindless 40-pound block, 640-pound block, and barrel cheese (Fig. 8.2) have evolved as the predominant forms and sizes in contemporary cheese manufacture for reasons of economy, ease of handling, and warehousing.

CHEESE HOOP SPECIFICATIONS

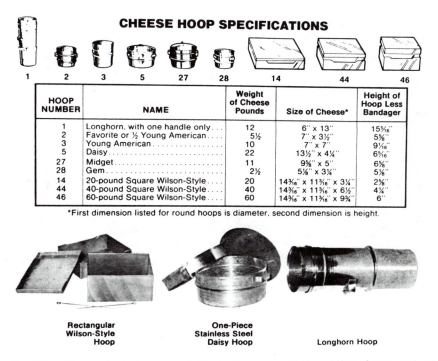

HOOP NUMBER	NAME	Weight of Cheese Pounds	Size of Cheese*	Height of Hoop Less Bandager
1	Longhorn, with one handle only	12	6" x 13"	15 5/16"
2	Favorite or ½ Young American	5½	7" x 3½"	5⅝"
3	Young American	10	7" x 7"	9 1/16"
5	Daisy	22	13½" x 4¼"	6 5/16"
27	Midget	11	9⅝" x 5"	6⅝"
28	Gem	2½	5⅛" x 3¼"	5⅛"
14	20-pound Square Wilson-Style	20	14 3/16" x 11 3/16" x 3¼"	2⅝"
44	40-pound Square Wilson-Style	40	14 3/16" x 11 3/16" x 6½"	4¼"
46	60-pound Square Wilson-Style	60	14 3/16" x 11 3/16" x 9¾"	6"

*First dimension listed for round hoops is diameter, second dimension is height.

Rectangular Wilson-Style Hoop **One-Piece Stainless Steel Daisy Hoop** Longhorn Hoop

Fig. 8.1. Examples of some of the hoops or molds used to form various shapes and sizes of Cheddar and related cheeses (Courtesy of Stoelting, Inc., Kiel, WI).

A "mammoth" is a large, oversized, attention-arresting Cheddar cheese. Such cheeses are formed for the express purpose of display, advertising, and a focus of interest for special occasions, such as the opening of a new supermarket or advent of a festival that features cheese or dairy products. The size of a mammoth cheese generally varies from 300 to 13,000 pounds. For many years the largest cheese on record was the 22,000-pounder made in Ontario, Canada and exhibited at the Columbian Exposition, Chicago, 1893. However, this one was exceeded by the 34,591-pound Wisconsin Cheese Foundation giant displayed at the 1964 New York World's Fair. Usually, these mammoths have excellent flavor and body and texture quality since the curd tends to cure quite well in a large cheese. In fact, since so much value is at stake, every precaution has been taken, from the selection of milk and curd handling to controlled curing for such a cheese.

At one time, the U.S. had a significant export market for Cheddar cheese, but this has eventually been lost to Canada, Australia, and

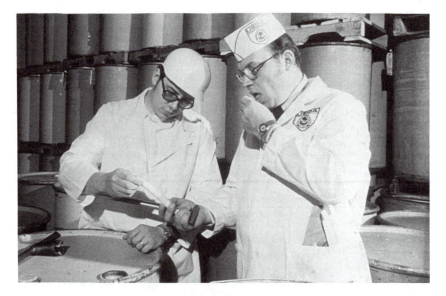

Fig. 8.2.   The manufacture of 600 pound (273 kg) barrels or 640 pound (291 kg) blocks has become a common size for Cheddar and several other cheese varieties due to some advantages of handling, transporting, warehousing, and economics (Courtesy USDA Grading Branch).

New Zealand. Two basic types of Cheddar cheese were once produced: "Export" and "Domestic," or "Home Trade" cheese. The former, made for sale abroad, had a relatively low moisture content, and hence, a firm body. The domestic version tended to possess a higher moisture content and a less firm body, which usually broke down and cured in a shorter period of time.

The above cheese types are now more commonly known as "long-hold" and "short-hold" Cheddar cheese. The former can generally be held to mature in a year or more and develop the typical sharp flavor characteristics of an aged or sharp Cheddar. Short-hold cheese, with a higher moisture content, is more appropriately intended for more immediate consumption (i.e., within two to six months).

The Cheddar Cheese Score Card

The quality score of cheese is determined by comparing the properties or characteristics of each cheese with their accepted standards of perfection. These standards of perfection, when assembled, form what is known as a score card for Cheddar cheese. It lists essential factors or items by which a cheese is evaluated; these items are assigned a point-weighting which reflects the relative importance of each factor in determining the overall sensory quality. For the novice cheese judge, the score card (Fig. 8.3) and associated scoring guide (Table 8.4) can be essential evaluation tools; as such, they should be studied in detail. The evaluator should keep in mind the relative values of the various score card items that are considered in the quality grading process. The American Dairy Science Association score card (Fig. 8.3) was de-

Table 8.4. Suggested Scoring Guide (ADSA) for Flavor and Body and Texture of Cheddar Cheese for Designated Defect Intensities.

	Intensity of Defect		
	Slight[b]	Definite	Pronounced[c]
Flavor Criticisms[a]			
High acid, sour	9	7	5
Bitter	9	7	4
Fermented, fruity	8	6	5
Flat, lacks flavor	9	8	7
Garlic, onion, weedy	6	4	1
Heated, cooked	9	8	7
Malty	8	7	6
Metallic	7	5	3
Moldy, musty	7	5	3
Rancid, lipase, putrid	6	4	1
Sulfide, skunky	9	7	4
Unclean, dirty	8	6	5
Whey-taint, sour whey	8	7	5
Yeasty	6	4	1
Body and Texture Criticisms[d]			
Corky, dry	4	3	2
Crumbly, friable	4	2	1
Curdy, rubbery	4	3	2
Fish eyes, slits, yeast holes	3	2	1
Gassy, pin holes	3	2	1
Greasy, salvy	3	2	1
Mealy, grainy	4	3	1
Open, mechanical holes	4	3	2
Pasty, sticky	3	2	1
Short, brittle, flaky	4	3	2
Sweet curd holes, Swiss eyes	4	3	2
Weak, soft, spongy	4	3	2

[a] "No criticism" for flavor is assigned a score of "10." Normal range is 1 to 10 for a salable product.
[b] Highest assignable score for defect of slight intensity.
[c] Highest assignable score for defect of pronounced intensity.
[d] "No criticism" for body and texture is assigned a score of "5." Normal range is 1 to 5 for a salable product.

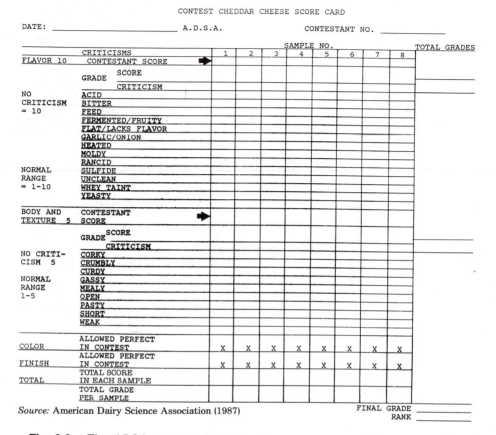

CONTEST CHEDDAR CHEESE SCORE CARD

DATE: _____ A.D.S.A.   CONTESTANT NO. _____

	CRITICISMS	1	2	3	4	5	6	7	8	TOTAL GRADES
FLAVOR 10	CONTESTANT SCORE ➡									
	GRADE SCORE									
	CRITICISM									
NO	ACID									
CRITICISM	BITTER									
= 10	FEED									
	FERMENTED/FRUITY									
	FLAT/LACKS FLAVOR									
	GARLIC/ONION									
	HEATED									
	MOLDY									
	RANCID									
NORMAL	SULFIDE									
RANGE	UNCLEAN									
= 1-10	WHEY TAINT									
	YEASTY									
BODY AND TEXTURE 5	CONTESTANT SCORE ➡									
	GRADE SCORE									
	CRITICISM									
NO CRITI- CISM 5	CORKY									
	CRUMBLY									
	CURDY									
NORMAL	GASSY									
RANGE 1-5	MEALY									
	OPEN									
	PASTY									
	SHORT									
	WEAK									
COLOR	ALLOWED PERFECT IN CONTEST	X	X	X	X	X	X	X	X	
FINISH	ALLOWED PERFECT IN CONTEST	X	X	X	X	X	X	X	X	
TOTAL	TOTAL SCORE IN EACH SAMPLE									
	TOTAL GRADE PER SAMPLE									

Source: American Dairy Science Association (1987)

FINAL GRADE _____
RANK _____

Fig. 8.3.   The ADSA contest Cheddar cheese score card for sensory defects.

veloped for use in training students in sensory evaluation of Cheddar cheese, as well as serving for many years as the official card for the Collegiate Dairy Products Evaluation Contest. Appendix XII contains an illustration of the collegiate contest Cheddar cheese score card designed for electronic grading of contestants' performance.

Tempering Cheese

Before evaluation, cheese samples should be tempered at 10°C to 15.5°C (50°F to 60°F) for a sufficient length of time to ensure a uniform temperature throughout the cheese. This usually requires 1–2 hours for the smaller styles (≤ 5 pounds), and 3–5 hours for larger ones. Generally, a cheese plug taken from a warm (overtempered) cheese appears weak-bodied; by contrast, a cold plug may appear brittle or corky. Actual body and texture characteristics cannot be determined readily unless cheese samples are properly tempered before evaluation.

Preparation for Evaluation

Appropriate facilities for cheese tempering, sampling, proper disposal of waste cheese, and cleaning of triers should be provided for evaluators. Prior to sampling, one's hands should be washed and dried, since they directly contact exposed cheese surfaces. As soon as the cheese samples to be evaluated are arranged in order and numbered or coded for proper identification, the sensory evaluation process may begin.

SEQUENCE OF SENSORY OBSERVATIONS

Appearance

Typically, the first procedure in grading Cheddar cheese is visual examination of surface finish or packaging material. The judge should note whether the sample appearance is generally clean, neat, attractive, and symmetrical, or whether the surfaces might be uneven, nonparallel, or rounded. Next, the evaluator should look more closely at the surfaces and observe whether the coating of plastic film (or paraffin) is smooth and free from holes, tears, or wrinkles. Finally, a close examination of the surface for possible mold growth should be undertaken by the judge; a mental record of all observations of the sample appearance should be made.

Obviously, this technique of evaluating appearance cannot be followed entirely when cheese is encased in opaque wrappers. Laminated paper-pliofilm or foil wrappers serve to obscure the cheese from the critical eye of the judge. About the only recourse the evaluator has in noting the appearance of such cheese is to note the cleanliness of the wrapper, the evenness and tightness of adherence, and freedom from breaks and tears. However, in the instance of transparent film-encased cheese, the judge can easily note the presence or absence of mold growth.

Sampling

Cheese samples are usually obtained with a double-edged, curved blade instrument known as a cheese (or butter) trier. For best service, the edges of a cheese trier need to be sharp. A trier that cuts a larger plug has an advantage over one of a smaller diameter since the extent of "openness" and possible color defects are easier to detect with a larger plug. A cheese trier with a 5 in. (127 mm) cutting edge, ⅝ in. (15.8 mm) diameter at the base (top) and ⁹⁄₁₆ in. (14.3 mm) diameter at the tip is recommended.

The trier should be inserted into the top surface of the cheese, preferably about half way between the center and the outer edge of the cheese sample. After insertion, the trier should be turned one-half way around to cut a sample core. The plug is withdrawn, which produces a long tapered cylinder of cheese. The upper 1 in. (2.54 cm) of the cheese plug is immediately broken off and replaced, flush with the surface of the original hole. This partially protects the cheese from developing mold contamination and retards drying and cracking of the cheese surface surrounding the hole. Various wax-like polymers of plastic or gels have been developed to seal trier plug holes to restrict the access of oxygen to the center of the cheese. The "surface plug" should be replaced as quickly as possible, but not until the plug has been passed slowly under the nose to ascertain its aroma. *The practice of smelling the plug of cheese immediately after sampling should automatically become a part of the evaluation technique.*

The evaluator should carefully examine the cheese plug and note whether the plug has a clean-cut surface (with no loose particles) or whether it is rough (with a feather-like edge as though the cheese had been cut with a dull knife). The evaluator should make a mental note of these observations.

Color

The evaluator should observe the color of the cheese and determine whether the appearance is bright and clear or dull and lifeless. It should be noted whether the color is uniform (free from mottled or light and dark portions), or whether there are curd seams or faded areas (surrounding any mechanical openings). The cheese judge should reexamine the plug and observe whether the cheese appears to be: (1) translucent, which is desirable, or (2) opaque, wherein it is difficult for the eyes to observe beyond the surface. The evaluator should especially note whether the color is uniform throughout the sample. In quality evaluation, color uniformity is generally more important than the shade of color. Some cheese consumers apparently prefer an uncolored product (no added annatto coloring). Uncolored (or lightly colored cheese) generally results in a light cream shade; this depends upon the milkfat and/or carotene content of the cheese milk. Other groups of consumers seem to prefer an intense deep-orange color for Cheddar cheese. Table 8.5 summarizes several color and appearance defects, their probable causes, and some remedial measures.

Openness

The judge should observe the nature and extent of the mechanical openings in the cheese. Their shape or configuration should be examined closely to see whether they are regular, angular, rounded, large, and/or small. It is also helpful to observe the luster or sheen of the inner surfaces of these openings and note whether the surfaces appear dry (preferable) or wet. Free moisture within these openings is sometimes indicative of certain flavor defects or potential quality shortcomings. See Fig. 8.4 for examples of the open defect.

Body and Texture

The evaluator should take the ends of the cheese plug by the forefingers and thumbs of both hands and bend the plug slowly into a semicircle, and observe when the sample breaks, as well as the nature of the break. It should be determined whether the cheese plug: (1) shows a definite resistance toward any bending and finally breaks abruptly (short); (2) bends until the plug ends nearly touch (weak), if it breaks apart at all; or (3) bends into approximately one-third to one-half of a full circle before it breaks apart (preferred elasticity, see Fig. 8.10).

Next, the judge should take one of the broken pieces of cheese between the thumb and the forefingers and attempt to manipulate it into a uniform mass. The relative resistance (or lack of resistance) offered by the cheese to applied pressure from the thumb and fingers should be ascertained. When the "worked cheese" sample is properly tempered (after an elapse of 10–20 sec), the evaluator should try to form a small ball or marble of the softened product. Formation of a cohesive sphere of cheese is generally indicative of an appropriate degree of waxiness or elasticity for a typical mild to medium-aged, as well as the highest-quality sharp or fully ripened, Cheddar cheese. Next, the formed "ball" of cheese should be placed into the depression between the tips of the first two fingers, and with gentle to moderate pressure, the evaluator should push the thumb (of the same hand) into the "manipulated" cheese. Then the thumb should be slowly pulled from the slightly depressed "cheese ball." If the cheese sample adheres or "sticks" to the thumb or feels "tacky" or wet to the thumb's touch, the cheese sample should be considered to demonstrate the pasty (sticky) defect. In stark contrast, if the cheese sample tends to fall apart in response to thumb pressure, either a curdy or crumbly defect is sug-

Table 8.5. Common Body and Texture, and Color and Appearance Defects of Cheddar Cheese, Their Probable Causes, and Remedial Measures.

Body and Texture Defects	Probable Causes	Remedial Measures
Corky, dry and hard	Lack of acid development.	Follow standard or recommended procedures for cheesemaking.
Crumbly, mealy/grainy	Excessive acid production and low moisture retention in cheese.	1. Avoid ripening at higher temperature. 2. Control acid development and moisture level in curd.
Curdy or rubbery	Inadequate curing conditions.	Optimize ripening temperature and time.
Pasty, sticky or wet	1. High moisture retained by curd. 2. Excessive acid development.	Control acid development in relation to time and temperature parameters.
Weak or soft	1. Excessive fat content. 2. High moisture in cheese. 3. Failure to develop "body" in cheese during cooking.	1. Standardize fat in cheese milk. 2. Cook curd to desirable firmness (higher temperature, longer time). 3. Avoid piling curd slabs too high or too soon while cheddaring curd.
Color and Appearance Defects	Probable Causes	Remedial Measures
Acid-cut, bleached or faded, or dull looking, (portions or entire cheese surface)	1. Excessive acid developing in the whey or at packing stage. 2. Nonuniform moisture distribution in the cheese.	1. Monitor acid development carefully. 2. Take precautions to insure consistent and uniform moisture retention in curd.

Defect	Cause	Prevention
Mottled appearance: irregularly shaped light and dark areas on cheese surface	1. Combining curds of different colors, batches, or moisture content. 2. Uneven acid development in curd. 3. Unwanted microbial growth: a) H_2O_2 production, and/or b) fruity off-flavor and c) pasty body.	1. Avoid adding starter culture after color incorporation. 2. Attempt to cut curd into uniform-sized particles. 3. Handle all curd carefully to avoid drying during matting, cheddaring, or "hold-overs."
Seamy: shows light colored lines around curd pieces	Exudation of milkfat from curd pieces due to excessive forking, too-warm temperatures, and lack of salt dissolution.	1. Wash "greasy" curd at 32°C (90°F) and thoroughly drain. 2. Avoid overforking of the curd. 3. Allow all of the salt to dissolve completely. 4. Press curd at 30°–32°C (86°–90°F).
White specks: granules or small hard mineral deposits	Generally occurs in aged cheese. Derived from proteolysis and crystallization of calcium lactate/tyrosinate complex	1. Ripen cheese at a higher temperature for a shorter time. 2. Reduce levels of $CaCl_2$ added to cheese milk.
Moldy appearance	Growth of mold on cheese surface	1. Insure airtight seals on cheese packages. 2. Avoid O_2 in the packages by vacuum, CO_2 or N_2 gas flushing.

Source: Compiled from Chandan (1980b); Van Slyke and Price (1979); Wilson and Reinbold (1965); Wilster (1980).

gested, respectively, depending on the advancing age of the cheese. The "worked cheese" should remain smooth, waxy, and somewhat pliable for an "ideal" Cheddar cheese. The tempered sample should exhibit a tendency to remain as a solid mass upon gentle finger manipulation (see Fig. 8.11).

An optional approach is to spread the cheese mass over the palm of the hand (with the thumb of the opposite hand) and determine whether the thin smear of cheese feels smooth, silky, waxy, and/or fine, or whether the sample variously appears to be sticky, pasty, mealy/grainy, or crumbly. The judge should then reassemble (or attempt to reassemble) the cheese particles and try to compress them into a compact "ball," and note the response of the cheese to this form of manipulation.

Aroma

By the time the sample has been worked into a semisoft ball, the temperature of the cheese mass should have increased from combined pres-

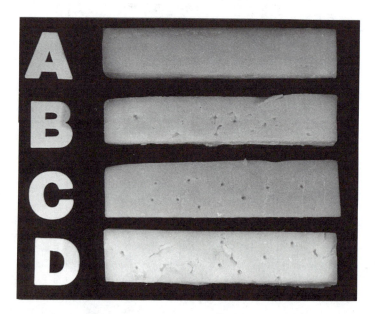

Fig. 8.4. Several examples of texture defects of Cheddar cheese compared to an ideal sample: A—An "ideal" texture; B—Definite "open" texture; C—Definite "gassy" texture; D—Combination "gassy" and "open" texture.

sure and hand warmth, and thus enable easier detection of any aroma. The evaluator should then place the tempered cheese sample directly under the nose and observe the aroma a second time. The judge should compare the aroma with that noted when the sample first was removed from the cheese. For tasting, the evaluator needs to place a small portion of an "unworked" plug into the mouth and chew until a semiliquid stage is reached. The judge should roll the macerated sample about in the mouth for sufficient time to determine both taste and aroma, then expectorate the sample and determine the overall flavor judgment(s).

As a rule, since many types of cheese tend to dull the sense of taste and smell, no more than 15–20 samples (ideally) should be tasted at one scoring, as they may eventually all tend to taste alike. For beginners, up to about 10 samples can be tasted successively with some assurance that the nerves controlling the sense of taste are functioning normally or are not overtaxed.

It can be helpful to rinse the mouth occasionally with a lukewarm saline solution to cleanse the mouth of previous cheese flavors (or off-flavors). A pinch of common table salt placed into the mouth and rinsed out with tepid water can be equally effective. Apple slices or grapes are also useful for cleansing the mouth between intense-flavored cheese samples. Experienced judges find it most helpful to "go back to the best sample in the lot" after experiencing a particularly poor-quality sample (i.e., rancid, garlic/onion, or intense sulfide/bitter). An experienced cheese judge can often grade cheese without actually tasting, on the basis of the color and appearance, amount and nature of openness, body and texture, and the perceived aroma of the worked sample mass. The experienced judge may taste an occasional sample simply to verify judgments ascertained by means of other sensory observations.

Smelling and tasting the cheese samples generally completes the evaluation process. All sensory observations should be recorded on a designated cheese score card or a form provided for this purpose. The less-experienced judge should strive to follow the aforementioned procedure quite closely. Deployment of a score card enhances accuracy when more than several samples are evaluated. The novice judge should also strive to keep in the "mind's eye" or "see" each cheese (once tasted) as though they were so many specimens. Once this ability is attained, it is unnecessary to continually reexamine the various samples. *The practice of reexamining, reworking, and retasting cheese is not conducive to the best evaluation performance.* Such a practice leads to vacillating judgment, which is just as apt to be wrong as to be correct. A calculated judgment should be made following the initial sampling, if possible.

REQUIREMENTS FOR HIGH-QUALITY CHEDDAR CHEESE

Color Evaluation

The color of Cheddar cheese, regardless of the chosen intensity, should always be uniform throughout the cheese. American cheese may be uncolored, light to medium colored, or high in color. For uncolored cheese, the most desired color is a light cream shade; for medium-intensity colored cheese, a deep cream color or a pleasant yellow-orange hue is acceptable. Deep, intense shades of yellow-reddish hues are generally discriminated against. Not only should the shade of color be appropriate and uniform for the given cheese, but the color should exhibit some luster. The cheese surface color should be slightly translucent; that is, it should appear as if one could actually see into the cheese interior for a short distance. The "translucent" quality of Cheddar cheese is closely associated with desirable body and texture.

Not only is cheese color one of the items capable of being most accurately evaluated, but when carefully observed and correlated, it may also serve as an index to other defects in body, texture, and flavor. On the ADSA score card for Cheddar cheese color is assigned a total point value of 5. The normal score range for the color of Cheddar cheese is from 4 to 5; serious color defects may occasionally occur and thus be assigned scores of less than 4.

Some color defects which may be associated with Cheddar cheese are:

Acid-cut (bleached, faded)   Mottled
Atypical color specks   Seamy (uneven, wavy)
Color too high (unnatural)   White specks.

Acid-cut (Bleached, Faded). These several color defects are quite similar, but differ primarily in their intensities. The color of "acid-cut" cheese generally appears dull and lifeless; it may be so slightly translucent that little light could be transmitted through even a thin slice. Quite often, a degree of bleaching may be noted more or less uniformly throughout the entire cheese (Fig. 8.5). In some cheese, acid-cut color may occur only within close proximity to mechanical openings. In such instances, the cheese may have a "mottled" appearance. Of these two defects, a uniform acid-cut color is less objectionable than a mottled one; however, neither is desirable. Evaluators should readily recognize the acid-cut color defect and be on the alert for the possible association with a given body and texture or a specific flavor defect. Generally, the

Fig. 8.5.   The acid-cut color defects of Cheddar cheese (excessive mechanical openings also evident).

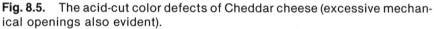

faded color of acid-cut may be associated with high moisture and high acid development in cheese, but it also may occasionally be observed in cheese with a dry body and a crumbly texture. Cheese showing this defect practically always has a distinctive high acid or sour flavor. The acid-cut color defect is becoming less common, due to better control of acid development by cheesemakers, improved lactic cultures, and better monitoring of the manufacturing process.

Atypical Color Specks. Atypical color specks take the form of either occasional white or black specks, rust spots, and/or red blotches. While there may be little or no association between foreign specks and a specific off-flavor, the presence of atypical color deposits generally reflects carelessness in the manufacturing process.

Color Too High (Unnatural). This defect is characterized by high color intensity, and often by an orange-yellow hue, especially when precut cheese is warmed to room temperature or higher. There is no association between this defect and flavor, since the defect stems from the use of an excessive amount of added colorant to the cheese milk. Intensely colored Cheddar cheese may be preferred in some specific markets, but in others it is often discriminated against.

Mottled. The "mottled" color defect appears as rounded, irregularly shaped areas of contrasting light and dark color, with one shade gradually blending in the other (Fig. 8.7). This defect may result either from certain physical causes during cheese manufacture, or be due to atypical microbiological activity during the curing process. Chief causes often ascribed to this defect are the combining of curd from two different lots of cheese or nonuniform development of acidity within the curd. When a mottled color results from unusual microbial growth, an associated yeasty, fruity, or acid off-flavor, and/or pasty body may sometimes accompany this appearance defect.

Seamy (Uneven or Wavy). The appearance defect "seamy" is portrayed when the cheese appears interlaced with light-colored lines around each original piece of curd (Fig. 8.6). This is particularly noticeable when one directly examines the surface appearance of freshly cut cheese (which possesses the seamy defect). The seamy appearance defect usually results from physically altered curd surfaces caused by exuded or crystallized milkfat, uneven salting, or moisture evaporation which probably occurred prior to curd pressing. Cheese exhibiting this color defect not only tends to lack color uniformity, but may also dem-

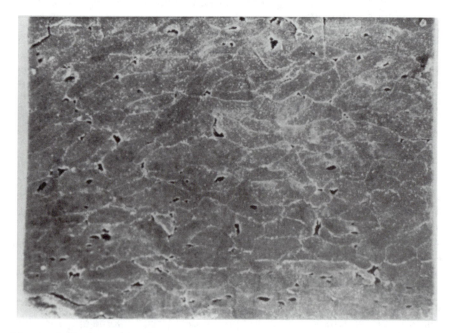

Fig. 8.6. The appearance defect "seamy" is interlaced with light-colored lines (definite mechanical openness also apparent).

onstrate a short-bodied, crumbly, and/or friable texture. The slight degree of seaminess that is occasionally noted in fresh or young Cheddar cheese is not particularly objectionable, since this form of seaminess generally disappears with additional aging. Occasionally, wider bands of discoloration may occur in cheese (without the seaminess lines); this condition may be described as uneven or wavy color.

White Specks. Any Cheddar cheese that has small "white specks" interspersed throughout its mass and/or on its surface is generally indicative of a mature cheese. These white particles are generally assumed to be an admixture of calcium lactate, tyrosine, and other components. Sometimes these specks are so small that they may be only noticeable when viewed from a close distance. This color (or appearance) defect (Fig. 8.7) is most commonly associated with aged cheese, but on rare occasions may also be noted in medium-cured cheese; curing at lower temperatures tends to favor formation of this insoluble complex. Accumulation of tyrosine should indicate to the evaluator that the cheese has been aged long enough for protein to partially break down and yield this amino acid. Some aged cheese that exhibits the combined appearance/texture characteristics of white specks also frequently exhibits a desirable "buttery"-like body.

Even an inexperienced judge should be able to associate the presence of white specks (and the possible associated mouthfeel) with an aged cheese; the cheese sample will most likely also have a fully developed intense flavor. White specks, on their appearance alone, should not be considered a serious color defect. Their presence may be noted, but a deduction in score should not be made unless an excessive grainy or objectionable gritty mouthfeel is present.

Finish and Appearance Evaluation

Cheese with a desired finish should generally show symmetrical, parallel ends; square, even edges; an evenly-folded, neat, close-fitting plastic film or wrapper free from wrinkles; a clean, thin, uniform, close-adhering coating of paraffin (if used) showing no blisters or scales; and freedom from pinholes, tears, breaks, cracks, mold, rot spots, or soiled areas.

The finish of the cheese is important during evaluation, as it furnishes an indication of the skill and care taken by the cheesemaker during manufacture of the cheese, and of the subsequent handling of the product. An ill-shaped, poorly formed and packaged cheese indicates carelessness in manufacture which may be correlated with undesirable sensory properties. Untidy, soiled, or moldy cheese does not present a pleasing appearance or full product utility. Defects in pack-

Fig. 8.7. The color/appearance defect of "white specks" or "surface precipitate" is generally associated with aged Cheddar cheese: A—"White specks" evident in a cross-section of cheese. Pronounced large gas holes and color mottling are also apparent; B—An example of "surface precipitate" of calcium lactate.

age finish are usually quite easy to observe and assess for their significance to maintaining product integrity.

The beginner judge should become familiar with the possible defects in cheese finish, and in turn correlate them, if possible, with other defects. The defects listed in the following paragraphs are closely associated with cheese wrapped with various types of protective coverings.

Rindless, Flexible-Wrapped, or Nonparaffined Cheese. Modern processing and merchandising has led to the introduction of new styles and packaging materials for Cheddar cheese. The 20- and 40-pound block (9.1 and 18.2 kg) has displaced the time-honored round "daisy" and "Cheddar," which were covered with a cotton bandage (cheese cloth) and paraffin. Taking their place are various flexible wrappers—coated kraft paper, cellophane, Cryovac, foil, glassine, nylon, Parakote, Pliofilm, Saran, and various other films. The Cryovac-like films are the most widely used coverings for Cheddar and Swiss cheese in the U.S. Some of these cheese coverings are used singly, but double or triple lamination films have provided greater tensile strength and bonding properties. In Cheddar cheese operations, these new packaging materials are generally applied directly to the pressed "wet curd" immediately after dehooping, with or without vacuum treatment, followed by heat sealing of the wrapper. The film-packed cheese may be overwrapped with a coated kraft paper, then placed in a fitted fiberboard box with a veneer reinforcement liner for storing and shipping. The cheese judge should be alert to possible flexible-wrapper defects listed in the following paragraphs.

Damaged Coverings. Torn or punctured wrappers readily permit air access and microbial contamination of bulk cheese and thus must be prevented, if at all possible. Careless handling contributes to the "damage" package defect. Hopefully, for economic reasons, damaged wrappers occur infrequently; but all wrapped bulk cheese warrants close inspection in this respect.

Loosened Coverings. For maximum protection against mold growth, air (oxygen) must be excluded insofar as possible from under the wrapper of cheese coverings. Some wrappers are bonded so tightly to cheese surfaces that loosening and removing of wrappers in cheese cutting and packaging operations may be difficult. All nonbonded wrappers must be pressure- or vacuum-sealed to void as much oxygen as possible. Usually, these wrappers cling to the cheese as though they were bonded. "Loosening" and "ballooning" of the wrapper is generally undesirable, as mold growth may occur within the air space provided if the integrity of the covering is lost. Loosened wrappers may be noted by sight, or by stroking the cheese with the hand. Cheese package edges and ends should be closely examined for any unnecessary looseness and air pockets.

Soiled Coverings. A "soiled (or greasy) wrapper" often denotes extreme carelessness in packaging, handling, and storage. Such a condition may suggest a general lack of concern for both cleanliness and good housekeeping. This defect is even more serious when it is accompanied by damaged wrappers.

Mold Growth. "Mold growth" can go unnoticed when the cheese is encased in opaque wrappers. If present, mold growth may be readily seen when transparent films are used. Any bluish-green mold spots are usually located in an air sac, as in a cheese opening, or along an edge where the film has not bonded closely to the cheese. Trier holes certainly are conducive to mold growth. A perforated film, or one having near-microscopic openings may be a source of contamination and/or air admission which can facilitate mold growth.

Pliable, Wax-Coated Cheese. This cheese covering is a form of microcrystalline paraffin which is modified to yield an adhesive, flexible, plastic-like protective coating when the surface-dried cheese is dipped into melted wax. This appealing, thick (yellow, amber, orange, or red) coating is semitransparent. The cheese must be handled with reasonable precautions so that the coating will not chip or flake. This type of flexible wax is often used as a cheese covering for any cheese that is subsequently cut into retail portions, or for small units cut from bulk cheese to be cured and sold as miniature-sized cheese. This coating is relatively free of defects if the proper form of wax is used.

Paraffined Cheese. Although paraffin currently finds limited use as a covering material for cheese, the cheese judge should be aware of the following defects related to its use.

Blistered. This defect manifests itself by areas of thin, loose paraffin, usually on the end of the cheese where cheesecloth may be absent. Such a condition readily lends itself to the possible entrance of mold and/or harboring cheese pests; therefore, blistering is quite objectionable in paraffined cheese.

Checked. A "checked" or cracked paraffin is denoted by breaks or formed cracks in the cheese covering. This defect is usually caused by the paraffin coating being heavier than necessary. Checked paraffin offers an opportunity for mold and pests to gain entrance to the cheese.

Rough. Rough paraffin is manifested by a lack of surface smoothness or paraffin finish. The paraffin surface seems to contain small hard particles; this leaves the impression that the surface of the cheese may have been covered with tiny particles of foreign matter prior to coating. Although not usually that serious, this defect is somewhat undesirable as a surface blemish for what may otherwise be a high-quality cheese. Roughness may be detected either visually or by running the hand over the surface.

Scaly. Loose or scaly paraffin offers poor protection for cheese; it permits moisture to escape and mold to gain entrance; hence, this represents a serious defect. In cutting cheese, particles of paraffin often become intermixed with the cheese itself, and thus produce an untidy, unappetizing cheese slice. Scaly-like paraffin should seldom occur if the cheese surface is predried sufficiently, then completely dipped in hot paraffin (not lower than 104.4°C [220°F] for at least 10 sec). The paraffin is then allowed to completely harden before subsequent handling occurs.

Workmanship of Cheese Finish

High Edges. Cheese showing this defect lacks square or symmetrical edges, such as desired in well-finished cheese. Sometimes, edges of the cheese may be so long that they tend to bend over (curl under) onto the end of the cheese, and thus form a protected area for mold growth or pests. These undesirable long edges are usually dry, do not cure properly, and thus represent waste.

Lopsided, Misshapen. A misshapen cheese is characterized by non-parallel ends or sides as a result of uneven distribution of curd in the hoops, possibly coupled with unequal pressure in the press. Such defects detract from a neat appearance of the cheese unit(s) under evaluation. This unwanted configuration may sometimes be correlated with weak-bodied cheese.

Uneven Edges. Heavy pressure against followers or press boards that are too small for the hoop may cause the curd to squeeze out around the edges and form a narrow raised edge or rim around the outer edge of the cheese, generally up to about one-half inch thick. The presence of these raised, uneven edges not only detracts from cheese appearance, but results in a waste of curd. The raised edge dries out and does not cure properly. Cheese should be pressed in a manner that ensures that the bottom edge of the cheese meets evenly with the sides.

Uneven Sizes. Cheese of a designated style should be well within a specified weight tolerance for that style of cheese; lack of size uniformity may result in an unattractive appearance. Carelessness in assuring even distribution of the curd among the various hoops is often correlated with other finish and/or appearance defects. An "uneven size" of cheese also may result in excess trim losses when blocks are cut subsequently into retail-sized pieces.

Surface

Bruised. A bruised surface is shown by slightly depressed areas over which the paraffin is broken. Cracks may radiate from the center of the break. Obviously, a bruised surface permits mold contamination and pest infestation.

Light Spots. A cheese that exhibits "light spots" has more or less irregular light and dark areas over the flat surfaces. Though this defect is quite noticeable, it is not a particularly serious one.

Moldy. "Mold growth" on cheese may occur on portions where the cheese covering has been broken by a cheese trier, or from holes or tears in the packaging material. The presence of even a slightly moldy portion not only substantially detracts from the appearance, but also may jeopardize the flavor and consumer acceptance of the entire cheese. As soon as the cheese is cut, mold mycelia usually have the opportunity to disperse across the entire cheese. Moldiness is considered a serious finish defect and a constant problem; annually it results in considerable waste and economic losses for the U.S. cheese industry.

Additionally, some mold contaminants can pose public health problems due to production of certain mycotoxins (carcinogenic aflatoxins). No absolutely successful method has as yet been found to prevent regrowth of mold from bulk forms of cheese onto cut and rewrapped cheeses. Even cheese that has been thoroughly cleaned, scraped, and repackaged, and possibly treated with approved mold inhibitors, may develop surface mold during storage or distribution.

Open. Short depressions on or near the surface are referred to as an "open" surface. This openness usually stems from insufficient curd pressing or a too-cold curd at the time of pressing. This open surface typically reflects an open-textured cheese; there tends to be many mechanical openings. Defects of surface openness are objectionable because these surface depressions and openings serve to: (1) increase the amount of cheese trimmings, and (2) provide sites for mold and/or cheese pests to establish themselves.

Rough. A "rough-surfaced" cheese exhibits severe irregularities of surface finish. This defect may result occasionally from: (1) the use of unclean press cloths to which particles of dried curd have adhered; (2) insufficient or improper pressing of hooped cheese; or (3) from rough and uneven shelving. Cheese that has this defect lacks the preferred neat and attractive appearance that facilitates marketing the product.

Soiled, Unclean. Most unfortunately, cheese takes on an untidy "soiled" or "unclean" appearance when dirt or soil adheres to cheese surfaces. Usually, soiled surfaces are due to carelessness on the part of the cheesemaker. This defect should not be tolerated in the manufacture of high-quality cheese.

Miscellaneous Factors

Huffed, Bloated. So-called "huffed" or "bloated" cheese results from gassy fermentation. A cheese suffering from this defect usually becomes rounded on sides and ends, producing a somewhat oval shape to the cheese unit. In occurrences of the huffed defect, the lower edges of the cheese may be raised slightly above the top plan of the shelf. Occasionally, a gassy condition may develop (within the cheese wrapper) to the extent that the general symmetry of the cheese unit is distorted and the packaging material may be ruptured. A huffed cheese usually yields a sample plug which is dominated by obvious gas holes. Plugs pulled from some bloated cheese may exhibit openings in the shape of narrow slits; these openings are commonly called "fish eyes." Huffed cheese generally portrays poor sensory qualities; serious off-flavors frequently accompany gassy fermentations.

Ink Smears. Occasional "ink smears" from careless cheese branding often detracts from the appearance of cheese. Generally, this is a relatively minor defect which is not correlated with other defects, other than careless workmanship.

Rust Spots. "Rust spots" on the surface of cheese are uncommon, since they are usually caused by corrosion of nails in wooden boxes, which are less frequently used for cheese storage. This defect may be caused by careless placement of oversized nails when closing wooden boxes. A rust spot is most objectionable because it represents a waste area and the puncture furnishes a focal point for possible mold contamination or cheese pest infestation.

Cheese Mites and "Skippers." The presence of "cheese mites" is usually manifested by a fine, loose, brown dust on the surface of aged cheese, cheese wrappers, or on the shelving. Microscopic examination has revealed this brown dust to consist of live and dead bodies, molted skins, excreta, and minute particles of cheese. In badly infested cheese (which has not been moved for some time), the brown dust may appear over extensive areas of the cheese; however, it is more generally localized in favorable harboring places (such as cracks, under a folded edge or under loose paraffin). "Skippers," the larvae of the cheese fly, may be infrequently noted; they only occur as the result of poor sanitation practices.

Body and Texture Attributes

Cheddar cheese with the most desirable body and texture displays a full, solid, close-knit plug (see Fig. 8.10) that possesses smoothness, meatiness, waxiness, and silkiness, and is entirely free from gas holes or mechanical openings. Cheddar cheese with the above-described quality attributes lends itself to uniform slicing into thin, intact pieces.

The term "body," as applied to cheese, usually refers to various physical attributes which primarily affect the relative firmness or softness of the cheese. By contrast, the term "texture" refers particularly to the structure and arrangement of the various parts which make up the whole (the cheese unit). Thus, texture in cheese is observed visually by the quantity, size, shape, and distribution of openings and by the sense of touch (as in mealy/grainy) to uncover internal particles. As a general rule a "close" (few or no openings in the cheese mass) or medium-close texture is preferred; however, a slightly open texture is not objectionable, providing the body possesses such properties that the open texture does not give rise to a weak-bodied cheese.

A close-textured cheese should yield a solid plug with practically no visible openings (Fig. 8.8). The plug, however, may gradually break apart along a curd seamline, especially in a young cheese. A plug withdrawn from a moderate close-textured cheese displays a few mechanical openings which may have been caused by insufficient matting (Cheddaring) or pressing of the curd, or both. An "open" cheese yields a plug which may contain numerous small or large irregularly shaped openings, referred to as "mechanical holes." Cheddar cheese may also

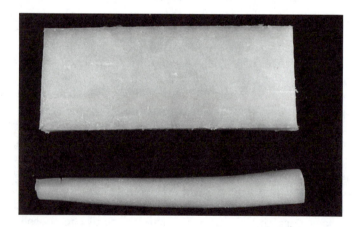

Fig. 8.8.   A Cheddar cheese that exhibits the preferred attributes of pliable body and close texture (no or few openings).

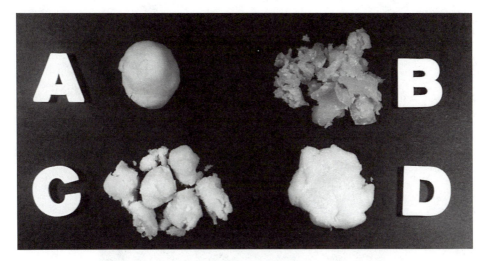

Fig. 8.9. "Worked" Cheddar cheese samples showing various body defects: A—Example of an "ideal" body. B—"Corky"-bodied curd; C—"Crumbly"-bodied cheese; D—"Pasty"-bodied cheese.

exhibit "gas holes" or "slits" as the result of CO_2 formation from microbial activity; these openings tend to be more symmetrical and are usually spherical or elliptical in shape. The so-called "late gas" defect may occur in close-textured cheese, but in this instance the plug will exhibit a split appearance. Worked plugs exhibiting various cheese body and texture characteristics are shown in Fig. 8.9.

Desirable Body and Texture Characteristics

Firm Body. Cheese with a firm body feels solid and offers some resistance to applied pressure. Firm-bodied cheese yields a clean-cut plug which generally tears apart slowly on bending, rather than breaking suddenly (Fig. 8.10). The preferred texture is close; the curd particles should be well-matted or fused together in a high-quality cheese. A slice of firm-bodied cheese tends to tear apart somewhat like the thoroughly cooked breast meat of a chicken. A firm-bodied cheese should not be confused with either a dry, corky, or curdy body; the latter cheese body products often resist pressure and seem excessively springy or quite rubber-like.

Waxy Body. A desirable "waxy body" is exhibited when a cheese plug responds to the combined pressure of thumb and fingers as would cold butter, tempered candle wax, or modeling clay. In "breaking

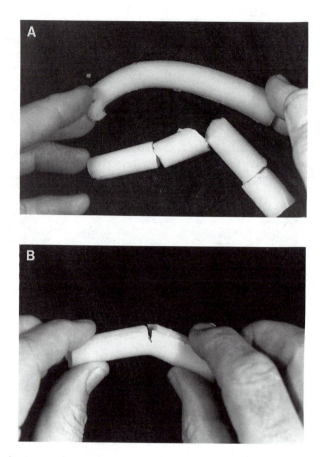

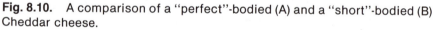

Fig. 8.10. A comparison of a "perfect"-bodied (A) and a "short"-bodied (B) Cheddar cheese.

down" a waxy-bodied sample by finger manipulation (Fig. 8.9), little resistance is offered other than the normal force required to mold the cheese into a cohesive "cheese ball." Preferably, a "malleable" cheese shows little tendency to "spring back" to the original position, but rather assumes or retains a new configuration as a result of applied finger pressure. A waxy body is generally associated with either medium-aged or aged (sharp) cheese. A pliable or waxy body is a good indicator of desired slicing properties and proper flavor development.

Silky, Smooth Body. A "silky, smooth-bodied" cheese exhibits fineness of grain and a continuous, slightly oily, silky-smooth film when the mass, worked between the thumb and fingers, is spread over the palm of the hand. The "worked cheese" usually spreads evenly without forming irregular patches in the hand. The spread-out cheese sample should readily reassemble into a small intact ball. The smooth, silky-like property of the cheese sample is generally indicative of proper cheese breakdown, flavor development, and desired mouthfeel.

Body Defects

Many duplicate terms are used in an effort to characterize undesirable body and texture defects of Cheddar cheese. The more common descriptors of cheese body defects are listed as follows:

Corky (dry, hard, tough)	Pasty (smeary, sticky, wet)
Crumbly (friable)	Short (flaky)
Curdy (rubbery)	Spongy
Greasy	Weak (soft).

Corky (Dry, Hard, Tough). This defect is generally associated with a low moisture, low fat, and/or young cheese. Difficulty is sometimes encountered in trying to sample dry, tough cheese, due to initial resistance against the trier during penetration. The drawn plug resists any form of pressure; when sufficient finger pressure is applied, the plug may resist breaking down and/or exhibits a distinct tendency to recover its original shape. The plug is stiff or rigid upon bending; it seems to have a rubber-like consistency. When a portion of a so-called corky cheese is worked between the thumb and forefingers, the desired smooth, silky, even distribution of cheese particles is notably lacking. The worked mass of cheese tends to curl up under sliding pressure of the thumb over the forefingers and is usually distributed in irregular patches. This defect may be associated with other body defects of which dryness is a closely related factor. A dry-bodied cheese generally has an opaque appearance. It may also exhibit the appearance defects of seamy or acid-cut color.

Crumbly (Friable). A "crumbly-bodied" cheese is one which tends to fall apart when sliced; it is difficult to cut a thin slice. A plug of such cheese may be extremely friable. This defect sometimes appears to be associated with curd mealiness (a texture defect) as well as with acid-cut and seamy color defects. A crumbly cheese may sometimes be quite dry, but more often will be normal in this respect. A crumbly, friable body is more likely to occur in aged cheese (≥ 10 months of aging).

Curdy (Rubbery). This body defect is quite characteristic of freshly made, "green," or uncured cheese. Such cheese usually seems firm, almost hard or rubbery. The plug resists finger pressure; when it does yield to pressure there is a tendency for the cheese to spring back to its original shape. A positive correlation exists between a curdy, rubbery body and fresh, "green," flat, or undeveloped flavor. Since curdiness is primarily a characteristic of young, uncured cheese, before the curd has had an opportunity to break down (undergo proteolysis), it is not usually considered an objectionable body defect in mild-aged cheese. Such cheese should eventually develop the desired body/texture characteristics upon additional aging.

Greasy. A "greasy" cheese is one which has free fat on the surface, as well as in and around openings within the cheese or surfaces of individual curds. The defect is easily recognized by an almost oil-like appearance or feel. Greasy-like cheese often exhibits marked seaminess or may eventually develop it upon additional aging.

Pasty (Smeary, Sticky, Wet). Cheese showing the "pasty" defect is usually characterized by the presence of high moisture. There is often difficulty in securing a full, well-rounded plug; the cheese shape is easily distorted. The cheese breaks down easily into a pasty, sticky mass which tends to adhere to the fingertips as the product is manipulated. This defect is often associated with a weak body and/or high acid, fruity, and/or fermented off flavors.

Short (Flaky). A "short" or "flaky" body is characterized by a lack of meatiness, waxiness, or overall homogeneity; the consistency of the cheese may appear loose-knit. The plug will not tear apart and will show a distinct lack of elasticity; it will break easily on bending a short distance. The sample piece may appear dull in color, but in many cases may exhibit a fairly even and somewhat glistening surface. A cheese having this body defect may be too acid and/or dry to exhibit more desirable body properties. Sometimes a short-bodied cheese is inclined to be mealy when a piece of a plug is worked between the thumb and forefinger (or by mouthfeel).

Spongy. A spongy-bodied cheese fails to yield a full, continuous plug, due to the presence of excessive gas or mechanical openings that prevent an adequate degree of firmness in the body of the cheese. When a spongy cheese is plugged, it tends to sink immediately next to the trier. Such cheese is distinctly springy when pressure is applied to the surface. This defect is commonly associated with gassy, high-moisture, weak-bodied cheese.

Weak (Soft). A weak-bodied cheese is noted particularly by the ease of cheese trier penetration, and/or by the relatively small amount of finger pressure necessary to break the structure. Weak-bodied cheese is soft and is closely associated with a high moisture content. An aged, weak-bodied cheese may demonstrate fruity/fermented, whey taint, and/or unclean flavor defects, enhanced presumably by a relatively high whey (moisture) content.

Texture Defects

The texture defects of Cheddar cheese may be listed as follows:

Mealy/Grainy (Gritty)

Slits (fish eyes, yeast holes)

Gassy (pin holes)

Sweet-curd holes (swiss holes,
  shot holes)

Fissures

Open (mechanical holes)

Mealy/Grainy (Gritty). A cheese that is worked between the thumb and forefingers and shows a lack of uniformity and smoothness, as well as irregularly shaped, hard particles of cheese, is criticized as being mealy/grainy (gritty), depending on the particle size. This physical condition often may be correlated with a dry, corky-bodied cheese. When the manually "worked cheese" feels like cornmeal, and the cheese tends to spread in irregular patches under sliding pressure of the thumb over the forefingers, the texture is described as mealy or grainy. A mealy/grainy cheese tends to exhibit dryness and seems to release fat readily. Often, a mealy textured cheese also exhibits a short body with little elasticity. Mealiness is most often associated with sharp or aged cheese; white specks may be obvious also. The cheese judge should also be able to detect a corn meal-like mouthfeel (mealy/grainy) when the cheese sample is masticated and pushed against the roof of the mouth.

Slits (Fish Eyes, Yeast Holes). Cheese made from poor-quality milk or starter culture that has been contaminated with yeast (or possibly coliform bacteria) may develop round, glossy-surfaced gas holes as the result of abnormal fermentation. Cheese which contains numerous yeast holes usually has a "spongy" body due to excessive gas production. During plugging, the cheese tends to sag immediately adjacent to the inserted trier. Such cheese usually yields a honeycomb-like plug. Yeast holes in cheese may flatten out as the cheese is cured, forming long narrow slits known as "fish eyes."

Gassy (Pin Holes). Gas holes in cheese may vary in size and are fairly uniform in distribution and shape. They are formed from gas produced by undesirable microorganisms within the cheese. The seriousness of these gas holes depends on the kind of organisms that form the gas and the relative size and frequency of the gas holes. Gas holes are referred to as "pin holes" when they are about the size of a pinhead, symmetrically rounded, evenly distributed, and/or show a tendency to be concentrated near the center of the cheese. "Pin" holes may result from the growth of undesirable bacteria from cheese milk, or a contami-

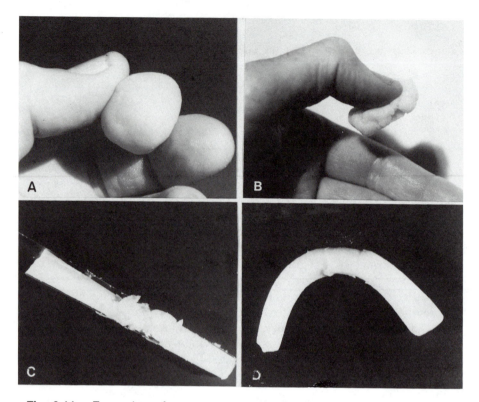

Fig. 8.11. Examples of some common body characteristics (defects) of Cheddar cheese: A—An "ideal" waxy body (practically forms a marble); B—A distinctly "pasty" or "sticky" body; C—A "crumbly" plug; D—A "weak" body.

nated culture, or a "gassy" culture (formed CO_2), that contains *S. diacetylactis* or *Leuconostoc* sp. This may also affect the flavor of the cheese; occasionally an objectionable fruity flavor may occur. The development of numerous "pin" holes and other gas holes may lead to a "huffed" cheese, especially if the cheese is cured at higher temperatures. If there are sufficient gas holes in the cheese to weaken the overall body structure, it is termed "spongy" cheese; undesirable flavor(s) are often associated with excess gas formation (see Fig. 8.12).

Sweet-curd Holes (Swiss Holes, Shot Holes). The large, uniformly distributed gas holes found occasionally in Cheddar cheese are usually the result of a particular bacterial growth. There is often a correlation between their occurrence and the flavor (or off-flavor) of the cheese. Large gas holes are often associated with a peculiar sweetish, pleasant flavor reminiscent of Swiss cheese; consequently, they are sometimes

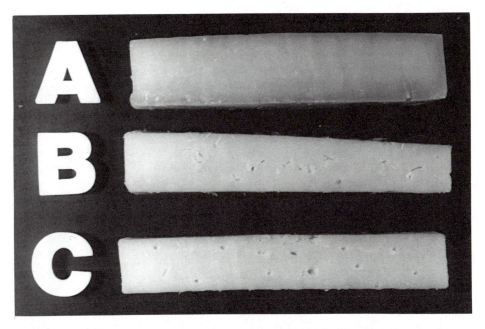

Fig. 8.12. A comparison of several texture defects in Cheddar cheese; A— No texture defect; B—Definite "open" texture; C—Definite "gassy" texture.

referred to as "Swiss holes," "sweet holes," or "shot" holes. The specific flavor defect that often develops may not be highly objectionable, but it is not typical of Cheddar cheese.

Open (Mechanical Holes). An "open," porous, or loose texture is traceable to the physical aspects of handling and pressing the cheese curd. Mechanical openings are characterized by their asymmetrical, angular shape and size, and by the dullness of their surface linings (Fig. 8.12). These irregular-shaped holes are derived from various conditions during the matting and pressing of the curd. There is little or no relationship between their presence and cheese flavor. In Cheddar cheese, as long as mechanical openings are not connected and are neither so numerous nor so large as to weaken the body or interfere with the integrity of the plug or slice, they should not meet with serious objection.

Fissures. A fissured texture is characterized by an elongated slit or extended separation of the curd particles. The curd lacks cohesion, and such defects may be associated with seaminess. This defect is not serious, but such an affected cheese often lacks the desired meatiness of body.

Scoring Body and Texture of Cheddar Cheese

Experience has always been a good guide for evaluation of the body and texture defects of cheese. However, gaining experience involves time and effort. A chart that summarizes the various intensities and assigned score values for defects is helpful. Such a guide for scoring cheese body and texture is shown in Table 8.4. In using this guide, one should keep in mind that combinations of several defects may lower the score to the minimum value of the range. Many of the body and texture defects can be corrected by simple modifications in cheese-making practices. The defects and their probable causes are given in Table 8.5.

EVALUATING THE FLAVOR OF CHEDDAR CHEESE

High-quality Cheddar cheese should possess the characteristic "Cheddar flavor," which is best described as clean, moderately aromatic, nutty-like, and pleasantly acidic. While the same general flavor qualities are desired in fresh, medium-cured, and aged cheese, the intensity of the characteristic Cheddar flavor will primarily depend upon the extent of curing and actual curing conditions. Usually, aged cheese has a sharp, aromatic, intense flavor that is entirely lacking in young cheese. The flavor of high-quality Cheddar cheese has been likened to that of freshly roasted peanuts or hazelnuts by various investigators (Davies and Law 1984; Kosikowski and Mocquot 1958; Van Slyke and Price 1979; Wilson and Reinbold 1965; Wilster 1980).

The flavor of Cheddar cheese is ascribed to a complex mixture of compounds, produced by bacteriological and enzymatic action during aging. The Cheddar flavor originates from: (1) protein breakdown to simpler and more volatile organic compounds; (2) acid developed in the curd; (3) milkfat; and (4) the small amount of salt added before the curd is pressed. Due to the relatively high degree of solids and the nature of the organic constituents, Cheddar cheese has a distinct, desirable flavor when the appropriate bacteriological, enzymatic, and chemical changes have occurred during controlled manufacturing and curing.

The beginner judge should try to appreciate that the finish, appearance, color, and body and texture characteristics reveal much regarding the flavor quality of the cheese. The evaluator should carefully study both the desirable and undesirable aspects of these quality criteria and note the flavor correlations that may be associated with them.

Once the physical properties of the cheese have been assessed, the flavor characteristics should be determined. This is accomplished by: (1) first noting the odor of the freshly drawn plug as it is passed slowly under the nose; (2) smelling the warm, semisoft cheese that results from the quick kneading of a portion of the plug between the thumb and forefingers; and (3) tasting a small piece of the cheese. Experienced judges may not even taste the cheese, but evaluate and grade samples on the basis of the noted odor characteristics and their association with certain observed physical attributes. The novice judge, however, should taste the sample not only to verify the odors previously noted, but also to perceive the nonvolatile taste sensations—bitter, salty, sour, and sweet, which could otherwise go undetected. When a larger number of samples are being tasted, an occasional rinse of the mouth between samples is helpful. This prevents any nonliquified portions, which may lodge between the teeth, from obscuring the flavor characteristics of subsequent samples.

Flavor Defects and Their Characteristics

Off-flavors in Cheddar cheese show wide variation and may be listed as follows (see Table 8.6 for probable causes and remedial measures):

High acid (sour)	Metallic (oxidized)
Bitter	Moldy (musty)
Fruity/fermented	Rancid (lipase)
Flat (lacking flavor)	Sulfide (Skunky)
Garlic/onion (weedy)	Unclean (dirty aftertaste)
Heated (cooked)	Whey taint (sour whey)
Malty ("Grape Nuts®")	Yeasty.

High Acid (Sour). Lactic acid is a normal component of Cheddar cheese flavor; however, an excessive acid or sour taste is undesirable. Depending on age, the normal pH range of Cheddar cheese should be 5.15–5.45. The "high acid" (sour) defect generally results from a too-rapid or excessive lactic acid production in the curd. High acid is by far the most frequently encountered flavor defect of Cheddar cheese.

When a portion of high acid cheese is placed into the mouth, a "quick" taste sensation is noted on the top and front sides of the tongue. This taste soon disappears (usually), leaving the mouth free of any off-flavor sensations. High acid flavor may sometimes be associated with a dull, faded, or acid-cut color defect. For some individuals,

Table 8.6.   Common Flavor Defects of Cheddar Cheese and Their Probable Causes and Remedial Measures.

Flavor Defect	Probable Causes	Remedial Measures
Bitter	1. Excessive moisture.	1. Use carefully selected cultures.
	2. Low salt level.	2. Reduce amount of starter.
	3. Proteolytic starter culture strains.	3. Monitor salting levels and method of adding
	4. Microbial contaminants	
	5. Excessive acidity	4. Upgrade milk quality
	6. Poor milk quality	5. Improve sanitation
	7. Plant sanitation problems	6. Control acid and rate of development
High Acid (Sour)	1. Development of excessive lactic acid	1. Reduce ripening time
	2. Excessive moisture	2. Reduce starter amount
	3. Use of too much starter	3. Monitor milk acidity
	4. Use of high-acid milk	4. Cook to slightly higher temperature
	5. Improper whey expulsion from curd	5. Follow a standardized procedure for cutting, cooking, draining, cheddaring, and salting steps
	6. Low salt level	
Flat (Lacks Flavor)	1. Lack of acid production	1. Check starter activity
	2. Use of milk low in fat	2. Increase starter amount
	3. Excessively high cooking temperature	3. Increase curing temperature
	4. Use of too low a curing temperature	4. Lengthen curing period
	5. Too short a curing period	5. Standardize cheese milk for fat content

Defect	Causes	Corrective Measures
Fruity/fermented	1. Certain strains of S. *lactis* or S. *diacetylactis* 2. Low acidity 3. Excessive moisture 4. Low salt level 5. Poor milk quality	1. Eliminate lactic strains that produce ethanol 2. Monitor starter activity 3. Check salting procedures 4. Upgrade milk quality
Rancid (Soapy)	1. Milk lipase activity 2. Microbial lipases from contaminants 3. Accidental homogenization of raw milk 4. Late lactation or mastitic milk	1. Check cheese milk for rancid off-flavor 2. Avoid excessive agitation, foaming, and severe temperature fluctuations 3. Improve sanitation 4. Monitor milk quality
Whey taint	1. Poor whey expulsion from curd 2. Improper Cheddaring techniques 3. Failure to drain whey from piles of curd slabs (especially between pieces)	1. Standardize the Cheddaring process 2. Constantly make sure expelled whey is free to drain away from Cheddaring curd 3. Wash curd with 32°C (90°F) water to remove excess whey
Unclean	1. Poor quality off-flavored or old milk 2. Unwanted microbial contaminants 3. Allowing off-flavored cheese to be "aged" 4. Improper technique of Cheddaring	1. Upgrade milk quality 2. Improve sanitation 3. Market marginal quality cheese as mild 4. Standardize the Cheddaring process

Source: Compiled from Chandan (1980a); Van Slyke and Price (1979); Wilson and Reinbold (1965); Wilster (1980).

the high acid off-flavor is sharp and puckery to the taste, suggestive of lactic acid. Frequently, in aged cheese, there is an associated bitter aftertaste. Numerous other off-flavors may occur in conjunction with a high acid note. It should be noted that *high acid is the most frequently encountered off-flavor in Cheddar cheese.*

Bitter. If volatile, other cheese off-flavors will be detectable by the sense of smell, but bitterness is noted only by the sense of taste. Bitter off-flavors may occur in mild cheese, but are found more frequently in aged cheese as an aftertaste. Certain lactic cultures, coagulating enzymes, and salt levels have been implicated in the development of this troublesome defect. Bitterness has been observed to develop in cheese made from both excellent-quality and poor-quality milk. "Sharpness" and the high flavor intensity of aged cheese should not be confused with a bitter taste. Sharpness gives rise to a temporary peppery sensation, whereas true bitterness is somewhat distasteful, resembling the taste of quinine or caffeine. The bitter sensation is somewhat delayed in terms of its initial perception, and tends to persist for some time after sample expectoration. Bitterness is observed by a taste sensation that occurs at the base of the tongue.

Fruity/Fermented. The "fruity" off-flavor is peculiarly sweet and aromatic; it resembles the odor of fermenting or overripe fruit, such as an apple or pineapple. This flavor defect is occasionally associated with high moisture cheese, and a weak, pasty body. The fruity/fermented defect intensifies as the cheese ages and may eventually lead to an unclean or combined fruity and unclean off-flavor. The fruity defect is attributed to the presence of ethanol-forming microorganisms in the cheese milk or certain cheese cultures. Esters formed from available ethanol and organic acids are responsible for the fruity note (Bills *et al.* 1965; Vedamuthu *et al.* 1966; Bodyfelt 1967). The fermented off-flavor in Cheddar cheese is more suggestive of acetic acid (vinegar-like).

Flat (Lacking Flavor). Cheese exhibiting this defect is practically devoid of any flavor. A flat flavor is particularly noticeable when the sample is initially tasted. Likewise, little odor is detectable. When associated with fresh or young cheese, the defect is not serious or objectionable, since full cheese flavor may eventually develop with additional aging. In an aged cheese, flatness (lacking flavor) represents a more objectionable defect.

Garlic/Onion. This flavor defect is relatively easy to detect because the off-flavor resembles that of garlic, onions, or leeks. Defective cheese usually shows a moderate odor, unless the sample has been stored at a high temperature. When the sample is tasted, the off-flavor is often quite pronounced and usually requires a thorough rinsing of the mouth prior to tasting additional samples.

Heated (Cooked). The heated (cooked) off-flavor of cheese differs from the clean, distinct cooked flavor of pasteurized milk; in cheese this defect more resembles the odor of old or spoiled milk, or the odor exhibited by melted Bakelite® forms of plastic. This off-flavor is somewhat suggestive of the unclean odor, in addition to whey taint. "Heated whey" is probably a more appropriate term to describe "heated" or "cooked" off-flavor in cheese.

Malty ("Grape Nuts®"). This off-flavor, which is seldom noted in Cheddar cheese, is easily distinguished by its harshness and distinctive character. The descriptors explain its flavor characteristics well. The growth of *S. lactis* var. *maltigenes* in cheese milk, and a subsequently produced malty flavor compound (3-methylbutanal), is responsible for this off-flavor.

Metallic (Oxidized). Occurrence of a metallic (oxidized) off-flavor in Cheddar cheese is quite rare, due to the reduction-oxidation potential of the cheese interior. This off-flavor, should it occur in cheese, is characterized by a flat, metal-like taste and a lingering puckery (mouthfeel) sensation. The sense of smell is of little or no value in detecting its presence. Oxidized (or metallic) cheese milk is the probable source for this cheese off-flavor when it infrequently occurs.

Moldy (Musty). A moldy or musty flavor defect often resembles the odor of a damp, poorly ventilated (potato) cellar. This defect is easily recognized by a characteristic smell. A slightly unclean off-flavor tends to persist after the tasted sample has been expectorated. The most frequent cause is mold growth on cheese surfaces, due to lost integrity of the cheese package and the admittance of air. In some cheeses where extensive mold contamination has occurred, a Penicillium-like mold (blue-green) growth may appear in the interior of the cheese, especially when it is open-textured. Serious economic losses, consumer dissatisfaction, and potential toxicological and allergenic consequences may occur from severe mold contamination of cheese.

Rancid (Lipase). A "rancid" off-flavor in cheese is characterized by: (1) a relatively slow reaction time; (2) a prominent odor that may be still noted after sample expectoration; and (3) an unpleasant, persistent aftertaste. The off-flavor is typically bitter, soapy, quite disagreeable, and usually somewhat repulsive. A rancid off-flavor in Cheddar cheese can usually be detected by the sense of smell. Rancidity is caused by activity of the enzyme lipase on milkfat; this yields volatile, unpleasantly flavored short-chain free fatty acids and their respective salts (or soap). When the concentrations of the free fatty acids from butyric (C_4)

to lauric (C_{12}) exceed levels desired for a balanced Cheddar cheese flavor, they impart on off-flavor variously described as goaty, cowy, unclean, bitter, or rancid. Rancid cheese usually results from abusive handling of cheese milk.

Sulfide (Skunky). The "sulfide" off-flavor of cheese is distinctive; it is similar to water with a high sulfur content. Sometimes an offensive sulphurous (skunky) or spoiled egg odor may be noted in aged cheese. Frequently, there is an associated bitter aftertaste, and/or a burning sensation within the mouth. Sulfide cheese often has a related sticky, pasty body. Usually, sharp or extra-sharp cheese is involved when this flavor defect is incurred. The cheese judge should keep in mind that a low to modest level of sulfide is an important component of aged Cheddar cheese flavor and aroma. Numerous sulfur-containing compounds can be formed during the aging process. However, when the sulfide note becomes dominant, to the point of obscuring other flavor characteristics, this is perceived as an off-flavor and a serious defect. In some regions of the world that produce Cheddar cheese, the sulfide flavor note is considered essential or highly desirable in sharp or extra-sharp cheese; hence, it is not criticized when it appears.

Unclean (Dirty Aftertaste). An "unclean" off-flavor is difficult to describe, since it often varies in intensity and lacks a definitive sensory description. This defect may suggest to the taster a general lack of cleanliness in producing the product, given the dirty, lingering, unpleasant aftertaste. This off-flavor persists long after the sample has been expectorated, and the mouth fails to "clean-up." An unclean off-flavor may occur in conjunction with other flavor defects such as high acid, bitter, and/or whey taint. Poor-quality or "old" milk used for cheese manufacture is a principal cause of the unclean flavor defect. Proteolytic and/or lipolytic enzymes, derived from psychrotrophic bacteria, may cause undesirable fermentations to occur within the cheese, and hence, result in an unclean off-flavor.

Whey Taint (Sour Whey). These terms describe various intensities of off-flavors in cheese associated with retained cheese whey. The slightly dirty–sweet/acidic taste and odor is characteristic of fermented whey. Ordinarily, the taste reaction of "whey taint" is perceived rapidly and is of short duration. The mouth tends to clean-up soon after sample expectoration. Some cheese authorities liken whey taint to the occurrence of a "fermented/fruity" off-flavor, with an "unclean" off-flavor superimposed over it. Whey taint cheese often has the body (rheological) characteristics of a high-moisture cheese. Also, whey taint is sometimes found in young Cheddar cheese that exhibits a seamy defect. Some judges may confuse whey taint and high acid off-flavors; however, only the former defect exhibits the distinctive aroma of fermented whey.

Yeasty. This off-flavor may be identified by its sour, bread dough, yeasty, or somewhat "earthy" taste and characteristic aroma. Yeastiness in cheese may be detected immediately after the sample has been put into the mouth. Since this defect is caused by yeast growth, the cheese will usually have numerous medium- to large-sized gas holes, which may be readily identified by their surface sheen, spherical or fish eye shape, and frequency.

Other Off-flavors. The off-flavors discussed above should be considered as the more common or frequently encountered ones in Cheddar cheese. However, the cheese judge should be alert to other possible flavor defects that may occur occasionally. Examples are an "atypical Cheddar flavor" and a "catty" (or cat-box odor) defect; the latter is possibly caused by low concentrations of mesityl oxide in cheese reacting with sulfides to produce the unpleasant aroma.

Guide for Scoring Cheddar Cheese Flavor

The Committee for Evaluation of Dairy Products of the American Dairy Science Association has established scoring guides for the evaluation of flavor and body and texture defects in Cheddar cheese, as shown earlier in Table 8.4.

Flavor Chemistry of Cheddar Cheese: A Brief Overview

Perhaps the best way to consider Cheddar cheese flavor in proper perspective is to briefly review some of the chemical characteristics responsible for the complex flavor of this product. A conservative estimate is that at least 50 different chemical compounds contribute to the flavor profile of Cheddar cheese. Kristoffersen (1963) stated that the above number of compounds would probably have to be multiplied several-fold when all varieties of cheese are considered.

The typical, desirable flavor of selected cheeses has been demonstrated, in numerous studies, to be associated with the volatile fraction of the flavor components. Low-temperature vacuum-distillation has been employed to separate volatile flavor components into water-soluble and water-insoluble fractions. The water-soluble fraction contains most of the cheese aroma; the water-insoluble portion retains only a bland, broth-like taste.

Cheddar cheese flavor develops as the result of bacterial action (lactic streptococci), enzymes (chymosin-rennin, bacterial enzymes, milk enzymes), and chemical reaction (oxidation and constituents reacting with each other). In addition to metabolites produced from lactose fermentation, proteolysis and lipolysis are of major significance in cheese flavor development. Methanethiol and other sulfur compounds probably originate from sulfur-containing amino acids. The bacterial flora of cheese milk contributes to the development of both the flavor of cheese and possible off-flavors.

Gas-liquid chromatography (GLC), mass spectrometry (MS), infrared spectrometry (IR), nuclear magnctic resonance (NMR), and other sophisticated procedures are used in flavor research. These techniques provide data which help to identify and quantify the chemical components of the flavor fraction. Examples of IR spectra obtained on flavor fractions of Cheddar cheese (Kristoffersen 1963) are shown in Fig. 8.13. This technique yields information on various types of chemical groups and structures which the compounds under observation possess.

Component Balance Theory. Cheese flavor intensity relates directly to the concentration of flavor components formed within the initial fermentation steps and the course of cheese ripening. Similar to wine, the quality of cheese flavor depends upon the formation of flavor components in a certain and desirable balance. Kosikowski and Mocquot (1958) derived what is known as the "component balance theory" of cheese flavor. Its premise states that a relatively small number of chemical compounds is responsible for cheese flavor. An example of proper and improper ratios (balance) of free fatty aids and hydrogen

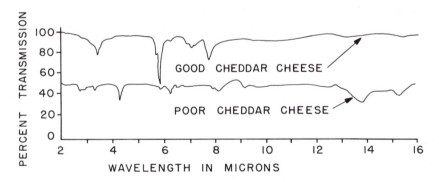

Fig. 8.13. Many techniques are used in research to obtain an objective description of flavor. This figure illustrates differences in the infrared spectra of flavor fractions from good and poor Cheddar cheese (from Kristoffersen, 1963).

Table 8.7. The Character of Cheddar Cheese Flavor Based on the Relative Concentrations of Free Fatty Acids (F.F.A.) and Hydrogen Sulfide (H₂S).

Flavor		Concentrations of		Molar Ratios		
Character	Intensity	F.F.A. mM/100g	H_2S mM/100g	F.F.A.		H_2S
Balanced	mild	10	0.7	14	:	1
	sharp	28	2.0	14	:	1
Sulfide-like	mild	7	1.0	7	:	1
(Unclean)	sharp	21	3.0	7	:	1
Fatty acid-like	mild	11	0.4	28	:	1
(Fermented)	sharp	42	1.5	28	:	1

Source: Kristoffersen (1963).

sulfide, and their respective effects on the character of Cheddar cheese flavor, is summarized in Table 8.7 (Kristoffersen 1963).

The component balance theory suggests that only when the amounts and relative proportions of key chemical compounds are correct (or within a certain range), is typical cheese flavor (of a given type) obtained. In the last several decades, a goal of Cheddar cheese flavor research has been to convert the complexity of flavor development into relatively simple manufacturing rules that would help insure or yield high quality cheese—consistently.

GRADING OF CHEDDAR CHEESE

Cheddar cheese can be graded at any stage between the time it is removed from the press and the time it is sold for consumption. Experienced cheese graders agree that Cheddar cheese ranging from only a few days to a few weeks old is more difficult to grade than a more mature product. In grading a young or "green" cheese, the grader should pay close attention not only to the flavor, but also to those conditions which might precede undesirable flavor development during ripening. There are occasions when a cheesemaker, cheese buyer, or processor would like to have fresh or "green" cheese graded, in order to: (1) sell it on a quality basis; (2) determine the best use of the cheese; (3) determine whether cheese quality will withstand storage; or (4) monitor the day-to-day quality of the cheese.

Different cheese-producing areas of the U.S. often grade cheese independently of each other; consequently, those assigned grades may differ slightly from Federal (USDA) cheese grade standards. Considering the purposes for which cheese is graded in different geographical regions, the variations in score cards or grading forms, and the wide interpretation of standards, there is little wonder that there is some lack of uniformity existing in grading Cheddar cheese. Conversely, re-markable agreement exists in what constitutes high-quality or low-quality cheese, regardless of the geographical region or the grading agency involved.

Grading of Young Cheese for Storage. Some Cheddar cheese is bought and sold when "green," or only a few days after removal from the press. Fresh, uncured cheese lacks the typical Cheddar flavor and body; hence, it must be graded on the basis of predicted quality development during early to midstages of the curing period. There is merit in grading fresh Cheddar cheese, in order to utilize the product to best advantage. However, some differences of opinion exist as to the value of judging "green" cheese to determine its future or "aged" potential. Since certain flavor, body, texture, and workmanship qualities have a bearing on the curing of cheese, a qualified cheese grader usually can reliably project or predetermine how a graded young cheese will develop with additional storage (curing time). Careful sensory evaluation of immature cheese (prior to storage) and records of manufacturing, moisture content, and of the relative quality of cheese milk are helpful factors in determining the probable success of cheese curing.

In grading young cheese for subsequent commercial use, Price (1943) suggested dividing Cheddar cheese into the following categories:

Long Hold—The quality level necessary for the most particular or discriminating use of the cheese.

Short Hold—Minor defects (slightly apparent) which will permit short storage periods without loss in commercial value.

Immediate Use Only—Distinct defects (easily detected, obvious) which require careful sorting of the cheese according to given markets; immediate utilization of the cheese is perhaps mandatory.

Limited Use—Major defects (quite serious faults) which restrict use of the cheese to a few markets, i.e., grinding purposes, process cheese, or immediate consumption as a "cooking cheese."

Culls—Inedible cheese, not to be used for human consumption.

The specific product defects which would necessitate placing cheese in the above respective classes are usually obvious and involve many of the defects listed on the cheese score card. Flavor is usually considered more critically than other factors, although body and texture, color, and appearance features of the cheese should not be overlooked. To possess the desired characteristics, the so-called "ideal" cheese should have: (1) a clean, delicate, pleasing aroma and, when cured, a nutty flavor; (2) a firm and springy body, showing smoothness and waxiness (if cured) when worked between the thumb and fingers, and slight curdiness if fresh; (3) a texture that reveals a smooth-bore (few or no openings); (4) uniform, translucent color, whether colored or uncolored (when fresh, it may be slightly seamy); and (5) a smooth finish that is clean, well-shaped, uniform in dimensions and overall size, with a complete, air-tight package, and mold-free.

Federal Grading of Cheddar Cheese

U.S. Cheddar cheese sold in central markets, or on contract, is usually sold on the basis of government grade. If sold on contract, the cheese age and style of package is generally specified. Such cheese is generally graded according to Federal standards by a United States Department of Agriculture grader; the cheese price is determined primarily on the basis of sensory quality. Generally, a college student who has mastered the evaluation of Cheddar cheese by the score-card system can, after a short apprenticeship with a Federal grader, become proficient in grading cheese according to Federal standards.

The Dairy Grading Branch of the Poultry and Dairy Quality Division, Food Safety and Quality Service, of the U.S. Department of Agriculture recognizes four grades of American Cheddar cheese. The nomenclature for these grades is as follows: (1) U.S. Grade AA; (2) U.S. Grade A; (3) U.S. Grade B; and (4) U.S. Grade C. A general description of Federal Cheddar cheese grades is summarized in Table 8.8. Detailed descriptions of the quality grades and U.S. Standards for grades of Cheddar cheese are summarized in Appendix VIII.

COLBY AND MONTEREY JACK CHEESE

Inasmuch as the general manufacturing procedures and bacterial fermentations occurring in Colby and Monterey Jack cheeses closely parallel those of Cheddar cheese, these three related varieties tend to share common defects. However, due to a higher moisture content, lower acid and salt content (sometimes), and higher microbial and enzymatic activity, some sensory defects may reach greater intensity and frequency in Colby and Jack cheese than in Cheddar. The above factors tend to limit the keeping quality of Monterey Jack and Colby cheese, compared to Cheddar.

Flavor. For cheeses two to three months of age, an acid flavor may be more apparent in Jack and Colby cheeses than in Cheddar. The likelihood that a typical, nutty, Cheddar flavor will develop in Colby or Jack cheese within several months is unlikely. The "acid flavor" tends to be more obvious in the two granular cheeses, since there is no partial

Table 8.8. A Summary of the U.S. Grades of Cheddar Cheese (USDA).

Grade	General description of medium-cured to aged Cheddar	Approximate score or score range[a]
AA	Flavor: Fine, highly pleasing, very slight feed flavor permitted. Body and Texture: Firm, solid, smooth, compact, close, translucent, few small mechanical or sweet holes permitted, no gas holes. Color: Uniform, tiny white specks if aged and very slight seaminess permitted. Finish: Sound rind well-protected and smooth, even-shaped.	93 or above
A	Flavor: Pleasing, may possess limited feed, or acid or bitter flavor (if aged). Body and texture: Reasonably solid, compact, close and translucent, few mechanical holes not large or connected, limited to two sweet holes per plug, no gas holes. Color: Slight white lines or seams. May be very slightly wavy. Finish: Sound firm rind, well protected but may possess to a very slight degree a soiled surface or mold growth; may be slightly lopsided, have high edges or rough, irregular surface.	92
B	Flavor: May possess certain limited undesirable flavors according to age. Body and texture: Texture may be loose and open and have numerous sweet holes, scattered yeast and other scattered gas holes, pinny gas holes not permitted. Color: May possess about the same defects as Grade A except to a greater degree. Finish: Rind sound, may be slightly weak, but free from soft spots, rind rot, cracks, or openings, bandage may be uneven, wrinkled but sound, surface may be rough, unattractive, but have good protective coating; paraffin may be scaly or blistered; no indication that mold has entered the cheese; may be huffed, lopsided, or have high edges.	90 to 91

Table 8.8. (*continued*)

Grade	General description of medium-cured to aged Cheddar	Approximate score or score range[a]
C	Flavor:	89
	May possess somewhat objectionable flavors and odors with a certain increase in tolerance according to age and degree of curing.	
	Body and texture:	
	May be loose with large connecting mechanical openings; have various gas holes and body defects with limitations varying with the degree of curing; must be sufficiently compact to permit drawing a full plug.	
	Color:	
	May possess various defects, but not to the extent that the color is unattractive.	
	Finish:	
	Rind may be weak, have soft spots, rind rot, cracks, and openings, with certain limitations varying with degree of curing. Bandage may be uneven, wrinkled, but not torn; may have rough unattractive appearance, paraffin scaly or blistered; mold permitted, but not evidence that mold has entered the cheese; may be huffed, lopsided, and have pronounced high edges.	

[a] These are the approximate numerical scores of each U.S. grade if scored by the score-card system. The U.S. grades are reported in letter grades only.
Source: United States Department of Agriculture (1985).

masking effect from a "Cheddar flavor." A notable exception is certain dry or low-moisture Monterey Jack cheeses, which can be aged nine or more months and often develop a distinct, full, nutty flavor. Frequently, when conventional Colby or Monterey Jack cheese exceeds 100 days of age, a distinct bitter taste may develp, which reflects a possible limitation for aging of these cheese types beyond three months.

Body and Texture Defects. Colby and Monterey Jack cheeses tend to have a weak body, due to their higher moisture content. This characteristic is anticipated and tolerated, up to a certain point. With respect to cheese texture, gas and/or mechanical openings are expected and more tolerated in these two granular (or stirred curd) forms of cheese, than in Cheddar. Occasionally, solid or "blind spots" occur in Colby and Monterey Jack cheese. These are usually related to the formation of curd lumps that developed before or during curd washing, cooling, or salting. The typical remedy is to try to continuously maintain the curd in a granular form by applying adequate agitation of the curd and uniform distribution of the salt. Applications of higher pressure to cheese hoops during pressing also account for the production of closed or blind Colby and Monterey Jack cheeses. Solid or blind cheese of these two types has apparently gained consumer acceptance; a granular or stirred curd appearance gradually has become a less and less common feature of Colby and Monterey Jack cheese.

SWISS CHEESE

Swiss cheese, also known as Emmental, Emmentaler, Schweizer, or Sweitzer cheese, is a type of hard cheese made from clean, fresh, whole milk. Specific processes of manufacture are used, which differ widely from those for Cheddar cheese. The utilization of thermophilic lactic bacteria and *Propionibacterium shermanii* for milk fermentation results in a cheese having flavor, body, texture, and appearance characteristics peculiar unto itself.

Correctly speaking, Swiss cheese made in the United States is called "domestic Swiss." The extent of manufacture and high consumer acceptance justifies the use of the general term, "Swiss" cheese. High-quality Swiss cheese is characterized by: (1) a cream-yellow color; (2) a solid, compact, slightly translucent body, interspersed with large, shiny-surfaced gas holes that are evenly distributed (preferably) throughout the center, but become less numerous near the edge of the cheese; and (3) a characteristic "sweet-hazelnut" flavor.

Swiss Cheese Score Card. Since Swiss cheese is an entirely different product from Cheddar cheese, the Cheddar score card is not readily adapted to scoring Swiss cheese. A score card for Swiss cheese recognizes the importance of proper hole or eye development. Three different score cards for Swiss cheese are summarized in Table 8.9. These score cards differ slightly in terms of the emphasis placed on various criteria.

Table 8.9. Alternative Swiss Cheese Scoring Outlines.

| | Possible Score | | |
Item	Mojonnier and Troy (1925)	Sammis (1937); Thom and Fisk (1938)	Contemporary
Flavor	40	35	10
Holes and Appearance	25	30	5
Texture	20	20	5
Salt	10	10	2
Style	5	5	3
Total	100	100	25

Quality Characteristics of Swiss Cheese

Flavor. A high-scoring Swiss cheese should have a clean, distinctive, pleasing, sweet-hazelnut flavor. The development of this desirable flavor is apparently associated with proper "eye" formation. During manufacture and curing, the lactic- and propionic-acid bacteria play an important role in converting lactose to lactates, and finally produce some propionic and acetic acids, as well as carbon dioxide. Carbon dioxide, diffused throughout the elastic curd, collects at foci to form individual eyes in the cheese, according to laws of physical chemistry. Appropriate eye formation in Swiss cheese is considered a good indication of typical Swiss-cheese flavor. Historically, any condition(s) that interfere with the fermentation that develops typical Swiss cheese "eyes" was assumed to jeopardize flavor development of the cured product. In contrast, some Swiss cheese authorities have stated that typical Swiss cheese flavor can be developed without "picture-perfect" eye size and distribution (Reinbold 1972; Foster *et al.* 1957).

The more common flavor defects found in Swiss cheese are as follows:

Lacks desirable Swiss flavor   Unclean (dirty aftertaste)
Rancid (bitter   Unnatural (atypical).
"Stinker" (foul)

Lacks Desirable Swiss Flavor. This defect is characterized by a partial or complete absence of the sweet-hazelnut, Swiss flavor. A "lack of desired Swiss flavor" is noticeable when the sample is first placed into the mouth. The defect may or may not be correlated with a lack of, or insufficient, eye formation. This defect is not considered as serious as some other flavor defects, since additional aging of the cheese may develop the characteristic flavor.

Rancid (Bitter). The rancid off-flavor and associated bitter aftertaste that may occasionally develop in Swiss cheese is usually the result of the mishandling of cheese milk, wherein lipolysis was induced, or due to microbial lipases. The formation of volatile fatty acids (C_4 through C_{12}) can be presumed responsible for this flavor defect in Swiss cheese.

Stinker (Foul). The "stinker" or "foul flavor" defect may be noted readily upon smelling the sample; the odor of hydrogen sulfide or spoiled eggs may be noted. This defect is often localized within the cheese, but may also extend over wide sections. An abnormal fermentation is generally the cause of this highly objectionable flavor defect.

Unclean (Dirty Aftertaste). Swiss cheese with this flavor defect leaves an undesirable, persistent taste after the sample has been expectorated. The defect may be correlated with undesirable eye formation, particularly with a niszler-type (pinhole defect) cheese.

Unnatural (Atypical). A cheese exhibiting an "unnatural Swiss" off-flavor generally possesses a relatively clean flavor, but the overall sensory perception is atypical for Swiss cheese. The unnatural flavor experienced by the evaluator may be similar to that of Cheddar cheese or some other type of cheese. An atypical Swiss flavor may emulate several possible off-flavors, which may be chemical, enzymatic, or bacteriological in origin.

Eye Development in Swiss Cheese

The shape, size, and distribution of the "eyes" in Swiss cheese have an aesthetic appeal to consumers (Fig. 8.14), in addition to a possible association with typical Swiss-cheese flavor. Hence, much emphasis is placed on the "eye" characteristics in evaluating Swiss cheese. The majority of eyes in Swiss cheese should be 3/4 to 13/16 in. (19.0–20.6 mm) in diameter. Round, symmetrical eyes are preferred in Swiss cheese, but a slightly elliptical or oval shape may be accepted without criticism. The "ideal" frequency distribution of eyes in cheese tends to fade toward the edge of the cheese (where Swiss cheese is somewhat drier). If the Swiss eyes are so large and/or so numerous as to "predominate" the plug or slice of cheese, such a cheese would be criticized severely. Various defects in eye formation of Swiss cheese are designated as follows:

Blind	Niszler (pin-holey)
Bloats	Overset, cabbage, blow holes
Dull glossy (dead eye, shell eye)	Pressler
Glaesler (glassler, glass)	Small eyes.
Irregular eyes	

Blind. A "blind" Swiss cheese has little or no eye formation. Such a cheese may be solid throughout without any openings (eyes) whatsoever, or may have a few small eyes scattered randomly; referred to as "partially blind." A blind or partially blind cheese often lacks the characteristic Swiss cheese flavor.

Bloats. The "bloat" defect is easily recognized; the cheese seems huffed, and in some cases so extensively that the sides of the cheese may be rounded. In extreme cases, the cheese piece may be severely cracked. The bloat defect is often the result of marked yeast outgrowth in a high-moisture cheese.

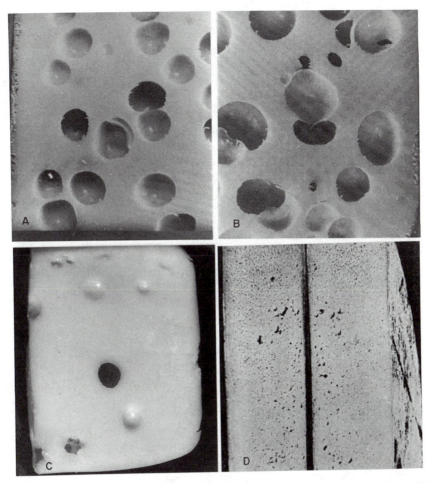

Fig. 8.14.   Cross sections of Swiss cheese of various levels of quality with respect to eye formation: A—Highest quality (good uniformity of size and distribution of "eyes"); B—Mid-quality (lacking uniformity of "eye" size); C—Inferior quality ("blind" or lacking presence and adequate size of "eyes"); D—Inferior quality ("pin-holey" or niszler-type eyes).

Dull Glossy (Dead Eye or Shell Eye). Swiss eyes that lack a glossy luster are designated as "dull glossy," "dead eye," or "shell eye." The inner lining of the cheese eye has a dull appearance similar to that of a nut shell. The eyes may be normal in shape, but have a distinct rough inner lining. Though this defect is not serious, it is not tolerated in the highest-scoring Swiss cheese.

Glaesler (Glassler, Glass). These terms describe a Swiss cheese that displays sizeable parallel clean-cut cracks within the body of the cheese. Cheese with a short-textured body apparently fails to respond to the normal development of round eyes, characteristic of an elastic curd. However, in the instance of a less elastic curd, the holes "split out" sideways, resulting in parallel layers of cracks, which are noticeable on the surface. This is a comparatively serious defect of Swiss cheese, since it substantially detracts from the typical appearance.

Irregular Eyes. This defect is characterized by distorted, somewhat elongated, walnut-shaped eyes. While a slight to medium defect is not considered serious, the presence of "irregular eyes" is not desired in the highest-quality Swiss cheese (Fig. 8.14). Frequently this defect is associated with oversetting (too many) eyes.

Niszler (Pin-holey). A "niszler" cheese contains numerous small pinholes (ranging in size from a pinhead to that of a puncture). This defect is due to an abnormal gassy fermentation within the cheese. The term "niszler" means "a cheese with a thousand eyes." Cheese suffering from this defect exhibit practically no Swiss eye development (Fig. 8.14). The flavor characteristics exhibited by "niszler" cheese are usually atypical, unclean, or otherwise undesirable.

Overset (Cabbage, Blow Holes). These terms describe a Swiss cheese in which the eyes are so large and so numerous that the quantity of cheese between them is nearly paper thin. Examination of a cross section of such cheese often reveals a cabbage-like appearance. This defect is progressively more accentuated toward the center of the cheese. Hence, a plug of cheese taken by trier from that part of the cheese may show a decided lack of cheese substance. Also, this defect may be localized, or it may extend throughout the center section of the cheese.

Pressler. A "pressler" cheese is one which develops too many eyes in the early stages of manufacture, particularly while the cheese is still in the press; hence, the name. This defect is apparently associated with an abnormal fermentation in the cheese; it also results in the development of an undesirable flavor, reminiscent of fermented whey.

Small Eyes. A preponderance of small eyes (smaller than 5/16 in. (7.9 mm) in diameter) in Swiss cheese is not desired; however, their presence does not represent a serious defect. More careful selection of the *Propionibacterium* culture and/or improved control of the manufacturing process should eliminate or minimize this defect.

Body and Texture of Swiss Cheese

The body and texture of high-quality Swiss cheese should be firm, closed, moderately flexible when bent, and free from such defects as "glass," pinholes, sponginess, or bloats. Occasional "picks" and "checks" may be tolerated, provided they are within 3/4 in. (19.0 mm) of the surface. "Picks" are small irregular or ragged openings within the body of Swiss cheese, somewhat like small mechanical openings in Cheddar cheese. "Checks" are short cracks within the body of Swiss cheese. Near the surface where the Swiss cheese is generally drier, the body may be slightly crumbly. A soft and pasty body is often associated with high moisture or abnormal eye formation (of one kind or another) and may be accompanied by poor flavor development.

Salt in Swiss Cheese

Swiss cheese is salted either by floating in the cheese salt brine or by rubbing salt on the outside, or both. Since salt brines migrate through cheese slowly, some difficulty may be encountered in securing uniform salting. Swiss cheese should be lightly and uniformly salted.

Finish and Appearance of Swiss Cheese

Swiss cheese should be symmetrical, with a smooth, even, clean, dry, and closed surface. The ends of cheese pieces should be parallel, neither bloated nor sunken, with surfaces free from cracks, and all edges square. It is most undesirable for cheese edges to exhibit long tabs or a cracked, open edge (referred to as "frog mouth").

Style

Swiss cheese is made in two general styles: (1) drums or "wheels" (approximately 36 in. diameter and 6 to 8 in. thick (91.4 cm × 15.2 to 20.3 cm)), and (2) rectangular blocks of various sizes. Four rectangular portions 8 × 11 × 17 in. (20.3 × 27.9 × 43.2 cm) may be cut from these blocks and packaged after adequate eye formation. This facilitates handling, storage, and cutting cheese into consumer-size units. Generally, Swiss cheese is not made in smaller-sized units, though there are some exceptions, notably, Baby Swiss. Larger cheese sizes are usually considered more conducive to proper eye and flavor development. However, Iowa State University has developed and successfully marketed an excellent Swiss-type cheese in round, flat styles of 25 to 30 lb (11.3 to 13.6 kg), with an approximate 16 in. (40.6 cm) diameter and a 6 in. (15.2 cm) thickness. Even smaller sizes have been successfully made and marketed. Baby Swiss is a description frequently applied to smaller sizes of Swiss cheese marketed in the midwestern U.S. Baby Swiss is generally higher in moisture and milkfat content;

hence, it exhibits a softer or weaker body than more conventional Swiss cheese.

Examination of Swiss Cheese

Much can be learned about the overall quality of Swiss cheese by carefully observing its general appearance. An abnormal internal gas formation will cause cheese to "bloat" or "huff," and sometimes to crack and form a "frog-mouth" opening, due to internal gas pressure. A high-quality cheese should show none of these abnormalities.

The proper number, size, and distribution of "Swiss eyes" can be determined in a general way by tapping the cheese and carefully listening to the resultant sound. The tapping is done by placing the second finger over the back of the index finger and briskly snapping it down on the surface of the cheese, or by tapping the surface gently with the handle of a special Swiss cheese trier.

Cheese possessing a desirable body will demonstrate a certain hollow sound or unique ringing effect. An extremely gassy or blind cheese will respond quite differently to tapping. A dull, nonresonant sound generally indicates distinct lack of eye formation (blind). A gassy cheese will sound "empty," "hollow," or "lifeless."

The next step is to plug the cheese and immediately pass the plug under the nose to note the aroma. The upper 1 in. (2.54 cm) of the cheese plug should be replaced into the formed hole in the cheese (to help "seal" the surface). After carefully observing the size, distribution and appearance of the "eyes," a small portion of the plug should be worked between the thumb and forefinger. The worked cheese mass is smelled again. High-quality Swiss cheese will have a clean, fragrant aroma and a pleasing, distinctly sweet, hazelnut-like taste.

BRICK CHEESE

Brick cheese, a semihard sweet-curd cheese, is made by adding a small amount of lactic culture to fresh, uncolored whole milk, and coagulating with rennet (Wilson and Price 1935). When brick cheese is properly made and cured, the body is softer than Cheddar but firmer than Limburger cheese. Brick cheese can vary from mild to quite robust in aroma, but should be rather sweet (nonacidic) in taste. This cheese lacks the typical acid sharpness of Cheddar cheese and the intense flavor of Limburger; it may however possess some of the flavor characteristic of both of these cheese varieties. Brick cheese is basically a surface-ripened cheese which involves a surface growth of *Brevibacterium linens*.

Score Card. A brick cheese score card, shown in Table 8.10 (Thom and Fisk 1938) reflects the relative importance of different criteria that should be considered in evaluating brick cheese.

Flavor. The flavor of brick cheese should be relatively clean, sweet, and mild, but simultaneously suggest a faint trace of "Limburger cheese" flavor (Price and Buyens 1967). The most common flavor defect of brick cheese seems to be an intense unclean off-flavor that occurs from abnormal bacterial growth. This flavor defect is frequently associated with a gassy or "slit" texture. A second defect of brick cheese is the development of excessive acid in the curd, which results in a sour off-flavor. If brick cheese exhibits flavor characteristics too reminiscent of either Cheddar or Limburger cheese, the product is considered to manifest an "atypical brick cheese" flavor.

Body and Texture. The body of brick cheese should be firm, smooth, and feel slightly moist; it should break down similar to cold butter when rubbed between the thumb and forefingers. This cheese type should not be crumbly, mealy, pasty, or sticky. The appropriate moisture content of brick cheese should be $\leq 44\%$, and commonly is 39 to 42%; if it is too high in moisture, the cheese cures too rapidly, and the body tends to become quite soft and pasty. By contrast, if there is insufficient moisture, the cheese body may be brittle and mealy (Price and Buyens 1967). A cut surface (or trier plug) of brick cheese typically has small, irregular, and/or somewhat rounded openings. The presence of these so-called "shot holes" may vary in size from "BB-shot" to approximately 3/8 in. (9.5 mm) in diameter. An extremely gassy or "pinholey" texture is criticized seriously. Occasionally, a "late-gas" defect may develop; this may cause the cheese to split in the center. This swelling or "bloating" of the cheese is readily apparent when a cross-section of an affected cheese is examined.

Table 8.10. Alternative Brick Cheese Scoring Outlines.

Item	Possible Scores	
	Thom and Fisk (1938)	Contemporary
Flavor	40	10
Texture	40	5
Color	10	5
Salt	5	2
Style	5	3
Total	100	25

Color. Brick cheese may be either lightly colored or uncolored (most common). When color is added it should be clear and uniform throughout the cheese. The carotene content of milkfat will generally impart a slightly translucent to light yellow color to brick cheese. A faded, dull, or chalky-white color is discriminated against in brick cheese; any discolored spots or areas are also undesirable appearance defects.

Salt. Brick cheese usually contains only about 1.5% salt, which is applied by either directly rubbing onto the surface, or by immersing the young cheese into salt brine. The proper amount and even distribution of salt has a marked influence on cheese ripening and the subsequent "brick cheese flavor." Added salt also facilitates the formation of a smooth, protective rind on the cheese exterior. A salty flavor, *per se,* is undesirable; furthermore, excessive salt interferes with proper and full ripening of brick cheese.

Appearance. High-quality brick cheese should have a neat, attractive appearance, be clean, well-shaped, free from checks and mold, and have a "closed rind" (predominantly smooth surface). Brick cheese is made in one style only—rectangular (in the general shape of a brick, from which it derives its name). The sides and corners of brick cheese should be square and parallel with each other, not bulged; uniformity in size and shape should prevail among pieces of cheese from a given lot.

Sensory Evaluation of Brick Cheese

Similar to the scoring of other cheese types, the general appearance of brick cheese and a withdrawn sample plug should reveal something about its quality attributes. By closely examining the body and texture of brick cheese, sensory quality may be partially revealed.

When the surface is pressed with the fingers, brick cheese should have some spring to it; this is indicative of a pliable body (desirable). The body of brick cheese may seem comparable or slightly weaker than Monterey Jack or Colby cheese. After noting the aroma of a trier plug of the cheese, the evaluator should look for a distinctly open texture (i.e., the presence of small, irregularly shaped mechanical holes, that show a slight luster on their surfaces). The top or upper 1 in. (2.54 cm) of the cheese plug should be replaced in the trier hole. The brick cheese should not exhibit severe gassiness (pinholes) or any long, developed cracks in the interior.

The judge should work a small portion of brick cheese sample (plug) between the thumb and forefingers, and again observe the aroma of the "worked cheese." In a properly developed brick cheese, there should be substantial aroma; this is indicative of considerable protein break-down. Upon tasting the cheese, the perceived sensation should be pleasantly sweet (nonacidic) and the overall flavor should confirm the earlier-noted aroma. After the cheese is expectorated, the flavor should linger briefly, but no lasting off-flavor, such as "bitter," "proteolysed," or "ammoniacal," should be suggested.

LIMBURGER CHEESE

Limburger is a semisoft cheese made from fresh, whole, uncolored milk. Usually, the curd for Limburger is heated only to 35°C–37.7°C (95°F–100°F) (Nelson and Trout 1965). The curd still appears soft and shiny when it is dipped into hoops to drain. Since the curd is not pressed, it retains considerable moisture (≤50%). The cheese is formed into small rectangular blocks; usually, the cheese surfaces are rubbed with salt daily during the first days of curing. This serves to develop a rind and helps prevent mold growth. Curing room temperatures range from 14.4°C to 17.7°C (58°F–64°F) with a relative humidity of about 95%. Limburger cheese ripens from the exterior toward the interior due to the combined surface growth of *Brevibacterium linens* and a proteolytic yeast; the latter provides the required enzymes for protein hydrolysis. After two to four weeks of curing, proteolysis should be uniform throughout the cheese (soft to the core). At this point, the cheese is considered ripe and ready for consumption. Due to the combined conditions of low acidity, high moisture, and an anaerobic environment for the development of bacteria, this cheese develops an intense characteristic odor and taste (flavor). This unique flavor may be quite suggestive of uncleanliness or organic decay (possibly) to individuals not accustomed to it; hence, the typical flavor of Limburger cheese is not cherished by many persons.

The Score Card. The brick cheese score card (Table 8.10) may be applied to the sensory evaluation of Limburger cheese. The items considered in evaluating the respective sensory attributes of Limburger cheese have the same relationships as in the case of brick cheese.

Desirable Qualities. Limburger cheese should have a somewhat symmetrical shape and a smooth surface that is free of cracks or breaks. The sides of the cheese should be parallel, not bulged nor misshapen. In a cured cheese, the interior color should be a uniform egg-shell-white shade. In the instance of an uncured Limburger cheese, the color may be slightly lighter in the center than near the outer edges.

Limburger cheese should meet the desired composition and contain the appropriate level of salt to enable characteristic flavor development. The body of cured Limburger cheese should be soft, buttery, and uniform throughout. Typical high-quality Limburger cheese flavor is pronounced or intense, without any suggestion of bitterness, or offensive putrid or ammoniacal off-flavors. Once this cheese is completely cured, the keeping quality is limited; consumption should occur within one or two weeks of reaching its "prime."

Undesirable Qualities. Limburger cheese is subject to some defects common to other kinds of cheese. Its unique flavor is not easy to obtain without the occurrence of some flavor defect, since the conditions of low curd acidity and the relatively high curing temperature are so conducive to the outgrowth of undesirable bacteria that may have entered the milk or the process as contaminants. "Gassy cheese," caused by gas-producing cultures, is a common defect; this not only gives the cheese a bad off-flavor, but often causes it to "bloat" as well. Too much acid development may result in slow curing, as well as sour and/or bitter off-tastes.

Any Limburger cheese that has too dry a curd, and subsequently a dry cheese body, will not cure normally. Limburger cheese with an initial high moisture content generally leads to a too-rapidly cured cheese, and an associated weak and pasty body; the finished cheese tends to flatten out during curing.

BLUE-VEINED CHEESE

The general class of cheeses known as the blue-veined type includes, among other lesser-known varieties: U.S. Blue, French Roquefort, Danish Blue (Danablu), English Stilton, and Italian (and U.S.) Gorgonzola. These cheeses are so-called because, in addition to curd produced by lactic acid organisms, the milk or curd is inoculated with selected species of blue-green mold, (i.e., *Penicillium roqueforti*) (Morris 1981). Following growth, this mold imparts blue-green streaks, providing a distinct marbled appearance. Other types of blue (or white) mold-based cheeses have been developed commercially or at agricultural experiment stations in the U.S. The American types are generally called "blue cheese" with the name of the state, district, locality, or company in which they have been developed prefixed to the name. "Iowa Blue," "Nauvoo," "Oregon Blue," "Treasure Cave," and "Maytag" are several examples. Some imported blue-veined cheese from Europe may be spelled "bleu." Blue (bleu) cheese may not be called or labeled "Roquefort," although "roquefort-type" is sometimes incorrectly used in labeling. Three prerequisites characterize genuine Roquefort cheese, thus making it distinctive and entitling it to uniquely bear the name "Roquefort:"

1. Roquefort cheese must be made from whole, pure, sheep milk,
2. the milk must be raw, and
3. the cheese must be ripened in the natural caves of Roquefort, Aveyron, France.

A Minnesota-created cheese, closely akin to blue cheese, but nonpigmented, has been named "Nuworld." It is ripened by spores of *P. roqueforti* that have been irradiated; and consequently, the blue-green marbling does not occur (Foster *et al.* 1957; Morris 1981). The ripened Nuworld cheese exhibits a bone-white to faint-brown colored mold growth. More recently, several combined blue and white mold cheeses have been developed in Europe and marketed successfully in the U.S. (e.g., "Bavaria-Blu", "Camemb-Blu," "Saga Blue," and "Brese Bleu."

American blue-veined cheese is made from fresh whole cows' milk with added lactic starter culture and coagulated by milk-clotting enzymes. Just before the curd is hooped, prepared spores of *P. roqueforti* are mixed with the curd. Within several weeks the cheese must be punctured with skewers to admit air to support the aerobic mold growth. The cheese is ripened at 10°C to 12.7°C (50°F–55°F) under high humidity. Mold grows readily during the ripening period; a cut cheese should show a pattern of openings lined with mold which gives the cured product a marbled blue-green and cream effect. The proteolysis and fat hydrolysis due to mold growth and metabolism, and other chemical reactions, impart to the cheese a typical, savory, peppery, and piquant flavor.

Evaluating Blue-veined Cheese

Nelson and Trout (1964) proposed the score card shown in Table 8.11 based on the popularity and increased consumption of American blue-veined cheese.

The body and texture of blue cheese is closely allied with its flavor characteristics. The color and appearance of blue cheese also has some bearing on flavor, since it is indicative of the particular distribution and growth of the blue-green mold which is such an important factor in proper ripening of this cheese. A lack of distinct mold marbling, indicated by a predominant creamy-white or speckled white color is generally undesirable or atypical for most types of blue cheese. The finish of blue cheese should be uniform and appealing, and the outer

Table 8.11. Blue-Veined Cheese Scoring Outline as Proposed by Nelson and Trout (1964).

Item	Possible Scores	
	Nelson and Trout (1964)	Contemporary
Flavor	45	10
Body and Texture	35	5
Color	15	5
Finish	5	5
Total	100	25

surface should be covered with a tight, smooth covering of aluminum foil or an equivalent material to protect the cheese from moisture loss.

Desirable Qualities. Most types of high-quality blue cheese have a distinct, characteristic aroma. Blue cheese flavor should be sharp, peppery, pronounced, and "linger" for some time after the cheese has been tasted and expectorated. However, the lingering flavor should impart and leave an overall pleasant taste sensation.

The characteristic flavor of blue cheese is largely due to some proteolysis and critical hydrolysis of milkfat. Free and combined forms of the short-chained fatty acids and various methyl ketones (e.g., 2-heptanone) are essential to development of the desired sharp peppery flavor of blue cheese.

The body of American blue cheese should be moist ($\leq 46\%$ but typically 40 to 43% H_2O) and slightly sticky (has a tendency to stick to the knife when cut), as well as somewhat crumbly (a tendency to crumble when cut). The ideal color and appearance of blue cheese should be a marbled cream with blue-green mold, indicating a good distribution of the mold development throughout the cheese.

Undesirable Qualities. Commonly, blue cheese may lack the typical flavor imparted by ketones (such as 2-heptanone). In such instances, the primary flavor note is often rancidity. Without proper mold development, a typical blue-veined cheese flavor cannot be achieved. A predominance of butyric acid odor (rancid-like) and a strong or pronounced bitter off-taste are not desired. Other undesirable qualities that are sometimes exhibited are too dry a body and a lack of sufficient mold growth.

Occasionally a black discoloration, accompanied by a musty flavor (due to the unwanted growth of *Hormodendrum olivaceum*) may become a serious quality problem (Bryant and Hammer 1940). Occasionally, gray discoloration and a mousy, ammonia-like off-flavor, that later may become soapy (accompanied by a pH increase), can be a serious defect. Variations in normal ripening apparently cause the aforementioned defects, presumably as the result of the formation of basic end products from proteolysis. Another physical defect of blue cheese is corner or edge softening.

SOFT, MOLD-RIPENED CHEESE

Cheeses classified as soft, mold-ripened types owe their softness to a combination of relatively high moisture content and extensive protein breakdown. These soft cheeses are generally manufactured as either small or thin-sized pieces to permit enzymes (formed by microbial activity on the surface) to diffuse throughout the interior of the cheese. Practically all chemical and physical changes occurring in soft ripened cheeses occur as a result of various microorganisms growing on the surface. Combinations of white mold, certain types of yeast, and bacteria are synergistically involved in the ripening of Camembert and Brie cheeses, which are the most popular and common of the soft, mold-ripened cheeses. The curd in these cheeses is not cooked or pressed during the manufacturing process.

Compared to other cheese types, Brie and Camembert cheeses ripen rather quickly, given the relatively high moisture content (48 to 52%), high humidity (75% RH), and the extensive microbial activity that occurs on the surface. These cheeses generally ripen in three to four weeks (or less) and the keeping time (prior to overripening) is usually only one to two months, at most (unless frozen). An interesting commentary about their relative perishability states: "Brie and Camembert are like flowers which bloom in their own time and then for a short time only" (Battistotti et al. 1984).

Camembert

Camembert (kam'-em-bare) is one of the oldest and most popular of the soft, surface-ripened cheeses. According to documented records, this cheese originated in the small community of Orne-Camembert, France circa 1780. Currently, Camembert is made in many countries of the world. There are several highly automated, continuous methods of Camembert manufacture in Europe; in several instances, the continuous approach is coupled with preconcentration of cheese milk by ultrafiltration. Brie and/or Camembert manufacture in the U.S. has been undertaken in at least the states of Illinois, Wisconsin, Minnesota, California, and Oregon. A current California manufacturer started production of Brie and Camembert as early as 1880.

Camembert cheese is made exclusively from cows' milk in the traditional circular or disk shape measuring about 4.5 in. (11.43 cm) in diameter and 1 in. (2.54 cm) thick. After curing, each cheese weighs approximately 8 oz (0.23 kg).

Within the first 8 to 10 days of curing, Camembert cheese is expected to develop an obvious, thick coat of white mold mycelia over the entire surface. This helps to develop an edible rind or crust, which in turn lends some needed rigidity to the ripened, soft cheese. At maturity, the body of the cheese should display a smooth, near-creamy, almost "flowing" consistency. The typical flavor of a matured Camembert is best described as slightly salty (initially), sweet, aromatic, possibly slightly bitter, and slightly mushroom-like (aftertaste), yet in a pleasing combination. The odor of ammonia and/or possibly a slightly unclean or fishy aroma may be barely noticeable in fully ripened (or overripened) Camembert cheese.

Composition-wise, good-quality Camembert and Brie cheese are usually 48 to 52% moisture; 25 to 27% milkfat; 18 to 20% protein, about 2.5 to 3.0% salt; and at least 50 to 52% fat in dry matter.

Brie

France produces numerous types of Brie, but the two major ones are Brie de Meaux (soft body and distinctive hazelnut flavor) and Brie de Meuse (firmer body and richer flavor). Original Brie were in the form of circular or thin wheels, varying between 15 and 18 in. (38–46 cm) wide. Brie cheeses generally exhibit a snowy white, velvety mold growth over the entire surface, but may also show irregular patches of tan-yellow or tan-red color due to the associated growth of desirable yeast and/or bacteria on the cheese exterior.

Score Card for Brie and Camembert Cheese

The following score card, based on essentially the same quality parameters as other specialty cheeses, is suggested by the authors as an evaluation tool for the sensory evaluation of Brie or Camembert cheese (Fig. 8.15).

Flavor. The desired flavor of Brie is delicate, subtle, slightly salty when young, and still modestly firm (two to three weeks of age). As Brie cheese matures further (three or four weeks), a more intense and moderately aromatic flavor should develop. This flavor may range from slightly nutty and sweet to a faint mushroom-like flavor-note. One French cheese authority has described a typical Brie as having "a richness, yet mellowness, and the flavor notes could be part mushroom, part cream, part cognac, and part earth" (Battistotti *et al.* 1984). When Brie exceeds five weeks of aging, the flavor is likely to become piquant or pungent, especially noted by the readily apparent, devel-

Item/Defects	Possible Score
FLAVOR	10
Acid	
Ammonia-like	
Bitter	
Flat, lacks flavor	
Harsh, earthy	
Moldy, musty	
Salty	
Unclean	
Whey Taint	
Yeasty	
Other	
BODY AND TEXTURE	5
Dry, corky	
Curdy, rubbery	
Gassy, pin holes	
Lacks uniform consistency	
Mealy, grainy	
Pasty, sticky	
Powdery	
Too soft, milky	
Other	
COLOR AND APPEARANCE	5
Atypical color development	
Dull color	
Uneven color	
Mold growth atypical	
Asymmetrical shape	
Atypical rind (crust)	
Other	
SALT	5
Lacks salt	
Too high salt	

Fig. 8.15. A suggested score card for soft, white mold, surface-ripened cheese (i.e., Brie and Camembert, etc.).

oped aroma. Once Brie develops an ammonia-like aroma or fishy-like odor, it is generally considered overripe. At this point an objectional bitter aftertaste may also appear.

By contrast, Camembert cheese is generally intended to have a somewhat more intense flavor than Brie at comparable stages of ripening and maturation. This higher flavor intensity is primarily due to the substantially higher ratio of surface area to volume per cheese represented by the smaller-sized Camembert. This permits a significantly more rapid and thorough diffusion rate for microbial enzymes from the cheese surface throughout the cheese interior. The enzymes catalyze protein hydrolysis, which leads to characteristic flavor development and body softening. The characteristic flavor of Camembert is somewhat more robust than the more delicate Brie; in fact, at maturity (four to six weeks) it is expected to be pungent (or piquant), slightly earthy, or definitely mushroom-like. As the Camembert matures further, it may display an aroma of ammonia. This is due to deamination of amino acids by the active microbial enzymes that have diffused through the cheese.

The accompanying suggested score card for soft, mold-ripened cheese should facilitate the process of evaluating either Brie or Camembert cheese for sensory characteristics. Standards for these two cheese types are not highly refined in the U.S., but the descriptors are self-explanatory for possible flavor defects. Other descriptive terms for off-flavors can be added by users of the score card as circumstances dictate.

Body and Texture. Brie and Camembert cheese, when freshly made, are definitely firm and exhibit hard, chalky but finely grained cores at the cheese center. After further aging, the center should become "flowable," with development of a uniform light creamy consistency and a white-yellow color. At the time of consumption, these two soft, ripened cheeses should be devoid of any curdiness or dryness, graininess, powdery or chalky consistency, whey pockets, or areas of extreme fluidity (milkiness). It is undesirable to observe a firmer, lighter colored center core "floating" in the soft, nearly flowable major mass of the cheese (Fig. 8.16). This is a natural occurrence of Brie (or Camembert) ripening. As proteolytic enzymes diffuse into the cheese, the center core is the last part to soften (undergo hydrolysis). Once the core is broken down, the cheese should manifest its full, typical flavor, but also experience a quite limited shelf-life.

The temperature for determining flavor and body/texture characteristics of Brie and Camembert cheese is important. At storage temperatures below 7.2°C (45°F), these cheeses may seem relatively firm and give the impression of underripening. When tempered at room temperature, the cheese may show its "true colors"—and flowable consistency. A properly ripened Brie or Camembert, following 45–60 minutes of tempering at room temperature, should slice and spread as readily as tempered butter. Consumption of the outer mold-crust is a matter of personal choice. The flavor, however, of this edible portion is generally more intense and different from the more delicate flavor of the cheese interior.

Color and Appearance. Depending on the species of mold employed (which varies geographically according to Kosikowski 1982), the outer

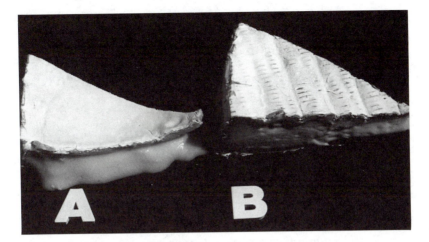

Fig. 8.16. A comparison of Brie cheese of varied maturity; A—Body soft and nearly "flowable," generally preferred; B—Body too firm (not spreadable), lacks sufficient maturity.

crust or rind may be snow white or a pleasant, mottled white color. The mottled effect is derived from the irregular growth of either red-pigmented bacteria and/or certain yeasts (tan-yellow) that grow in association with the *Penicillium candidum* (or *P. camemberti*) mold. Dark grey, black, or blue discoloration due to other mold contaminants is undesirable. There should be as much symmetry to the cheese wheels as possible, but inasmuch as they are soft cheeses, some misshapening is tolerated. Each wheel should be fairly level across the top. A severe concave appearance is probably indicative of possible overripening and extensive flow within the cheese interior.

ITALIAN CHEESE VARIETIES

Few people probably realize or appreciate Italy's unique heritage as a significant cheese-manufacturing country. Today's Italian cheeses are the result of an ancient culture. They may be made from the milk of sheep, goats, buffalo, and/or cows. At least 50 different cheese varieties claim Italy as their country of origin—too many to discuss in these limited pages. The following are typical examples of Italian cheeses, grouped according to type (Reinbold 1963):

1. Very hard (grating)
   a. Ripened by bacteria
      Asiago, Parmesan, Reggiano, and Romano
2. Hard
   a. Ripened by bacteria, no eyes
      Caciocavallo and Provolone
   b. Ripened by bacteria, with eyes
      Asin, Bitlo, and Fortina
3. Semisoft
   a. Ripened by bacteria principally
      Fresh Asiago
   b. Ripened by bacteria and surface microflora
      Taleggio
   c. Ripened primarily by blue mold—interior
      Castelmagno, Gorgonzola, Pannarone
4. Soft
   a. Ripened
      Bel Paese, Crescana, Formaggini, and Italico
   b. Unripened
      Braccio, Mozzarella, and Scamorze

Most typical Italian cheese, as known and made in the U.S. include the following groups or types: Parmesan, Pasta filata (plastic curd, e.g., Mozzarella, string, etc.), Provolone, and Romano. Interestingly enough, these cheeses (as a group) are second in popularity only to Cheddar cheese in the U.S.

Cheese Defects

Unfortunately, Italian cheese varieties are occasionally subject to some of the same shortcomings in manufacture and spoilage defects common to many other cheeses.

Flavor. Flavor defects found in several varieties of Italian cheese may be of an absorbed, microbiological, or chemical origin. The production of milk, or storage of cheese, may incur certain absorbed odors from the environment (i.e., cowy, barny, feed, and foreign off-flavors). Off-flavors of microbiological origin include acid, bitter, fermented/fruity, malty, moldy, musty, putrid, unclean, or yeasty. The more common off-flavors derived from chemical activity include bitter, metallic, oxidized, tallowy, and rancid. Other possible off-flavors in Italian cheese varieties include a flat or lack of typical flavor, harsh (due to the wrong component balance of chemical compounds), unclean aftertaste, soapy (excessive lipase activity), and whey taint.

Body, Texture, and Finish. Invariably, the same body, texture, and finish defects that occur with other cheeses are also found in Italian types. Rapid moisture evaporation from the cheese surface often leads to checked or cracked rinds, especially in Provolone. Control steps include: (1) increasing the relative humidity; (2) reduction of storage temperature; (3) reducing the rate of air circulation; and (4) earlier waxing, paraffining, oiling, or packaging. Any uneven or solid surfaces in the cheese are a poor reflection of the workmanship and handling procedures. If cheese texture of granular types is weak or too smooth, the causes may be related to unusually high concentrations of milkfat, improper cutting and curing of the curd, too high a moisture content, or other causes.

Other Defects. Unwanted mold growth may be found on some Italian cheese varieties, especially in higher-moisture types. Mold detracts from the appearance and product acceptance, and causes undesirable off-flavors and considerable trim losses, plus the potential presence of carcinogenic mycotoxins. Control of mold starts and ends with rigid and continuous plant housekeeping and sanitation, and humidity control in the manufacturing and curing rooms. The use of modern packaging materials and sorbic acid or its salts are beneficial in the control of mold. Undesirable gas formation is common among several of the higher-moisture varieties of Italian cheese, such as Mozzarella. Control measures for gas formation are similar for these varieties of cheese as for others—minimize or prevent sources of unwanted microbial contamination in milk, cultures, and the manufacturing process.

CREAM CHEESE

Cream cheese is a soft, unripened cheese made from cream, coagulated either by microbial development of lactic acid (aided by milk-coagulating enzymes) or by direct acidification. This is followed by collection of the formed soft curd by centrifugation or pressing in cloth bags. This cheese is creamy-white in color, has a fine, smooth, spreadable texture, and a full, rich cream-like flavor with a slight acidic taste.

Cream cheese is generally made from a cream base that contains 12 to 20% milkfat; the fat content of the finished cheese may vary from the typical minimum of 30% to as high as 40%. The moisture content will vary in inverse proportion. Neufchatel is a similar cheese made from whole milk of high fat content and hence has a correspondingly lower milkfat content (20–25%) in the final product.

In recent years, a U.S. manufacturer (Vitex Divison of Carlin Foods) has developed a direct-acidification process for converting a cream or milk base into cream cheese or Neufchatel cheese. Glucono-delta-lactone (GDL) and phosphoric acid are the acidulants used for coagulating the milk protein. Gluconic acid is formed when GDL is added to an aqueous system such as the cream or milk base; the resulting pH decrease induces the clotting of casein. Milk or cream is usually preacidified with phosphoric acid. The GDL requires heating of the milk for its conversion to gluconate.

Flavor

Cream cheese should have a full, rich, clean, mild acidic flavor, in accordance with the high percentage of milkfat present (30 to 40%). The lower fat content cream cheese of the Neufchatel type may have a moderate acid taste. A sweet, cream-like flavor is generally preferred over a distinct acid flavor in cream and Neufchatel cheese. The more common flavor defects in various types of cream cheese are flat (lacking flavor), sour or too high acid, metallic, an unclean aftertaste, and yeasty. All of these flavor defects should be readily recognized; the descriptors are self-explanatory.

Body and Texture

Cream cheese should have a soft body, yet be sufficiently firm to retain its shape at serving temperature. The texture should be somewhat buttery, silky-smooth, and simultaneously provide excellent spreading and slicing properties. This is reflective of the popular use of cream or Neufchatel cheese as a cracker, bread, or pastry spread, or topping.

Research on cream cheese manufacture by Roundy and Price (1941) noted that, when the cheese was made from cream containing about 16% milkfat, then cheese with 50 to 54% moisture and 37 to 42% fat exhibited the most desirable spreading and slicing properties. They also observed that a reduction of the fat content of the cream base for cheese manufacture to less than 16% caused a grainy texture and a crumbly body, and that an increase in the milkfat content above 20% tended to cause excessive smoothness (slick-like) and stickiness. The body and texture defects of cream cheese may be listed as: coarse (rough or nonhomogenous), crumbly, excessively smooth (slick), grainy (gritty), sticky, too firm (hard), and too soft (lacking viscosity). Most of these defects are easily recognized upon close observation of the product.

Flavored Cream Cheese

In recent years, consumers have experienced a wide range of both domestic and imported versions of flavored cream cheese. There are two basic types of product: (1) a naturally made cream cheese with added flavoring material, and (2) a processed form of cream cheese with added flavorings. The flavoring materials consist of: fruit purees, pieces, and/or extracts; nut pieces and/or extracts; herbs and spices; and alcoholic-based flavorings such as rum, brandy, kirsch, port wine, and liqueurs. The incorporation of the above additional flavoring materials into cream cheese has served to provide a broad spectrum of appealing flavor combinations and enhance the use and acceptance of cream, processed cream, and Neufchatel types of cheese. Some of the more popular flavored cream cheeses evident in the U.S. marketplace are: peach, pineapple, orange, strawberry, raspberry-melba, chocolate, kirsch, walnut, nut-raisin, rum raisin, brandy, port wine, amaretto, onion, garlic, pepper, mixed herbs and spices, and mixed vegetables.

OTHER CHEESE

From time to time, the person who is responsible for the sensory evaluation of cheese and various other dairy products may be called upon to examine cheese other than those types described in this chapter. Unfortunately, space does not allow for a detailed description of the sensory features of all the known cheese types available to U.S. consumers. The information presented in Table 8.6 may help the individual who wishes to address some of the sensory characteristics of such other kinds of cheese.

The true connoisseur of cheese tends to have developed a general or thorough knowledge of the origin, history, methods of manufacture and curing, uses, and some of the desired flavor, body/texture, color, and appearance characteristics of many, many kinds of cheese. It may be especially helpful to become familiar with several kinds of cheese from each major category. This is an ongoing, lifetime process of education of each "cheese authority," because exciting new varieties or interesting modifications of conventional cheese appear continuously. A genuine sense of satisfaction is felt by the cheese connoisseur who finds the right, or the near-"ideal," cheese of any kind. Thus, the search, discovery, and eventual appreciation of fine cheese becomes a near-avocation. Typically, the cheese authority might feel compelled to share this adventure into good eating, based on his or her knowledge and familiarity of technical aspects of manufacture and the intricate quality parameters of numerous cheese varieties.

A study of the origin, development, and primary sensory characteristics of the leading cheese varieties of many countries can be most rewarding. The cheese judge or authority will enjoy knowing and sharing with others something about the background and the desired qualities of such prominent cheeses, to name a few, as Bel Paese, Blue-Vein, Cheddars, Dana-Blu, Gorgonzola, Roquefort, Stilton, Brie, Camembert, Edam, Gouda, Emmentaler, Fontina, Gruyere, Jarlsberg, Swiss, Raclette, Feta, Kasseri, Parmesan, Romano, Mozzarela, String, Provolone, Monterey Jack, Muenster, Leyden, Noekkelost, Port du Salut, Havarti, Tilsit, Kirsch or Walnut Gourmandise, and Sap Sago.

REFERENCES AND BIBLIOGRAPHY

Agricultural Research Service. 1978. Cheese Varieties and Descriptions. Agric. Handbook No. 54. U.S. Dept. of Agric. Washington, D.C. 151 pp.

Battistotti, B., Bottazzi, V., Piccinardi, A., and Volpato, G. 1984. *Cheese: A Guide to the World of Cheese and Cheesemaking.* Facts on File, Inc. New York. 168 pp.

Bills, D. D., Morgan, M. E., Libbey, L. M. and Day, E. A. 1965. Identification of compounds responsible for fruity flavor defect of experimental Cheddar cheeses. *J. Dairy Sci. 48:*1168.

Bodyfelt, F. W. 1967. Lactic streptococci and the fruity flavor defect of Cheddar cheese. M. S. Thesis, Oregon State University, Corvallis, OR. 118 pp.

Bryant, H. W. and Hammer, B. W. 1940. Defects in blue (Roquefort-type) cheese. IA. Agri. Expt. Sta. Res. Bul. No. 283. Ames, IA.

Chandan, R. C. 1980a. Flavor problems in Cheddar cheese varieties. *Dairy Record. 81*(4):117.

Chandan, R. C. 1980b. Texture problems in Cheddar cheese. *Dairy Record. 81*(6):94.

Code of Federal Regulations. 1985. Title 21-Food and Drugs. Part 133. Cheese and Cheese Products. U.S. Government Printing Office, Washington, D.C.

Davis, F. L. and Law B. A. (Editors). 1984. *Advances in the Microbiology and Biochemistry of Cheese and Fermented Milk.* Elsevier Applied Sci. New York. 260 pp.

Eekhof-Stork, N. 1976. *The World Atlas of Cheese.* Paddington Press, Ltd. Amsterdam, Netherlands. 240 pp.

Emmons, D. B. and Tuckey, S. L. 1967. *Cottage Cheese and Other Cultured Milk Products.* Pfizer Cheese Monographs. Chas. Pfizer & Co., Inc. New York. 143 pp.

Feeley, R. M., Criner, P. E., and Slover, H. T. 1975. Major fatty acids and proximate composition of dairy products. *J. Am. Diet Assoc. 66:*190.

Foster, E. M., Nelson, F. E., Speck, M. L., Doetsch, R. N., and Olson, J. C., Jr. 1957. *Dairy Microbiology.* Prentice-Hall, Inc. Englewood Cliffs, NJ. 492 pp.

Hammer, B. W. and Babel, F. J. 1957. *Dairy Bacteriology.* (4th Ed.) John Wiley and Sons. New York. 614 pp.

Kosikowski, F. V. 1977 and 1982. *Cheese and Fermented Milk Foods.* (2nd Ed.) Edwards Brothers. Ann Arbor, MI. 709 pp.

Kosikowski, F. V. and Mocquot, G. 1958. Advances in Cheese Technology. FAO Agr. Studies No. 38. Food and Agricultural Organization, United Nations. Rome, Italy. 263 pp.

Kristoffersen, T. 1963. Cheese flavor in perspective. *Manufact. Milk Prod. J. 54*(5):12.

Kristoffersen, T. 1967. Interrelationships of flavor and chemical changes in cheese. *J. Dairy Sci. 50*:279.

Lacroix, D. E., Mattingly, W. A., Wong, N. P., and Alford, J. A. 1973. Cholesterol, fat and protein in dairy products. *J. Amer. Diet. Assoc. 62*:276.

Mojonnier, T. and Troy, C. 1925. *The Technical Control of Dairy Products.* Mojonnier Bros. Co. Chicago, IL.

Morris, H. A. 1981. *Blue Cheese.* Pfizer Cheese Monographs. Chas. Pfizer & Co., Inc. New York. 65 pp.

National Cheese Institute. 1982. Unpublished data. Arlington, VA.

National Dairy Council. 1978. Nutritive value and composition of cheese. *Dairy Council Digest, 46*(3):15.

Nelson, J. A. and Trout, G. M. 1964. *Judging Dairy Products.* Avi Publishing Co. Westport, CT. 463 pp.

Olson, N. F. 1980. New approaches to cheese quality evaluation. *Dairy Field 163*(9):78.

Posatim, L. P., Kinsella, J. E., and Watt, B. K. 1974. The fatty acid composition of milk and eggs. *Proc. 57th Meeting Amer. Dietetic Assoc.* Philadelphia, PA.

Price, W. V. 1943. Comments on tentative cheese grades. *Cheese Reporter.* Nov. 5. Madison, WI.

Price, W. V. and Buyens, H. J. 1967. Brick cheese: pH, moisture and quality control. *J. Dairy Sci. 50*:12.

Reinbold, G. W. 1963. *Italian Cheese Varieties.* Pfizer Cheese Monographs. Chas. Pfizer & Co., Inc. New York. 43 pp.

Reinbold, G. W. 1972. *Swiss Cheese Varieties.* Pfizer Cheese Monographs. Chas. Pfizer & Co., Inc. New York. 187 pp.

Roundy, Z. D. and Price, W. V. 1941. The influence of fat on the quality, composition and yield of cream cheese. *J. Dairy Sci. 24*:235.

Sammis, J. L. 1937. *Cheese Making.* The Cheese Maker Book Co. Madison, WI. 249 pp.

Thom, C. and Fisk, W. A. 1938. *The Book of Cheese.* Macmillan Co. New York. 403 pp.

Scott R. 1981. *Cheesemaking Practice.* Applied Science Publishers Ltd. London, U.K. 475 pp.

United States Department of Agriculture Food Safety and Quality Service. 1985. Specifications for U.S. Grades of Cheddar cheese. U.S. Government Printing Office, Washington, D.C.

Van Slyke, L. L. and Price, W. V. 1979. *Cheese.* Ridgeview Publish. Co. Reseda, CA. 522 pp.

Vedamuthu, E. R., Sandine, W. E., and Elliker, P. R. 1966. Flavor and texture in Cheddar cheese. II. Carbonyl compounds produced by mixed-strain lactic starter cultures. *J. Dairy Sci. 49*:151.

Wilson, H. L. and Price, W. V. 1935. The manufacture of brick cheese. U.S. Dept. Agr. Cir. No. 359. Washington, D.C.

Wilson, H. L. and Reinbold, G. W. 1965. *American Cheese Varieties.* Pfizer Cheese Monographs. Chas. Pfizer & Co. Inc. New York. 67 pp.

Wilster, G. H. 1980. *Practical Cheesemaking.* OSU Bookstores Inc., Corvallis, OR. 360 pp.

Sensory Evaluation of Butter

Standards for U.S. grades of butter are addressed in Title 7 of the Code of Federal Regulations (1987), Part 58, Subpart P:

> For this purpose, 'butter' means the food product usually known as butter and which is made exclusively from milk or cream or both, with or without common salt, and with or without additional coloring matter and contains not less than 80% by weight of milkfat. . . .

The term "cream" (as used in Subpart P) refers to a concentrated source of milkfat separated from milk that is produced by healthy cows.

> The cream shall be pasteurized at a temperature of not less than 74°C (165°F) and held continuously in a vat at such temperature for not less than 30 minutes or pasteurized at a temperature of not less than 85°C (185°F) for not less than 15 seconds, or it shall be pasteurized by other approved methods giving equivalent results.

Butter is generally marketed wholesale according to its quality grade in the U.S., but may also be retailed according to quality grade in many states. These butter grades are based on sensory quality and are assigned by competent "official" judges who conduct prescribed sensory examinations of the product. Although there are known regional preferences for certain flavor characteristics, body and texture properties, salt levels, color intensity, and shape and style of package, the basis for scoring or judging butter quality remains remarkably uniform across the U.S.

In addition to milkfat, butter contains moisture, curd (milk proteins, milk ash, lactose, and other minor constituents), and common salt (usually). Thus, the possible off-flavors of butter are not necessarily limited to those associated with milkfat, but flavor defects may also result from the action of microorganisms on milk proteins and/or lactose.

Farm-churned butter was once a major source of the U.S. butter supply, but for all practical purposes this form of product (formerly referred to as "dairy butter") is nearly extinct. Currently, all commercial butter in the U.S. is creamery- or factory-made butter (Wilster 1968). The primary method of manufacturing butter has gradually changed from the traditional batch process to the continuous method of churning. Industry trends are for an increasingly higher proportion of butter churned by the more efficient continuous process and for "lightly salted" butter ($\leq 1.5\%$ added NaCl).

Ingredients for Buttermaking

A typical butter manufacturing plant (creamery) starts with either fresh milk or cream separated at the plant as the raw material for butter making. Only a few creameries in isolated parts of the U.S. still accept farm-separated cream (which is commonly sour); this is primarily due to the substantial inefficiency of handling cream in this manner and the poor-quality butter that generally results from such practice. An accepted truism of the butter industry was that "the quality of fresh milk is much superior to collected, farm-separated cream" (Hunziker 1940; Wilster 1968). During the first half of this century, farmers typically sold cream to cream-buying stations, which in turn supplied the butter manufacturing plants. At the creamery receiving platform, this cream had to be carefully graded, since most of it came from small producers who produced the cream over a period ranging from several days to a week. Frequently, only slight attention was given to the cleanliness of the cream separator, utensils, containers, or to the storage temperature of the raw cream.

The vastly improved quality of current U.S. butter supplies is primarily due to the "fresh milk system" of creamery operation. As the overall quality of the U.S. milk supply continued to improve, low-grade butter can be expected to practically disappear (USDA 1983).

KINDS OF BUTTER

Sweet Cream Butter. The majority of the butter on the U.S. market is of the "sweet cream" type. The "sweet cream" designation implies that the titratable acidity of the churning cream did not exceed 0.20% (calculated as lactic acid). Most of the cream probably had no "developed acidity" whatsoever. Bulk forms of sweet cream butter that are free of off-flavors normally receive U.S. Grades AA or A (when and if graded). Higher acidity development (>0.20% as lactic acid) in cream mandates production of "sour cream" butter. This type of butter generally exhibits reduced shelf life and storage properties, even though cream acidity may have been neutralized to conform approximately to that of sweet cream.

Cultured Cream Butter. "Cultured cream butter" is made from either sweet cream or slightly soured cream in which a pleasant, delicate aroma was developed by added lactic culture just prior to churning. The cream (or a portion of the cream) was inoculated with a carefully selected lactic culture (starter) for the production of certain desired aromatic compounds. Cultured cream butter can usually be distinguished by its distinct aroma of diacetyl and other pleasant volatile compounds. Properly made, cultured cream butter has a delicate flavor which is sometimes referred to as "real butter flavor." Some cultured butter is made by adding either starter or starter distillate to the butter at the time of salting, and working it directly into the butter.

Other Butter Types. The addition of salt to butter is optional. The salt intensity of butter can vary over a wide range (0.75–2.5%). Most of the butter on the U.S. market is salted, but the trend in recent years has been toward more lightly salted (~ 1.5%) butter. The various types of butter available to U.S. consumers (although not necessarily in all markets) include: sweet cream salted, lightly salted, and unsalted butters; cultured cream salted and unsalted butters; whipped butters; and flavored butters (e.g., spices, herbs, honey, fruit, and berries).

Whipped butter is available for both institutional and home use. The U.S. grade of quality is assigned a given butter prior to whipping, but sensory evaluation of whipped butter can be undertaken. The air or gas (preferably nitrogen) incorporated by a whipping process changes the body characteristics and generally improves product spreadability.

Miscellaneous Spreads. Other products which emulate butter are margarine, butter-margarine blends, and "lowfat spreads" made from either milkfat and/or vegetable oil. Vegetable oils with differing melting points are the principal constituents of margarine. Although sensory properties vary widely for all products in this group of "spreads," some generalities still apply for their sensory evaluation. The general prerequisites for high-quality spread-type products are: desirable flavor and appearance, the absence of off-flavors, quality of workmanship, and product performance in terms of intended functional properties.

Grading Milk and Cream for Buttermaking

Alternative raw materials for butter manufacture include whole milk, sweet cream, whey cream, and sour cream. Butter made from fresh, sweet cream (or fresh milk) usually grades higher in sensory quality than that made from other cream sources.

The cream used for buttermaking is generally graded in accordance with the grade of butter that can be made from it (if properly processed). A close review of flavor defects (and their respective intensities) associated with various butter grades reveals that most of the frequently encountered butter off-flavors are derived from the cream. Some off-flavors may result from faulty cream processing or churning of the cream; certain other flavor defects may develop in the finished butter. Butter flavor defects that may be derived wholly or in part from cream, include:

Acid	Musty
Barny	Old cream
Bitter	Onion/Garlic
Cheesy	Rancid (lipase)
Coarse	Smothered
Feed	Stale
Foreign	Unclean/Psychrotrophic (utensil)
Fruity	Weedy
Metallic/Oxidized	Yeasty

There is little or no advantage in mixing cream (or milk) of different grades; the most probable result is a reduction in quality of the better raw material equivalent to the poorer one. Segregation of cream into various grades is a recommended procedure. Appropriate segregation into quality groups may be simply and practically accomplished by the sense of smell. The same precautions that apply to the sensory evaluation of raw milk are applicable to raw cream. Due to the potential health hazard of tasting raw products, a laboratory pasteurization procedure should be designed and used for small samples that insures adequate destruction of potential pathogens. The acidity of sour cream samples may require neutralization prior to pasteurization to prevent curdling.

Title 7, Code of Federal Regulations (1987), Part 58.322 addresses quality specifications for cream in part as follows:

§58.322 Cream.—Cream separated at an approved plant and used for the manufacture of butter shall have been derived from raw material meeting the requirements as listed under §58.132 thru 58.138 of this subpart.

The inspection of farm-separated cream to be used for manufacturing or processing into dairy products under this part shall be based on organoleptic examination and quality control tests to determine suitability of cream at the time of delivery thereof at the receiving plant or substation.

Organoleptic Examination. —Cream received at an approved receiving plant or substation shall be identified as to the producer, seller or shipper from whom received. Each can of cream in each shipment shall be examined for physical characteristics, off-taste and odors including those associated with developed acidity. The condition of the cream shall be wholesome and characteristic of normal cream. The organoleptic examination and segregation of the cream which is used in the manufacturing or processing into butter, shall be consistent with the applicable flavor classification of butter set forth in the U.S. Standards for Grades of Butter. Any cream having pronounced or offensive off-taste or odors or which is in an abnormal condition (including, but not being limited to surface mold, foamy, yeasty, fruity or containing extraneous matter), or which is otherwise unwholesome, shall be rejected to the producer, seller or shipper and shall not be used in the processing or manufacturing of dairy products.

Sediment Content Classification. —For the purpose of quality control and establishing a rejection level of cream to the producer, seller, or shipper, the following classifications of cream for sediment shall be applicable using a USDA Sediment Chart (7 CFR 58.2726) as the basis for classification.

Sediment (Off-the-Bottom Method):
No. 1—USDA Sediment Standard (not to exceed) 0.50 mg.
No. 2—USDA Sediment Standard (not to exceed) 1.00 mg.
No. 3—USDA Sediment Standard (not to exceed) 2.50 mg.

Sediment (Mixed-sample Method):
No. 1—USDA Sediment Standard (not to exceed) 0.20 mg.
No. 2—USDA Sediment Standard (not to exceed) 0.30 mg.
No. 3—USDA Sediment Standard (not to exceed) 1.00 mg.

Since high-quality raw materials are essential for the production of excellent butter, vigorous milk and cream quality improvement programs should be conducted continuously. The objectives are "clean flavored," low bacteria count, low sediment milk (or cream), which is of normal composition (low somatic cell count) and contains no extraneous, inhibitory, foreign, or unwholesome substances. Such a quality program must address general farm sanitation in the production of milk, the sanitary condition and handling of equipment, prompt and adequate cooling, and sufficiently frequent delivery of milk (or cream) to prevent age deterioration.

GRADES OF BUTTER

Since April 1, 1977, the U.S. Department of Agriculture has recognized only three consumer grades of butter; namely U.S. Grade AA, U.S. Grade A, and U.S. Grade B (CFR 1987). The U.S. Grade C designation was deleted at that time in recognition of the substantial improvements made in cream and butter quality. Handling and production practices for cream and whole milk improved so markedly by the mid-1970s that only insignificant quantities of butter of U.S. Grade C were made any longer. U.S. Grade designations based on numerical values (previously used as an alternative to the letter grade) were also deleted in 1977. Thus, the highest-quality butter was known as U.S. Grade AA or "U.S. 93-score" butter. Other equivalent designations were Grade A and "92-score," Grade B and "90-score," and Grade C and "89-score." Numerous butter samples entered in state fair competitions (or other educational exhibits or contests) were frequently scored higher than 93 points. Sometimes the scores approached, though probably never quite achieved, the mythical "perfection of 100 points." Some butter judges or contest officials may still employ the former 100-point scoring system in butter exhibits or competitions.

Since so much butter is evaluated by government graders prior to marketing, the USDA grading system for butter should be examined at this point. A review of Tables 9.1, 9.2, and 9.3 can provide the reader

Table 9.1. U.S. Grade Classification of Butter According to Flavor Characteristics.

Identified Flavors by Grading[a]	Grade Classification by Flavor[b]		
	AA	A	B
Feed	S[c]	D	P
Cooked	D		
Acid		S	D
Aged		S	D
Bitter		S	D
Coarse		S	
Flat		S	
Smothered		S	D
Storage		S	D
Malty			S
Musty			S
Neutralizer			S
Scorched			S
Utensil			S
Weed			S
Whey			S
Old cream			D

Source: Code of Federal Regulations (1987).
[a] When more than one flavor is discernible in a butter sample, the flavor classification of the sample should be established on the basis of the flavor that carries the lowest classification.
[b] U.S. Butter Grade as determined by official USDA grading standards.
[c] Defect intensity: S = slight; D = definite; P = pronounced.

Table 9.2. Characteristics and Disratings for Body, Color, and Salt for U.S. Butter Grades.[a]

Butter Characteristics	Disratings[b]		
	S	D	P
Body			
Crumbly	0.5	1	
Gummy	0.5	1	
Leaky	0.5	1	2
Mealy or grainy	0.5	1	
Short	0.5	1	
Weak	0.5	1	
Sticky	0.5	1	
Ragged boring	1	2	
Color			
Wavy	0.5	1	
Mottled	1	2	
Streaked	1	2	
Color specks	1	2	
Salt			
Sharp	0.5	1	
Gritty	1	2	

Source: Code of Federal Regulations (1987).
[a] U.S. Butter Grade as determined by official USDA grading standards.
[b] Defect intensity: S = slight; D = definite; P = pronounced.

with an overview of the USDA butter-grading process. For example, to merit U.S. Grade AA, a given butter may exhibit a slight feed or a definite cooked flavor, but cannot exhibit any other off-flavors. In the workmanship category (pertains to body, color, and salt content and distribution), a concept known as a "disrating" is used. For Grade AA butter, the total permissible disrating for a "workmanship fault" is only 0.5 point. Thus, for a given butter, the flavor classification may actually be "AA," but the assigned U.S. Grade may be lower due to assigned disrating(s) for product workmanship. Examples of this concept are illustrated in Table 9.3.

SOME TECHNIQUES OF BUTTER SCORING

The Butter Score Card. The USDA grading system for butter may be inappropriate for some quality assurance activities or for those situations wherein the quality of one product is compared with that of others. A group of products may include some samples for which a U.S.

Table 9.3. Examples of the Relationship of U.S. Butter Grades to Flavor Classification and Total Disratings for Body, Color, and Salt Characteristics.

Example No.	Flavor Classification	Disratings[a]			Total Disratings	Permitted Total Disratings[a]	Disratings in Excess of Total Permitted	U.S. Grade
		Body	Color	Salt				
1	AA	0.5	0	0	0.5	0.5	0	AA
2	AA	0.5	0.5	0	1	0.5	0.5	A
3	AA	0	1	0	1	0.5	0.5	A
4	AA	0.5	1	0	1.5	0.5	1	B
5	A	0.5	0	0	0.5	0.5	0	A
6	A	0	0.5	0.5	1	0.5	0.5	B
7	A	0	1	0	1	0.5	0.5	B
8	B	0.5	0	0	0.5	0.5	0	B

Source: Code of Federal Regulations (1987).
[a] Maximum disratings permitted for any U.S. grade is 0.5. The given examples illustrate how disratings of varying magnitudes affect the assigned U.S. grade of butter.

grade is not assignable, but which solicit identification of defects and assignment of a score that is reflective of problem seriousness. Useful instruments for assisting in this quality assurance endeavor are score cards and scoring guides, such as those shown in Fig. 9.1 and Tables 9.4, 9.5, 9.6, and 9.7, respectively. The butter score card and scoring guides were adopted (with some modifications) from those proposed by the American Dairy Science Association (1987).

Condition of the Judging Room. The room used for scoring butter should always be clean and well ventilated. Ideally, the temperature of the room should be 15.5° to 21°C (60°F–70°F). There should be no strong odors within the room or from nearby areas. The use of tobacco

Table 9.4. A Suggested Scoring Guide for the Flavor of Butter.

Flavor Defect	Scores for a Given Intensity[a]				
	Slight[b]	Moderate	Definite	Strong	Pronounced[c]
Acid	7	6	5	4	—[d]
Bitter	7	6	5	4	0–3
Cheesy	3	2	1	0	0
Coarse	8	7	6	—	—
Cooked	10	9	8	—	—
Scorched	6	5	4	3	0–2
Feed	9	8	7	6	0–5
Fishy	2	1	0	0	0
Flat	9	8	7	6	—
Foreign[e]	0–3	0–2	0–1	0	0
Garlic/onion	3	2	1	0	0
High salt (briny)	8	7	6	5	0–4
Malty	6	5	4	3	—
Musty	5	4	3	2	0–1
Neutralizer	5	4	3	2	—
Old cream	6	5	4	3	0–2
Oxidized/metallic	4	3	2	1	0
Rancid	4	3	2	1	0
Storage (aged)	7	6	5	4	0–3
Stale	4	3	2	1	0
Tallowy	2	1	0	0	0
Unclean (utensil)	6	5	4	3	0–2
Weedy	6	5	4	3	0–2
Whey	6	5	4	3	0–2
Yeasty	4	3	2	1	0

[a] "No criticism" is assigned a numerical score of "10." Normal range is 1 to 10 for a salable product.
[b] Highest assignable score for a defect of slight intensity.
[c] Highest assignable score for a defect of pronounced intensity. However, a sample may be assigned a score of "0" (zero), if the product is deemed to be unsalable.
[d] A dash (–) indicates that the defect is unlikely to occur at this intensity level.
[e] Due to the wide variation of possible foreign flavors, establishing a fixed scoring range does not seem appropriate. Some foreign off-flavors warrant a zero ("0") score, even when intensity is slight (e.g., gasoline, pesticides, lubricating oil).

Table 9.5. A Suggested Scoring Guide for Body and Texture and Color and Appearance in Butter.

	Intensity				
	Slight[b]	Moderate	Definite	Strong	Pronounced[c]
Body and Texture Defect[a]					
Crumbly	4	3	2	1	—[d]
Gummy	4	3	2	1	—
Leaky	4	3	2	1	0[e]
Mealy or grainy	4	3	2	1	
Ragged boring	4	3	2	1	0
Short	4	3	2	1	—
Sticky	4	3	2	1	—
Weak	4	3	2	1	—
Color and Appearance Defect[a]					
Color specks	3	2	1	0	0
Foreign material	0	0	0	0	0
Mold	0	0	0	0	0
Mottled	3	2	1	0	0
Streaky	3	2	1	0	0
Surface faded/ high	4	3	2	1	0
Unnatural	0	0	0	0	0
Wavy	4	3	2	1	0

[a] "No criticism" is assigned a numerical score of "5." Normal range is 1 to 5 for a salable product.
[b] Highest assignable score for a defect of slight intensity.
[c] Highest assignable score for a defect of pronounced intensity. However, a sample may be assigned a score of "0" (zero) (unsalable product).
[d] A dash (—) indicates that the defect is unlikely to be present at this intensity level.
[e] When a product is determined to be unsalable for a given sensory defect, a "0" (zero) numerical score is assigned to the sample for the quality attribute(s) in question.

in any form should always be prohibited within any judging room, since tobacco odors are so readily absorbed by butter. Although the judging room may be "aired out," tobacco odors can nearly always be detected as soon as the room is "closed" again. There should be adequate lighting so that the uniformity and shade of butter color can be readily determined. Daylight or fluorescent light is superior to incandescent light for the examination of color.

Tempering Butter. The delicate aroma of butter is more readily detected, and the body and texture characteristics are more easily determined, when butter is at the appropriate temperature for judging. Butter stored in temperatures colder than 10°C (50°F) should be placed into the grading or sensory evaluation room in advance of judging to allow tempering to 10°C (50°F). Adjusting the temperature of butter samples is commonly referred to as "tempering." Guidelines for federal (USDA) graders state that the temperature of butter at the time of

Table 9.6. The ADSA Scoring Guide for Sensory Defects of Butter (Suggested Flavor Scores for Designated Defect Intensities).

Criticisms[a]	Score for Given Intenstiy of Defect		
	Slight[b]	Definite	Pronounced[c]
Flavor			
Acid (sour)	6	5	4
Bitter	6	5	4
Briny (high salt)	7	6	5
Cheesy	3	2	1
Coarse	8	7	6
Cooked	9	8	6
Feed	9	8	6
Fishy	2	1	0[d]
Flat	9	8	7
Garlic/onion	3	2	1
Malty	6	5	4
Metallic	4	3	1
Musty	5	4	2
Neutralizer	5	4	3
Old cream	6	5	4
Oxidized	4	3	2
Rancid	4	2	1
Storage	6	5	4
Tallowy	2	1	0
Unclean (utensil)	4	3	1
Weedy	5	4	3
Whey	6	5	3
Yeasty	4	3	2

Source: American Dairy Science Association (1987).
[a] No criticism is assigned a score of "10". Normal range is 1 to 10 for a salable product.
[b] Highest assignable score for a defect of slight intenstiy.
[c] Highest assignable score for a defect of pronounced intensity.
[d] An assigned value of "0" (zero) is indicative of an unsalable product.

grading is quite important in determining the true characteristics of body; it should be between 7.2°C and 12.8°C (45°F to 55°F). The lower the temperature at which the butter is stored, the longer the time required for tempering in the judging room. The required tempering time also depends on the relative size of the butter samples and the temperature of the judging room. One-pound prints will temper in a relatively short time (1–3 hours), while larger wholesale cubes (approximately 68 lbs [36.4 kg]) require a much longer time; possibly overnight under some conditions. Flavor may be evaluated satisfactorily at a temperature above 15.5°C (60°F) but the body is likely to appear atypical at this higher temperature.

Use of the Butter Trier. Samples are taken by a two-edged, curve-bladed tool known as a trier (Fig. 9.2). Facilities for cleaning the trier

BUTTER SCORE CARD

Product: _____   Date: _____

SAMPLE NO.

		1	2	3	4	5	6	7	8
FLAVOR 10	CRITICISM SCORE ➡								
	ACID/SOUR								
	BITTER								
NO CRITICISM	CHEESY								
10	COARSE								
	COOKED								
	FEED								
	FISHY								
UNSALABLE	FLAT								
0	FOREIGN								
	GARLIC/ONION								
	HIGH SALT								
NORMAL	MALTY								
RANGE	METALLIC								
1-10	MUSTY								
	NEUTRALIZER								
	OLD CREAM								
	OXIDIZED								
	RANCID								
	STORAGE (AGED)								
	TALLOWY								
	UNCLEAN (UTENSIL)								
	WEEDY								
	WHEY								
	YEASTY								
BODY AND									
TEXTURE 5	SCORE ➡								
	CRUMBLY								
	GREASY								
NO CRITICISM	GUMMY								
5	LEAKY								
	MEALY/GRAINY								
UNSALABLE	RAGGED BORING								
0	SALVY								
	STICKY								
NORMAL RANGE	WEAK								
1-5									
COLOR AND									
APPEARANCE 5	SCORE ➡								
	COLOR SPECKS								
NO CRITICISM	FOREIGN MATERIAL								
5	MOLD DISCOLORATION								
	MOTTLED								
UNSALABLE	STREAKY								
0	SURFACE FADED/HIGH COLOR								
	UNNATURAL								
NORMAL RANGE	WAVY								
1-5									
SALT 3	SCORE ➡								
NO CRITICISM	EXCESSIVE (TOO HIGH)								
3	GRITTY								
UNSALABLE	UNEVEN DISTRIBUTION								
0									
PACKAGE AND									
FINISH 2	SCORE ➡								
	EXPOSED PRODUCT								
NO CRITICISM	PACKAGE AND LINER								
2	CARELESS								
	DAMAGED								
UNSALABLE	DIRTY/UNSANITARY								
0	NOT PROTECTIVE								
	PRINTING DEFECTIVE								
	UNATTRACTIVE								
	ROUGH FINISH								
	TOTAL SCORE OF								
TOTAL 25	EACH SAMPLE SCORE ➡								
LABORATORY	FAT CONTENT (%)								
PARAMETERS[a]	WEIGHT (LB)								
	PROTEOLYTIC COUNT (PER G)								
	YEAST AND MOLD COUNT (PER G)								
	COLIFORM COUNT (PER G)								
	FREE FATTY ACID								
	7-DAY, (21°C) KEEPING QUALITY								

SIGNATURES OF EVALUATORS: _____   _____   _____

[a] The laboratory parameters are not scored; they provide information that helps determine
the legal status and company specifications of the product.

Fig. 9.1.   Suggested score card for the sensory evaluation of butter.

387

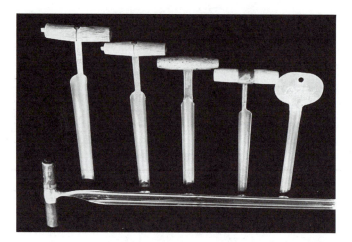

Fig. 9.2. Various sizes of triers used for sampling butter and cheese.

and disposal of waste butter should be provided. The trier should not be washed in warm water (prior to use) but should be wiped with soft tissue or absorbent paper. Washing the trier in warm water often results in a melted, greasy surface on the first plug of butter taken. This obscures the true condition of the body and makes observation of the color more difficult. Paper napkins or other disposable, sanitary paper products are convenient and economical materials for wiping triers. Either disposable or washable containers (plastic pans or metal pails) are suitable for the disposal of discarded samples, the paper used to wipe the trier, and for sample expectoration, if necessary. Disposal of the refuse should be made promptly after the evaluation is completed or when the container is full.

Obtaining the Sample. Since the hands will usually come in direct contact with the butter during the sampling, they should be thoroughly washed before commencing judging. For protection and sanitary reasons, a clean, light-colored frock or apron should be worn (Fig. 9.7). The trier should be cleaned by wiping it with soft tissue or absorbent paper. The number or code of the sample is recorded on the score card or scoring record and the evaluation process is commenced. The judge(s) should stand squarely in front of the sample and observe the cleanliness and neatness of the package. Next, the cover or packaging material is removed and the sample observed for evenness of the liner (if present) and/or the squareness of the wrapping material.

The judge should grasp the butter trier firmly in hand, and diagonally insert the sampling device near the center of the butter sample.

The trier should be turned one-half around and the plug (core sample) withdrawn.

Sequence of Observations. Immediately after withdrawing the plug, and before making any observations for color, the judge should pass the butter sample slowly under the nose, slowly inhale and note any aroma present. A "mental record" of any observed odor should be made by the evaluator. The next step is an examination of the color, especially for uniformity. At this point the judge should examine the body and texture by pressing the ball of the thumb against the sides of the butter plug (core sample) until it shows a break (see Fig. 9.3). The judge should note the presence or absence of free moisture (or "beads" of water) and their relative clarity (see Fig. 9.4). The judge should also be concerned with the nature of the break; that is, whether it is smooth or jagged.

Up to this point, evaluation of the butter has been performed primarily by the sense of sight. Now the judges' senses of smell and taste are "brought into action." The evaluator should break off approximately a 0.5–1 in. piece from the end of the butter plug and place it into the mouth. When a long trier (5 to 10 in. [12.7–25.4 cm]) is used (as in official USDA grading) a sample is typically secured from the rounded, exposed portion near the center of the plug. This sample is generally obtained by means of a stainless steel knife or spatula (cleaned and prepared in the same manner as the trier). A disposable plastic knife or spoon would serve this purpose in both a functional and sanitary

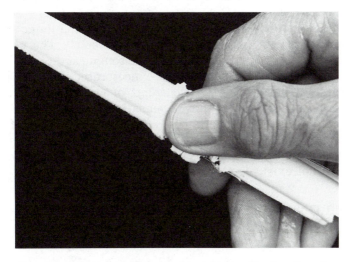

Fig. 9.3. By pressing the butter sample (plug) with the thumb, waxiness and/or leakiness may be determined.

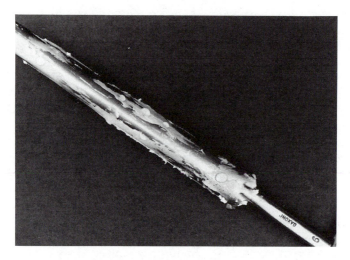

Fig. 9.4.   Leaky-bodied butter usually exhibits free moisture in the form of small water droplets on the back of the trier.

manner. The sample should be chewed until melted. The melted butter is then rolled around in the mouth until it attains approximately body temperature. Meanwhile, the judge should consciously try to feel for the possible presence of "grit" (undissolved salt) between the teeth or between the tongue and the roof of the mouth. The evaluator should also note the manner in which the butter melts; a homogenous melting process is desired.

Simultaneous to these other sensory processes the judge should be experiencing various sensations of taste and smell. The melting (or melted) butter should be rolled around the tongue and to the back and roof of the mouth; then, the sample is expectorated. Finally, the judge should carefully observe for the occurrence of a possible aftertaste; and note whether or not any off-flavor sensation(s) persist. The physical scoring of the butter sample is now complete and the set of sensory observations should be recorded on a butter score card. Less-experienced judges must be especially careful to avoid "imagining a flavor which does not exist" in the butter samples.

OTHER CONSIDERATIONS IN BUTTER QUALITY EVALUATION

Package

The package in which butter is sold should be neat, clean and tidy in appearance and show a good "finish" (smooth, attractive surfaces).

This is important regardless of the type of butter package, whether a quarter-pound or one-pound (0.11 or 0.45 kg) print, a five-pound (2.27 kg) carton, or a larger wholesale unit of product. Fingerprints should not be in evidence on any packaging materials. All butter packages should appear fresh and unsoiled, as if the wrapper or container had not been in previous service; the closure should be fastened firmly and neatly. Any inner linings should impart an impression of neatness and reflect a pride in workmanship. In the instance of one-pound (0.45 kg) prints, removal of an outer carton should always reveal a uniform, neatly wrapped print (piece) of butter.

USDA graders will frequently comment on the general condition of bulk containers, but packaging is not one of the criteria used for determining the U.S. grade of butter. However, this should not detract from the importance and efforts of providing sound, attractive butter packages for facilitating quality assurance and merchandising. Butter packages serve to adequately protect the product and, simultaneously, must be clean, neat, and attractive in order to appeal to and invite purchase by consumers. See Table 9.7 for a suggested scoring guide to help assess the appearance and condition of packaging and product finish.

Salt

Individuals differ in their preference for the amount of salt in butter. Some consumers prefer a highly salted butter ($\geq 2\%$), some desire a lightly salted butter ($\leq 1.5\%$), while other people use unsalted butter exclusively. Different butter markets seem to demand varying percentages of salt. The same score card, however, is used in scoring butter that has a wide range in percentage of salt. Salt is not generally criticized in butter grading regardless of whether the butter is high or low in salt (or even unsalted), providing any salt is completely dissolved (not gritty) and not too "sharp" (high intensity). However, within the quality assurance efforts of a butter manufacturer, a criticism for "briny or high salt" may occasionally be warranted if the saltiness intensity is objectionable or does not meet a buyer's specifications.

Butter should be examined for possible undissolved salt when first placed into the mouth; otherwise salt will quickly go into solution with saliva; hence, it may not be readily detected. The presence of "grittiness" or "grit" (undissolved salt) can be detected most easily by placing some butter between the molars and pressing together gently. If undissolved salt is present, a gritty effect is usually noticed at once. Undissolved salt on the surface or wrapper of an exposed sample does not necessarily indicate presence of undissolved salt in butter. If butter is not "worked" sufficiently during the manufacturing process, then

Table 9.7. A Suggested Scoring Guide for Package, Product Finish, and Salt in Butter.

Package or Product Finish Defect[a]	Score for a Given Intensity				
	Slight[b]	Moderate	Definite	Strong	Pronounced[c]
exposed product	0[d]	0	0	0	0
Package and liner					
careless	1.5	1	0.5	0	0
damaged	1.5	1	0.5	0	0
dirty/unsanitary	1	0.5	0	0	0
not protective	0	0	0	0	0
defective printing (code)	1.5	1	0.5	0	0
defective seal	0	0	0	0	0
unattractive	1.5	1	0.5	0	0
Rough finish	1.5	1	0.5	0	0
Salt Defect[e]					
gritty	2	1.5	1	—[f]	—
uneven distribution	2	1.5	1	—	—

[a] "No criticism" for package or product finish is assigned a score of "2."
[b] Highest assignable score for a defect of slight intensity.
[c] Highest assignable score for a defect of pronounced intensity. However, a sample may be assigned a score of "0" (zero) (unsalable product).
[d] When a product is determined to be unsalable for a given defect, a "0" (zero) score is assigned to the sample for the quality attribute(s) in question.
[e] "No criticism" for salt is assigned a numerical score of "3."
[f] Defect is unlikely to be present at this intensity level.

water droplets that contain salt may reside on the surface. As the water evaporates, salt in the form of white crystals remains on the surface of the butter. In order for butter to merit a perfect score, salt in the interior of the butter must be completely dissolved. A sharp, salty taste sensation usually indicates excessive salt in the butter, particularly when the butter is well "worked" (blended). This is generally indicated by no visible water droplets on the trier or butter plug (the product is void of "leakiness"). Table 9.7 includes a suggested guide for scoring the condition and distribution of salt in butter.

Color and Appearance

Just as the amount of added salt varies with the market for butter, so does the color intensity. In some geographic regions of the U.S., a moderately high color seems to be preferred, while in other areas a lighter color is more desirable (Wilster 1968). A uniform light straw color seems to most often meet the demand or expectations of U.S. consumers. As a rule, the shade of butter color is of little consequence in scoring, providing the color is a natural shade of yellow. Such a yellow color is commonly associated with milkfat, especially if the intensity is no higher than the natural color of the butter produced when cows consume grass as the sole source of roughage (higher carotene content imparted to milkfat). Greenish-yellows or reddish-yellows tend to be discriminated against as a butter color. The primary item to observe in scoring butter for color is the uniformity of color throughout the product.

The butter judge should be aware of the following possible color and appearance defects in butter:

Black, green, red, white,
  or yellow specks
Bleached, dull, pale, lifeless
Faded surface
High-colored surface
Lack of uniformity between
  churnings

Mold discoloration
Mottles
Streaks
Unevenness
Unnatural
Waviness

Faulty workmanship, particularly over- and underworking of butter during the manufacturing process, is responsible for most color and appearance defects. The size, number, and distribution of water and air droplets markedly influence the color of butter. There are several possible causes of color specks, including solidified pure butter oil (yellow, curd particles (white), copper salts (green), iron salts (black), and undissolved butter coloring (yellow). Microorganisms, including molds, can cause serious quality deterioration problems of butter. Butter that is inadequately protected against moisture evaporation tends to exhibit an intense or high-colored surface. There have been instances of escaped refrigeration gases reacting with the color pigments of butter. Contamination with extraneous or foreign substances poses serious problems of aesthetics that go beyond color or sensory effects; occasionally, even questions of wholesomeness and toxicity may be raised due to product adulteration.

The more common color defects of butter can essentially be eliminated by proper working at the time of manufacture. Generally, the flavor of poorly worked butter is not as good as the flavor of the same butter, had it been properly worked. Furthermore, butter with color defects due to insufficient working usually does not keep as well as butter that is adequately worked. Therefore, a color defect may serve as a hint to the judge to be more on the alert for possible flavor defect(s) that may be associated with the cause of the appearance shortcoming.

"Mottles" are spots of lighter and deeper shades of yellow, caused by an uneven distribution of salt and/or moisture due to insufficient working. "Streaks" are recognizable as an area of light color surrounded by more highly colored portions; "waviness" is an unevenness that appears as waves of different shades of yellow. Their common cause—insufficient working, may be aggravated by a poor mechanical condition of the churn and/or careless reworking of remnants from previous churnings.

Body and Texture

Immediately after examining a trierful of butter for aroma and color, the body should be examined before it is affected by exposure to ambient temperatures. The judge should notice the plug surface for the possible presence of "beads" of water, for smoothness, for solidity, and for firmness. Next, the evaluator should press the ball of the thumb (good sanitary practices must be observed) against the sample surface and notice how the plug "breaks" or responds (see Fig. 9.3). If beads of water appear (Fig. 9.4), the judge should note whether this water is clear or milky. A milky appearance often furnishes a clue to an off-flavor that may be present. By means of a spatula, spoon, or knife, (preferably the single-service type, for sanitary reasons) the judge should cut off a small piece of the plug and place it in the mouth, and note how it melts. The evaluator should determine if the physical features of the plug seem to disappear. *High-quality butter should melt evenly and disappear slowly.* The evaluator should note the mouthfeel characteristics of the sample with the tongue and the palate as it is melting. *The body* of good-quality butter *should be firm, and exhibit a waxy, close-knit texture.* When broken, butter should present a somewhat jagged, irregular, wrought iron-like surface.

The physical-chemical system that determines the characteristic body and texture of butter is quite complex. Since milkfat is a mixture of triglycerides which melt at different temperatures, butter at normal handling temperature is a mixture of both crystalline and liquid fat. The type of feed which cows consume influences the proportion of high-to-low melting triglycerides in the milkfat. The fat of butter also exists in the form of globules and free fat. Both the size of fat crystals and the diameter of fat globules may influence butter body and texture. Seasonal differences in milkfat composition, primarily due to different feeds, may be partially compensated for by varying some manufacturing steps. In much of the U.S., butter tends to be harder (firmer) in the winter season due to an elevated content of higher-melting triglycerides. Generally, milkfat is softer in the summer because it contains a larger proportion of lower-melting triglycerides; hence, the butter body may tend to be weaker and/or leaky. Butter is a water-in-fat emulsion, in which milk proteins and possibly milk salts may play a "stabilizing" role. Manufacturing steps which influence the body and texture of butter include: (1) tempering of the cream; (2) churning temperature; (3) butter granule washing temperature; (4) extent of working; (5) the method of adding salt; and (6) the manufacturing equipment and churning methods used.

BODY AND TEXTURE DEFECTS OF BUTTER AND THEIR GENERAL CAUSES

Briefly, the terms "body and texture" refer to the physical properties of butter. These physical properties primarily depend upon the composition of milkfat, structure of fat globules, rate of fat crystallization in cream and butter, amount of liquid fat, and size of the fat crystals in butter. Although the term "body" refers to the general makeup or consistency of the butter mass, and the term "texture" relates to the arrangement of the particles that make up the mass, they are so closely related that they are not considered separately in evaluating the physical properties of butter. The major body and texture defects of butter are:

Crumbly	Ragged boring
Greasy	Salvy
Gummy	Short
Leaky	Sticky
Mealy/grainy	Weak or spongy

Each defect will be discussed in more detail in the following paragraphs.

Crumbly. The fat particles in a "crumbly" or "brittle"-textured butter lack cohesion and do not hold together. Some of the butter usually adheres to the trier and reflects a rough appearance. As the term "crumbly" suggests, the butter appears dry and readily falls apart, rather than appearing waxy and homogenous when pressure is applied to the plug (Fig. 9.5). A crumbly texture of butter suggests that it has been underworked; however, if it is worked more, the body usually becomes sticky (salvy).

Crumbliness in butter seems to be the result of large fat crystals and a deficiency in liquid fat. The defect is usually observed during late fall and winter months; it has been associated with feeding cottonseed

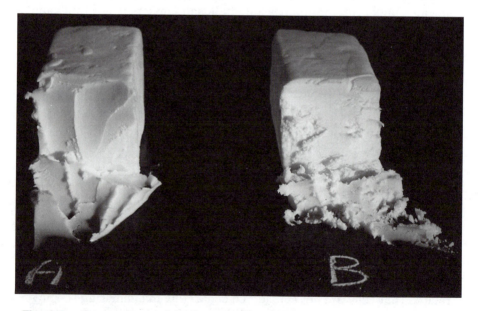

Fig. 9.5. Comparison of butter samples for body characteristics: A—A more desirable smooth, waxy body; B—A crumbly, nonhomogeneous body (poor spreadability).

meal and alfalfa hay. The temperature to which cream is cooled after pasteurization, length of the holding period, churning, and wash water temperatures are factors to be considered in overcoming this defect (Coulter and Combs 1936; Wilster 1958; Wilster *et al.* 1941; Zottola 1958). The temperature of samples during the evaluation is an important factor in detecting crumbliness, since a normal body may appear crumbly at a low temperature, while a crumbly butter may appear normal at a markedly higher temperature. Crumbly butter is always difficult to cut into neat, attractive patties.

Greasy. A "greasy" butter consistency may be noted by evidence of extreme smoothness and immediate melting when a sample of such butter is placed into the mouth. Also, this defect may be suggested by the extreme ease with which a trierful of sample may be removed from the product. Instead of a clean, clear feeling in the mouth after expectorating (as when a desirable "waxy" sample has been tasted) the oral cavity may be left with a distinct sensation of greasiness. The most likely cause of greasiness is overworked butter, particularly when the body of the butter is already too soft. A higher proportion of low-melting triglycerides is the physical-chemical factor responsible for this defect.

Gummy. "Gummy"-bodied butter tends to stick to the roof of the mouth and may leave a gum-like impression. This defect is more prevalent in areas where cottonseed products are fed as the protein supplement in the dairy ration. Likewise, the defect is more prevalent during the winter months; cottonseed products are probably fed more extensively then.

Gumminess in butter is considered to be due to an abnormally high percentage of high-melting-point glycerides, which causes a firmer or harder milkfat. The relative hardness of such butter is often first noticeable by the additional time and high temperature required to properly temper the butter for evaluation. This defect is generally not that objectionable, but it can interfere markedly with butter spreadability. A reduced holding time of the cream prior to churning, slower cooling of the cream, higher churning temperature, higher temperature of wash water, and longer working time are some of the manufacturing steps which have been found to aid in control or minimization of this defect.

Leaky. Butter that exhibits beads or droplets of moisture on the plug and/or the back of the sampling trier is criticized as being "leaky." Such butter fails to retain moisture within the product mass due to the larger size of water droplets. Leakiness is usually caused by insufficient working; the butter has not been worked to the point where the water droplets are reduced sufficiently in size to be well-distributed

CONTEST BUTTER SCORE CARD

DATE: _____ A.D.S.A.   CONTESTANT NO. _____

	CRITICISMS					SAMPLE NO.								TOTAL GRADES
		1	2	3	4	5	6	7	8	
FLAVOR 10	CONTESTANT SCORE ➡									
	GRADE SCORE									
	CRITICISM									
NO CRITICISM = 10	ACID (SOUR)									
	BITTER									
	BRINY (HIGH SALT)									
	CHEESY									
	COARSE									
	FEED									
	FISHY									
	FLAT									
	GARLIC/ONION									
NORMAL RANGE = 1-10	MALTY									
	METALLIC									
	MUSTY									
	NEUTRALIZER									
	OLD CREAM									
	OXIDIZED									
	RANCID									
	STORAGE (STORAGE)									
	TALLOWY									
	UNCLEAN (UTENSIL)									
	WEEDY									
	WHEY									
	YEASTY									
BODY AND TEXTURE	ALLOWED PERFECT IN CONTEST	X	X	X	X	X	X	X	X	
COLOR	ALLOWED PERFECT IN CONTEST	X	X	X	X	X	X	X	X	
SALT	ALLOWED PERFECT IN CONTEST	X	X	X	X	X	X	X	X	
PACKAGE	ALLOWED PERFECT IN CONTEST	X	X	X	X	X	X	X	X	
TOTAL	TOTAL SCORE OF EACH SAMPLE ➡									
	TOTAL GRADE PER SAMPLE									

FINAL GRADE _____
RANK _____

Fig. 9.6.   The ADSA contest butter score card.

throughout the butter mass (Hunziker 1940; McDowall 1953; Totman *et al.* 1939; Wilster 1968). Replacement of the plug into leaky butter is sometimes accompanied by a "slushing" sound (refer to Fig. 9.4).

Leaky butter is routinely discriminated against due to significant economic losses that are likely to be incurred in handling and printing (packaging). The leaky defect is considered substantially more serious if the droplets of moisture appear milky or cloudy; this indicates either insufficient washing or insufficient working of the butter. Use of too high a churning temperature, which results in softer granules and retention of excessive amounts of buttermilk within the granules, is also conducive to a milky or cloudy brine in the resultant butter. A milky or cloudy brine not only causes economic losses in handling and printing, but this condition is also detrimental to butter keeping-quality. The flavor is also often jeopardized, due to the lack of appropriate blending of fat, water, salt, and curd; the four major constituents of butter. An uneven salt distribution may cause migration of water towards the salt; leakiness is a possible consequence.

Mealy or Grainy. A "mealy" or "grainy" texture is easily recognized when a sample of partially melted butter is compressed between the tongue and palate and a mouthfeel reminiscent of cornmeal mush or a distinct "grainy" sensation is perceived. This is considered a somewhat serious texture defect; such butter lacks a smooth, waxy texture for which good butter is noted. Carelessness on the part of the buttermaker is assumed in the case of this defect. A mealy (grainy) texture may be caused by improperly neutralized sour cream or allowing milkfat to "oil-off" at some stage of the buttermaking process. Improper melting of frozen cream, allowing milkfat in cream to separate or "oiloff" in the pasteurizer or remelting butter scraps in a pasteurizer vessel may result in a grainy textured butter. When this defect is due to frozen cream or remelting of butter scraps in the pasteurizer, the consequent severe graininess is considered one of the most serious texture defects of butter. The buttermaker is in a position to prevent or control the mealy/grainy defect by proper selection and processing of cream and application of appropriate churning techniques.

Ragged Boring. Usually a full trier of butter cannot be readily drawn from butter that has a sticky-crumbly texture; it is also difficult to replace the ill-shaped plug into the formed trier hole. The butter simply seems to roll from the trier, rather than the trier cutting a distinctly formed plug. Butter that exhibits this sampling difficulty is referred to as "ragged boring." This is considered a fairly serious body defect, since it can be projected that this condition would interfere with cutting butter into individual serving-size patties. This defect also unfavorably affects butter spreadability. (See Fig. 9.7.)

Factors that cause ragged-boring butter include the rate of cream cooling after pasteurization, the holding temperature of cream prior to churning, the churning temperature, wash water temperature, or any processing condition that tends to interfere with the formation of a well-made, close-knit texture in butter.

Salvy. "Salvy" texture in butter is denoted by a definite lack of smoothness and an uneven surface; it appears as though the product is an admixture of two or more different lipid substances. This defect practically defies description beyond the term "salvy." Butter-processing conditions that cause a salvy texture also frequently provoke a sticky-bodied butter. These processing shortcomings are enumerated in subsequent discussion.

Short. A "short" body in butter refers to a product that lacks the desirable characteristics of plasticity and waxiness. This defect is noted when the plug has a tendency to break sharply when moderate thumb pressure is applied. At less-than-typical temperatures ($\leq 7.2\,^\circ$C [$\leq 45\,^\circ$F]) of scoring, a short-textured butter exhibits marked brittleness. Other factors which may be involved in short-textured butter are: (1) high-melting point fats (that contain relatively small fat globules); (2) an extremely low curd content in the butter; (3) manufacturing processes wherein part of the milkfat is melted (hence, normal butter granules are not formed); and (4) rapid cooling of recently made butter to an extremely low temperature.

Sticky. As the term implies, a "sticky"-bodied butter adheres (sticks) to the trier and appears to be quite dry. Usually it is difficult to secure a uniform, smooth-surfaced plug from such butter; it appears "ragged" or "rough." This is particularly the case when the trier is cold. Since the market generally demands butter that slices and/or spreads relatively easily, a sticky body is quite undesirable. As stated earlier, when crumbly or brittle-textured butter is overworked, the entire mass tends to become sticky. In fact, a sticky body and crumbly texture are often present concurrently in butter. A sticky body (as is crumbly or brittle-textured butter) is observed most frequently in late fall and winter when there is a greater predominance of high-melting-point fatty acids in churning cream. Hence, a sticky body is primarily a feed-related defect; it appears to be more prevalent in dairy areas where alfalfa is the major roughage fed to milk cows. Various temperature treatments of cream and butter, as well as churn working conditions, seem to markedly affect the occurrence of the sticky defect.

Weak or Spongy. A "weak" or "spongy" body is typically indicated by a quick meltdown or an exaggerated softness of the butter when it is exposed to ordinary room temperatures. This is not a particularly serious defect, but it is a butter characteristic which is, overall, not in good favor with most butter graders or buyers. A weak-bodied butter often produces an imperfect plug; there is a tendency for the trier to "cut in" on the plug. When the ball of the thumb is pressed against a plug of "weak" butter, difficulty is often encountered in defining a distinct "breaking point" for the plug. Supposedly, a weak body is due to a state of incomplete milkfat crystallization, which results in an excess of milkfat in the liquid form within such butter. Incomplete crystallization of the milkfat may be due to inadequate cooling of pasteurized cream, or due to a relatively high proportion of low-melting-point glycerides in the cream.

Based on natural variations in the composition of milkfat in different butter-producing regions of the country, it would seem appropriate that butter judges allow for some leeway or range in butter body and texture characteristics. Consequently, a weak body in butter is not generally considered one of the more serious defects. Churning at too high a temperature or incorporating too much air may also lead to a weak-bodied butter.

Frequency of Sensory Defects in Butter

Apparently, no statistics are available to quantify or document the improvement in butter quality in recent decades. However, anyone who has been involved in the sensory evaluation of butter for a decade or longer, would most certainly conclude that the overall quality of U.S. butter has vastly improved. The one development most responsible for this significant flavor improvement is the marked change from farm-gathered cream to plant-separated cream (from fresh milk). Simultaneously, technological advances in butter manufacturing have substantially reduced defects that were previously due to workmanship. Continuous churns have served to significantly reduce "personnel errors" through semiautomation and better process control.

Annual summaries of the frequency of defects encountered by USDA graders provide only an approximate assessment of current butter quality, since not all butter is graded. Furthermore, official USDA grades are placed on butter prior to printing into retail-sized

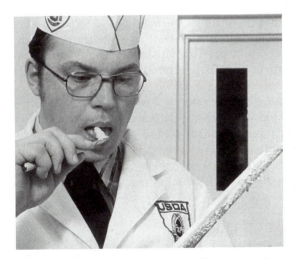

Fig. 9.7. This USDA grader is applying a prescribed technique for assessing the sensory properties of a butter sample which will result in determination of an assigned grade. An apparent defect is "ragged boring" for this sample, based on the butter particles adhering to the back of the trier.

packages. A summary prepared by the USDA (1983) indicated that about 68% of the samples from 256,275 churnings exhibited no sensory defects of any kind. The most serious flavor criticisms (assigned to about 16% of these churnings) were: bitter (4.7%), and either aged, coarse, whey, acid, feed, or old cream (0.4%). About 15% of the samples were criticized for body and texture. The most frequent criticisms were "short" (11%), leaky (3.5%), and sticky (0.6%). The churnings criticized for color were infrequent (about 0.6%); wavy color occurred in about 0.5% of the churnings and mottled, streaked, and color specks accounted for the remainder of the color/appearance defects. Only 0.03% of the churnings were criticized for salt content or adverse condition; "sharp" salt was the most prevalent criticism. Only three churnings were found to manifest a "gritty" salt defect. In about 0.08% of the samples, USDA graders noted at least one of the following product shortcomings: mold present, poor condition of container, foreign material evident, exposed butter surfaces, rough finish, unnatural color, too low milkfat content, and/or an unsatisfactory keeping-quality test result.

Correlation of Body and Texture Defects. Sometimes, two samples of butter may have distinctly different body and texture characteristics, but due to regional preferences or grading interpretations, each sample may be given a similar or perfect (if warranted) body and texture score in the grading process. As a rule, the body and texture of butter from different butter-producing regions will not be exactly the same, though made in the same season of the year. Tolerances in grading are allowed for these different characteristics. If a body/texture defect is noted when grading butter, it should be either sufficiently pertinent, intense, or readily obvious to be recorded on the score card. Also, it is not unusual to have two or more body and/or texture defects occur in the same butter sample. For instance, butter with a leaky defect may also exhibit a mealy texture; sticky-bodied butter may also have a crumbly texture. Due to the occurrence of these dual defects, two criticisms are sometimes noted. However, in such cases both of the defects must be sufficiently obvious, intense, or serious (demonstrated beyond any doubt), for the dual defects notation to be recorded.

FLAVOR OF BUTTER

The ability to consistently detect various off-flavors and assess their intensity is probably the most difficult part of evaluating butter. With few exceptions, almost any person (with a little experience) can detect and evaluate the previously mentioned defects related to packaging, salt, color, and body and texture. The senses of sight and touch are primarily relied on for assessing the aforementioned butter defects.

For evaluating butter flavor, the judge should recall the aroma which was mentally recorded at the time the trierful of butter was obtained. The evaluator must be ready to correlate, if possible, this perceived aroma with the taste sensation which is about to be experienced. The judge should then remove about 1 in. (2.54 cm) of the end section of the butter plug with a knife, spoon, or spatula. If taken properly, this portion of butter represents the approximate center of the butter sample. The judge places this small quantity of butter in the mouth and chews it enthusiastically, thereby bringing the butter into a liquid state as soon as possible. The evaluator should continue manipulating the sample with the tongue and jaws until the butter sample approximately reaches body temperature.

It is most important that the butter judge particularly notice the first hint of a taste or smell to make an appearance. The evaluator needs to observe whether this first taste sensation disappears or not.

Fig. 9.8. Students participating in the sensory evaluation of butter in the Collegiate National Contest.

The judge should mentally record, as the sensory procedure progresses, whether there is a succession of flavors; that is, do the first flavors pass off and others appear? Some judges employ a technique of tilting the head back and with an action of the tongue, attempt to draw the warm, melted butter to the back of the mouth. The evaluator should bear in mind that the sense organs of taste and smell are quite delicate, and with certain flavor sensations the sensitivity of these delicate organs is easily dulled (sensory adaptation). In this event, the flavor notes are either less readily perceived or may no longer be observed. To help prevent sensory adaptation, a butter sample should not be kept in the mouth too long.

After the judge notes the various flavor sensations that may be present, the sample is expectorated into a container or sink provided for that purpose. This generally completes the task or sequence of observations with the butter sample. However, it is most important that any aftertaste be carefully noted and studied. The evaluator should observe any taste sensation which remains in the mouth and note the relative degree of pleasantness or unpleasantness as well as the extent of persistence. Having done this, the evaluator should replace the remainder of the plug into the same hole from which it was obtained; the plug should be reinserted level with the butter surface. Next, the trier hole should be smoothed with a knife or spatula, which will help keep the butter surface neat in appearance and restrict access of air to the sample interior. The authors want to emphasize that the process of evaluating butter samples requires the application of sanitary precautions throughout the various steps of securing samples, tasting, replacing core plugs, and the discarding of sample portions and other refuse.

All sensory observations should be recorded on the score card, by assigning each item an appropriate numerical value. A typical order of sensory procedures followed by most experienced butter judges has just been outlined. Experience might reveal several efficient steps or short-cuts in the evaluation routine which might enable an evaluator to proceed faster. However, the systematic approach suggested here is useful in developing correct scoring habits, which lead to more proficient work.

Characteristics of the Various Flavor Defects. High-quality butter should have a mild, sweet, clean, and pleasant flavor and a delicate aroma. A characteristic feature of such high-quality butter is that the appetite seems to "crave more of the product." To manufacture butter of "first-class" flavor, the raw materials must definitely be free of objectionable flavor defects. This is also true of cultured cream butter, which should exhibit a distinct starter flavor (an aroma with diacetyl as the principal component).

In summary, butter should have a subtle, delicate flavor. This delicate flavor is somewhat reminiscent of the best "buttered popcorn" that a person has ever consumed. If pronounced flavor notes are observed in butter, then at least one or a combination of several flavor defects are surely present. In the instance of a combination of flavor defects, the most obvious, serious, or objectionable off-flavor should draw the eventual criticism. If two or more flavor defects are present, with each one equally prominent, then more than one flavor criticism should be assigned. A suggested scoring guide for the various flavor defects of butter is shown in Table 9.4. Illustrated in Table 9.8 are the major divisions of a 100-point score card. This older version of a butter score card may still be used for specific applications such as state fair or other competitions.

Acid/Sour. An acidic or sour off-flavor in butter usually stems from either high-acid (sour) cream, overripened cream, excessive use of lactic culture, or excess retention of buttermilk in the butter. When buttermilk is retained (frequently indicated by a milky brine), it is often designated as a "buttermilk flavor" defect. The term "sourness" denotes a marked intensity of acid taste in the product. An acid off-flavor in butter is characterized by a sharp, sour taste on the tip of the tongue, as well as an associated "sour" aroma, due to the presence of certain volatile components. This sour taste is easily and quickly detected when the butter is placed into the mouth; however, the flavor sensation usually clears up quickly, and leaves no aftertaste.

Aged. An "aged" off-flavor in butter is probably best described by the term "lacks freshness." The lacks freshness sensation can typically

Table 9.8. The 100-Point Contest Butter Score Card[a]

Quality Attribute	"Perfect" Score	No Criticism[b,c] Score
Flavor	45	38
Body and Texture	30	30
Color	10	10
Salt	10	10
Package	5	5
Total Score	100	93

[a] A form of butter score card frequently used for fair competitions, exhibits, and/or clinics.
[b] Samples may have a definite cooked and/or slight feed flavor and still be assigned a flavor score of "38."
[c] Traditionally, a rather narrow range of scores separates samples which are not criticized from those which are unsuitable for table use. To illustrate, a butter which receives a total score of 88 is an extremely poor product, perhaps only suitable for cooking. When several excellent butter samples (i.e., samples which cannot be criticized) are encountered in competitive events, the 100-point score card provides a mechanism for distinguishing between them. Flavor scores between 38 and 45 are used to comparatively rate the quality of these exceptional products. A "perfect" sample, if encountered, would be assigned a total score of 100.

be detected by smelling the sample or by noting a moderately persistent aftertaste. The aged off-flavor may be confused with either "storage" or "old cream" off-flavors. The aged flavor defect is caused by either holding butter for extended time periods at relatively low temperatures, or for short time periods at relatively high temperatures. If butter, especially "printed" butter, is to be held for an extended time, it should be stored at $-17.7\,°C$ ($0\,°F$) or lower to minimize the development of aged and/or oxidized off-flavors. Failure to promptly process milk or cream (even if it is of high quality) can result in a loss of freshness or the aged butter flavor defect. USDA graders distinguish between the aged and storage off-flavors in determining U.S. grades; however, only the "storage" criticism appears on the score card recognized by the Committee on Evaluation of Dairy Products of the American Dairy Science Association (1987). In the case of the ADSA score card, both defects are considered to be various stages of the same process of product deterioration.

Bitter. In butter, bitterness has been attributed to a number of causes, including the action of certain microorganisms, enzymes, specific feeds or weeds, late lactation milk, impurities in butter salt, and inappropriate use of some neutralizer compounds. The particular bitter defect produced by the action of lipases is more appropriately described as a rancid or lipolytic off-flavor (discussed later). Conventional (or generic) bitterness is recognized by the sense of taste alone. Once the butter sample is melted in the mouth, it should be rolled to the back center of the tongue, where the taste buds sensitive to the bitter sensation are located. Bitterness resembles the taste sensation exhibited by quinine; it persists as a distinct, lingering aftertaste, even after the sample has been expelled from the mouth.

Briny/High Salt. Government graders identify this defect as "sharp salt" under the category of salt, rather than noting a "high salt' ("briny") problem as a flavor defect *per se.* Regardless of the category for designating this defect, a distinct to pronounced salty taste in butter tends to prevail to an extent beyond a "range of ordinary acceptability." Usually, the cause is simply the use of too much salt, though uneven distribution of salt and water can also provoke this defect. An intense high salt taste in butter will probably receive more attention or criticism in the future, given the recent focus on the sodium content of foods and its possible relationship to human health. As a guideline for the butter judge, a salty taste should not be the first, last, and only taste sensation perceived in a butter sample. If such is the case, then it should probably be criticized for high salt (briny).

Cheesy. A "cheesy" off-flavor in butter has a striking resemblance to the aroma and taste of Cheddar cheese. The presence of this off-flavor is easily detected from an initial sensory observation, due to both the intensity and peculiar cheesy characteristics. From the instant of placing the sample in the mouth, through manipulation of the sample and subsequent expectoration, to the last lingering aftertaste, this flavor defect is unique and readily noticeable. A cheesy off-flavor is persistent; the mouth definitely fails to "clean-up."

When soured cream is held too long at nonrefrigeration temperatures, proteolytic microorganisms (e.g., psychrotrophs) frequently hydrolyze the protein (curd), which results in a cheesy off-flavor. When such abused cream is churned, a most unpleasant, cheesy off-flavor often prevails in the butter. This off-flavor can also be formed in butter made from initial, good-quality cream, but as the result of contamination by psychrotrophic bacteria in churning or butter handling (especially when there is poor moisture and/or salt distribution). Lightly salted and unsalted types of butter are especially vulnerable to this defect.

In some extreme cases, a cheesy off-flavor in butter may somewhat resemble the odor of limburger cheese or putrid meat. Cheesy is usually considered to be an extremely serious defect. Quite often a bitter aftertaste will accompany the cheesy flavor defect, due to proteolysis and some of the resultant chemical end products (peptides). Butters that have a higher curd content (above 0.9%) are more prone to the development of cheesy (or unclean) off-flavors than butter with 0.7% or less curd. Complete removal of buttermilk and thorough washing of butter granules are important processing steps for minimizing the cheesy off-flavor.

Coarse. Butter which lacks that sweet, pleasing, delicate flavor that is generally associated with fresh milkfat is generally criticized as being "coarse" in flavor. The lack of flavor refinement, or a slight harshness of flavor, is typically noticed when the butter sample is first placed into the mouth. A coarse off-flavor does not particularly give rise to a pronounced, undesirable flavor sensation; the butter just seems to lack that pleasant flavor sensation or the balanced taste and aroma characteristics that is anticipated in the highest-quality product.

The term "coarse" has been part of the flavor vocabulary of butter for several decades, but it has not always had the same meaning. Obviously, the meaning of this term should be fixed by definition. Considering the state of the art of present day buttermaking, the term "coarseness" has become associated with butter made from commingling some fresh, high-quality cream with a cream source of somewhat lower quality or from cream that has a slight acid development. From a practical standpoint, however, whenever a butter is found to lack a fine, delicate, smooth flavor, *the coarse criticism is employed when no other criticism appears justified or appropriate.* Thus, the criticism "coarse" for butter is similar to the criticism "lacks fine flavor" which is applied to other dairy products. "Coarse" is primarily reserved for that butter that has reasonably good sensory properties but just seems to fall short of the top or best-quality product.

Cooked. A "cooked flavor" is generally associated with high-quality (best-grade) butter. This flavor note in butter should be readily recognized when the core sample (within the trier) is passed under the nose or when a portion of the sample is first placed into the mouth. Unless the flavor is intense, its presence, as noted by tasting or smelling, is of relatively short duration. Provided that other off-flavors are not present, butter exhibiting a "cooked" flavor "cleans up" completely and leaves absolutely no aftertaste.

A cooked flavor in butter, which can be described as a smooth, nutty, custard-like character, is produced by pasteurizing sweet cream at a relatively high temperature. It is not unusual (and frequently desirable) to have a definite cooked flavor in freshly churned butter. If the butter is free of an associated "coarseness," and it is not "scorched" *per se,* this flavor sensation in butter is not objectionable; in fact, it is generally considered delightfully aromatic and pleasing (i.e., buttery). United States butter grades allow a *definite* cooked flavor in the highest grade (AA) of butter. Typically, much of this flavor note dissipates from the product before the butter reaches the consumer. Pasteurization at higher temperatures also enhances the keeping-quality of butter; the high heat-treatment serves to significantly counteract the onset of oxidized off-flavors (cardboardy, metallic, and/or tallowy). Reducing compounds, such as sulfhydryls, formed from the high-temperature heat-treatment of cream, are effective antioxidants.

A "scorched" off-flavor in butter is considered objectionable. Causes include pasteurization at severely high temperatures, and/or long holding times in the presence of developed acidity, and product "burn-on" that may occur on heating surfaces due to inadequate agitation or extended processing times of HTST pasteurizers. For improperly neutralized cream, a defect may develop that is known as "scorched-neutralizer," which resembles the off-flavor of old nut meats.

Feed. The presence of different "feed" off-flavors can usually be detected in the aroma and verified on the palate when the butter is melted. The mouth cleans up quite soon after the sample is expectorated, in the case of the feed flavor defect.

Most forms of dry feeds, such as hay, many of the grain concentrates, silage, green alfalfa, and the various grasses generally lead to no worse than what is referred to as a "normal" feed flavor note in butter (when heat treatment of cream is at minimum temperature). Even when fed in large quantities, these feeds usually only have a slight objectionable effect on butter flavor. Green alfalfa tends to produce a characteristic, mild, sweet flavor (with a possible instantaneous bitter-sweet tinge). This is common for much of the butter produced in the irrigated valleys of the Rocky Mountain and Pacific Coast states, where alfalfa is fed so extensively to dairy cows. When cows are placed on fresh grass pasture in spring or early summer, the butter produced may exhibit a characteristic "grassy" off-flavor. A slight or "normal" feed flavor is actually allowed in U.S. Grade AA butter. Occasionally, some feed sources may impart an objectionable "bitter" off-taste to butter.

Proper feeding routines for dairy cows can do much to eliminate or minimize feed off-flavors in butter. Generally, if cows are not fed within three hours of milking time, feed off-flavors are substantially minimized in subsequently produced butter. If large quantities of highly aromatic feeds are fed, the period of time between feeding and milking should be increased. When a cooked flavor is imparted to cream (and the resultant butter), it tends to mask any feed off-flavors in butter (for at least the first several months after product manufacture).

Fishy. As this term indicates, a "fishy" butter may have flavor and aroma characteristics similar to codfish, cod-liver oil, or fish meal. This is one of the most serious, most pronounced, and objectionable flavor defects of butter; it is an off-flavor which is persistent and the mouth distinctly fails to "clean-up." The development of fishiness in butter is favored by the combined conditions of high acid, high salt, overworking, and elevated levels of metallic salts in the cream. A fishy off-flavor is believed to develop from the chemical decomposition of lecithin, a trace compound of butter. Fishiness is primarily a storage-incurred defect, but it occasionally occurs in fresh or reasonably fresh butter, if appropriate catalysts trigger it. In summary, fishy off-flavor in butter represents a most severe degree of auto-oxidation of lipid components.

Flat. Butter that simply lacks a characteristic, full, pleasing "buttery" flavor is criticized as being "flat." The absence of typical butter flavor is noted when the butter is first placed into the mouth. The lack of flavor character is most readily noted as the butter melts in the mouth upon tasting. The flat defect should not be confused with the lower flavor profile of lightly salted or unsalted butter, though there are similarities. Unsalted butter may exhibit several flavor notes in sufficient intensities for detection, but the lack of salt generally suppresses rather than enhances the flavor notes. By contrast, in a product with a flat flavor defect there is little or no characteristic butter flavor. A flat defect is generally caused by an apparent lack of volatile acids and/or low content of diacetyl and other volatile compounds that are partially responsible for a desirable "buttery" flavor.

Dilution of churning cream with water or excessive washing of butter granules during manufacture may result in a flat flavor. Certain feeds may also be more conducive to production of milkfat with less characteristic flavor. The imparting of a cooked flavor to churning cream is probably the simplest expedient for masking the flat flavor defect in butter.

Foreign. Off-flavors derived from the careless use of cleaning and sanitizing chemicals, absorption of combustion products, odors absorbed from direct or indirect contamination with gasoline, kerosene, fly spray, paint, varnish, etc. are unacceptable in butter. Unfortunately, since milkfat can function as an excellent solvent for many of these materials, any cream or butter contamination must be absolutely avoided. Even atmospheric vapors from these kinds of compounds can be a serious problem in terms of possibly imparting foreign (atypical) or chemical-like off-flavors.

Garlic or Onion. "Onion" or "garlic" are most objectionable off-flavors occasionally found in butter. They are easily detected from the distinctive odors suggestive of their names. Both of these off-flavors are most pronounced when samples are heated to body temperature. Interestingly, the flavor taints of garlic and onion are surprisingly similar when detected in butter by tasting and/or smelling. Both are quite odorous, as well as distinctly persistent in aftertaste, both are equally objectionable and out-of-place in either fresh or stored butter.

Malty. The "malty" off-flavor that is occasionally encountered in butter resembles the odor of malted milk or "Grape Nuts®" cereal. Sometimes this off-flavor may be suggestive of black walnuts. The flavor sensation extends throughout the entire tasting period and generally persists after the sample has been expectorated. The malty off-flavor results from the outgrowth of *Streptococcus lactis* var. *maltigenes* in either milk or cream that has been cooled inadequately. Storage temperatures of milk or cream were probably in excess of 13°C–16°C (55°F–60°F). Increased acidity of the milk or cream may subsequently occur, hence, a combined malty and high acid off-flavor is probable.

Metallic. As the name indicates, a "metallic" off-flavor is suggestive of metal. This flavor defect conveys a slightly astringent and puckery feeling to the mouth interior. If a person holds an iron nail in the mouth until saliva contacts it, the resultant sensory perception is analagous of the typical characteristics of a metallic off-flavor. The metallic note may be detected as soon as the butter is placed in the mouth; the sensation perceived by the palate generally becomes more intense as the sample melts and is liquified. To some persons, the initial taste perception experienced with the metallic defect seems flat. This off-flavor persists after the sample has been expectorated; a somewhat bitter taste or other objectionable aftertaste may appear at the end of the tasting period.

A metallic off-flavor is often caused by storing cream, particularly soured cream, in direct contact with metals such as tin-coated cans, copper, or iron until a metallic salt is formed, which can be conveyed into the finished butter. Rusty cream cans, cans from which the tin has been abraded and the use of milk can steamings (heated rinsings) can cause metallic-flavored butter. Since chemical compounds which impart a metallic off-flavor are formed during the course of lipid auto-oxidation, this defect may also be considered to be a distinct stage in the progression of off-flavors that comprise the oxidized off-flavor complex.

Musty. The "musty" off-flavor in butter resembles the odor of a poorly ventilated cellar, potatoes, or a swamp. The aroma of musty butter may also resemble musty or poorly cured hay. When musty off-flavored butter is first placed into the mouth, actual perception of the off-flavor seems to lag or be delayed; it may not be apparent until the sample has melted. Usually, a musty off-flavor is most noticeable when the sample has been expectorated; the mouth definitely fails to clean-up.

The primary cause of musty off-flavored butter can be attributed to the growth of a specific spoilage microorganism (psychrotroph). Morgan (1976) reported that the production of 2-methoxy-3-alkylpyrazine by *Pseudomonas taetrolens* was apparently responsible for a distinct musty odor in cream. A musty off-flavor may also be the consequence of storing cream in a damp, musty-smelling space or poorly ventilated room. Improperly cleaned cream separators or unwashed cream cans (held for several days, with the lids in place in warm weather) may cause a musty off-flavor in stored cream. This can result in a musty flavor defect in the butter subsequently produced from the cream. This off-flavor is also associated with milk from cows that consume musty smelling feeds, slough grass, and stagnant water.

Neutralizer. The presence of a "neutralizer" off-flavor in butter can be observed immediately after the sample has melted in the mouth. However, this defect is often more readily perceived just after the sample has been expectorated and air is inhaled through the mouth. The aftertaste of added neutralizer in butter is persistent. This flavor note, depending on intensity, may be soda cracker-like or somewhat alkaline, suggestive of bicarbonate of soda or similar compounds. The soda neutralizers may also produce an associated bitter-like aftertaste, sometimes referred to as "limey." A strong neutralizer off-flavor in butter generally stems from the incorporation of highly concentrated solutions of neutralizer or excessive quantities of neutralizer (necessitated by high levels of formed lactic acid in the cream).

Old Cream. Cream which is fresh, sweet, clean, and without production or handling defects is certainly preferred for making butter that will exhibit a clean, appealing, "buttery" flavor. As cream ages, it seems to lose the desirable, delicately balanced flavor characteristics that should be transmitted to butter. After reaching several days of age, some cream sources will exhibit a typical "old cream" off-flavor, which usually carries through into the resultant butter. The old cream defect may also be caused by exposing cream to improperly washed cans and utensils, unclean storage and processing equipment and/or inadequate cooling rates. Lactic acid development frequently accompanies old cream off-flavor. Butter manufactured from old cream is characterized by staleness or lack of freshness and a characteristic aroma that is somewhat reminiscent of the unpleasant "background" odor noticed in a creamery or dairy plant that has not practiced the best sanitation. When a butter sample with this defect is first placed into the mouth, the flavor seems to lag, not making "an appearance" until the sample is melted. Usually, the old cream defect is most noticeable when the sample has been eliminated from the mouth; the off-flavor lingers and does not clean-up readily.

Oxidized. The oxidation of unsaturated fatty acids in dairy products causes a series of different off-flavors which are frequently grouped under the generic term "oxidized." However, since different flavor sensations are perceived in various samples or stages of development of oxidized butter, associative terms such as metallic, oily, tallowy, painty, and fishy have been used to describe the various observed defects. The butter score card (Fig. 9.1 and Appendix XI) currently approved by the Committee on Evaluation of Dairy Products of the American Dairy Science Association (1987) contains three different entries for oxidized off-flavor: metallic, oxidized, and tallowy. The term "oxidized" best describes the metal-induced form of oxidized flavor which is most common to milk and other dairy products. A cardboardy character and often an associated puckery mouthfeel are the usual distinguishing features. The oxidized off-flavor is probably most frequently noted as a "surfact taint" in butter, whereas the metallic and tallowy defects may be more commonly observed within the mass of a butter sample. Preventive measures for oxidized defects were discussed earlier in this text. The so-called "oily stage" of oxidized off-flavor in butter is quite uncommon in current butter supplies.

Rancid (Lipase). The "rancid" off-flavor of butter is strong, objectionable, soapy, and/or bitter. Rancidity of butter somewhat resembles the strong, disagreeable off-flavor of darkened, decayed nut meats. The odor is pungent and resembles that of volatile fatty acids, hence the odor may generally be noted from carefully smelling the contents of the withdrawn trier. When this off-flavor is prominent, the impression is one of soapiness and, frequently, pronounced bitterness. A rancid off-flavor is the result of hydrolysis of milkfat through the enzymatic action of lipase, which liberates fatty acids. A rancid off-flavor is attributed to the formation of free, short-chain fatty acids and salts of fatty acids; the latter are technically referred to as soaps.

Pasteurization of cream that contains high levels of free fatty acids does not eliminate the rancid off-flavor (Woo and Lindsay 1984). A characteristic of the rancid off-flavor (which is useful for recognition) is a certain astringent mouthfeel, perceived at the base of the tongue and upper throat. This mouthfeel persists after the sample has been expectorated. Those individuals who may have a relatively high threshold for the characteristic odor of fatty acids may still be able to recognize rancid butter by this mouthfeel sensation; otherwise they are advised to wait for the delayed bitterness and the unclean-like aftertaste.

Storage. Butter held for considerable time (months to several years) in cold storage may gradually undergo some deterioration of the protein and fat and/or absorb odors from the storeroom environment. Under these circumstances the delicate flavor characteristics of high-quality butter are lost; the consequent flavor deterioration is referred to as the "storage" defect. After extended storage, butter made from fresh, clean-flavored, sweet cream seems to undergo this chemical change much more slowly (exhibit less flavor deterioration) than butter that was made from lower-quality cream.

The particular off-flavor that results from this overall loss of product freshness is difficult to describe, since a storage off-flavor appears to be a composite of several deteriorative processes. The desirable sensory characteristics that are attributed to "product freshness" are distinctly absent in butter that exhibits the storage flavor defect. Even butter of the highest sensory quality will gradually deteriorate during storage, especially if odorous foods or materials are stored in close proximity to the butter or storage temperatures are too high.

Tallowy. As the term suggests, this off-flavor of butter resembles the odor and taste of tallow. There are varied intensities of this form of milkfat oxidation. A "tallowy" off-flavor may be detected by carefully noting the surface odor of the butter sample. The tallowy odor appears quite prominent immediately after the sample has been expectorated. In some extreme cases, butter manifesting a tallowy off-flavor also has a bleached color, especially on the surface layers.

A tallowy off-flavor is caused by an extensive degree of oxidation of the unsaturated fatty acids in milkfat. Its development is favored by holding butter at high storage temperatures in the presence of light, and by contamination with certain metallic salts (divalent cations). Copper and iron are the two metals most commonly involved. The addition of sufficient neutralizer to cream, to the point that it becomes alkaline, may accelerate oxidation of milkfat to the extent that it causes the resultant butter to become tallowy.

Butter with a tallowy off-flavor prevailing throughout the bulk of the block or cube has practically been eliminated. However, the surface of butter, particularly in consumer packages (where the ratio of surface to volume is high), rather frequently exhibits a tallowy or strongly oxidized defect. The logical term for this condition is "surface taint," since the butter in the interior may be of acceptable quality. Unfortunately, many years ago the cheesy-putrid flavor defect of butter, whether it occurred only on the surface or throughout the butter mass, was inappropriately referred to as surface taint. This dual use of terminology increases the possibilities of confusion and misunderstanding in interpretation, especially as related to cause, effect, and remedies.

Unclean/Utensil. As the term implies, the "unclean/utensil" off-flavor is characterized by an "off" flavor indicative of poor cream handling conditions and/or improper sanitary care of the storage and production equipment with which the cream and butter came in contact. Possibly, slow cooling rates of the milk or cream, and/or elevated storage temperatures, may have facilitated the outgrowth of spoilage bacteria (psychrotrophs), which produce end products that are responsible for causing this rather unpleasant off-flavor. Sometimes this flavor defect is referred to as an "unclean," "dishrag," or "dirty socks" off-flavor. This butter flavor defect manifests itself as a most unpleasant odor which intensifies as the sample is melted. This off-flavor persists for some time after the sample has been expectorated. The generic term "unclean" allows inclusion of butter flavor defects formerly described as "barny," "cowy," and "smothered," even though the occasional presence of those defects may have developed in a different way. Thus, barny may be an absorbed flavor from foul-smelling stables, and cowy may result from a physiological disturbance in the animal, while smothered is generally attributed to growth of psychrotrophic bacteria due to improper sanitation and/or delayed cooling of cream.

The term "utensil" still appears in the USDA grade classification, but its use should be discontinued. It represents an anachronism in that it is no longer relevant to current methods of cream handling and butter manufacture. Furthermore, this defect is caused by spoilage bacteria, not by "utensils."

Weedy. "Weedy" off-flavors in butter result from churning cream that has an absorbed weed taint, which sometimes occurs with seasonal feeding patterns. Some weeds are more common in early spring, when cows may be placed on weed-infested pastures (before grass is sufficiently developed to furnish sufficient sustenance), while other weeds seem more prevalent in late summer or fall. Specific weeds cause characteristic (often slightly bitter) off-flavors in butter, which are readily detected when encountered. Weed off-flavors are more pronounced after samples are heated to body temperature; usually the flavor note that is typical of the weed remains in the mouth after the sample has been expectorated.

In the past, a distinction was made between common and obnoxious weeds in identifying this defect. Obnoxious weeds are those that produce a particularly unpleasant flavor. The present practice is to allow the assigned sample score to reflect the degree of seriousness of the imparted off-flavor. Wild onion or wild garlic are examples of possible weed off-flavors.

Whey. Butter made from cream separated from cheese whey tends to exhibit flavor characteristics that are suggestive of the given type of cheese whey that was the source of the churning cream. The nature and intensity of the "whey" off-flavor depends on the freshness and quality of the whey, as well as the proportion of whey cream to regular cream that may have been blended to produce the butter. Since the so-called "whey flavor" of butter may not be familiar to some evaluators, practice with known or authentic samples is usually required to insure correct identification of this defect. A whey off-flavor is somewhat similar to the combined coarse/acid flavor defect of butter, plus an associated moderate odor and/or aftertaste suggestive of the given cheese whey. Some butter judges describe a particular "whang" or aftertaste for "whey" butter. A whey off-flavor is somewhat similar to the "old cream" defect; however, flavor notes of both "coarse" and "acid" are prevalent in this flavor defect. Some manufacturers label "whey cream butter" as "old-fashioned style" butter, or may employ another fanciful product name.

Yeasty. A "yeasty" off-flavor is detected in the early stages of development by its typical fruity, vinegary, yeasty, and slightly fragrant aroma, which is apparent when the sample is first taken into the mouth. As the sample melts, the odor becomes more and more distinctly yeasty (ethanol-like).

This flavor defect of butter occurs infrequently, but when it does happen, it is most often noted in butter produced during the hot summer months. By-products formed by yeasts that have grown in poorly handled, abused cream are responsible for this off-flavor. Old, yeasty cream may also impart a bitter flavor to the resultant butter. Comments from some evaluators are that yeasty butter has an off-flavor reminiscent of yeast-raised bread. A yeasty off-flavor is a serious defect, since the cream from which the butter was made had undergone considerable decomposition. Rejection of such cream before churning would be the desirable, sensible approach.

Other Butter Off-flavors. At various times, other off-flavors have been described in butter, and a few are summarized here. A "fruity" off-flavor may be the result of psychrotrophic bacterial growth (*Pseudomonas sp.*) in milk or cream.

"Stale" butter may result when cream of marginal or poor quality is held too long prior to churning. Faulty sanitation in production and handling will enhance a stale off-flavor. This flavor defect may be suggestive of both protein decomposition and fat oxidation, to some degree. This off-flavor is noted for both its foretaste and aftertaste. A type of staleness results from aging or storing unsalted butter that was made from unripened cream which had been held at room temperature. The stale defect develops slowly if the butter is held at 4.5°C (40°F), and even more slowly at lower cold-storage temperatures. It is not uncommon for a stale off-flavor to develop in unsalted butter after it is removed from frozen storage. Unfortunately, the stale off-flavor may develop into a cheesy off-flavor if the butter is held at temperatures approaching 15.6°C (60°F) or above. This particular off-flavor might be referred to as a form of the old cream defect.

A "woody" off-flavor, detected both by the tasting and smelling of butter, has become quite rare. This off-flavor becomes more pronounced toward the end of the tasting routine. The odor resembles the fragrant, sometimes piney, odor of a new wooden churn. The aromatic properties of this off-flavor vary from that of the fresh odor of new hardwood to the somewhat musty odor of water-soaked or partially decayed wood. This infrequent off-flavor may be caused by cream or butter that has absorbed a woody-like odor from a new wooden churn (practically extinct in the U.S.) which had not been properly treated prior to use, or from unlined or poorly lined wooden butter molds for bulk butter. Fortunately, wooden churns have been almost exclusively replaced by either aluminum alloy or stainless steel churns; a wooden churn would be an oddity in the U.S. today.

The feeding of strong-flavored vegetables and other strong-flavored feeds may cause the milk and the subsequent cream and butter to acquire the same flavor as that of the vegetable or feed consumed by the cows. The feeding of cabbage, turnips, potatoes, or rape within the period of 30 min. to 3 hr. prior to milking is likely to result in "tainted cream," the off-flavor of which is intensified in the resultant butter. These flavor notes are so typical of each vegetable that when encountered, they are easily recognized to both taste and smell. A "vegetable" off-flavor in butter is actually a form of the more commonly recognized feed flavor defect.

REFERENCES AND BIBLIOGRAPHY

American Dairy Science Association. 1987. Committee on Evaluation of Dairy Products. Champaign, IL.

Code of Federal Regulations. 1987. Title 7, U.S. Standards for Grades of Butter. Part 58, Subpart P, paragraphs 58.2621–58.2635. U.S. Government Printing Office. Washington, D.C.

Coulter, S. T. and Combs, W. B. 1936. A study of the body and texture of butter. Minn. Agr. Expt. Sta. Tech. Bul. No. 115. St. Paul, MN.

Hunziker, O. F. 1940. *The Butter Industry*. Published by the Author. LaGrange, IL. 821 pp.

McDowall, F. H. 1953. *The Buttermaker's Manual*. Vol. I and II. New Zealand University Press. Wellington, NZ. 1589 pp.

Morgan, M. E. 1970. Microbial flavor defects in dairy products and methods for their simulation. I. Malty flavor. *J. Dairy Sci. 53:*270.

Morgan, M. E. 1976. The chemistry of some microbiologically induced flavor defects in milk and dairy foods. *Biotechnol. and Bioeng. 18:*953.

Totman, C. C., McKay, G. L., and Larsen, C. 1939. *Butter*. John Wiley and Sons. New York. 472 pp.

United States Department of Agriculture. 1983. A summary of defects in USDA graded butter. USDA Food Safety and Quality Service. Washington, D.C.

United States Department of Agriculture. 1985. Details for butter grades and grading. USDA Food Safety and Quality Service. Washington, D.C.

Wilster, G. H. 1968. *Practical Buttermaking*. Oregon State University Bookstores, Inc. Corvallis, OR. 275 pp.

Wilster, G. H. 1958. Smooth spreading butter. *Milk Prod. J. 49*(4):10.

Wilster, G. H., Jones, I. R., and Haag, J. R. 1941. Crumbliness and stickiness in butter: Physical and chemical properties of the milkfat. *Nat. Butter and Cheese J. 32*(1):2.

Woo, A. H. and Lindsay, R. C. 1984. Characterization of lipase activity in cold-stored butter. *J. Dairy Sci. 67:*1194.

Zottola, E. A. 1958. Effect of certain manufacturing methods on the physical characteristics of butter. M.S. Thesis. Oregon State Univ. Corvallis, OR.

10

Sensory Evaluation of Concentrated and Dry Milk

The common characteristic of various types of concentrated milk products is the reduced water content. Generally, the water is removed by boiling under reduced atmospheric pressure (a partial vacuum) at relatively low temperatures (in the approximate range of 43°C–60°C [110°F–140°F]). The various products in this category vary with respect to: (1) the degree of concentration; (2) a skim milk or whole milk source; (3) whether preserved or perishable; and (4) the method of preservation (if preserved). Some forms of concentrated milks are intended for beverage consumption, while others are primarily used as ingredients for the formulation of various food products.

Currently, only small quantities of concentrated milk are produced for beverage use. Considerable research has contributed to our understanding of the technical problems encountered in freezing or sterilizing milk concentrates; however, consumers have not responded favorably to these product forms. With pasteurized fluid milk of high quality so readily available at reasonable prices, U.S. consumers tend to resist trying alternate forms of beverage milk. This consumer attitude could change in the future, when and if appropriate incentives for utilizing concentrated milks occur.

CONCENTRATED MILK PRODUCTS

The Grade A Pasteurized Milk Ordinance, 1978 recommendations of the U.S. Department of Health and Human Services (USDHHS), defines the following beverage quality, concentrated products:

Concentrated Milk. Concentrated milk is a fluid product, unsterilized and unsweetened, resulting from the removal of a considerable portion of the water from milk, which when combined with potable water in accordance with instructions printed on the container, results in a product conforming with the milkfat and milk-solids-not-fat levels of milk as defined in Title 21, CFR, Section 131.110.

Concentrated Milk Products. Concentrated milk products shall be taken to mean and to include homogenized concentrated milk, concentrated skim milk, concentrated lowfat milk, and similar concentrated products made from concentrated milk or concentrated skim milk, and which when combined with potable water in accordance with instructions printed on the container, conform with the definitions of the corresponding milk products as defined in the CFR.

Frozen Milk Concentrate. Frozen milk concentrate is a frozen milk with a composition of milkfat and milk-solids-not-fat in such proportions that when a given volume of concentrate is mixed with a given volume of water, the reconstituted product conforms to the milkfat and milk-solids-not-fat requirements of whole milk. In the manufacturing process, water may be used to adjust the primary concentrate to the final desired concentration. The adjusted primary concentrate is pasteurized, packaged and immediately frozen. This product is stored, transported, and sold in the frozen state.

Concentrated milk of beverage quality is defined also in Title 21, Code of Federal Regulations Section 131.115 (1987a) in part as follows:

Description. Concentrated milk is the liquid food obtained by partial removal of water from milk. The milkfat and total milk solids contents of the food are not less than 7.5 and 25.5 percent, respectively. It is pasteurized but is not processed by heat so as to prevent spoilage. It may be homogenized.

Vitamin Addition (Optional). If added, vitamin D shall be present in such quantity that each fluid ounce of the food contains 25 International Units (I.U.) thereof, within the limits of good manufacturing practice.

Optional Ingredients. The following safe and suitable optional ingredients may be used:

1. Carrier for vitamin D;
2. Characterizing flavoring ingredients with or without coloring as follows:
   a. Fruit and fruit juice, including concentrated fruit and fruit juice;
   b. Natural and artificial food flavoring.

The minimum prescribed composition of 7.5 percent milkfat and 25.5 percent total milk solids in concentrated milk implies a minimal concentration ratio of 2:1. Those milk products concentrated by a ratio of 3:1 incur greater processing cost; however, this is partially compensated for by lower distribution and packaging costs. In the latter instance, one container of milk concentrate would typically yield three equivalent size containers of beverage milk after the addition of water.

Concentrated milk intended for use as beverage milk is generally evaluated for sensory properties in a manner similar to its counterpart, unconcentrated or regular milk. For sensory evaluation, only good-quality potable water or distilled water should be used in diluting a milk concentrate back to the original composition. Milk concentrates may also be tasted without prior dilution. Sometimes, flavor perception may be more difficult, but with experience, most off-flavors can be detected (particularly the more serious defects). Since the background flavor of milk concentrates usually tastes sweeter and saltier than conventional, unconcentrated milks, dilution of the concentrate to the original composition provides a more typical set of test conditions for sensory evaluation of the products. Basically, the reconstituted product should emulate its counterpart in flavor, mouthfeel or consistency, and appearance. In addition to flavor defects, any visible evidence of discoloration, thinning, or thickening and other abnormalities should be noted as defects (Hammer 1919; Hunziker 1949; Summer and Hart 1926).

In the mid-1960s, considerable interest was generated in the market potential of a 3:1 sterile concentrated milk, although only relatively small quantities were actually produced. The major problems of sensory quality involved shortcomings of both taste and mouthfeel. The off-flavors which regularly developed in these products during storage were difficult to describe. Judges commonly labeled these off-flavors of sterile milk concentrates as "stale," "caramel-like," or a combination stale/caramel defect. This particular off-flavor could be associated with the browning reaction of heated milk (Arnold *et al.* 1958; Muck *et al.* 1963). Any possible future success of sterile milk concentrates would depend on processors' ability to prevent flavor deterioration during storage. Initially, objectionable product gelation posed a serious problem; however, preventive measures were subsequently discovered and successfully applied.

Descriptive terminology as applied to concentrated milk products is somewhat confusing; hence, a review of several key terms should be helpful. For example, what is the difference between concentrated, condensed, and evaporated milk when the products' composition in all three cases may be identical? Evaporated and concentrated milk are clearly defined in the CFR, as is sweetened consensed milk, but what kind of product is referred to by the term "unsweetened condensed milk?" This confusion may be eased somewhat if it is assumed that "evaporated milk" represents a special type of sterile concentrated product, of which the composition and processing are clearly defined. A reasonable suggestion (by the authors) is to reserve the term "concentrated" for *products of beverage quality* and use the word "condensed" when the milk product in question is primarily *intended as an ingredient* in cooking, baking, candymaking, or food manufacture.

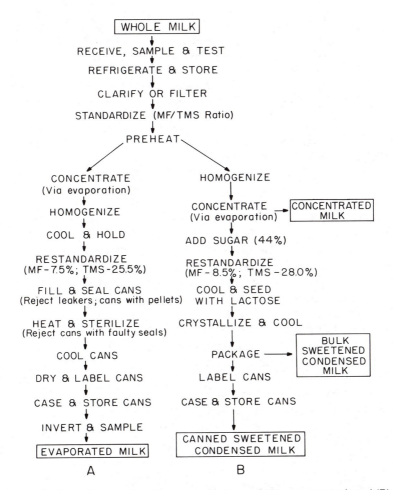

Fig. 10.1. Flow diagram for the manufacture of (A) evaporated and (B) condensed milks.

Evaporated Milk. Title 21 CFR, Section 131.130 (CFR 1987a) describes evaporated milk in part as follows:

Description. Evaporated milk is the liquid food obtained by partial removal of water only from milk. It contains not less than 7.5 percent by weight of milkfat and not less than 25 percent by weight of total milk solids. Evaporated milk contains added vitamin D as prescribed by paragraph (b) of this section. It is homogenized. It is sealed in a container and so processed by heat, either before or after sealing, as to prevent spoilage.

Vitamin Addition. Vitamin D shall be present in such quantity that each fluid ounce of the food contains 25 I.U. thereof within limits of good manufacturing practice. Addition of vitamin A is optional. If added, vitamin A shall be present in such quantity that each fluid ounce of the food contains not less than 125 I.U. thereof within limits of good manufacturing practice.

Optional Ingredients. The following safe and suitable ingredients may be used: (1) carriers for vitamins A and D; (2) emulsifiers; (3) stabilizers, with or without dioctyl sodium sulfosuccinate as a solubilizing agent (when permitted by and in compliance with stated provisions); and (4) characterizing flavoring ingredients, with or without coloring and nutritive carbohydrate sweeteners, as follows: fruit and fruit juice, including concentrated fruit and fruit juice; natural and artificial food flavoring.

In addition to meeting the legal composition and chemical requirements for the product, high-quality evaporated milk should be creamy in color, have a relatively viscous body, be uniformly smooth in texture and possess a relatively mild, pleasant flavor. Furthermore, the container should present an attractive appearance and exhibit a neat, well-applied label; and the ends of the can should appear polished and show no evidence of bulges or dents. The overall examination of the product includes flavor, body and texture (viscosity, break, gelation, and homogeneity), and appearance (color, fat separation, and serum separation). A comprehensive examination of evaporated milk should include the following tests and observations:

Coffee whitening properties	Flavor
Color	Gelation
Container	Sedimentation
Curd tension	Serum separation
Fat separation	Viscosity
Fill of container	Whipping quality
Film formation (protein "break")	

Objective tests which yield measurable data for concentrated milks, require the use of color charts, viscosimeters, electric mixers, and other special laboratory equipment. The subjective tests, based on sensory evaluation, may employ score cards or rating scales. A modified hedonic scale for quality evaluation of flavor in evaporated milk may take the following form:

Sensory Reaction and Scores for Relative Quality

Excellent—9	Slightly desirable—6
Very desirable—8	Neither desirable nor
Moderately desirable—7	undesirable—5

Slightly undesirable—4   Extremely undesirable—1
Moderately undesirable—3
Very undesirable—2

Examination Procedures for Evaporated Milk

Establishment of a definite routine for examining cans of evaporated milk can facilitate the evaluation of numerous samples. The steps outlined in the following paragraphs have been found most helpful in evaluating sample sets of evaporated milk.

Undue agitation should be avoided when cans of the product are transported to the laboratory. The product should be carried in an upright position and be placed vertically on the table to avoid remixing any possible precipitates (sediment) or fat layers into the product.

Examination of the can appearance should be done without lifting the can from the table. The upper end of the can should be noted for the degree of polish; the attractiveness of the label and the evenness of its application should be observed. The evaluator should insert a knife under the label and cut it from top to bottom. After partially or completely removing the label, the judge should note the condition or integrity of the can, especially with respect to freedom from rust spots.

With an edge-cutting can opener the evaluator should almost cut around the entire periphery of the upper end of the can and turn back the lid. By opening a can in this manner, both the container and the contents may be examined carefully.

Color. Evaporated milk should display a light, uniform cream color, but may tend toward a light brown color. In case of brown discoloration, the exact shade of the color may be determined either instrumentally or by visual comparison with color charts (using a numerical or graphical intensity scale), or by noting and recording the relative intensity of darkening as follows: −, none; +, slight; + +, distinct; and + + +, pronounced.

Uniformity. Evaporated milk should be uniform or homogeneous as evidenced by any absence of a cream layer, curd, or butter particles; there should be no evidence of any fat separation. Product uniformity may be more readily determined with the assistance of a spatula. Results of the examination for product uniformity may be verified when the product is examined for body and texture. In the macroscopic examination of the product for uniformity, the evaluator should notice particularly the under surface of the "turned-back" lid for possible adherence of cream or precipitated salts.

Study the Body and Texture. The contents of the can should be poured slowly into a clean glass beaker; the judge should note the flow properties of the product. A smooth, relatively viscous evaporated milk should pour in a similar manner to a thin cream (without marked splashing action). The can is allowed to drain completely; when the container is empty, the evaluator should look for any possible types of deposits on the can bottom. If the bottom metal surface cannot be seen through the remaining film of evaporated milk, the can bottom should be scraped with a spatula to determine whether a firm, tenacious deposit is present. The can is set aside for later examination; the observer should proceed with an examination of the evaporated milk for viscosity and texture. This is done by "spooning up" some of the milk with a plastic or hard-rubber spatula and allowing it to drip back into the beaker. The evaluator needs to note the relative thickness and uniformity of the film that adheres to the spatula.

An additional test for examining the "grain" of evaporated milk is to study a film of the milk through which a light source has been transmitted. By means of a 1.27 to 1.9 cm (1/2 to 3/4 in.) wire loop (or a cut-away spoon), the milk film is observed for surface "evenness" or uniformity. This is done by dipping the loop into the product and withdrawing it carefully to form a film across the face of the loop. Next, the milk film is held up to the light source and the observer looks for curd particles of pinpoint size. The appearance of small grains throughout the film indicates protein "break" or denaturation. If the milk appears rough, grainy, or lacks uniformity, these conditions may be associated with excessive viscosity and could also provoke the "feathering" defect in coffee.

Should evaporated milk "lack uniformity" of body/texture, the evaluator should try to determine the possible cause. Contributing factors may be "destabilized" fat or protein, or the presence of precipitated salts or foreign material. If de-emulsified milkfat is responsible, the defect generally will appear as a cream layer or as butter-like particles on the product surface. When denatured protein is the cause, the defect usually appears as either various-sized curds (distributed throughout) or as a form of gelation of different intensities. Salt deposits are responsible for formation of a hard, gritty precipitate which may have settled on the can bottom. Foreign material is the probable cause if the sediment is evident as a smudge-like discoloration on the can bottom; this is only evident when the last traces of the product contents are decanted.

Observe the Condition of the Container. The observer should especially look for either "spangling," blackening of the seams or container corrosion (rustiness). Next, the container should be rinsed and the inner surfaces observed for any evidence of chemical activity. Spangling refers to the appearance of alternate clean, bright and dark, overlapping blotches on the surface (as though the tin were attacked by acids). Typically, any such blotches are well-distributed over the inner surfaces of the can. Discoloration and rusting may occasionally be noted on any part of the can, but it tends to occur particularly at the milk-air interface.

Determine the Product Reaction in Coffee. Though use of evaporated milk as a coffee whitener has declined, there is still merit in checking its color reaction and miscibility in coffee. Evaporated milk should impart a rich, golden-brown color to coffee. The coloring power of evaporated milk may be readily determined by adding approximately 10 ml of the product to 100 ml of test coffee. Occurrence of an iron contamination of the product may be indicated by the development of a greenish-dark, muddy, slate-like discoloration in coffee. Thus, this off-color in an evaporated milk–coffee mixture can often be associated with container rust formation. "Feathering" in coffee is not a common defect, but protein denaturation may manifest itself as finely divided, serrated curds shortly after a susceptible evaporated milk has been added to extremely hot coffee.

Determine the Flavor. For flavor determination, evaporated milk should be mixed with distilled water in a 1:1 ratio. Sampling and flavoring is conducted by the same procedure employed in evaluating fluid milk. High-quality evaporated milk (made by a conventional process) tends to have a specific, "milk/cream flavor," which some individuals find reminiscent of a delicate, high-quality mushroom soup.

The evaluator should bear in mind that the source of added water might have an adverse effect on the flavor of evaporated milk. Some experienced judges of evaporated milk prefer direct tasting of the final sterile concentrate rather than evaluating a diluted product. This method of sensory evaluation requires keen perception, but it has the advantage of eliminating the "flavor diluting" effect of the water used for product reconstitution.

The declining demand in the U.S. for evaporated milk has served to discourage the development of product forms. Contemporary systems of sterilization are probably capable of producing sterile products of improved sensory characteristics, with flavor attributes comparable to regular pasteurized (market) milks, in the judgment of some authorities. There is still a remote possibility that various forms of concentrated, sterilized milk could gain a larger share of future milk markets in this country. Such products may be processed with new or as yet undeveloped technologies, as well as be packaged in a most unique manner. The body characteristics of conventionally processed evaporated milk have been markedly improved through the use of stabilizers that prevent physical separation during storage and help keep the product smooth and creamy throughout typical distribution cycles (Graham *et al.* 1981).

The evaluator should be aware that evaporated milk is intended to be a shelf-stable product; any evidence of bacterial growth or spoilage is unacceptable. The defects which will be subsequently discussed are the result of physical causes and/or chemical activity, which proceed in the absence of any viable microorganisms.

Specific Sensory Defects of Evaporated Milk

Flavor. The flavor defects which usually occur in evaporated milk are unlike those commonly encountered in fresh beverage milk, due to its concentration under vacuum (which removes volatile off-flavors), and to the extent of the applied heat of sterilization.

Probably the most common storage defect of evaporated milk results from the progressive age-darkening or browning of the product. No single term seems to describe this off-flavor adequately. Such terms as "old," "strong," slightly acid, sour, and "stale coffee" may suggest the nature of the defect. The term "caramel," which is probably suggested by the brownish milk color, is not appropriately descriptive in this instance; however, it does suggest the chemical origin of the off-flavor. A caramel flavor, as in certain confections, generally connotes a pleasant, appetizing taste sensation; however, this agreeable reaction is definitely lacking when this flavor occurs in evaporated milk. A caramel off-flavor is associated with the age-darkening of evaporated milk. When a caramelized sample is first placed into the mouth, the flavor sensation is not particularly different from that of normal evaporated milk, but soon a distinctly old or slightly acid off-flavor is evident. This flavor defect may persist for some time, even after the sample has been expectorated. This off-flavor may be accompanied by an odor that suggests staleness. The underlying taste reaction of age-darkened evaporated milk is "acidic." The extent of staleness is primarily a function of product age and storage temperature.

A study by Sundararajan *et al.* (1966) determined flavor changes which occurred during the storage of evaporated milk produced by the: (1) conventional (long-hold retort); (2) high temperature, short-time (HTST) (short-hold retort); and (3) aseptic (ultra-high temperature— UHT) methods of processing. These workers concluded that the type of heat processing had a significant effect on the initial flavor score. The aseptic process employed yielded the best-flavored product initially, and the flavor remained the best when the product was stored at 10°C (50°F) or 27°C (80.6°F) for about two months. After storage for one year, flavor scores of the HTST and aseptically-made products were similar, but the flavor of conventional evaporated milk was significantly lower in quality. Flavor ratings of the conventionally processed product scored the lowest of the three product forms throughout the storage study. These investigators employed a fluid milk score card with a 40 point scale for flavor. The evaporated milk samples were evaluated after appropriate dilution. The initial flavor of the conventionally manufactured product was described as "cooked" and "caramel." The off-flavors which developed during subsequent storage were variously described as acidy, stale, storage, bitter, astringent, and puckery (mouthfeel).

Body and Texture. Contemporary manufacture of evaporated milk is subject to adequate quality control; this has resulted in marked product uniformity from batch-to-batch, as well as among processors. Currently, fresh evaporated milk is remarkably free of body and texture defects. However, when evaporated milk is held for extended time periods, or under adverse conditions, some body and texture defects may be encountered, such as the following:

Buttery, fat separation	Grainy
Curdy	Low viscosity
Feathering	Sediment
Gassy	

Buttery, Fat Separation. The "buttery defect" appears as a 0.64 to 1.27 cm (1/4 to 1/2 in.) layer of heavy cream at the top of the can. The cream layer may be so dense and tenacious that it is not miscible with the remainder of the milk. Under such conditions, the shaken milk appears curdy with floating masses of creamy or buttery particles within a liquid of relatively low viscosity. Several alleged causes of this defect are: (1) inadequate homogenization; (2) high storage temperature; (3) extended storage period; and (4) improper handling while in storage. The incorporation of stabilizing agents has helped to control this serious defect. Consumers object to this defect, since such milk fails to pour readily and thus creates the suspicion that the product "may have spoiled." This body defect is not associated with any particular flavor defect. The occasionally noted thin film and/or surface streaks of cream are undesirable product features, but do not adversely alter the functionality of the product for the user.

Occasionally, discs of free fat, from 0.08 to 0.32 cm (1/32 to 1/8 in.) diameter may appear on the surface of evaporated milk; these rarely encountered droplets of hydrophobic milkfat in the product are referred to as "moon spots." The fat appears yellowish, crystal clear and as flattened spheres scattered sporadically on the surface. The defect seems to be associated with inadequate homogenization, destabilized protein, and low viscosity, which is probably accelerated by high-temperature storage. Such evaporated milk lacks the homogeneity of a high-quality product.

Curdy. "Curdy" evaporated milk may be noted by the presence of many coagulated particles interspersed throughout the milk or by a continuous mass of coagulum. This condition differs from the buttery defect in that it is associated more with protein than with fat. This defect is primarily due to the distinct low heat-coagulation point of the end product; such evaporated milk simply could not withstand the "severity" of the sterilization process. With modern processing and technical control, this defect is observed rarely. Nonetheless, a tendency toward age-gelation should be watched closely.

Feathering. The "feathering" of evaporated milk in hot coffee cannot be predicted by macroscopic examination; it must be ascertained by actually testing the milk sample in hot coffee. Such a test was proposed by Whitaker (1931), wherein he found that, upon examination of 52 cans of commercial evaporated milk, feathering in hot coffee was not a common defect (one-half century ago). In addressing another coffee-whitening problem, Mojonnier and Troy (1925) found that curd formation (when evaporated milk was added to coffee) was due entirely to excessive viscosity of the product.

Gassy. Fortunately, "gassy" evaporated milk is uncommon. This defect is manifested by bulged cans, and sometimes, by a hissing sound of escaping gas when the can is punctured on opening. This defect can be due occasionally to certain physical-chemical causes, but microbial fermentation is the more typical cause if gassiness occurs.

Grainy. "Graininess," like curdiness, is related to the relative heat stability of milk proteins. A grainy evaporated milk is one that lacks smoothness and uniformity throughout; such a product appears coarse. If this defect is present, a film across a loop or an open-bottom spoon will transmit light unevenly. Grainy evaporated milk is often associated with an excessively heavy, viscous body. The evaluator should recognize that grainy evaporated milk does not actually contain "grains" of sediment. The presence of curd particles of pinpoint size may be noted when a light source is transmitted through a film of the product; hence, the visible grain is indicative of protein "break" or denaturation.

Low Viscosity. A "low-viscosity" evaporated milk may be noted by its milk-like consistency; such milk lacks creaminess and pours from the container as readily as fresh milk. The viscosity of evaporated milk is related to heat stability. Highly stable milk and technical efforts to achieve high heat stability tend to produce low viscosity; by contrast, low heat stability leads to high viscosity in the finished product. The viscosity attained immediately after sterilization may change, depending on several factors (storage temperatures, especially). Thinning or thickening (even to the point of gelation) may occur as the result of product aging; this depends on such factors as solids content, preheating temperatures, type of sterilization process, milk quality, and initial viscosity. In conventional evaporated milk, the addition of stabilizers has simplified the control of viscosity.

Sediment. Sedimentation, as observed in evaporated milk, may be of two distinct kinds; each type of precipitation may arise from entirely different causes. The "sediment" resulting from the settling of somatic cells (leukocytes), denatured protein, and/or foreign material (of possible colloidal nature) is usually darker in color than the product itself. Since these forms of sediment are readily miscible, they may only be seen when an undisturbed can is emptied slowly. This infrequent defect is not readily experienced by the consumer, since evaporated milk is subject to some agitation, especially when decanted through small puncture holes in the can top.

The second type of sedimentation that may occur in evaporated milk results from the crystallization of specific calcium and magnesium salts as tricalcium phosphate $Ca_3(PO_4)_2$, magnesium phosphate $Mg_3(PO_4)_2$, and tricalcium citrate $Ca_3(C_6H_5O_7)_2$. This form of gritty-like sedimentation frequently accompanies the "aging" of evaporated milk. The rate at which crystals form seems to be influenced by the nature of the milk, conditions of manufacture, and storage temperature. Sato (1923), Mojonnier and Troy (1925), and Gould and Leininger (1947) found these white, gritty, sand-like particles to be chiefly lime salts of citric acid or tricalcium citrate, $Ca_3(C_6H_5O_7)_2 \cdot 4H_2O$. Their rather bland, chalky taste suggests a form of calcium salt. These crystals vary from the size of a pinpoint to the size of a wheat kernel. They are usually found on the container bottom and may be noted when the contents are emptied.

Color. The principal color defect of evaporated milk is "browning." This color defect results from the Maillard reaction which involves chemical interactions between lactose and milk proteins (and between their hydrolysis products) upon severe heat treatment and subsequent storage. Numerous flavor compounds are also produced during the course of the browning reaction, which in this case leads to characteristic flavor defects. The degree or intensity of the brown discoloration is related to the intensity (time and temperature) of the sterilization process and the storage temperature. Aseptic and HTST sterilization systems generally yield a lighter-colored product than the conventional retort (long hold) process. However, additional darkening may occur during storage in all cases, as a function of age and the storage temperature of the product. There is evidence that the color produced during storage is qualitatively different from that imparted by the heat of sterilization (Webb and Holm 1930).

SWEETENED CONDENSED MILK

Description. A description of sweetened condensed milk can be found in 21 CFR 131.120 (CFR 1987a):

> Sweetened condensed milk is the food obtained by partial removal of water only from a mixture of milk and safe and suitable nutritive carbohydrate sweeteners. The finished food contains not less than 8% by weight of milk-fat, and not less than 28% by weight of total milk solids. The quantity of nutritive carbohydrate sweetener used is sufficient to prevent spoilage. The food is pasteurized and may be homogenized.
> *Optional Ingredients.* The following safe and suitable characterizing flavoring ingredients, with or without coloring and nutritive carbohydrate sweeteners, may be used:
> Fruit and fruit juice, including concentrated fruit and fruit juice.
> Natural and artificial food flavoring.

Since sweetened condensed milk contains a sufficiently high percentage of sugar for preservation, the flavor sensation is predominantly (or overwhelmingly) sweet. However, beyond this intense sweetness, the flavor of this dairy product should be clean and pleasant with a slight or trace aftertaste of milk caramel. The body of the product should be smooth and uniform; the color should be a light, translucent-yellow.

Whether sweetened condensed milk is used in the home kitchen or in a food processing plant, its primary function is as an ingredient in candy, cookies, pies, and ice cream—not as a beverage. Hence, its sensory properties are nearly exclusively evaluated in the research or quality control laboratories of processors. Careful consideration must be given to the functional properties of this product, but sensory characteristics are also important in the overall process of the quality evaluation of sweetened condensed milk.

Examination Procedures for Sweetened Condensed Milk

The particular precautions and steps that were applicable in the evaluation of evaporated milk are not as germane to the examination of sweetened condensed milk (in consumer-size containers). However, a specified routine enables the evaluator to best utilize the available time, with greater assurance that the examination is complete. Hence, the following recommended procedure can be helpful.

The evaluator should place a representative container on a table for examination. The can should be in exactly the same (upright) position that it had assumed prior to examination. This readily enables the judge to open the container and make an initial examination of the top surface and product contents. Next, the evaluator should cut and turn back the container lid so that the milk surface may be closely examined and the contents easily decanted from the container. The recommended order of visual examination is listed in the following paragraphs.

Appearance of the Container. The sweetened condensed milk container should appear to be in good condition. Since the container has not been subjected to high heat treatment, as in retorting (which dulls container surfaces), the can ends should be as bright as new tin. It is advisable that the evaluator develop the habit of carefully scrutinizing or observing the relative condition of all containers.

Appearance of the Product Surface. The product surface should have the same color intensity as various underlayers of the condensed milk. The product should be uniform in consistency with no indication of lumps, free fat, or film formation.

Color. With a spatula, the judge should "spoon up" some of the product and note the relative translucency of a milk layer. The color should be uniform throughout, rather than have a lighter-colored layer at the container bottom. The evaluator should determine whether the sweetened condensed milk has a creamy, or a less desirable brownish color.

Viscosity. Next, the evaluator should tilt the container, and note the relative ease with which the product tends to "seek its own level." The product is poured into a beaker. The observed pouring (flow) characteristics should resemble those of a medium-heavy molasses; this is indicative of the desired viscosity. There definitely should be no indication of a gel or custard-like formation. Flow characteristics (viscosity) can also be determined objectively by physical measurement.

Sediment. After the can has been emptied, the evaluator should scrape the bottom and note the presence or absence of a thickened layer (which may be a crystalline, granular material). The color of the granules should be compared with the bulk of the milk and the size of any precipitated crystals measured against any suspended in the liquid.

Flavor. After the above macroscopic examination has been completed, the judge should note the flavor characteristics. A small teaspoonful of the sweetened condensed milk should be placed into the mouth; the evaluator needs to observe both the mouthfeel and taste sensations. The relative smoothness of the product and the grain fineness can be noted by pressing some of the sample against the palate with the tongue. By this time, the evaluator may have experienced a secondary taste reaction—a perceived flavor other than sweetness. This "delayed" flavor note usually represents a blend of the sensory perception of the added sugar and dairy ingredients.

Defects of Sweetened Condensed Milk

Flavor. Sweetened condensed milk, due to its concentration under vacuum, tends to have none of the volatile off-flavors which may occur in fresh milk. Since this product is preserved by sugar rather than by heat, it should not exhibit those off-flavors that result from the higher heat treatments applicable to evaporated milk and certain other milk products. Hence, when this product is properly manufactured, it is remarkably free of flavor defects. However, several off-flavors in sweetened condensed milk have been noted to develop with increased storage time, as indicated below:

Metallic   Strong
Rancid   Tallowy

Metallic. The "metallic" off-flavor of sweetened condensed milk distinctly is chemically induced; it is usually traceable to copper contamination. Hunziker (1949) stated that "sweetened condensed milk may have a pronounced, disagreeable metallic flavor—suggesting the puckery, copper-like taste of copper salts." Copper contamination should be encountered infrequently due to near ubiquitous use of stainless steel equipment.

Rancid. Fortunately, a "rancid" off-flavor occurs most infrequently in sweetened condensed milk. As discussed earlier, rancidity results from milkfat hydrolysis due to enzymes secreted by spoilage bacteria or indigenous milk lipase, which may not have been heat inactivated. If the milk source was rancid, the peculiar, offensive odor associated with hydrolytic rancidity may be readily noted when the can is first opened.

Strong. The term "strong" or "strong-caramel" is often used to describe the off-flavor that accompanies the progressive thickening and browning of condensed milk. While this particular flavor sensation must be classified as a defect, it is not usually a serious one. Unfortunately, a caked or gelled product with its associated deep brown color often suggests that the product may have undergone serious flavor impairment. However, such milk occasionally may develop a rather pleasant caramel-like taste. When this off-flavor is not accompanied by a strong after-taste, this defect is not considered that serious or objectionable.

Tallowy. Rice (1926) observed in the instance of "tallowy" condensed milk that "on opening a tin, the sample appears sometimes, but not always, a little paler than normal. The tallowy off-flavor of the freshly opened sample remains even after exposure to the air for several days." Tallowiness has become a rarely encountered oxidation defect in sweetened condensed milk. Elimination of copper contamination and prevention of exposure of milk to light and air are the most likely reasons why this off-flavor is practically extinct.

Off-flavors Caused by Microorganisms. Certain osmophilic microorganisms, including yeasts, molds, and bacteria can tolerate high sugar concentrations and under certain conditions can cause spoilage in sweetened condensed milk. The growth of these microorganisms may be accompanied by characteristic physical and appearance changes, gas production, off-flavors, and odors. Depending on the type of microorganism involved, the resultant odor may be acidic, stale, cheesy, unclean, or yeasty. Any products that show evidence of microbial activity should be considered unsalable.

Body and Texture. Due to the relatively high percentage of sugar required for preservation, sweetened condensed milk exhibits a relatively heavy body (somewhat like molasses). Also, this product usually has a fine-grained, smooth, and uniform texture. However, the following body and texture defects may be encountered:

Buttons	Sandy (rough, grainy, granular)
Lumpy	Settled
Fat separation	Thickened
Gassy	

Buttons. Although they generally change the consistency of a portion of the product, formed "buttons" are visually observed as round, firm, cheesy curds at the product surface. These buttons result from the proteolytic activity of certain molds. Product losses due to button formation can be eliminated by preventing contamination by molds and other microorganisms.

Lumpy. Occasionally, a product may exhibit pronounced differences in viscosity ("lumpiness") within portions of the container contents. Sometimes, portions of the product may have actually gelled. It should be determined whether this problem is due to possible microbiological contamination or some other cause.

Fat Separation. "Fat separation" in sweetened condensed milk seldom occurs. This defect may be noted by either an off-color, fatty film at the surface and/or floating droplets of free fat.

Gassy. Condensed milk that has developed gassiness may be recognized by a "bloated" or "huffed" can. This defect results from contamination by and subsequent outgrowth of gas-producing microorganisms. Hammer (1919), studied the formation of gas in sweetened condensed milk and found the causative agent to be a yeast, which he named *Torula lactis condensi.* A yeasty odor was associated with this gaseous condition. Today, the defect is rarely noted.

Sandy (Rough, Grainy, Granular). All of these terms are used interchangably to describe sweetened condensed milk which contains detectable or oversized lactose crystals. The solid lactose particles are sufficiently large enough to impart a distinct grittiness and general lack of product smoothness, which is readily noticeable as the sample is tasted. This defect can be readily detected by the consumer. The condition referred to as "sandiness" is due to the presence of relatively large lactose crystals. So-called "smooth" condensed milk has minute-sized lactose crystals, which seem to appear like a fine "flour" mixed into the condensed milk. If manufacturing conditions are not conducive to the formation of small lactose crystals, then large, coarse crystals are likely to form (sandiness). The sandy defect may also be caused by sucrose crystals, when the concentration of this sugar exceeds the saturation level.

Settled. The term "settled" is used to describe a condensed milk in which a distinct settling of sugar crystals has occurred. The syrup which settles out forms a thick sugary layer on the container bottom. This sugar sediment consists primarily of lactose crystals, according to Hunziker (1949). Key measures for prevention of this are efforts to insure small crystals and development of an adequate product viscosity to retard sedimentation.

Thickened. "Overly thickened" condensed milk is one of the more common defects that can be encountered in sweetened condensed milk. The defect is manifested by an extensive gel formation, which leads to a product appearance more suggestive of a solid than a liquid. Excessively thickened condensed milk is usually associated with browning; both undesirable conditions become progressively more intense upon additional storage (especially at elevated temperatures). This defect varies markedly in intensity from a slight jelly to a firm custard consistency. A high-quality sweetened condensed milk should pour like molasses. When the product is poured, it should seek its own level and leave no traces of "folds" on the surface. The formation of a gel, even a soft gel, is not desired. Both physical and chemical factors are commonly responsible for thickening of sweetened condensed milk, but certain microorganisms may also cause product thickening.

OTHER CONCENTRATED MILK PRODUCTS

The evaluation of other concentrated milk products differs little from that of the products just described. Evaporated skim milk and sweetened condensed skim milk should be evaluated in a similar manner to their fat-containing counterparts. Obviously, one must allow for the absence of fat in evaluating both the flavor and tactile properties. Some products are produced to provide certain functional properties for specific applications. A good example is superheated condensed milk (or skim milk) for use as a milk ingredient in ice cream manufacture. This product should possess a desirable flavor and an appealing color, as well as impart the desired "bodying" properties to ice cream. Instrumental measurements of viscosity should supplement sensory-derived assessments of product consistency. As a general principle, when a concentrated milk product is intended for beverage purposes, sensory evaluation should ascertain how closely the product quality approaches that of its unconcentrated, high-quality, fresh milk counterpart. When a concentrated milk is used as an ingredient, the primary question becomes "how will it influence the quality and desired properties of the end product?"

DRY MILK PRODUCTS

Since its commercial origin, dry milk has been graded on the basis of bacteria, moisture, and certain physicochemical properties. More recently, flavor and other sensory properties have become important criteria in grading dry milk products. Despite the possible attainment of perfection in physical and chemical properties, dry milk must also have good flavor characteristics if it is to gain consumer acceptance. The relative importance of flavor characteristics is governed to a large extent by the intended use of the product. The evaluator of dry milk should be familiar not only with the product standards and the associated laboratory tests, but also with the appropriate flavor standards and potential flavor defects.

Methods of Drying Milk

There have been two principal methods of drying milk: (1) the roller process (nearly extinct in the U.S.), and (2) the spray process. Numerous technical developments in milk drying have materially improved certain properties of dried milk and facilitated the drying of several milk product forms which would not have been possible otherwise. One development has served to virtually revolutionize the drying of nonfat milk—the process known as *instantizing*. This involves hydrating and dehydrating previously dried milk to attain a more soluble particle form. Other drying processes include foam drying, freeze-drying, and fluidized-bed drying, although these methods have had a greater impact on foods other than dairy products.

The actual concentration of fluid milk that occurs at the instant of drying and the type of drying process substantially influence the physicochemical properties of the resultant dry milk. Thus, certain qualities of the finished product provide clues to the method of product manufacture. A descriptive outline of several milk-drying methods and some characteristic qualities of the respective dry products are given in the following paragraphs:

Atmospheric Roller. In this process, milk is dried in the open air on the surface of revolving, internally heated drums. The dried milk film is shaved from the drums and pulverized. The end product is characterized by a relatively heavy body, coarse texture, and comparative insolubility when it is initially added to water. Under the microscope, the solid particles appear angular, flaky, and irregular; seldom are spherical-shaped grains or particles noted.

Vacuum Drum. This drying process is similar to the atmospheric roller process except that the drum rolls are enclosed within a vacuum chamber and thus permit drying at a lower temperature. This is advantageous from a product quality standpoint. Vacuum-drum-dried powder readily solubilizes when added to water (similar to spray-process powder), but it may be easily distinguished from the latter by its appearance under the microscope. Grains of spray-process powder are generally spherical, whereas particles from the vacuum drum process tend to be distinctly angular and fragmented.

Spray Process. In this process, milk is sprayed under high pressure into a current of filtered, heated air in a drying chamber. Spray-process powder is fine grained, rather fluffy, and readily soluble. Under the microscope, the grains appear bead-like or spherical, and are of relatively uniform size.

Instantizing. "Instantizing" is a unique modification of the spray process of drying, which is generally applied to the drying of nonfat milk, but the process may also be adapted to whole or lowfat milks. The instantizing process substantially increases the particle size of the given milk powder, which significantly minimizes the tendency to "ball-up" when dried milk is mixed with water. This markedly improves the dispersibility and reliquification characteristics of dried milks. Since the introduction of instantized milk products in the 1950s, a number of patents have been issued that cover two basic processes, the two-step and one-step processes (Graham *et al.* 1981; Hall and Hedrick 1971). The two-step process, which appears to be the most commonly employed method, consists of bringing previously spray-dried milk in contact with water or steam (under appropriate conditions). The moistened particles adhere to each other and form a distinctly porous, agglomerated particle of larger size, which is then redried to the desired moisture content. In a typical one-step instantizing process, the drying is conducted in such a manner to enhance particle clustering. The larger agglomerates that are formed are subsequently separated and the final drying step occurs in a secondary dryer.

Foam Drying. In "foam drying," the product is dried after a liquid slurry is converted to a foam state. Two basic processes can be applied: (1) foam drying, and (2) foam-spray drying. In the former process, a nitrogen-gassed, whole milk concentrate (50% solids) is initially foamed, and the foam is then applied to a continuous belt that leads into a vacuum-drying chamber. In the foam-spray drying method, compressed air is injected into concentrated milk through a mixing device, which is located between a pressure pump and the spray nozzle. The gas-injected milk subsequently forms a foam upon sudden ejection into a heated air chamber. The thin air-cell films that are formed tend to dry as a fragile, eggshell-type of particle.

Freeze Drying. "Freeze drying" consists of removing moisture from a frozen product by sublimation under high vacuum. A food product dried by this method retains many of its initial, natural qualities. However, freeze drying and some of the other drying processes have enjoyed only limited application to dairy products. This is due primarily to rather substantial economic constraints related to energy inputs compared to spray processes for milk drying.

Kinds of Dry Milk Products

The various dry milk products may be enumerated as follows:

Dry whole milk	Lowfat dry milk
Nonfat dry milk	Dry cream
Dry buttermilk	Dry ice cream or ice milk mix
Dry whey	Malted milk
Edible dry casein	Miscellaneous dry milk products.

Dry whole milk and nonfat dry milk have standards of identity promulgated by the Food and Drug Administration (CFR 1987) and quality standards (administered on a voluntary basis) by the U.S. Department of Agriculture (CFR 1987b; USDA 1986). Only, USDA quality standards (CFR 1987b) exist for dry buttermilk, dry whey, and edible dry casein. Lowfat dry milk and dry cream have applicable FDA standards of identity. No standards of identity or USDA quality standards exist for the other dry-milk-based products listed above. Occasionally, state or local regulations apply to the manufacture and use of these dried milk products. In certain instances, a definition may not exist for the dry form of a product, but when it is reconstituted, the final product may have to comply with the definitions of its liquid counterpart. For example, dried ice cream mix has no definition (or standard of identity), but ice cream does. When dehydrated products are made into and sold as ice cream or ice milk, the final product form must comply with the existing regulations that pertain to the respective type of frozen dairy dessert.

In the ensuing discussion, the major emphasis will be placed on the sensory properties of dried milk products, although some details or other pertinent facts will also be provided. Some limited information from the Code of Federal Regulations and several other documents related to dried milk will be cited. Since federal regulations may change from year to year, the reader is urged to consult the most recent edition(s) of the Code of Federal Regulations (CFR) for authoritative and current information. Absolute compliance with USDA quality standards does not excuse failure to comply with certain rigorous provisions of the Federal Food, Drug and Cosmetic Act.

DRY WHOLE MILK

The Food and Drug Administration has defined dry whole milk in 21 CFR 131.147 (CFR 1987a) as follows:

> *Description.* Dry whole milk is the product obtained by removal of water only from pasteurized milk, as defined in §131.110(a), which may have been homogenized. Alternatively, dry whole milk may be obtained by blending fluid, condensed, or dried nonfat milk with liquid or dried cream or with fluid, condensed, or dried milks as appropriate, provided the resulting dry whole milk is equivalent in composition to that obtained by the method described in the first sentence of this paragraph. It contains the lactose, milk proteins, milkfat and milk minerals in the same relative proportions as the milk from which it was made. It contains not less than 26% but less than 40% by weight of milkfat on an as is basis. It contains not more than 5% by weight of moisture on a milk-solids-not-fat basis.

Other provisions include the optional addition of vitamins A and D (when added, the content is regulated) and incorporation of the following safe and suitable optional ingredients: carriers for vitamins A and D, emulsifiers, stabilizers, anticaking agents, antioxidants, characterizing flavoring ingredients with or without coloring and nutritive carbohydrate sweeteners (including fruit, fruit juice, fruit juice concentrates, and natural and artificial food flavoring).

Grading standards of the USDA are published in Title 7, CFR, in paragraphs 58.2701–58.2710. They pertain primarily to basic dry whole milk, which optionally may be fortified with vitamins A and D, or both vitamins. Two USDA grades are recognized: (1) U.S. Extra Grade, and (2) U.S. Standard Grade. They are determined on the combined basis of flavor, physical appearance, bacterial estimate, coliform estimate, direct microscopic count, milkfat content, moisture content, scorched particle content, and solubility index. Tables 10.1, 10.2, and 10.3 summarize the requirements for the above two grades of dry whole milk. Definitions of the terms used in these tables will be discussed in a later segment of this chapter.

Testing for certain other quality parameters may also be done at the option of the USDA (1986) or when examination is requested by an interested party. These optional quality parameters (requirements) are:

(1) Copper content—not more than 1.5 ppm.
(2) Iron content—not more than 10 ppm.
(3) Titratable acidity—not more than 0.15%.

Table 10.1. U.S. Grade Classifications Based on Flavor and Odor of Dry Whole Milk (Reliquified Basis).

Flavor Characteristics[a]	U.S. Extra Grade	U.S. Standard Grade
Cooked	Definite	Definite
Feed	Slight	Definite
Bitter		Slight
Oxidized		Slight
Scorched		Slight
Stale		Slight
Storage		Slight

Source: CFR (1987b).
[a] The flavor applies to the reconstituted product. In general, it shall be sweet, pleasing, and free of undesirable flavors. It may contain off-flavors as indicated.

Table 10.2. U.S. Grade Classifications of Dry Whole Milk Based on Physical Appearance Characteristics.

Physical Appearance Characteristics[a]	U.S. Extra Grade	U.S. Standard Grade
Dry product:		
Unnatural color	None	Slight
Lumps (break up under:)	Slight pressure	Moderate pressure
Visible dark particles	Practically free	Reasonably free
Reconstituted (reliquified) product:		
Grainy	Free	Reasonably free

Source: CFR (1987b).
[a] In general, the dry product shall be a white or light cream color, and shall have other characteristics as indicated.

Table 10.3. U.S. Grade Classifications of Dry Whole Milk According to Laboratory Analyses.

Laboratory Tests (or Parameters)	U.S. Extra Grade	U.S. Standard Grade
Bacterial estimate, SPC/g	$\leq 50{,}000$	$\leq 100{,}000$
Coliform estimate/g	≤ 10	≤ 10
Milkfat content, %	Not less than 26.0 but less than 40.0	Not less than 26.0 but less than 40.0
Moisture content, %[a]	≤ 4.5	≤ 5.0
Scorched particle content, mg:		
Spray process	≤ 15.0	≤ 22.5
Roller process	≤ 22.5	≤ 32.5
Solubility index, ml:		
Spray process	≤ 1.0	≤ 1.5
Roller process	≤ 15.0	≤ 15.0

Source: CFR (1987b).
[a] Milk-solids-not-fat basis.

(4) When vitamins are added:
    vitamin A—not less than 2000 IU/qt.
    vitamin D—not less than 400 IU/qt.

Failure to meet "standard grade" or optional quality requirements (when the tests are performed), or a direct microscopic clump count in excess of 100 million/g, suffice to deny a given product the assignment of a USDA grade. Deficiencies in so-called "good manufacturing practices" by a processor may also disqualify appropriate products from eligibility for USDA grade assignment.

Procedures for all of the applicable quality tests may be found in *Methods of Laboratory Analysis*, DA Instruction No. 918-103 (dry milk products series), Dairy Grading Branch, AMS, U.S. Department of Agriculture, Washington, D.C. 20250, as well as in *Official Methods of Analysis of the Association of Official Analytical Chemists*, 13th Edition, or the latest revision.

Flavor Properties of Dry Whole Milk

Flavor. Ideally, dry whole milk should have flavor characteristics which are clean, rich, sweet, fresh, and pleasant, not unlike that of fine pastry. Sensory defects may be due to either poor-quality raw material, handling and processing of the fluid milk, the drying method, and extended or inappropriate storage. The development of storage defects in dry whole milks are most difficult to control or eliminate. The more serious quality defects encountered in dry whole milk are scorched, stale, and oxidized.

Scorched. A "scorched" off-flavor is likely to occur in those products which have been subjected to excessive heat (during the drying stage) or have remained in the drying chamber too long. This product defect is usually accompanied by numerous scorched particles; sometimes dark discoloration occurs, hence, suggestive of overheating of product particles in processing.

Stale. A "stale" off-flavor develops during storage, even in products that have been nitrogen packed and/or contain an extremely low oxygen concentration in the headspace of the container. Dry whole milks stored with a moderately high level of oxygen in the headspace generally develop an oxidized off-flavor. The discovery of effective preventive measures against the development of a stale off-flavor has been eluding researchers for decades. However, if and when a remedy is found, dry whole milk could conceivably make a giant step forward in general consumer acceptance.

Oxidized, Tallowy. The "oxidized, tallowy" off-flavor is certainly a troublesome sensory defect of dry whole milk. This off-flavor, suggestive of old tallow, simply makes much dry whole milk unpalatable. Frequently, various stages of oxidation may be noted. Numerous factors seem to affect the development and rate of oxidation; notably, (1) temperature; (2) light; (3) product acidity; (4) metallic salts; (5) moisture condensation; (6) headspace oxygen content; and (7) the type of packaging.

Other Properties of Dry Whole Milk

Tactual properties of dry whole milk vary with the method of manufacture, the degree of concentration prior to drying, and the particle size (Hall and Hedrick 1971; Hunziker 1949). Dry whole milk manufactured by the spray process may be extremely fine and uniform throughout, but two powder defects may occasionally be noted; lumpy and caked.

Lumpy. Lumpy powder definitely lacks homogeneity in appearance. Hard lumps the size of wheat grains or larger may form. This defect is found more frequently in spray-process forms of dry milk product. The lumps can result from insufficient drying, dripping spray nozzles or particle exposure to moisture-laden air. Dry whole milk, because of its relatively high fat content, may contain so-called "soft lumps." This condition is particularly characteristic of cold-stored products. It stems from the agglomeration of powder particles. This defect should not be confused with a "hard lumpy" product, wherein the formed particles (lumps) feel firm when they are pressed between the fingers.

Caked. Usually, the "caked" defect is not encountered in dry whole milk. However, when it does occur, dry whole milk loses it powdery consistency and becomes "solid as a rock." When this solid mass is broken up, the product remains as chunks, and thus fails to regain the original powdery state. This defect is considered most serious, since such an altered dry milk has lost sales value for human use.

Color. Dry whole milk is typically light yellow in color, but it can vary seasonally with the amount of carotene in the milkfat. The color can range from a creamy white to a deep yellow.

The possible defects of color in dry whole milk are:

Browned or darkened   Lack of uniformity
Scorched

Browned or Darkened. This color defect of dry whole milk is associated with product age. When this defect occurs, the typical creamy color has been replaced by a distinct brown shade. Furthermore, this color defect is usually associated with a distinctive old or stale off-flavor.

Scorched. Discoloration due to burning (scorching) of milk solids is more commonly associated with roller-processed powders than spray-processed products. The powder color may vary from light to dark brown; rarely will burnt particies be so dark as to appear black. Milk powders that exhibit discolored particles or foreign sediment are severely discriminated against.

Lack of Uniformity. This defect may be due to either partial discoloration (browning) that may develop after product packaging or it may be the result of partial scorching during the manufacturing process.

NONFAT DRY MILK (NDM)

The Food and Drug Administration has two definitions for nonfat dry milk, as noted in 21 CFR 131.125 and 131.127. The only difference in the second definition is that the product is fortified with vitamins A and D. Nonfat dry milk (NDM) is defined as follows:

Description. Nonfat dry milk is the product obtained by removal of water only from pasteurized skim milk. It contains not more than 5% by weight of moisture and not more than 1.5% by weight of milkfat unless otherwise indicated.

Optional ingredients. Safe and suitable characterizing flavoring ingredients (with or without coloring and nutritive carbohydrate sweetener) as follows:

Fruit and fruit juice, including concentrated fruit and fruit juice.

Natural and artificial food flavorings.

The following is the additional language for nonfat dry milk fortified with vitamins A and D:

Vitamin addition. (1) Vitamin A is added in such quantity that when prepared according to label directions, each quart of the reconstituted product contains 2000 I.U. thereof.

(2) Vitamin D is added in such quantity that when prepared according to label directions, each quart of the reconstituted product contains 400 I.U. thereof.

(3) The requirements of this paragraph will be deemed to have been met if reasonable averages within limits of good manufacturing practice are present to ensure that the required levels of vitamins are maintained throughout the expected shelf-life of the food under customary conditions of distribution.

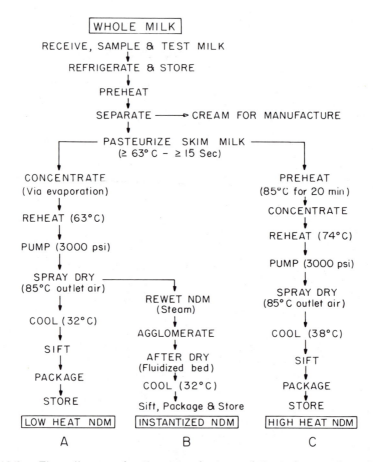

Fig. 10.2. Flow diagram for the manufacture of three forms of nonfat dry milk: (A) low heat, (B) instantized, and (C) high heat.

The USDA has quality standards for three types of NDM; namely, spray-process, roller-process, and instant. These may be found in Title 7, CFR, paragraphs 58.2525, 58.2551, and 58.2750, respectively. A summary of the requirements is given in Tables 10.4, 10.5, and 10.6. The products covered by these standards must not contain buttermilk or any added preservative, neutralizing agent, or other chemical. Conditions under which a "U.S. Grade" is not assignable vary for the different types of nonfat milk. Only the "Extra Grade" is recognized for instant nonfat milk. For spray- and roller-process nonfat milk, failure to meet the requirements for U.S. Standard Grade and/or a direct microscopic clump count in excess of 100 million/g results in nonassignment of a grade. One or more of three reasons may be responsible for

Table 10.4. U.S. Grade Classifications of Nonfat Dry Milk (Reliquified Basis) Based on Flavor and Odor.[a]

Flavor Characteristics	Flavor Classification	
	U.S. Extra Grade	U.S. Standard Grade[b]
Bitter		Slight
Chalky	Slight	Definite
Cooked (Spray and Instant)	Slight	Definite
Feed	Slight	Definite
Flat	Slight	Definite
Oxidized		Slight
Scorched		
Roller	Slight	Definite
Spray and Instant		Slight
Stale		Slight
Storage		Slight
Utensil		Slight

Source: CFR (1987b).

[a] In general, the flavor shall be sweet, pleasing, and desirable, but certain off-flavors are permitted in intensities as indicated.

[b] Applies only to spray and roller process. Only one grade, "U.S. Extra," is recognized for instant nonfat dry milk.

Table 10.5. U.S. Grade Classifications of Nonfat Dry Milk Based on Physical Appearance Characteristics.

Physical Appearance Characteristics[a]	Classification	
	U.S. Extra Grade	U.S. Standard Grade[b]
Dry Product		
Lumpy	Very slight[c]	Slight
Unnatural color		Slight
Visible dark particles		
Spray	Very slight	Slight
Roller	Slight	Definite
Instant		
Reliquified		
Grainy		
Spray		Slight
Roller	Slight	Slight
Instant		

Source: CFR (1987b).

[a] In general, the dry product shall be white or light cream in color and shall not exceed the intenstities of other characteristics as indicated.

[b] Applies only to spray and roller process. Only one Grade, "U.S. Extra" is recognized for instant nonfat dry milk.

[c] Instant product must be reasonably free-flowing (i.e., pours in a fairly constant, uniform stream from the open end of a tilted container or scoop).

Table 10.6. U.S. Grade Classifications of Nonfat Dry Milk According to Laboratory Analyses.

	Classification[a]	
Laboratory Tests (or Parameters)	U.S. Extra Grade	U.S. Standard Grade
Bacterial estimate, standard plate count/g		
Spray and Roller	≤ 50,000	≤ 100,000
Instant	≤ 30,000	
Butterfat content, %	≤ 1.25	≤ 1.5
Moisture content, %		
Spray and Roller	≤ 4.0	≤ 5.0
Instant	≤ 4.5	
Scorched particle content, mg		
Spray and Instant	≤ 15.0	≤ 22.5
Roller	≤ 22.5	≤ 32.5
Solubility index, ml		
Spray	≤ 1.2	≤ 2.0
U.S. high heat[b]	≤ 2.0	≤ 2.5
Roller	≤ 15.0	≤ 15.0
Instant	≤ 1.0	
Titratable acidity, %	≤ 0.15	≤ 0.17
Coliform count/g		
Instant	≤ 10	
Dispersibility, %		
Instant	≥ 85	

Source: CFR (1987b).
[a] Instant nonfat dry milk may be assigned only one grade, "U.S. Extra."
[b] Heat classification is as follows:
    1. U.S. high heat ≤ 1.5 mg undenatured whey protein nitrogen/g dry product
    2. U.S. low heat ≥ 6.0 mg undenatured whey protein nitrogen/g dry product
    3. U.S. medium heat 1.51 to 5.99 undenatured whey protein nitrogen/g dry product

nonassignment of grade for instant NDM: (1) failure to meet requirements for U.S. Extra Grade; (2) a direct microscopic clump count greater than 75 million/g; and (3) a phosphatase test that shows more than 4 μg of phenol/ml of reconstituted nonfat milk.

When NDM (especially the instantized form), is used as a beverage, a sensory comparison with fresh fluid skim milk is inevitable. Under ideal conditions, the sensory difference may not be that significant; even expert evaluators may find little to criticize in reconstituted nonfat dry milk of high quality. However, there are several points to keep in mind. Fresh, liquid skim milk (or another liquid product) is not guaranteed to be free of flavor defects; in some instances, fresh skim milk may be inferior to the dehydrated product. Generally, there is no logical basis for comparing a good-quality fluid product with a poor-quality dry product or vice versa. Each product form should be evaluated for its own merits and defects. Fresh, fluid skim milk (as do other highly perishable milk products) deteriorates with age, generally due to microbial activity. On the other hand, flavor deterioration in a dry product is most commonly due to chemical mechanisms—such as the browning reaction, oxidation, and the process of staling. Also, since dry products may be in storage for months or years (as opposed to a maximum of about two weeks for fluid products), certain gradual chemical reactions generally have adequate time to manifest themselves. Thus, a sample of one-year-old nonfat milk may exhibit flavor characteristics inferior to that of fresh, fluid skim milk. However, a year-old dry product may be substantially more acceptable in flavor than a three-week-old fluid skim milk.

Sensory Properties of NDM

Flavor. The flavor of high-quality nonfat dry milk should be similar, when reconstituted, to that of fresh fluid skim milk. Due to the extremely low milkfat content, NDM does not possess the rich "pastry" flavor of products of higher fat content. The flavor should be clean, sweet, and pleasant, but NDM may possess a slight cooked or heated flavor. Likewise, the off-flavors found in reconstituted nonfat dry milk have much in common with those of dry whole milk, but differ in their relative importance. The common flavor defects of nonfat dry milk are as follows:

Stale, storage, old   Oxidized, tallowy
Scorched

Stale, Storage, Old. This flavor defect is frequently encountered in nonfat dry milk (NDM). This particular off-flavor is even more "quick" to occur and distinct in NDM than in dry whole milk. Usually, this flavor defect is accompanied by a slight to definite darkening of the powder color. However, some staleness may frequently be detected before any change in color is noted. As pointed out elsewhere in this chapter, there are some reasons for considering "stale" and "storage" off-flavors as separate entities. Many graders of milk powders do not attempt or even make the effort to distinguish between these two off-flavors. In old, darkened products, a sharp, slightly sour taste may be detected after the first sensation of staleness has completely disappeared. This slightly sour taste is quite similar to that noted in darkened evaporated milk, which may have resulted from storage at a high temperature for an extended time. Lea and coworkers (1943) variously described this off-flavor as "burnt," "stale," or "glue-like." They reported that the so-called "burnt flavor" may have stemmed from a blend of the "toffee" flavor (derived from milkfat) and slight lactose caramelization, and that quite possibly the stale off-flavor was derived from protein deterioration.

Scorched. As in the instance of dry whole milk, a "scorched" off-flavor is also developed in NDM's which have been subjected to abnormally high heat during processing. This defect is usually accompanied by an excessive number of scorched particles in the product; sometimes, a darker color may be observed also.

Oxidized, Tallowy. This off-flavor is less frequently encountered in NDM's than in dry whole milks. Since tallowiness is a fat-associated off-flavor, it develops when appreciable fat constitutents are present. Nonfat dry milk should contain a negligible amount of milkfat available to undergo auto-oxidation; nonetheless, under certain conditions an objectionable oxidized or tallowy off-flavor can develop. Of particular note for the dried milk products judge, a "tallowy" product tends to have a pronounced odor, whereas a "stale" powder does not exhibit an intense odor.

Physical Characteristics of NDM

Fineness and Homogeneity. The grain fineness of high-quality NDM is dependent upon the size of spray nozzle(s) and the extent of milk concentration prior to spray drying, and upon the extent of pulverization and the mesh of the bolting when the product is roller-dried. Nonfat dry milk manufactured by the spray process usually exhibits a fine, uniform particle size (Fig. 10.3). The dried product made by the roller process is much coarser and less homogeneous, unless it is extensively pulverized after drying.

Instant nonfat dry milk is usually quite granular; the product should pour as readily as corn meal. By contrast, normal spray-dried NDM is most hygroscopic, light, dusty (nearly airborne) and has "flow" characteristics similar to flour (see Fig. 10.3).

Color. Nonfat dry milk, like dry whole milk, should be uniform in color and be free of foreign specks and burnt particles. NDM should exhibit a creamy white or light yellow color, though it may vary slightly in intensity with season of the year. Under certain conditions, nonfat dry milk tends to darken in color with aging; the light yellow color darkens to a distinct brown. This appearance defect is usually associated with an old or stale off-flavor. For reasons not well understood, spray process products seem to be more susceptible to "age darkening" (and to a greater intensity) than roller process powders. However, dry powders made by both processes are susceptible to this defect.

Fig. 10.3. A comparison of the relative particle size of (A) conventional and (B) instantized nonfat dry milks.

DRY BUTTERMILK

The definition and standards for dry buttermilk (as established by the USDA) are found in 7 CFR 58.2651–58.2678. This product is defined as follows:

> Dry buttermilk (made by the spray process or the atmospheric roller process) is the product resulting from drying liquid buttermilk derived from the manufacture of sweet cream butter to which no alkali or other chemical has been added and which has been pasteurized either before or during the process of manufacture at a temperature of 161°F for 15 seconds or its equivalent in bacterial destruction.

The two U.S. grades of dry buttermilk, "U.S. Extra" and "U.S. Standard," are determined on the basis of flavor and odor, physical appearance, bacterial estimate, milkfat content, moisture content, scorched particle content, solubility index, and titratable acidity. The U.S. grade requirements for dry buttermilk are summarized in Table 10.7.

The flavor of dry sweet cream buttermilk should be clean, sweet, and pleasant; it should have a somewhat richer flavor than nonfat dry milk. Whereas NDM contains less than 1.5% of milkfat, dry sweet cream

Table 10.7. U.S. Grade Classifications of Dry Buttermilk Based on Flavor, Physical Appearance, and Laboratory Analyses.

Quality Attributes (or Laboratory Tests)	Classification[a]	
	U.S. Extra Grade	U.S. Standard Grade
Flavor[b]	Free from nonbuttermilk flavors and odors	Not more than slight unnatural flavors and odors, and has no offensive flavors and odors
Physical Appearance	Cream to light brown color; free from lumps that do not break up under slight pressure; and practically free from black and brown scorched particles.	Cream to light brown color; free from lumps that do not break up under moderate pressure; and contains brown and black scorched particles to not more than a moderate degree.
Bacterial estimate per g	≤50,000	≤200,000
Butterfat content, percent	≥4.50	≥4.50
Moisture content, percent	≤4.0	≤5.0
Scorched particle content, mg		
Spray	≤15	≤22.5
Roller	≤22.5	≤32.5
Solubility index, ml		
Spray	≤1.25	≤2
Roller	≤15	≤15
Titratable acidity, %	0.1–0.18	0.1–0.2
Optional test		
Alkalinity of ash, ml of 0.1 N HCl per 100 g	≤125	≤125

Source: CFR (1987b).
[a] A U.S. grade is not assigned when these requirements are not met.
[b] Applies equally to the reliquified form.

buttermilk usually contains from 4.8–8.4% milkfat (Davis 1939). This fat percentage appears compatible with the 4.5% milkfat minimum advocated by the American Dry Milk Institute (since 1986 renamed: American Dairy Products Institute 1971). With this much milkfat present in sweet cream buttermilk, the product should possess a richer, fuller flavor than NDM. On the other hand, the evaluator should remember that buttermilk is rich in lecithin, a lipid constituent which is quite susceptible to auto-oxidation. Thus, dried buttermilk powders are frequently quite vulnerable to rapid flavor deterioration. Off-flavors noted in dry buttermilk (stored under adverse conditions) in a study by Davis (1939) included various intensities of: stale, old, musty, sharp, bitter, soapy, coarse, cheesy, rubbery, acid, fruity, tallowy, and putrid. A wider range of off-flavors will probably be noted in evaluating dry sweet cream buttermilk than when judging NDM for flavor.

The American Dairy Products Institute (1971) has established the minimum flavor requirements for dry buttermilk. For Extra Grade, "the reliquified product shall have a desirable flavor, but may possess the following off-flavors to a slight degree: chalky, cooked, feed and flat"; also slightly scorched for roller-dried powder. For Standard Grade, "the reliquified product shall possess a fairly desirable flavor, but may possess the following off-flavors to a slight degree: bitter, oxidized, stale, storage, utensil, and scorched; and chalky, cooked, feed and flat flavors to a definite degree." As far as quality guidelines other than flavor, the American Dairy Products Institute's requirements are nearly identical to those of the USDA.

DRY WHEY

The U.S. standards for dry whey are cited in 7 CFR 2858.2601–2858.2610. Whey and dry whey are therein defined as follows:

Whey is the fluid obtained by separating the coagulum from milk, cream and/or skim milk in cheesemaking. The acidity of the whey may be adjusted by the addition of safe and suitable pH adjusting ingredients. Salt drippings (moisture removed from cheese curd as a result of salting) shall not be collected for further processing as whey.

Dry whey is the product resulting from drying fresh whey which has been pasteurized and to which nothing has been added as a preservative. It contains all constituents, except moisture, in the same relative proportions as in the whey.

Only a single grade of dry whey, "U.S. Extra," is recognized; compliance is determined on the basis of flavor, physical appearance, bacterial estimate, coliforms, milkfat content, moisture and the optional tests for protein content, alkalinity of ash, and scorched particle content (see Table 10.8).

The flavor characteristics of dry whey will vary with the whey acidity and the drying process. The flavor of good-quality dry whey is usually pleasantly sweet, with a subtle or slightly subdued acid aftertaste. The flavor may change markedly during storage toward a stale, slightly sour flavor, which is accompanied by a definite browning of the product. In one stability study, samples of commercial dry whey were stored at room temperature for three years. During that period, the dry whey underwent a marked change in flavor and color. Within a year, the color changed from a light-yellow to a dark-brown. The initial pleasant aroma changed to an odor that resembled that of corn gluten

Table 10.8. Requirements for U.S. Extra Grade Dry Whey.[a,b]

Category	Requirement
Flavor[c]	Shall have a normal whey flavor free from undesirable flavors, but may possess the following flavors to a slight degree: bitter, fermented, storage, and utensil; and the following to a definite degree: feed and weedy.
Physical appearance	Has a uniform color and is free-flowing, free from lumps that do not break up under slight pressure, and is practically free from visible dark particles.
Bacterial estimate	Not more than 50,000 per g standard plate count.
Coliform	Not more than 10 per g.
Milkfat content	Not more than 1.5%.
Moisture content	Not more than 5%.
Optional tests:	
Protein content (N × 6.38)	Not less than 11%.
Alkalinity of ash (sweet-type whey only)[d]	Not more than 225 ml of 0.1N HCl per 100 g.
Scorched particle content	Not more than 15 mg.

Source: CFR (1987b).

[a] No U.S. grade shall be assigned to a product which fails to meet any of the required or optional criteria or one which was produced in a plant found on inspection to be using unsatisfactory manufacturing practices, equipment, or facilities, or to be operating under unsanitary plant conditions.

[b] All required tests and optional tests when specified shall be performed in accordance with the following methods: "Methods of Laboratory Analysis," DA Instruction series 918-103-2, 918-103-5, 918-109-2, and 918-109-3, Dairy Grading Branch, Poultry and Dairy Quality Division, Food Safety and Quality Service, U.S. Department of Agriculture, Washington DC 20250, or the latest revision thereof.

[c] Applies to the reliquified form.

[d] Acidity classification is not a U.S. Grade requirement. The dry whey is classified as follows:
   1. Dry sweet-type whey— ≤0.16% titratable acidity on a reconstituted basis.
   2. Dry whey __% titratable acidity—Dry whey over 0.16% but below 0.35% titratable acidity (on a reconstituted basis); the blank is filled with the actual acidity.
   3. Dry acid-type whey—Dry whey with 0.35% or higher titratable acidity on a reconstituted basis.

(feed) or of roasted cereal. To several evaluators, the formed odor suggested a poor grade of coffee or chicory.

Bodyfelt *et al.* (1979) studied the quality impact of dried wheys of various degrees of age and flavor quality on vanilla ice cream mix. Gasliquid chromatography analyses indicated several pyrazines and 2-furfural to be partially responsible for mix off-flavors, variously described by the investigators as lacks freshness, stale, and "whey flavor."

The initial flavor quality of whey depends on such factors as: (1) the quality of the milk from which the cheese was made; (2) the type of cheese manufactured; (3) the method of whey handling immediately after curd draining; (4) the elapsed time between draining and pasteurization; and (5) the extent of adherence to "good manufacturing practices." The manufacture of cheese requires the combined activity of microorganisms and enzymes, but these biochemical activities must be suddenly terminated in the whey to prevent off-flavor(s) development.

EDIBLE DRY CASEIN

The USDA definition of edible dry casein (acid) is cited in 7 CFR 58.2800–58.2808:

> *Edible Dry Casein (Acid).* (1) For the purposes of these standards, edible dry casein (acid) is the pulverized or unpulverized product resulting from washing, drying or otherwise processing the coagulum resulting from acid precipitation of skim milk which has been pasteurized before or during the process of manufacture in a manner approved by the Administrator.
>
> (2) The product shall have been produced in a plant under conditions suitable for the manufacture of human food and packaged in a container which will prevent contamination, deterioration and/or development of a public health hazard under normal conditions of storage and transportation.

Two grades of edible dry casein are recognized, "U.S. Extra" and "U.S. Standard," which are assigned on the basis of flavor and odor, physical appearance, bacterial estimate (based on the standard plate count and coliform count), protein content, moisture content, extraneous material, free acid, particle size (optional), ash (phosphorus fixed), copper content, lead content, iron content, yeast and mold count, thermophile count, reducing sugars (as lactose), and absence of significant *Staphylococcus* (coagulase positive) and *Salmonella* counts. The requirements and recommended criteria for edible dry casein are summarized in Table 10.9.

Table 10.9. U.S. Grade Classifications of Edible Dry Casein (Acid) Based on Flavor and Odor, Physical Appearance, and Laboratory Analyses.[a,b]

Category	Classification	
	U.S. Extra Grade	U.S. Standard Grade
Flavor and odor	Bland natural flavor and odor, and free from offensive flavors and odors such as sour and cheesy.	Not more than slight unnatural flavors or odors, and free from offensive flavors and odors such as sour or cheesy.
Physical appearance	White to cream colored physical appearance; if pulverized, free from lumps that do not break up under slight pressure.	White to cream colored physical appearance; if pulverized, free from lumps that do not break up under moderate pressure.
Bacterial estimates:		
Standard plate count/g	≤30,000	≤100,000
Coliform count/0.1 g	Negative	≤2
Protein content, N × 6.38, dry basis, %	≥95	≥90
Moisture content, %	≤10	≤12
Milkfat content, %	≤1.5	≤2
Extraneous materials	Scorched particles not more than 15 mg and free from foreign materials in 25 g.	Scorched particles not more than 22.5 mg and free from foreign materials in 25 g.
Free acid	Titrated to not more than 0.2 ml of 0.1N NaOH per g.	Titrated to not more than 0.27 ml of 0.1N NaOH per g.

Optional tests (recommended criteria):

Ash (phosphorus fixed) %	≤2.2
Copper, ppm	≤5
Lead, ppm	≤5
Iron, ppm	≤20
Yeast and mold, per 0.1 g	≤5
Thermophiles, per g	≤5000
Reducing sugars (as lactose) %	≤1
Staphylococcus (coagulase positive)	Negative
Salmonella in 100 g	Negative
Particle size—30, 60, 80, or other specified mesh	
30 mesh	100% must pass 30 ASTM screen, 10% may pass 60 ASTM screen.
60 mesh	99% must pass 50 ASTM screen, 10% may pass 80 ASTM screen.
80 mesh	100% must pass 60 ASTM screen, 85% may pass 80 ASTM screen.

Source: CFR (1987b).

[a]Testing methods contained in "Methods for the Analysis of Edible Dry Casein (acid)," Dairy Division Inspection and Dairy Branch Laboratory, Consumer and Marketing Service, U.S. Department of Agriculture, or latest revision thereof, shall be used.

[b]Edible dry casein (acid) which fails to meet the requirements of U.S. Standard grade; or of *Salmonella* or *Staphylococcus* tests when such tests have been made; or product manufactured, packaged, or transported under conditions unsuitable for human food; or otherwise found to be unwholesome, shall not be assigned a U.S. grade.

Caseinates. Acid casein is commonly converted to a more usable form such as sodium caseinate or calcium caseinate. In the salt form, caseinates have found wide application as a food ingredient, principally in nondairy foods such as bakery products, dairy product analogs, processed meats, and coffee whiteners. A blend of caseinate and whey solids may be made to emulate the composition and functional properties of nonfat dry milk. In various food applications, caseinates perform specific functions. How adequately a given lot or source of caseinate performs the various food ingredient functions should be a primary criterion of the quality evaluation process for these milk-derived products.

Casein and caseinates are subject to variations in sensory quality either during the manufacturing process or as a result of deteriorative changes that occur during storage. The USDA standards specify freedom from offensive flavors and odors; off-flavors such as sour and cheesy are identified. A "stale" off-flavor may develop during storage, which may be related to a similar off-flavor that occurs in stored dry milk, sterile milk, and evaporated milk. More research is needed to better chemically characterize this off-flavor and to learn the mechanism(s) of its formation.

In the process of manufacturing casein, the curd is washed to remove impurities. Lactose is one of several compounds which may be retained in excessive concentration if the casein curd is not adequately washed. The USDA standards establish 1% as the upper limit for lactose in casein. The presence of lactose in casein products unfortunately potentiates the Maillard (browning) reaction, especially when casein has been converted to a more alkaline caseinate. A brown pigment need not appear for off-flavors to manifest themselves, because pigment formation occurs in later stages of the reaction, after numerous actual and potential flavor compounds have been formed. Thus, low residual lactose levels should be sought in dry casein products.

A frequent and serious flavor defect in caseinates is referred to as "gluey." As the term implies, this off-flavor is suggestive of protein degradation. Under alkaline conditions (as with caseinates), protein degradation occurs at an accelerated rate.

LOWFAT DRY MILK

The standard of identity for lowfat dry milk differs from that of dry whole milk in a few details. The description for lowfat dry milk is included in 21 CFR 131.123:

Description.—Lowfat dry milk is the product obtained by removal of water only from pasteurized lowfat milk, . . . , which may have been homogenized. Alternatively, lowfat dry milk may be obtained by blending fluid, condensed or dried nonfat milk with liquid or dried cream or with fluid, condensed or dried milk, as appropriate, provided the resulting lowfat dry milk is equivalent in composition to that obtained by the method described in the first sentence of this paragraph. It contains not less than 5% but less than 20% by weight of milkfat on an "as is" basis. It contains not more than 5% by weight of moisture on a milk-solids-not-fat basis. Lowfat dry milk contains added vitamin A as prescribed . . . (below).

Vitamin Addition.—Vitamin A shall be present in such quantity that when prepared according to label directions, each quart of the reconstituted product contains not less than 2,000 I.U. thereof.

Provisions relating to vitamin D and other optional additives are the same as those for dry whole milk. The sensory properties of this product, depending on the milkfat content, resemble those of NDM at one extreme and dry whole milk at the other. There are no specific U.S. grade standards for this product. The expected sensory defects are those due to heat, Maillard reaction, storage and stale off-flavor development, oxidation, and vitamin addition (under some circumstances).

DRY CREAM

The FDA standard of identity for dry cream may be found in 21 CFR 131.149. Following is the description and list of optional ingredients for dry cream:

Description.—Dry cream is the product obtained by removal of water only from pasteurized milk or cream or a mixture thereof, which may have been homogenized. Alternatively, dry cream may be obtained by blending dry milks as defined in §§131.123(a), 131.125(a) and 131.147(a) with dry cream as appropriate: *Provided* that the resulting product is equivalent in composition to that obtained by the method described in the first sentence of this paragraph. It contains not less than 40% but less than 75% by weight of milkfat on an "as is" basis. It contains not more than 5% by weight of moisture on a milk-solids-not-fat basis.

Optional ingredients.—The following safe and suitable optional ingredients may be used:

Emulsifiers

Stabilizers

Anticaking agents

Antioxidants

Nutritive carbohydrate sweeteners

Characterizing flavoring
  ingredients, with and without
  coloring, as follows:

Fruit and fruit juice, including concentrated fruit and fruit juice.
Natural and artificial food flavoring.

No specific classification for grades of dry cream has been issued by
the USDA. Off-flavors in dry cream products parallel those that develop in dry whole milk (i.e., stem from auto-oxidation of lipid components during storage). In addition to lipid oxidation, browning reactions and staling are significant quality problems of dry cream.

DRY ICE CREAM AND DRY ICE MILK MIX

These products differ from the other dry products in that mere reconstitution with water does not yield the finished product, in this case,
frozen ice cream or ice milk. The reconstituted mix generally requires
added flavoring and this mixture is then frozen. Thus, evaluation of
the dry mix following reconstitution may not be adequate, since the
sensory properties of the resultant frozen ice cream are of paramount
interest. As a rule, a mix that has inferior flavor characteristics can be
expected to yield an ice cream of poor flavor quality. Freezing characteristics, body and texture, and color/appearance are additional quality
considerations for the product.

Dry ice cream or ice milk mix may be made by spray-drying the liquid mix, although a portion of the sweetener may be withheld prior to
drying to avoid excessive browning. The remaining required sugar can
be subsequently dry-blended with the dry mix. Alternatively, the entire dry mix may be assembled by dry-blending all of the various ingredients, such as nonfat dry milk, dry cream, sugars, and any stabilizer/
emulsifier. The question as to whether the reconstituted mix can then
be frozen without pasteurization (after reconstitution) can only be an-
swered by regulatory officials of the particular state or local jurisdic-
tion in which use of the product is proposed.

Dry ice cream (and ice milk) mixes are subject to the development
of exactly the same defects as dry whole milk and dry cream. These
defects result from heat treatment, browning reaction, staling, and oxidation processes.

MALTED MILK

"Malted milk" is a product made by combining whole milk with a liquid separated from a mash of ground barley malt and wheat flour, with
or without the addition of sodium chloride, sodium bicarbonate, and
potassium bicarbonate. This is done in such a manner as to secure the
full enzymatic action of the malt extract and is followed by the removal

of water. The resultant product should contain not less than 7.5% milk-fat and not more than 3.5% moisture.

Each pound (0.455 kg) of malted milk contains the total solids of approximately 2.2 lb (1.0 kg) of whole milk. If a whole milk testing 13% total solids were used in malted milk manufacture, the final product would then contain only about 29% milk solids. Thus, it appears that the term "malted milk" is a misnomer. In reality malted milk is a grain-based product, which is made principally from malted wheat and barley flour, and fortified with milk solids.

Flavor. Since malted milk is composed in large part of maltose and dextrose, it definitely manifests a sweet taste. In addition, it has the distinct flavor of malt, which is derived from barley during the mashing process. This "malty flavor" is highly desired or sought in malted milk. Flavor defects in malted milk are rare, but the following have occasionally been encountered:

Lack of Malt Flavor. Lack of a definitive malt flavor occasionally occurs in malted milk. Such a product usually exhibits the desired level of sweetness, but will have an insufficient intensity of characteristic flavor.

Tallowy. Occurrence of a tallowy, oxidized off-flavor is rare in malted milk; nevertheless, moderate degrees of tallowiness are possible in this product. This flavor defect in malted milk has similar sensory characteristics of tallowiness in dry milk products.

Relative to the keeping quality of malted milk, Hunziker (1949) stated that: "Those who have given the manufacture, properties and marketing of malted milk careful study appear to agree that the dependable keeping quality of malted milk is in all probability due to a film or layer of gluten, sugars and salts that protects the surface of fat globules against the quality deteriorating influence of contact with air."

Texture. Malted milk is usually quite coarse and grainy, unlike the finer texture of spray-dried milk or the corn meal-like texture of instant NDM. This product has a high affinity for water; when exposed to moist air it initially becomes sticky and then tends to "cake" together in large lumps. However, instantization has essentially overcome this problem, since a more readily soluble, free flowing powder is produced by this process.

MISCELLANEOUS DRY PRODUCTS

A partial list of miscellaneous dry milk products includes instant chocolate drink, instant hot cocoa mix, instant breakfast drinks, dry cheese, casein-whey solids blends, and nondairy coffee whiteners. Products of this type are generally formulated according to proprietary specifications; in fact, some are covered by patents. Sensory quality control of dry-milk-based foods depends on maintaining a high level of consumer acceptibility; this embraces flavor, physical appearance, rehydration characteristics and product functionality. Some of these products are manufactured by evaporating moisture initially and then drying from a liquid slurry state, while other dry products may be "assembled" as the result of dry-blending various ingredients.

SCORING AND GRADING DRY MILK

Several sensory terms have been applied in an attempt to classify flavor defects of various dry milk products. Unfortunately, these particular descriptors have not been used that consistently between technologists or researchers involved with dry milk products. As early as 1957, a committee of the American Dairy Science Association (Thomas 1958) proposed definitions for the flavor and appearance characteristics, as well as for the packaging of dry milks. A suggested dry milk products score card is presented in Fig. 10.4 and a suggested scoring guide for flavor is offered in Table 10.10.

Flavor Descriptors of Dry Milks

Acid. The term "acid" is used to describe the odor and taste (primarily) that results from the action of lactose-fermenting bacteria in milk and milk products. The principal fermentation end-product is lactic acid, which typically exhibits a clean, distinct sour taste.

Astringent. "Astringent" refers to a puckery type of mouthfeel sensation similar to that produced by a chemical such as alum. There is an associated tactile sensation to the "astringent" off-flavor; the mucous membranes of the palate and/or tongue tend to shrink (pucker).

Bitter. The "bitter" defect resembles the taste sensation imparted by bitter substances, such as quinine, caffeine, or certain peptides. This defect is often associated with rancidity (lipolysis) and/or the growth of either lipolytic or proteolytic microorganisms in milk (certain psychrotrophs and some spore-forming bacteria).

The USDA (1987b) employs comparable definitions of bitter for several dry milk products. For instance, in describing bitterness in dry whole milk, the USDA states: "Similar to taste of quinine and produces a puckery sensation." For dry whey, the USDA states: "Distasteful, similar to taste of quinine." And for nonfat dry milk, bitter is succinctly stated: "Similar to taste of quinine."

Product: _____

SAMPLE NO.   DATE: _____

FLAVOR 10	CRITICISM SCORE								
	ACID								
	ASTRINGENT								
NO CRITICISM	BITTER								
10	CHALKY								
	COOKED								
	FEED								
	FERMENTED								
UNSALABLE	FLAT								
0	FOREIGN								
	GLUEY								
	METALLIC								
NORMAL RANGE	NEUTRALIZER								
1-5	OXIDIZED/TALLOWY								
	RANCID (LIPOLYSIS)								
	SALTY								
	SCORCHED								
	STALE								
	STORAGE								
	UNCLEAN/UTENSIL								
	WEEDY								

PHYSICAL APPEARANCE 5	SCORE								
	DRY PRODUCT:								
	CAKED								
	DARK PARTICLES								
NO CRITICISM	LUMPY								
5	UNNATURAL COLOR								
	RECONSTITUTED								
UNSALABLE	PRODUCT:								
0	CHURNED PARTICLES								
	DARK PARTICLES								
NORMAL RANGE	GRAINY								
1-5	UNDISPERSED LUMPS								

PACKAGE 5	SCORE								
NO CRITICISM	RUPTURED VAPOR BARRIER								
5	SOILED								
	UNSEALED								
UNSALABLE 0									
NORMAL RANGE 1-5									

LABORATORY TESTS 5	SCORE								
	FAT (%)								
	MOISTURE (%)								
NO CRITICISM	TITRATABLE ACIDITY								
5	(% LACTIC ACID)								
	SOLUBILITY INDEX (ML)								
	BACTERIAL ESTIMATE								
	(PER GRAM)								
UNSALABLE 0	COLIFORM (PER GRAM)								
	DIRECT MICROSCOPIC								
	CLUMP COUNT (PER G)								
	SCORCHED PARTICLES (MG)								
	DISPERSIBILITY (MODIFIED								
	MOATS-DABBAH METHOD, %)								
	PHOSPHATASE TEST								
	MICROGRAMS PHENOL/ML								
	UNDENATURED WHEY PROTEIN								
	NITROGEN (MG/G)								
	OXYGEN CONTENT (%)								
	COPPER (PPM)								
	IRON (PPM)								
	VITAMIN A (I.U.)								
	VITAMIN D (I.U.)								
	ALKALINITY OF ASH								
	(ML/100 G)								
	PROTEIN CONTENT (%)								
	MESH (SCREEN %)								
	ASH, PHOSPHORUS FIXED (%)								
	LEAD (PPM)								
	YEAST AND MOLD (PER 0.1 G)								
	THERMOPHILES (PER G)								
	REDUCING SUGARS (AS								
	LACTOSE %)								
	STAPHYLOCOCCUS								
	(COAGULASE POSITIVE)								
	SALMONELLA (IN 100 G)								

SIGNATURES: _____   _____   _____

Fig. 10.4.   A suggested dry milk products score card.

Table 10.10. A Suggested Scoring Guide for the Flavor of Dry Milk (Reliquified Basis).

Defect	Scores for a Given Intensity[a]				
	Slight[b]	Moderate	Definite	Strong	Pronounced[c]
Acid	2	1	0	0	0
Astringent	8	7	6	5	0–4
Bitter	6	5	4	3	0–2
Chalky	8	7	6	5	0–4
Cooked	9	8	7	6	5
Feed	8	7	6	5	0–4
Fermented	6	5	4	3	0–2
Flat	9	8	7	6	—[e]
Foreign[d]	2	1	0	0	0
Gluey	2	1	0	0	0
Metallic	4	3	2	1	0
Neutralizer[f]	0	0	0	0	0
Oxidized/tallowy[g]	4	3	2	1	0
Rancid (lipolysis)	5	4	3	2	0–1
Salty	7	6	5	4	0–3
Scorched	4	3	2	1	0
Stale	4	3	2	1	0
Storage	7	6	5	4	0–3
Unclean/utensil	5	4	3	2	0–1
Weedy	3	2	1	0	0

[a] "No criticism" is assigned a score of "10." Normal range is 1 to 10 for a salable product.
[b] Highest assignable score for a defect of slight intensity.
[c] Highest assignable score for a defect of pronounced intensity. However, a sample may be assigned a score of "0" (zero) if the defect makes the product unsalable.
[d] Due to the variety of foreign off-flavors, suggesting a fixed scoring range is not appropriate. Some foreign off-flavors warrant a score of "0" (zero) even if their intensity is slight (i.e., gasoline, pesticides, lubricating oil).
[e] The defect is unlikely to be present at this intensity level.
[f] The use of neutralizers is not authorized except in whey. However, dry, sweet-type whey must have an alkalinity of ash not to exceed 225 ml of 0.1 N HCl per 100 g.
[g] When an oxidized off-flavor has progressed to the tallowy stage, the assigned flavor score should be "0" (zero).

A direct statement such as "resembles the taste of quinine or caffeine" seems to be an adequate definition of bitterness. The fact that this off-flavor may be caused by lipolysis, proteolysis, bacterial growth or, occasionally by the consumption of certain weeds by cows is helpful information, but it does not realistically describe the sensation of bitterness.

Chalky. This descriptor of a common off-flavor in concentrated milk products suggests the inclusion of fine, insoluble, chalk (powder) particles. The USDA definition for "chalky" is: "A tactual type of (off-)flavor, lacking in characteristic milk flavor." The chalky off-flavor is as much an objectionable mouthfeel sensation as it is an off-taste. The

Fig. 10.5.   Collection of nonfat dry milk samples for composition and quality analysis by a USDA inspector/grader.

chalky defect frequently tends to manifest itself as a delayed mouthfeel—an aftertaste response of the evaluator.

Cooked. "Cooked" has an odor and flavor resembling that of milk that has been heated to 73.8°C (164.8°F) or higher. The USDA definition for cooked flavor in dry milk products is: "Similar to a custard flavor and imparts a smooth aftertaste."

Feed. A milk off-flavor that is usually characteristic of the roughage (feeds) consumed by milk cows, is simply referred to as a "feed" defect. Several USDA definitions state: "Feed flavors (such as alfalfa, sweet clover, silage, or similar feed) in milk that are carried through into the dry whole milk," or "characteristic of the feed flavors found in milk" (instant nonfat milk), and "characteristic of the feed flavors in milk that are carried through into the nonfat dry milk."

Flat. The descriptor "flat" implies a lack of fullness of flavor; this flavor defect is suggestive of added water. It is not detectable by odor perception. The listed USDA definitions for flat are: "Lacking characteristic flavor" and "lacking characteristic sweetness or full flavor."

Fermented. The following definition for "fermented" is taken from the USDA standards for dry whey: "Flavors such as fruity or yeasty produced through unwanted chemical changes brought about by microorganisms or their enzyme systems."

Foreign. "Foreign" refers to any atypical or objectionable off-flavor that is not ordinarily associated with good-quality milk; sometimes a chemical- or medicinal-like off-flavor may have occurred. This flavor defect usually stems from the fluid milk used as a raw material to produce the dry milk.

Metallic. The off-flavor, "metallic," is quite suggestive of the presence of copper or iron in the raw material used to produce the dried product. Metallic is usually regarded as a phase of oxidized (metal-induced) off-flavor.

Neutralizer. The "neutralizer" off-flavor is an alkaline taste generally derived from alkaline substances used to neutralize any developed acidity in milk. The USDA has made provisions for the pH adjustment of dry whey, but acid neutralization of other dry products is not permitted.

Oxidized. Milkfat oxidation is responsible for the "oxidized" off-flavor in dry milk products. The perceived sensation in an oxidized off-flavor resembles wet cardboard, oily substances, or tallow, depending on the defect intensity. The USDA definition simply paraphrases the previous sentence.

Rancid. "Rancidity" in dry milk products usually exhibits a pungent odor, which is frequently accompanied by a soapy and/or bitter taste. These sensory properties are primarily due to the formation of volatile fatty acids, that have resulted from lipolysis of the milk components used to produce the dried product.

Salty. A "salty" taste defect in dry milk products is simply a perceived primary taste; it resembles a milk product that has had salt added to it. Perception of a salty taste on the front tip and sides of the tongue is relatively rapid, compared to other experienced taste sensations.

Scorched. This flavor defect is produced when milk powder has been subjected to excessive heat in the drier; it is generally suggestive of burnt protein. The USDA definition for "scorched" is: "A more intensified flavor than cooked," plus an additional statement that this flavor defect is generally characterized by having a "burnt" aftertaste.

Stale. "Stale" generally implies a lack of product freshness. This flavor sensation in dried milk products is ordinarily associated with deterioration of milk protein, rather than milkfat. Some dairy product evaluators tend to use the descriptor "lacks freshness" in lieu of the term "stale" while other evaluators use both of the aforementioned descriptors interchangeably. The terms "stale" or "lacks freshness" are commonly applied when the flavor is not as refreshing as expected by the evaluator.

There is an apparent anomaly in use of the terms "stale" and "storage" as flavor descriptors. The USDA provides guidelines for various intensities of both stale and storage off-flavors, but their singular definition treats them as one and introduces some element of confusion for product evaluators: (i.e., *"Stale, storage.* Lacking in freshness and imparting a 'rough' aftertaste").

The authors feel that a logical argument can be made for the acceptance of separate meanings of the terms "stale" and "storage." It is true that a stale off-flavor in dry milk can develop during storage, but so can the oxidized off-flavor. Analogous to the oxidized off-flavor, stale is a distinctively recognizable off-flavor which can develop at low oxygen levels. Unfortunately, thus far, research has not conclusively pinpointed the chemical precursor or the actual chemical entity that is responsible for the stale off-flavor. The precursor could be any of the following: (1) a protein; (2) a product of the Maillard reaction; or (3) some compound(s) derived from milkfat. The chemical compound(s) produced from potential precursor(s) may require that the substance(s) undergo oxidation to eventually produce the stale off-flavor. The salient point is that the stale off-flavor is a distinct entity, whereas the designation "storage off-flavor" is somewhat more generic. Hence, the descriptor "storage" more appropriately encompasses those particular off-flavors that dry milk products may acquire during a period of storage. These shortcomings may range from absorbed off-flavors (from the storage environment) to flavor defects that develop from slow, gradual chemical reactions in the product, which can be appropriately designated as a "lacks freshness" and/or "storage" off-flavor.

Unclean (Utensil). Typically, the "unclean" flavor defect in dry milks refers to an unpleasant odor and lingering aftertaste that is suggestive of organic decomposition products. The sensation of "uncleanliness" may vary from an odor that resembles "barny" or "barnyard-like," to that of spoiled feed or the decay of organic matter. These objectionable sensory characteristics are usually due to proteolytic or lipolytic activity by spoilage bacteria in milk. The unpleasant aftertaste is often "dirty-like," persistent, and generally objectionable, if not obnoxious.

The USDA definition is somewhat more general and only relies on the antiquated term "utensil." Hence, unclean (utensil) is described by

USDA terminology thusly, "A flavor suggestive of improper or inadequate washing and sterilization of milking machines, utensils, or factory equipment."

Due to its questionable relevance, the term "utensil" should probably no longer be used in describing this off-flavor. The activity of spoilage microorganisms (psychrotrophs) in residual milk soils that remain on the equipment is responsible for the defect, not the equipment and/or utensils themselves.

Undesirable. The USDA uses the term "undesirable" to describe certain off-flavors which are in excess of the permitted intensity in specific grades of dried milk products or for those miscellaneous off-flavors that are not otherwise listed.

Weedy. "Weedy" is a flavor characteristic of certain weeds which may be consumed by cows that produced some of the raw material used for manufacture of the dried product.

Terms Describing the Appearance of Dry Products

The reader is advised to review Table 10.11 for a suggested scoring scheme for physical appearance characteristics of dry milks.

Caked. "Caked" means a hardened mass of powder that results from lactose crystallization. It usually disintegrates into smaller hard chunks, which are practically nondispersible in water.

Lumpy. "Lumpy" refers to a nonhomogeneous appearance of dry

Table 10.11. A Suggested Scoring Guide for the Physical Appearance Characteristics of Dry Milk.

Defect	Scores for a Given Intensity[a]				
	Slight[b]	Moderate	Definite	Strong	Pronounced[c]
Dry product:					
Caked	2	1	0[d]	0	0
Dark particles	3	2	1	0	0
Lumpy	4	3	2	1	0
Unnatural color	4	3	2	1	0
Color not uniform	4	3	2	1	—[e]
Reconstituted product:					
Churned particles	3	2	1	0	0
Dark particles	3	2	1	0	0
Grainy	3	2	1	0	0
Undispersed lumps	3	2	1	0	0

[a] "No criticism" is assigned a score of "5." Normal range is 1 to 5 for a salable product.
[b] Highest assignable score for a defect of slight intensity.
[c] Highest assignable score for a defect of pronounced intensity.
[d] A score of "0" (zero) is assigned if the product is determined to be unsalable.
[e] The defect is unlikely to be present at this intensity level.

milk which is due to sizeable lumps of agglomerated powder particles. The USDA definition for lumpy is: "Loss of powdery consistency but not caked into hard chunks."

Unnatural Color. "Unnatural color" refers to an abnormal or atypical color of the product due to either caramelization of lactose, nonenzymatic browning, or added color. The USDA defines unnatural color for dry whole milk and nonfat dry milk as follows: "A color that is more intense than light cream and/or is brownish, dull or grey-like."

Visible Dark Particles. Scorched powder particles or visible extraneous matter is termed "visible dark particles." A similar definition is offered by the USDA: "The presence of scorched or discolored specks readily visible to the eye."

Of note, the American Dairy Products Institute (1971) includes this statement within their *General Grading Requirements:* "The dry milk product shall be free from extraneous matter as described under Sec. 402(a) of the Federal Food, Drug, and Cosmetic Act."

Terms Describing the Appearance of Reconstituted Product

Churned Particles. Masses of coalesced fat and/or coagulated protein which may float to the surface (and eventually adhere to the side wall of the container) are generally called "churned particles."

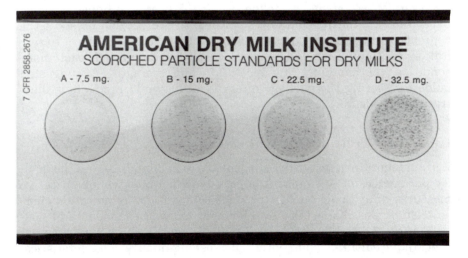

Fig. 10.6. The American Dry Milk Institute (American Dairy Products Institute) and USDA photograph standard for scorched particles in nonfat dry milks (from 7 CFR 2858.2676).

Grainy. "Grainy" refers to visible insoluble particles in reconstituted milk products which distinctly appear granular. This is the only appearance term defined by the USDA for reconstituted dry products. The USDA definition reads: "Minute particles of undissolved powder appearing in a thin film on the surface of a glass or tumbler."

Undispersed Lumps. Masses of caked or lumpy powder that do not readily dissolve in water are called "undispersed lumps."

Terms to Describe Packaging Defects

Ruptured Vapor Barrier. Any visible mechanical opening in the product package is referred to as a "ruptured vapor barrier."

Soiled. The unsightly appearance of the package exterior due to adherence of dried product or any foreign substance is simply called "soiled."

Unsealed. "Unsealed" refers to a closure that is not secured in such a manner to guarantee that access to the product is impossible without breaking or tearing a visible seal on the product container. The product package is not a defined criterion in the USDA grading system. However, a note in the standards for instant nonfat milk specifically directs the graders to perform as follows: "All tests to be determined upon samples drawn from sound, undamaged packages."

The American Dairy Products Institute (1971) has addressed the packaging requirements for dried milk products under its *General Grading Requirements* as follows: "The dry milk product shall be packed in substantial containers suitable to protect and preserve the contents without significant impairment of quality with respect to sanitation, contamination and moisture content under various customary conditions of handling, transportation and storage." A suggested guide for scoring the package integrity of dry milk products is illustrated in Table 10.12.

Laboratory Tests of Dry Milk Products

Certain laboratory tests are indispensible in helping to assess the quality parameters of dry milk. Analyses provide objective, quantitative measures of hygienic quality, product composition, rehydration characteristics, possible acidity development (as well as evidence of neutralization of excessive acidity), compliance with minimum pasteurization requirements, and potential keeping-quality characteristics. Standard procedures are used for conducting these tests. They may be found in: *Standards For Grades of Dry Milks—Including Methods of Analysis,* published by the American Dairy Products Institute, Inc.,

Table 10.12. A Suggested Scoring Guide for the Packaging of Dry Milk.

Parameter	Score Range[a]
Soiled package, graded and scored proportional to the nature and quantity of soil	0–5
Unsealed package and/or ruptured or defective vapor barrier	0
Any packaging that fails to meet the requirements of regulatory agencies	0

[a] A score of zero ("0") is assigned if the defect is so serious (or pronounced in intensity) as to render the product unsalable.

Chicago, IL (1971 Edition, 6th printing, [or latest revision]). *Standard Methods for the Examination of Dairy Products* published by the American Public Health Association, Inc., New York, N.Y. (1985 or latest revision); *Official Methods of Analysis of the Association of Official Analytical Chemists* (1984), and *Methods of Laboratory Analysis,* DA Instruction No. 918-103 (dry milk products series), Dairy Grading Branch, AMS, U.S. Department of Agriculture, Washington, D.C. 20250 (1986).

Methods of Reconstituting Dry Milk for Flavor Examinations

Limited quantities of reconstituted dry milks are used as beverage products in the U.S. However, even if they are used only as ingredients in dairy products or other foods, the sensory properties of reconstituted dry milks must meet desired standards and favorably contribute to the desired quality criteria of finished product(s). Therefore, a standardized procedure should be devised by each user for evaluating dry milk products for determining their suitability as a food product ingredient. For example, if a poor quality (off-flavored) nonfat dry milk is used in ice cream manufacture, the off-flavor(s) will most likely carry through into the ice cream. On the other hand, a highly lipolyzed (rancid) dry milk may sometimes be desired in certain milk chocolates.

Two types of test situations may arise with dry milk products to be consumed as a beverage. In acceptibility testing using a consumer panel, the product should be reconstituted in exactly the same manner as the consumer is instructed to do by the "user directions" on the container. The temperature at which the reconstituted product is served in the test should be the typical consumption temperature for the product. In grading or quality evaluation (discrimination) by trained evaluators, conditions are chosen, when appropriate, in order to optimize detection of off-flavors but not exaggerate their intensity. Since the perceived intensity of flavor characteristics varies with temperature, comparative judgments should be made with reconstituted samples at the same temperature.

Normally, grading or quality evaluation should be performed on dry milk products which are intended as ingredients for other foods. The odor perceived immediately after the containers are opened should be carefully noted, since it provides an immediate clue to a possible flavor problem. Precautions should be taken to avoid inhaling powder. The powder should be reconstituted and evaluated under standardized conditions, including a specified ratio of powder to water, source of water, manner of mixing, temperatures, and time interval between reconstitution and actual testing. The reliquified product should be evaluated in practically the same manner as its fluid counterpart. The evaluators should know, learn, and "fix-in-mind" the desirable flavor characteristics of whey, sweet cream, buttermilk, skim milk, etc. to which they must mentally compare the flavor of the reconstituted product.

Distilled water is commonly used for reconstituting dry milk for flavor evaluation, even though tap water is more likely to be used in the home, as well as in the plant. Since tap water varies in hardness and flavor in different locations, there is a rationale for specifying distilled water. However, since distilled water may also vary in sensory properties (depending on residual impurities), a good precaution is to ensure that the water is relatively tasteless and odorless.

Directions for determining the taste and odor of products derived from reconstituted milk is prescribed by the USDA as follows:

> Reconstitute with an electric mixer 6.5 g of whey, 10 g of nonfat dry milk, or 13 g of dry whole milk in 100 ml of distilled water. Allow samples to stand one hour, stir thoroughly and taste at room temperature. Observe odor and taste in a room free of disturbance and off-odors. Report the flavor as satisfactory or report the off-flavor in accordance with the apropriate U.S. grade standards.

The American Dairy Products Institute (1971) provides a similar set of directions: "Immediately after opening the container, examine the samples of dry milk for flavor and odor. The flavor and odor should be determined on the reliquified sample as follows:

> Add 10 g nonfat dry milk or dry buttermilk or 13 g dry whole milk to 100 ml distilled water and mix thoroughly (electric or other type mixer and mixing container). Permit sample to stand 1 hour, stir gently, and determine flavor and odor. Determination should be made with sample at 75°F (24°C).

The American Dairy Products Institute (1971) publication also describes those products that might be unfit for human consumption: "Any nonfat dry milk, dry whole milk, or dry buttermilk failing to meet requirements for Standard Grade may be considered an ungraded product. However, any dry milk product which has a flavor, odor or other characteristic indicative of decomposition or neutralization, or which fails to meet the General Requirements shall be termed 'Unfit for Human Consumption.'"

REFERENCES AND BIBLIOGRAPHY

American Dairy Products Institute, Inc. 1971. *Standards for Grades of Dry Milks, Including Methods of Analysis* (Manual). Chicago, IL. 53 pp.

American Public Health Association. 1985. *Standard Methods for the Examination of Dairy Products* (15th Edition). New York. 412 pp.

Arnold, R. G., Libbey, L. M., and Day, E. A. 1968. Identification of components in the stale flavor fraction of sterilized concentrated milk. *J. Food Sci. 31*:566.

Association of Official Analytical Chemists. 1984. *Official Methods of Analysis* (14th Edition). Washington, D.C. 1141 pp.

Bodyfelt, F. W., Andrews, M. V., and Morgan, M. E. 1979. Flavors associated with the use of Cheddar cheese whey powder in ice cream mix. *J. Dairy Sci. 62*(I):51.

Code of Federal Regulations. 1987a. Title 21. Part 131. Milk and Cream Products. U.S. Government Printing Office, Washington, D.C.

Code of Federal Regulations. 1987b. Title 7. Part 58. Dairy Products Grading Standards. U.S. Government Printing Office, Washington, D.C.

Davis, R. N. 1939. Some properties of milk powders with particular reference to sweet buttermilk powders. *J. Dairy Sci. 22*:179.

Gould, I. A. and Leininger, E. 1947. Composition of the solids which deposit from evaporated milk during storage. *Mich. Agr. Expt. Sta. Quart. Bul. 30*(1):54.

Graham, D. M., Hutton, J. T., and McIntire, J. M. 1981. Concentrated and dry milks and wheys in the third quarter of the 20th century. *J. Dairy Sci. 64*:1055.

Hall, C. W. and Hedrick, T. I. 1971. *Drying Milk and Milk Products* (2nd Edition). AVI Pub. Westport, CT. 338 pp.

Hammer, B. W. 1919. Studies on formation of gas in sweetened condensed milk. IA. *Agr. Exp. Sta. Res. Bul.* 54. Ames.

Hunziker, O. F. 1949. *Condensed Milk and Milk Powder.* Published by the author. LaGrange, IL. 502 pp.

Lea, C. H., Moran, T., and Smith, J. A. 1943. The gas-packing and storage of milk powder. *J. Dairy Res. 13*(2):162.

Mojonnier, T. and Troy, H. C. 1925. *The Technical Control of Dairy Products.* Mojonnier Bros. Co. Chicago, IL.

Muck, G. A., Tobias, J., and Whitney, R. M. 1963. Flavor of evaporated milk. I. Identification of some compounds obtained by the petroleum ether solvent partitioning technique from aged evaporated milk. *J. Dairy Sci. 46*:774.

Rice, R. E. 1926. Sweetened condensed milk. VI. Tallowiness. *J. Dairy Sci. 9*:459.

Sato, M. 1923. Sediments of evaporated milk. *Proc. World's Dairy Cong. II*:1284. Washington, D.C.

Sommer, H. H. and Hart, E. B. 1926. The heat coagulation of evaporated milk. Wis. *Agr. Expt. Sta. Tech. Bul.* 67.

Sundararajan, N. R., Muck, G. A., Whitney, R. McL., and Tobias, J. 1966. Flavor of evaporated milk. II. Changes in flavor on storage of evaporated milk made by three processes. *J. Dairy Sci. 49*:169.

Thomas, E. L. 1958. Dry milk products score card including scoring guide and definition of terms; Report of subcommittee to develop a score card for nonfat dry milk. *53rd Ann. Meeting, Amer. Dairy Sci. Assoc.* Raleigh, NC.

U.S. Department of Agriculture. 1986. *Methods of Laboratory Analysis.* DA Instruction No. 918-103. Dry milk products series. Dairy Grading Branch, AMS. Washington, D.C.

U.S. Department of Health and Human Services. 1978. *Grade A Pasteurized Milk Ordinance.* Food and Drug Administration. Washington, D.C.

Webb, B. H. and Holm, G. E. 1930. Color of evaporated milk. *J. Dairy Sci. 13*:20.

Whitaker, R. 1931. The feathering of evaporated milk in hot coffee. *J. Dairy Sci. 14*:177.

Preparation of Samples for Sensory Training

Individuals responsible for providing the instruction or training of people in the recognition of dairy foods off-flavors are interested in methods for preparing and demonstrating various examples of product defects. Several dairy products lend themselves to the preparation of special samples that simulate common or classic off-flavors. The purpose for such prepared samples is to simulate actual flavor defects, or various body, texture, color, and appearance shortcomings of dairy products. The preparation of so-called "doctored samples" is necessary, since it is impractical or may be impossible to locate examples of a majority of the typical sensory defects in the local marketplace, within a reasonable expenditure of effort and cost.

In the U.S., various studies (Bandler and Barnard 1974; Dunkley 1968; Nelson and Trout 1964; Shipe et al. 1978; USDA 1975) have been conducted over several decades in an effort to develop "recipes," "formulations," or laboratory treatments that would provide specific reference standards for milk off-flavors. Such flavor reference standards can do much to facilitate the recognition of off-flavors and the training of personnel in sensory evaluation.

Dairy foods off-flavors are usually caused by a mixture of flavor compounds which are difficult, if not impossible, to duplicate artificially. Though synthesized mixtures of chemicals may simulate natural food flavors, these combinations of "flavor components" seldom truly duplicate them. For flavor training purposes, it is not necessary to exactly duplicate a given flavor sensation. However, there is the risk of improper flavor identification if an "atypical standard" is used.

For the off-flavor recognition of fluid milk, the most recent and comprehensive treatment of reference standards is the publication of Shipe et al. (1978). This paper summarized methods for preparing milk off-flavors for the following flavor categories: (1) heated milk; (2) light induced; (3) lipolyzed; (4) microbial; (5) oxidized; and (6) transmitted. The authors of this standard paper on the preparation of "doctored milk" samples also provided an extensive bibliography for each of the aforementioned off-flavor categories. The quality control director, instructor, or coach of dairy products evaluation teams is advised to consult this publication for determining the most appropriate method(s) for preparing various reference standards for given milk off-flavors.

At this point, the authors will share some of their knowledge and experience in the preparation and selection of dairy products samples, to assist in the ongoing effort of training and developing future judges of dairy products. Prepared samples may be used to train students: (1) in university or community college courses in sensory evaluation, quality control, or preparation for judging competitions; (2) vocational education (Future Farmers of America) activities; (3) within the dairy industry (clinics for milk haulers, fieldmen, receivers, and quality assurance personnel); and (4) personnel in product development.

PREPARATION OF MILK OFF-FLAVORS

Inasmuch as whole milk is the basic raw material for dairy products, it is logical to examine this item initially in the typical progression of the development of sensory skills. The authors express the philosophy that most initial educational efforts related to the recognition of sensory defects of dairy foods should "center around" an intensive exposure to the sensory elements of fluid milk flavor. It is probably advisable to "accentuate the positive" and, hence, try to demonstrate with a reference sample what either an "ideal" or "near-ideal" milk sample is like in terms of taste, aroma, and mouthfeel characteristics.

Coaches for most of the teams that compete in the Collegiate Dairy Products Evaluation Contest generally expend substantial time and effort on the sensory examination of fluid milk, compared to the other five dairy foods evaluated in the competition. Currently, the Dairy Foods Quality Contest of the Future Farmers of America is limited to the sensory evaluation of fluid milk. This focus on the sensory attributes of fluid milk seems justified for the following reasons:

1. The traditional rationale—judging milk is simply where one should start!
2. If the evaluator can recognize and name the "classic" sensory defects of milk, the judge should likewise be capable of identifying the occurrence of the same sensory faults in other dairy products.
3. The selection or "doctoring" of milk samples for off-flavors is generally easier to accommodate than the modification or preparation of "manufactured" dairy products.

4. The physical handling, storage, coding, and tempering of milk samples is less complicated than is the case for other dairy products.

Commonly, milk samples are prepared and presented to participants in either 1/2-pint, quart, or 1/2-gallon size containers, depending on the number of evaluators involved in the training activity or clinic. The containers used most frequently are clean, dry, reusable glass, or plastic bottles (with tight-fitting closures). For the annual Collegiate Contest, single-service plastic containers have been used since 1981, due to the unavailability of glass milk bottles in certain locales. Glass bottles are probably the most convenient and readily available containers in many classroom and many industry laboratory settings. The features of glass include: (1) easy tempering in water baths; (2) cleanability; (3) inert contact surface; (4) easy marking or coding by indelible pens or pressure-sensitive labels; (5) easy capping; and (6) no background odors.

It is advisable to prepare many of the "doctored" samples of milk or cottage cheese at least 24 to 48 hours in advance of when they may be needed for training purposes. Some off-flavors may require this amount of time to fully "develop" or have the flavor characteristics reach equilibrium (e.g., oxidized (metal-induced), rancid, and some microbial-caused off-flavors). The "recipes" provided by the authors in Table 11.1 (for milk) should permit an instructor's assistant or laboratory helper to prepare the needed samples before the designated class time. It is recommended that an experienced dairy products judge (the instructor) closely supervise the preparation of off-flavor samples, as well as, conduct a final check of each off-flavor (present or absent), the defect intensity, assign a score (if appropriate) and check the coding (numbering) of each sample.

At this point, the authors would like to emphasize some precautions regarding the preparation or "doctoring" of dairy products samples. Whenever samples are "doctored" with chemicals and/or other additives there is a possibility of inducing a toxic or allergic response in some people. Prudent judgment must be exercised in deciding whether or not to use such samples. If "doctored" samples are used, the instructor must inform the student tasters of what was done to the treated products. Furthermore, the instructor should provide directions or advice as to whether the samples may actually be tasted or merely smelled. In no instance should any such samples be swallowed. When the risk to human safety or health appears too great, the instructor should endeavor to locate samples with desired specific defects from commercial products.

Table 11.1. Procedures for Preparing Samples of the Common Off-Flavors of Fluid Milk.[a]

No. Off-flavor	Procedure[b] (Quantities for 600 ml sample)
1. Acid (sour)	Add 25 ml fresh cultured buttermilk to 575 ml fresh past./homog. milk. Vary ± 5–10 ml to alter intensity. Should prepare 24–48 hr ahead.
2. Cooked	Heat 600 ml fresh past./homog. milk to 80°C (176°F) and hold for 1 min and cool.
3. Feed (alfalfa)	Add approx. 2–3 g of alfalfa hay to about 100 ml of fresh past./homog. milk and hold for approx. 20 min. Then strain the milk through cheese cloth (in a funnel) into another container. It is highly advisable to pasteurize this "stock solution" of milk by heat treating it at 70°C (158°F) for ≥ 10 min. Next, for each 575 ml of milk: (a) Add ~20 ml of this "alfalfa" milk for—*slight*. (b) Add ~30–35 ml of this "alfalfa" milk for—*definite*. *Note:* Other roughages may be used to prepare feed off-flavors in a similar manner. Pasteurize any prepared "stock solutions" of feeds in milk.
4. Flat	Add approx. 75–100 ml of distilled or good quality tap water to 525–500 ml of fresh past./homog. milk (*slight* intensity); add 110–120 ml water to 485 ml milk for a *definite* intensity.
5. Garlic/onion	Add 0.15 g garlic or onion salt *or* 2 drops garlic or onion extract to 600 ml past./homog. milk (*definite* intensity).
6. Malty	Add 15 g Grape Nuts® or Grape Nuts Flakes® breakfast cereal to 100 ml milk and hold for 20–30 minutes. Strain through cheese cloth, then add 13 ml of the "stock" to 590 ml past./homog. milk (*definite* intensity). Alternatively, a direct addition of 1/2 to 1 teaspoon of unflavored malted milk to 600 ml past./homog. milk may be employed.
7. Malty/acid (sour)	For a typical, combined malty/acid off-flavor, add 12 ml cultured buttermilk and 15 ml of "malty stock" to 575 ml fresh past./homog. milk.
8. Oxidized (metal-induced)	Prepare 100 ml of 1% $CuSO_4 \cdot 5H_2O$ solution and keep refrigerated. Add the following amounts of "stock copper" solution to 600 ml past./homog. milk: *Slight*—0.75 ml 1% $CuSO_4$ *Definite*—1.2 ml 1% $CuSO_4$ *Pronounced*—1.8 ml 1% $CuSO_4$ *Note:* Highly advisable to prepare 24–48 hr ahead of use. Alternatively, if pasteurized creamline (unhomogenized) milk is available, exposure of this milk to sunlight will produce the oxidized off-flavor without addition of copper ions.
9. Oxidized (light-induced)	Add 600 ml past./homog. milk to a clear glass or plastic milk container. Expose milk to bright, direct sunlight for the following exposure times: *Slight*—8 to 9 minutes *Definite*—10 to 11 minutes *Pronounced*—12 to 15 minutes *Note:* Plan to use such prepared samples for only 1 or 2 days; the generic oxidized (metal-induced) off-flavor may develop within 36–48 hours after light exposure.

476

Table 11.1. (*continued*)

No. Off-flavor	Procedure[b] (Quantities for 600 ml sample)
10. Rancid	Agitate 100 ml raw milk with 100 ml past./homog. milk in a Waring ⓦ blender or similar mixer (or milkshake maker) for 2 min. Extend to 600 ml total volume with past. homog. milk. *Notes:* (a) Prepare at least 24–36 hr ahead, if possible. (b) This prepared sample should be pasteurized prior to presentation to tasters. Heat to 70°C (158°F) for $\geq$ 10 min and cool. If facilities permit, homogenize raw milk at 32°C (90°F) and hold until desired intensity of rancid off-flavor develops (5–20 min). Pasteurize, cool, bottle, and store the milk. Since relatively large quantities can be prepared in this manner, the rancid milk can be incorporated into ice cream, cottage cheese, and other products as needed for training purposes.
11. Fruity/fermented	Prepare a "stock ester" aqueous solution of 1% ethyl hexanoate (food grade). Obtain this additive from a supplier of food additives. *Slight*—Add 1.0 ml of stock solution to 600 ml of past./homog. milk *Definite*—add 1.25 ml of stock solution to 600 ml milk.
12. Lacks freshness	(a) Add 10–15 g nonfat dry milk powder to 600 ml past./homog. milk. (b) Select past./homog. milk samples that are approaching end of a 10–15 day "pull date." (c) Sometimes, 2% lowfat milks when compared to whole milk will exhibit the "lacks freshness" off-flavor.
13. Unclean	(a) Select past./homog. milk samples that have exceeded "pull date" by several days. (b) If above selected sample(s) do not exhibit "unclean" off-flavor, then incubate 4–12 hr at room temp., and reexamine for "unclean" off-flavor. (c) Can maintain an "unclean" milk sample in refrigerator, and add 5–50 ml of it to 600 ml of fresher milk.
14. Bitter	Prepare a 2% aqueous solution of quinine sulfate. *Warning:* Some individuals may be allergic to quinine compounds. *Slight*—Add 1.0 ml quinine solution/600 ml milk *Definite*—Add 2.0 ml quinine solution/600 ml milk
15. Foreign	(a) Chemical (chlorine). Add 1.0 ml 5% chlorine bleach (commercial sanitizer type approved for dairy plants)/600 ml milk. It is advisable to only smell, and *not* taste, such prepared samples. *Notes: Definite intensity.* Prepare fresh daily. (b) Vanilla (*definite*). Add 3–4 ml of 2-fold vanilla extract/600 ml milk.

[a] *Precautionary Note:* Whenever chemicals, other additives, or special treatments are used to "doctor" samples, coaches or instructors are firmly obligated to:
  a) Inform all participants of the fact that certain samples (preferably tagged with a warning label) have been treated with designated chemicals (or in another specified manner);
  b) Provide instructions as to whether samples may be tasted or whether they should merely be smelled (*all samples—not merely those for which precautions appear in this table*);
  c) Emphasize that in no instance should any of the given samples be swallowed.
[b] Base milk (pasteurized/homogenized whole) should be of good quality, free of any off-flavors (i.e., only slight cooked or feed permissible).

PREPARATION OF COTTAGE CHEESE SENSORY DEFECTS

The authors have summarized some of the techniques for preparation of the more common off-flavors of creamed cottage cheese in Table 11.2. If, in the course of preparing product defects, it is determined that certain samples exhibit excessive intensity of off-flavor, high-quality cottage cheese can be added to the sample(s) to dilute the defect intensity.

The various body and texture defects of creamed cottage cheese are difficult to simulate or prepare in the laboratory. It is advisable to survey various available brands of commercial cottage cheese and critique them (Connolly *et al.* 1984) in an effort to find a range of possible defects that are determined by mouthfeel. However, many color and appearance defects of cottage cheese can be staged or simulated by the coach (see Table 11.3). Occasionally, it is helpful to critique samples of large curd cottage cheese; this product type frequently exhibits a "high acid" flavor note and is often rated as "too firm/rubbery," due to its comparably larger mass.

PREPARATION OF BUTTER OFF-FLAVORS

The selection or preparation of butter samples for sensory training can be approached in four different ways. First, the instructor can purchase various brands and types (kinds) of butter in the local marketplace (e.g., sweet cream, lightly salted, unsalted, whey cream, cultured cream, and whipped). This approach will generally provide a fairly broad spectrum of qualities and flavor characteristics. However, one

Table 11.2. Procedures for Preparing Samples of the Common Off-Flavors of Creamed Cottage Cheese.

No. Off-flavor	Procedure[a,b,c] (Quantities for 400 g sample)
1. Acid (high)	*Slight* intensity—Add 15–20 ml fresh buttermilk to 385 g good quality cottage cheese. *Definite* intensity—Add 30–40 ml fresh buttermilk to 365 g cheese.
2. Bitter	*Slight*—Add 2.0 ml of 1% aqueous quinine sulfate solution to 400 g cottage cheese. *Definite*—Add 3.5–4.0 ml of 1% aqueous quinine solution to 400 g cheese. *Note:* Advisable to disperse quinine solution in half-and-half or whole milk. *Warning:* Some individuals may be allergic to quinine compounds.

Table 11.2. (continued)

No. Off-flavor	Procedure[a,b,c] (Quantities for 400 g sample)
3. Diacetyl	*Slight*—Add 0.1 ml diacetyl (food grade) to 400 g cottage cheese.
	Definite—Add 0.2 ml diacetyl (food grade) to 400 g cottage cheese.
	Notes: (1) Use food grade diacetyl or starter distillate obtainable from suppliers of food additives. (2) Can disperse diacetyl with milk or cream (~5–10 ml).
4. Fruity/fermented	*Slight*—Add 1.0 ml of 1% aqueous solution of food grade ethyl hexanoate to 400 g cottage cheese.
	Definite—Add 1.25 ml of 1% aqueous food grade ethyl hexanoate to 400 g cheese.
	Note: Alternatively, can add 1/2–1 tsp. of banana, pineapple, or pina colada yogurts to 400 g of cottage cheese. Restrict addition of fruit pieces. Food grade ethyl hexanoate is available from suppliers of food additives.
5. High salt	*Slight*—Add 2.0 ml saturated NaCl (table salt) solution (23% w/w) to 400 g cottage cheese.
	Definite—Add 3.0–3.5 ml saturated NaCl solution to 400 g cheese.
6. Oxidized (metallic)	*Slight*—Add 2.0 ml of 1% aqueous $CuSO_4 \cdot 5H_2O$ solution to 5 ml milk or half-and-half (H&H) and add to 400 g cheese.
	Definite—Add 3.0 ml of 1% $CuSO_4 \cdot 5H_2O$ to 5 ml milk or H&H and add to 400 g cheese.
	Note: It is advisable to prepare about 24 hrs in advance.
7. Rancid	Add 10–15 ml of known rancid milk, half-and-half, or cream to 400 g cottage cheese. Be sure rancid milk (cream) has been pasteurized prior to addition to cottage cheese.
8. Lacks fine flavor (acetaldehyde, green)	Add 1/2–1 tsp. of plain yogurt to 400 g of cottage cheese.
9. Lacks freshness	Select samples of cottage cheese that have approached pull date (or exceeded it). Additional temperature abuse may develop "lacks freshness" off-flavor if not already evident. Selected lowfat cottage cheese samples may also demonstrate this defect.
10. Unclean	Select samples of cottage cheese that have exceeded pull date. "Heat shock" for 4–12 hr at room temperature if "unclean" off-flavor is not evident in samples.

[a] The base cottage cheese should be of relatively good flavor quality, free of any off-flavors (i.e., only slight acid, flat, lacks fine flavor permissible).

[b] Alternatively, many of the off-flavors in creamed cottage cheese may be produced by adding "off-flavored" dressing to fresh dry curd. Thus oxidized dressing (exposed to sunlight), rancid dressing (prepared by homogenizing as described in Table 11.1), etc. may be added to dry curd to prepare the samples with the desired off-flavors for training purposes.

[c] *Precautionary Note:* Whenever chemicals, other additives, or special treatments are used to "doctor" samples, coaches or instructors are firmly obligated to:
   a) Inform all participants of the fact that certain samples (preferably tagged with a warning label) have been treated with designated chemicals (or in another specified manner);
   b) Provide instructions as to whether samples may be tasted or whether they should merely be smelled (*all samples—not merely those for which precautions appear in this table*);
   c) Emphasize that in no instance should any of the given samples be swallowed.

Table 11.3. Procedures for Simulating or 'Staging' Common Color and Appearance Defects of Creamed Cottage Cheese.

Defect	Procedure for Developing Defect
Free cream	Add sufficient half-&-half or whole milk to creamed cottage cheese to obtain a "zone" of free dressing ≥ 0.3 cm (1/8 in.) around periphery of product mound on display plate.
Free whey	Add sufficient whey from a "wheyed-off" buttermilk, sour cream, or plain yogurt (or tap water) to obtain a "zone" of separated clear or translucent liquid ≥ 0.3 cm (1/8 in.) around periphery of product mound on plate.
Lacks cream	Either (1) rinse cottage cheese free of cream dressing with warm water (45°C (113°F)), or (2) add sufficient quantity of dry curd to provide distinct impression of an "underdressed" cottage cheese.
Matted	"Search" through several samples of commercial cottage cheese to locate and retrieve large curd pieces (lumps) that are at least 3–4 times the size of typical curd particles. Place "lumps" at edge of product mound on the plate.
Shattered curd	A vast majority of U.S. cottage cheese exhibits this defect. The greater problem will be encountering a cottage cheese sample that is void of this defect. Curd size and shape uniformity is especially sought. When this defect is called, the evaluator should observe at least 4 or more quite small, shattered pieces of curd ("curd dust") per typical curd particle in the sample. Another approach is to note 6 or more particles of "curd dust" within the cream dressing on the back side of a teaspoon that has been used to stir the cheese sample.

unfortunate limitation may be a frequent occurrence of a storage (and/ or absorbed) off-flavor in butter samples secured in this manner. Secondly, the instructor or coach can ask various butter manufacturers or packers to be on the "look out" for "interesting" samples (i.e., butter with distinct flavor defects) and save them for subsequent classroom examination. Third, the sensory trainer can "treat" small quantities of butter (~ 1 kg [~ 2.2 lb]) to simulate certain flavor defects (e.g., cheesy, garlic/onion, musty, neutralizer, unclean, or yeasty). This approach can be accomplished by either: (1) blending appropriate substances with tempered butter (≥ 25°C [≥ 77°F]) or (2) placing butter in an airtight jar or vessel adjacent to the appropriate volatile material for 12–24 hr (to allow for odorant absorption). Fourth, small batches of butter may be churned (in a one or two gal [3.6–7.2 l] home-size churn). Especially treated (or abused) cream can be converted to small lots of butter that exhibit flavor defects (e.g., acid/coarse, cheesy, cooked, metallic, musty, malty, old cream, neutralizer, rancid, or unclean).

The ingenuity or creativity of the coach in "tainting" the cream is the prime determinant of the effectiveness of this approach. Some of the treatments of milk (Table 11.1) can be extended to heavy (whipping) cream, which can then be churned in an effort to develop butter samples with various flavor defects.

SELECTION OF CHEDDAR CHEESE SAMPLES FOR SENSORY DEFECTS

Generally, there is a wide selection of Cheddar cheese in the U.S. marketplace, thus making it quite simple and easy (though relatively expensive, perhaps) to obtain an outstanding assortment of cheese samples. This large choice bodes well for finding a vast cross-section of sensory properties in this dairy product. This situation is partially due to the range in age of Cheddar cheese (fresh, mild, medium, sharp, and extra sharp) as well as the availability of local, regional, and national brands. Some collegiate team coaches purposely examine other cheese varieties (but in a comparative way with Cheddar cheese) in an effort to establish certain reference points, formulate word descriptors of sensory characteristics, or develop points of emphasis (or exaggeration). Some of the other cheese varieties thus examined might be washed curd, Colby or Monterey Jack (flat or acid); Swiss (low acid, whey taint, or unclean); Havarti or brick (unclean); baby Gouda (whey taint); Parmesan, Romano, or Kasseri (rancid); aged club (sulfide); etc. In addition to flavor defects, this approach may also assist in characterizing certain body, texture, color, and appearance defects of Cheddar cheese.

SELECTION AND PREPARATION OF ICE CREAM SENSORY DEFECTS

A wide variety of sensory properties can be found in the U.S. marketplace for ice cream. In the unlikelihood that the sensory instructor cannot purchase a "range of qualities" from retailers or processors, it is possible to "doctor" ice cream mixes in an effort to predetermine flavor defects in the frozen product. Home-style ice cream freezers or batch freezers facilitate the conversion of abused or treated mixes into desired defect samples for training purposes. Use of mellorine, ice milk, or frozen vanilla yogurt can be useful to illustrate certain body and texture or flavor characteristics.

Table 11.4. Self-Test Sheets for ADSA Score Guides.

	Slight	Definite	Pronounced
MILK			
Acid	_____	_____	_____
Astringent	_____	_____	_____
Bitter	_____	_____	_____
Cooked	_____	_____	_____
Feed	_____	_____	_____
Flat	_____	_____	_____
Fermented/fruity	_____	_____	_____
Foreign	_____	_____	_____
Garlic/onion	_____	_____	_____
Lacks freshness	_____	_____	_____
Malty	_____	_____	_____
Oxidized (light-induced)	_____	_____	_____
Oxidized (metal-induced)	_____	_____	_____
Rancid	_____	_____	_____
Salty	_____	_____	_____
Unclean	_____	_____	_____
BUTTER			
Acid	_____	_____	_____
Cheesy	_____	_____	_____
Coarse	_____	_____	_____
Cooked	_____	_____	_____
Feed	_____	_____	_____
Flat	_____	_____	_____
Garlic/onion	_____	_____	_____
Metallic	_____	_____	_____
Musty	_____	_____	_____
Neutralizer	_____	_____	_____
Old cream	_____	_____	_____
Oxidized	_____	_____	_____
Rancid	_____	_____	_____
Unclean/utensil	_____	_____	_____
Whey	_____	_____	_____
Yeasty	_____	_____	_____
COTTAGE CHEESE			
Flavor			
Bitter	_____	_____	_____
Diacetyl	_____	_____	_____
Fermented/fruity	_____	_____	_____
Flat	_____	_____	_____
Foreign	_____	_____	_____
High acid	_____	_____	_____
High salt	_____	_____	_____
Lacks fine flavor	_____	_____	_____
Lacks freshness	_____	_____	_____
Oxidized	_____	_____	_____
Rancid	_____	_____	_____
Unclean	_____	_____	_____
Yeasty	_____	_____	_____

Table 11.4. (*continued*)

	Slight	Definite	Pronounced
Body and Texture			
Firm/rubbery			
Mealy/grainy			
Pasty			
Slick			
Weak/soft			
Appearance and Color			
Free cream			
Free whey			
Lacks cream			
Matted			
Shattered curd			
CHEDDAR CHEESE			
Flavor			
Bitter			
Fermented/fruity			
Flat			
Garlic/onion			
High acid			
Rancid			
Sulfide			
Unclean			
Whey taint			
Yeasty			
Body and Texture			
Corky			
Crumbly			
Curdy			
Gassy			
Mealy			
Open			
Pasty			
Short			
Weak			
ICE CREAM			
Flavor			
Acid			
Cooked			
Lacks flavoring			
Too high flavor			
Unnatural flavor			
Foreign			
Lacks fine flavor			
Lacks freshness			
Old ingredient			
Oxidized			
Rancid			

(*continued*)

Table 11.4. (continued)

	Slight	Definite	Pronounced
Salty	———	———	———
Storage	———	———	———
Lacks sweetness	———	———	———
Too sweet	———	———	———
Syrup flavor	———	———	———
Whey	———	———	———
Body and Texture			
Coarse/icy	———	———	———
Crumbly	———	———	———
Fluffy	———	———	———
Gummy	———	———	———
Heavy	———	———	———
Sandy	———	———	———
Weak	———	———	———
SWISS-STYLE YOGURT			
Flavor			
Acetaldehyde	———	———	———
High acid	———	———	———
Low acid	———	———	———
Bitter	———	———	———
Cooked	———	———	———
Foreign	———	———	———
Lacks fine flavor	———	———	———
Lacks flavoring	———	———	———
Lacks freshness	———	———	———
Lacks sweetness	———	———	———
Old ingredient	———	———	———
Oxidized	———	———	———
Rancid	———	———	———
Stabilizer	———	———	———
Too high flavoring	———	———	———
Too sweet	———	———	———
Unnatural flavoring	———	———	———
Unclean	———	———	———
Body and Texture			
Gel-like	———	———	———
Grainy	———	———	———
Ropy	———	———	———
Too Firm	———	———	———
Weak	———	———	———
Appearance			
Atypical	———	———	———
Color leaching	———	———	———
Excess fruit	———	———	———
Lacks fruit	———	———	———
Lumpy	———	———	———
Free whey	———	———	———
Shrunken	———	———	———

Table 11.5. ADSA Dairy Product Scoring Guides, the More Pertinent Defects.

MILK

Minor Defects	Sl	Def	Pron
Cooked (aroma)			
Feed (aftertaste)			
Flat (mouthfeel)			
Moderate			
Lacks freshness			
Salty			
Serious			
Bitter (delayed)			
Fermented/fruity			
Foreign, (Cl, vanilla)			
Garlic/onion			
Malty (Grape Nuts®)			
Oxidized/light-induced			
Oxidized/metal-induced			
Most Serious			
Acid (tingle)			
Rancid (soapy)			
Unclean (dirty)			

ICE CREAM

Flavoring	Sl	Def	Pron
Lacks flavor			
Too high flavor			
Lacks fine flavor			
Unnatural			

BUTTER

Minor Defects	Sl	Def	Pron
Cooked			
Coarse			
Flat			
Moderate			
Acid			
Acid/course			
High salt/briny			
Feed			
Whey			
Serious			
Old cream			
Old cream/neutralizer			
Metallic			
Musty			
Unclean			
Rancid			
Cheesy			
Yeasty			

CHEDDAR CHEESE

Minor	Sl	Def	Pron
High acid (frequent)			
Bitter (aftertaste)			
Flat (lacks flavor)			

(continued)

Table 11.5. (continued)

ICE CREAM CONTINUED

	Sl	Def	Pron
Sweetener			
Lacks sweetness		___	___
Too sweet		___	___
Syrup flavor (Karo®)		___	___
Dairy Ingredients/Processing			
Cooked (eggy)		___	___
Lacks freshness (stale)		___	___
Storage (stale)		___	___
Salty		___	___
Old ingredient		___	___
Whey (graham-cracker)		___	___
Acid		___	___
Oxidized (cardboardy)		___	___
Rancid (soapy, bitter)		___	___
Body and Texture			
Coarse/icy (too cold)		___	___
Weak (watery, quickmelt)		___	___
Gummy (taffy-like)		___	___
Crumbly (friable)		___	___
Sandy (hard crystals)		___	___
Fluffy (too high air)		___	___
Soggy (no air, pudding)		___	___

COTTAGE CHEESE

	Sl	Def	Pron
Minor Defects			
High acid		___	___
Flat		___	___
Diacetyl		___	___
High salt		___	___
Lacks fine flavor		___	___

CHEDDAR CHEESE CONTINUED

	Sl	Def	Pron
Moderate			
Whey taint (sweet, slight dirty)		___	___
Unclean (dirty)		___	___
Fermented/fruity (apple-like)		___	___
Sulfide		___	___
Serious			
Garlic/onion		___	___
Rancid (soapy, bitter)		___	___
Yeasty (earthy)		___	___
Body and Texture			
Weak (reduced bite resistance)		___	___
Short (brittle)		___	___
Open (mechanical)		___	___
Gassy (symmetrical)		___	___
Crumbly (falls apart)		___	___
Mealy (by mouthfeel)		___	___
Pasty (sticky, gummy)		___	___
Curdy (young cheese)		___	___
Corky (cork-like, older cheese)		___	___

STRAWBERRY YOGURT SWISS-STYLE

	Sl	Def	Pron
Acidity Level			
Acetaldehyde		___	___
High acid		___	___
Low acid		___	___
Sweetness Level			
Too sweet		___	___
Lacks sweetness		___	___

Moderate Defects
Lacks freshness ||||
Bitter ||||
Foreign ||||
Malty ||||

Serious Defects
Fermented/Fruity ||||
Oxidized ||||
Garlic/Onion ||||
Musty ||||
Rancid ||||
Unclean ||||

Body and Texture
Mealy/grainy ||||
Weak/soft ||||
Firm/rubbery ||||
Pasty ||||

Appearance/Color
Shattered curd ||||
Matted ||||
Free cream ||||
Free whey ||||
Lacks cream ||||

Flavoring System
Unnatural flavor ||||
Lacks fine flavor ||||
Lacks flavor ||||
Too high flavor ||||

Dairy Ingredients/Processing
Lacks freshness ||||
Old ingredient ||||
Unclean ||||
Bitter ||||
Oxidized ||||
Rancid ||||
Foreign ||||
Cooked ||||

Body and Texture
Weak ||||
Too firm ||||
Gel-like ||||
Grainy ||||
Ropy ||||

Appearance and Color
Atypical color ||||
Lacks fruit ||||
Color leaching ||||
Shrunken ||||
Free whey ||||
Excess fruit ||||
Lumpy ||||

SELECTION AND PREPARATION
OF YOGURT SENSORY DEFECTS

In the experience of the authors (Bodyfelt 1980), there is undoubtedly as wide a range of sensory properties for strawberry Swiss-style yogurt (Connolly *et al.* 1984) as one can find with any of the dairy products in the "select six" of the Collegiate Contest. No difficulty should be encountered in finding a good cross-section of sensory qualities of yogurt in most areas of the U.S. In the event that a satisfactory quality range of Swiss-style yogurt samples cannot be located, fruit on the bottom (Sundae-style) yogurt samples may provide an additional market source. If particular desired defects are still lacking, the trainer (instructor) can make small batches of yogurt and "build-in" the appropriate shortcomings.

SELF-TEST FORMS

For prospective competitors in Collegiate Dairy Products Evaluation Contests, several formats for self-testing the students' command of the ADSA scoring guides for six different dairy products are presented in Tables 11.4 and 11.5. Each form presents the range of criticisms from a different perspective—Table 11.4 in an alphabetical arrangement and Table 11.5 in a more logical grouping or categorization of the defects.

REFERENCES AND BIBLIOGRAPHY

Bandler, D. K. and Barnard, S. E. 1974. *Milk Quality Assurance Handbook.* Cornell University Press. Ithaca, NY.

Bodyfelt, F. W. 1980. Body, texture, color and appearance defects of strawberry Swiss-style yogurt, plus glossary of terminology (35 mm slide set). Oregon State Univ. Corvallis, OR.

Connolly, E. J., White, C. H., Custer, E. W., and Vedamuthu, E. B. 1984. *Cultured Dairy Foods Quality Improvement Manual.* Amer. Cult. Dairy Prod. Inst. Washington, D.C.

Dunkley, W. L. 1968. Milk flavour: I.—As it comes from the cow; II.—Changes during storage and processing; III.—Flavour quality control. *Dairy Ind. 33*(1,2,3):19, 91, and 162.

Nelson, J. A. and Trout, G. M. 1964. *Judging Dairy Products* (4th Ed.). AVI Pub. Co. Westport, CT.

Shipe, W. F., Bassette, R., Deane, D. D., Dunkley, W. L., Hammond, E. G., Harper, W. J., Kleyn, D. H., Morgan, M. E., Nelson, J. H., and Scanlan, R. A. 1978. Off-flavors of milk: nomenclature, standards and bibliography. *J. Dairy Sci. 61*(7):855.

United States Department of Agriculture. 1975. Judging and scoring milk and cheese. Farmer's Bull. No. 2259. U.S. Government Printing Office, Washington, D.C. 19 pp.

12

Sensory Testing Panels: An Overview

The subject matter of preceeding chapters has primarily focused on the training and development of qualified or "expert" evaluators. The so-called "expert" judge of dairy products is a person who should recognize both desirable and undesirable sensory characteristics of various dairy foods. As a rule, *natural* foods (as opposed to *fabricated* food products) more readily lend themselves to sensory evaluation by expert judges. This is the case since only limited technical modifications of so-called "natural foods" can be successfully undertaken. For example, if a person does not care for a particular apple variety, the alternatives are to find either a "better" representative of the given apple variety or select a different variety. For the intents and purposes of quality assurance, it would seem that the preferred approach would be to grade the apples by established standards of quality for the given variety. Ideally, quality grades should have a positive correlation with consumer acceptance; however, it should be recognized that the preferences of consumers and the sensory-based decisions of graders may not always be in agreement.

We might consider whole milk and chocolate milk as two examples of dairy products to be evaluated and compared for their respective sensory characteristics. First, in the instance of whole milk, the highest product quality is attained when milk is produced by healthy animals under desirable conditions. This includes the proper feeding of cows, sanitation, housekeeping, storage, and milk transportation to the plant, followed by "ideal" conditions of processing and distribution. Some milk from such sources (just described) should be free of any off-flavors when evaluated by the most competent and sensitive judge. Some samples of milk can thus serve as an image or model of the "perfect milk," against which other samples of milk can be compared. If the quality characteristics of a given milk sample fall short of the image for the so-called "perfect" milk, the sample is usually "scored down." An attempt is made to state the reasons for this "criticism" in meaningful terms of sensory description.

Although chocolate milk (or chocolate drink) only requires minimal modification from the original state of milk, this flavored product can be considered a form of fabricated food. Chocolate milk (as a fabricated food) possesses a number of specific attributes such as color, concentration and type of chocolate (cocoa), level of sweetness, and viscosity which may be varied at the processor's option. Judgment in this instance should be based on some assessment and/or perception of consumer preference (acceptance). By contrast, in the case of whole milk, the absence of off-flavors is presumed to be a major contributor to the hedonic (the psychological range of feelings from pleasant to unpleasant (or favorable to unfavorable)) aspects. For chocolate milk, the absence of any off-flavors generally contributes to overall quality; however, it does not determine the degree of consumer acceptance compared to the attributes discussed above (i.e., flavor level, sweetness, viscosity, and/or color).

In certain instances, some consumers may actually prefer milk which is not completely free of off-flavors. It could be suggested that those particular consumers should be enlightened as to what constitutes the more appropriate sensory characteristics for fluid milk. To better illustrate this point, would a milk producer or dairy processor be justified in producing or processing milk under less than adequate conditions of sanitation or temperature controls in order to permit bacteria to impart certain milk off-flavors, which conceivably might please some consumers? To proceed accordingly could pose health hazards for the public.

TYPES OF SENSORY PANELS

Objectives of sensory evaluation dictate the type of sensory panel which is most appropriate. In some instances, training is required; in other cases, it is paramount that there strictly be no training of the panelists. For so-called *"consumer panels," the participants are not trained;* this provides the intended realm of reactions to sensory attributes. For *"discrimination panels," different individuals may have to be selected for different products,* since some persons may perform effectively on one product (or on one particular sensory note), but not on another. The test objective(s) (i.e., what we are trying to find out) dictates the type of sensory panel that should be employed.

New product development (or product improvement) proceeds through several stages of evolution. It begins with the conception of a new product, or an idea for the improvement of an existing product. Product development continues through various evolutional phases until the final, critical assessment—product acceptance (or nonacceptance) by consumers. Various steps along the way require different types of sensory panels, which must be carefully chosen or designed. Critical to predicting the success of a food product is the selection of: (1) appropriate sensory procedures, and (2) correct interpretation of sensory test results.

The Individual Expert

Butter and cheese graders are typical examples of "individual experts," whose impartiality and technical authority or expertise may be widely recognized. Designated food products are assigned official U.S. Department of Agriculture grades, based on the sensory judgments of certified USDA graders. It has been traditional in the dairy industry to use individual experts for sensory evaluation of milk, ice cream, and cultured dairy foods, as well as butter and cheese.

Reliance on individual experts for the sole assessment of sensory quality is appropriate only when the quality attributes of a given food product are thoroughly defined. In this approach to sensory evaluation, any departure from the "fixed" mental image of a "perfect product" is visualized and described in terms of individual defects, which usually have two quantitative dimensions—intensity and the degree of seriousness. For example, a slight oxidized off-flavor is considered more serious than a slight feed off-flavor; hence the former is assigned a lower score. The two aforementioned dimensions of sensory defects are in effect "additive" as far as determining either the score or the product grade, within the expert grading format. While intensity scales are commonly used in sensory evaluation procedures, there is definite controversy among sensory authorities related to the appropriateness of "point deductions" that are based on the element of defect "seriousness." One view (O'Mahony 1979, 1986) is that point deductions based on the relative seriousness of a defect transgresses the area of consumer acceptance. Consequently, this professional view insists that a score card which is used for determining the grade of a product (or defect intensity), should not be used for the simultaneous assessment of product acceptance (relative quality score). Stated another way, the score card approach of grading products probably attempts to perform too many sensory assessment tasks.

In response, one can argue that this grading system has performed quite well for the improvement of dairy product quality for many years. Obviously, meaningful assessments of the relative "seriousness of a defect," as perceived by consumers and dairy product graders, should be based on research (appropriate comparative studies). However, certain defects such as "sour," "oxidized," "rancid," "foreign," and "unclean," for example, should be "scored down" in the opinion of the authors, regardless of consumer perceptions. Properly made dairy products that conform to both composition and sanitary regulations should be free of the just-mentioned objectionable defects or shortcomings. Recommendations as to how many points should be deducted (from a "perfect score") for a slightly rancid flavor, or any other defect, should be established after considering relevant data and the opinions of recognized authorities in the field. The American Dairy Science Association, through its Committee on the Evaluation of Dairy Products is responsible for performing this professional or authoritative consensus on the sensory characteristics of dairy products.

Expert judges should be able to: (1) recognize all of the defects that may be encountered; (2) identify those defects by an appropriate and meaningful verbal description; (3) assess the defect intensity; and (4) score the product(s) accordingly (with the aid of an accepted scoring guide). Some evaluators may not qualify as full authorities or experts (due to individual differences in sensitivity), but instead, may be trained for a given or specific task. For instance, raw milk in a dairy farmer's bulk tank may be rejected by the milk hauler, who has been trained to recognize, by smell, those flavor defects which render milk unacceptable. In this case, the mere presence of such off-flavors as oxidized, weedy, rancid, sour, foreign, etc. may be sufficient reason for rejection for Grade A milk, without regard to intensity.

The initial training of the individual expert is usually done by the "coach and pupil" method; this helps ensure (from the start) that relative uniformity is practiced in scoring. To help maintain uniformity, expert judges should periodically assemble for the purpose of scoring the same samples.

Traditionally, individual experts have been used to conduct sensory quality assessment of (1) raw materials; (2) intermediate production samples; (3) fresh and stored final products; (4) competitors' products; and (5) samples taken from various stages of new product development.

Trained Discrimination Panels

Generally, the expert evaluator or judge of dairy products is visualized as not only a sensory expert, but also as a person with training or experience in dairy technology; however, this is not necessarily the case with "discrimination panelists." The latter can be trained for specific sensory discrimination tasks because of their motivation, interest, reliability, and willingness to learn and improve their "sensory performance." In some situations, the availability of an individual is by itself a prime recommendation. Training is designed so that improvement is achieved in self-confidence and discrimination skills of designated sensory attributes. Furthermore, performance on carefully controlled tests provide the criteria, or a basis, for the rejection of less capable or "insensitive" panelists.

Essentially, discrimination panels are used as a laboratory tool to provide information on flavor, tactual, and color properties, which would either be impossible, more costly, or more time-consuming to obtain by other means. Basically, the type of tests performed by these sensory panels include: (1) simple difference; (2) difference, providing direction and extent; and (3) descriptive evaluation. In the simplest form of difference test, the panelists indicate whether they perceive (taste, smell, etc.) a difference between samples of the food or beverage product presented to them. Since a singular sensory note may provide the difference, a prerequisite is that panelists be selected, in part, because of their sensitivity to the sensory note(s) in question. Panelists for descriptive evaluation may also render judgments as to differences, but in a more sophisticated manner. They provide descriptive analyses of flavor, body and texture, or color/appearance (or combinations thereof) in a written or word descriptor format.

There is some similarity between the descriptive panelist and an individual expert in the depth to which both probe into sensory properties, but the end result of their work is not the same. In descriptive evaluation, the product is not graded; instead, its sensory properties are described and differences between products can be pinpointed.

Difference tests performed by discrimination panels are useful in: (1) product improvement; (2) new product development; (3) comparisons of competitive products; and (4) quality assurance. It should be emphasized that the results of difference tests have no bearing on consumer acceptance. An exception may be when no difference is observed between a well-accepted product and a test product; this may be an indication that the test sample may also be acceptable. In such a case, there is a definite implication, at least from the sensory standpoint, that the product should perform satisfactorily. Obviously, there are other important factors such as price, packaging, advertising, product or brand name, etc. that influence market acceptance.

Untrained Laboratory (In-House) Panels

People working within a given location, both office and plant workers (in-house), can provide a convenient source of panelists for hedonic tests (i.e., affective tests that measure sample preference or like and dislike levels). Though these persons are consumers, it must be stressed that these individuals are not fully representative of the target-market population. However, these panelists frequently can provide valuable input, particularly toward screening-out potential "losers." In-house sensory panels pose a number of serious bias problems; hence, they should be used only for screening purposes.

During the course of product development, a need may arise for various types of sensory testing for both trained and untrained panels, including difference, descriptive, preference, and acceptance evaluations. When an untrained panel is applicable (for preference and acceptance evaluation) an "in-house panel" may be activated: (1) quickly; (2) economically; (3) and as often as needed without a great deal of administrative preliminaries.

There are severe limitations or constraints related to the use of in-house (in-company) panels which must be considered when interpreting results. Persons directly involved in development or manufacture of the tested product definitely should not serve on the panel. People with prior knowledge of the products under test must be excluded because of their possible biased opinions. In performing sensory preference work, it is customary to ask panelists, "are you users" of the given product; this may pose various problems in-house (in-company). People may not use the product at all, or they may use it only because they can purchase it economically through a company outlet or "employee's store."

For statistical interpretation, the in-house panel must be large enough, preferably fifty or more people. This may present a problem in smaller companies. Whether duplicate evaluations by twenty-five people is equivalent to a single evaluation by fifty people is often a debatable point. Duplicate determinations measure the ability of individuals to repeat a previous result, whereas increasing the size of the panel provides a better sample of the population. In some situations, the use of the in-house panel would be inappropriate because it would not be representative of the target population. An example would be a food or beverage product intended for children or an ethnic group which is not adequately represented within the company's work force.

Consumer Panels

"Consumer panels" are larger untrained panels of more than a hundred individuals that are used to evaluate products on scales of like/dislike, preference, or acceptance (hedonics). Panelists should be chosen as "random representatives" of product users or potential consumers. If facilities at a manufacturing plant are available, panel evaluations may be conducted there. Commonly, the tests are conducted at a convenient central location (testing facility, mall, county fair, etc.) or in homes. The type of product and facilities needed to properly conduct the test determine the manner in which the test is conducted. Unless the product is specifically intended for a segment of the population, the panel should include proportionate representation from all demographic categories (i.e., age, sex, income bracket, etc.). Results of the consumer panel often may be analyzed for both: (1) overall acceptance, and (2) acceptance within specific demographic categories.

The judgment as to the number of required panelists for the desired test objectives depends on such factors as: (1) the minimum number of evaluations needed for statistical inference; (2) the nature of the product (e.g., new, improved, or modified product); (3) its regional and/or demographic preference, and occasionally; (4) product availability (all samples tested must be identical for statistical inference). Some new products may require complete plant retooling or new manufacturing facilities, so their marketing potential should be thoroughly determined by the use of a greater number of tests and panelists. Other product introductions (improvement or modification) merely involve the addition of a new flavor to an existing line of product flavors (e.g., ice cream) with no additional requirements, except for additional ingredient(s) and properly labeled packaging. In this instance, exhaustive testing is not especially critical, but in the case of a product's failure, substantial economic losses can be subsequently incurred from the inventory of unused packages. The actual number of panelists may consist of several hundred persons (for instance, one hundred families). In a major market study for a finalized new product, thousands of consumers may need to participate in the product evaluation to obtain a more reliable prediction of acceptance.

The consumer test may be monadic (one-sample test) when entirely new or different products are evaluated, or when consumer panelists are asked to make a mental comparison of the given test product with a product that they already have been using. When two samples are evaluated, the presentation may include both samples simultaneously, or one sample at a time. In an "in-home test," the consumer may be given one product. This test involves a request to use it for a period of time, followed by delivery of a second sample. Each of the different products should be presented initially to one-half of the families in the study. Judgment, based on test objectives, dictates whether the samples should be identified or unidentified when presented. There is a logical rationale that could be advanced for each of these variations, but in the final analysis the decision will be based on the best judgment of the test coordinator, supplemented by a knowledge of the principles of sensory evaluation and human behavior.

If the consumer test objective is the prediction of marketplace performance of a certain food item (a new yogurt style for example), it could be argued that the panelists should have all of the information available that normally motivates consumers to buy that type of product. By contrast, if the test objective is to determine whether there are superior sensory properties for one of the products (e.g., an "improved" blueberry flavor in yogurt), all other "choice factors" should be equalized or minimized, when the judgments are rendered. In this latter manner of conducting a consumer test, the panelists perform with minimum bias.

Precautions in Conducting Consumer Panels. To minimize unwanted bias and thus limit statistical inference and to enhance the opportunity of achieving objectives of the consumer test, certain precautions are mandatory in conducting sensory evaluation with consumer panels. The following discussion is an attempt to review some of those precautions.

The results of panel evaluations may be obtained by use of a simple ballot on which the panelists indicate their sample preference or rate the products on some form of preference (affective) scale. Sometimes, it may be advisable to include the question "why did you prefer the selected sample?" on the ballot. The solicited response can be an indication of the reasons that prompted a preference for a given product. In other consumer tests, a more elaborate questionnaire may be developed and used for a specific test or purpose. With a refined questionnaire, the consumer test coordinator may probe beyond sensory preferences and address matters that pertain to: (1) unit price; (2) frequency or forms of use; (3) consumption levels of different-aged family members; and (4) other questions which may be helpful in data interpretation, advance product development, promotion advertising, packaging, etc.

Personal interviews, face-to-face or by telephone, may be used either in lieu of, or as a follow-up to, the questionnaire. Regardless of whether the questions are written or oral, they must not be in the form of a "leading question." Questions must be worded and stated carefully, so that there is absolutely no suggestion (due to the wording or voice inflection) of a preferred response. Unless the answers to the questions contain only the unbiased views and true opinions of consumers, the exercise is invalid as a determination of product preference or acceptance. With care, skilled interviewers can probe with more depth or detail than is possible with a questionnaire.

More than two samples may be tested simultaneously, but the sample properties and other test considerations dictate the optimum sample number per test. When possible, it is advisable to restrict the number of samples that are compared, and keep test conditions as uncomplicated as feasible. It must be stressed again that any potential panelists who have prior knowledge of the test products, or a vested interest in them, should be excluded from participation.

The use of one type of preference/acceptance panel does not preclude the use of another test format for the study of a given product. Consumer panel methods may be sequential. It is conceivable to use testing formats that proceed from the least to the most complicated and expensive, as the product's attributes are refined through the product development process.

There is constant debate among sensory specialists as to whether consumer panelists should, for various reasons, be asked to provide judgments of a discriminatory or descriptive nature. This is a task that is normally reserved for trained panels to perform. The most common response of sensory authorities on this matter is to discourage the practice of descriptive judgments by consumer panelists; caution is urged in interpreting such generated data.

SENSORY TESTS

Triangle Test

The "triangle test" is probably the most commonly used form of sensory difference test. Basically, the triangle test is a sensory procedure that is used to determine a difference in the sensory characteristics between two samples. In conducting the triangle test, three samples are presented, two of which are the same and one is different; panelists are asked to identify the odd sample. Instructions to participants specify that the samples be tested in a certain order to avoid position errors. See Fig. 12.1 for an example of a ballot for the triangle test. The position of the odd sample is randomized so that this sample is tasted either first, second, or third an equal number of times (chances).

Although the extent of sensory difference for the sample is not measured, this test is relatively sensitive for small differences in sensory characteristics. Idealistically, all samples must be homogeneous, except for the one variable that is being measured.

The triangle test is a true difference test, and *any* difference that is there will be "used" by the panelist to differentiate between samples. Instructing people to evaluate a sample for "sweetness only" when

Name_____ Date_____ Product_____

Taste the samples in the order indicated on your ballot. Two of the
samples are identical, the third one is different. Place a check
mark (√) in the square corresponding to the ODD sample.

425 ☐

322 ☐

538 ☐

Comments:

Fig. 12.1. A sample ballot for a triangle test.

other differences are apparent is only deceiving the experimenter. Panelists frequently will use *any means* to "get it right." Panel coordinators should be aware of this component of human behavior and accept this fact. Color differences may be minimized or disguised by red lights or other forms of lighting. When it is the objective to measure a flavor difference, it is critical that differences in product consistency or appearance not give away the identity of the "odd sample."

In the triangle test, the probability is always 33.3% for the odd sample being correctly identified, when the panelist cannot discern any difference. Tables which take this 33.3% probability into account have been developed (Roessler *et al.* 1948, 1978). These tables specify the minimum number of correct identifications required for a statistically valid test result, for varying numbers of panelists in a triangle test, at various confidence levels. Since the percentage of required correct responses decreases as the number of evaluations increase, it is desirable to have a reasonably high number of evaluators or have the evaluators repeat the test one or more times. However, repeated tests should not follow one another in rapid succession. The test coordinator is dealing with two different components of variation: (1) between panelists, and (2) within panelist. The onset of sensory fatigue among the panelists can be a key performance factor.

Statistical significance at the 5% probability level implies that there is a 5% chance of reaching a conclusion that the samples are different, when actually they are not different ($\alpha = 0.05$, Type I error). Obviously, there is substantially more statistical confidence at either the 0.1% or 1% level that at the 5% or 10% significance levels. Exactly which level of significance to accept is often a matter of personal judgment. However, the 5% significance level is commonly chosen for the triangle test. Administrators at the upper echelons of an organization or firm frequently make the decision on the selected probability level.

Duo-Trio Test

The "duo-trio test" is another form of sensory difference method that is similar to the triangle test in that three samples are presented for evaluation. However, one of the duplicate samples is designated as the reference or control and the panelists are instructed to sample it first and acquaint themselves with it. The presentation of samples is balanced; one-half of the panelists receive sample A as the reference, the other half of the panelists recevie sample B as the reference. The benefit of this test is that the reference sample serves as a "warm-up" exercise for panelists. Then, each panelist is asked to determine which one of the remaining samples is identical to the control (or reference sample). In this case, the chance probability of guessing correctly is 50%; consequently, to demonstrate a significant difference, a higher percentage of correct responses is required than in the triangle test. Tables are available to assist the sensory specialist in determining the statistical significance for duo-trio tests (Roessler et al. 1956, 1978).

Experimental guidelines and precautions discussed for the triangle test also apply to the duo-trio test. There is evidence to indicate that if the "odd" sample is greater in intensity, then it is easier to detect sensory differences than if it is less intense than the other samples (ASTM 1968; O'Mahony 1986; Stone and Sidel 1985). Panelists should be made aware that the position of the samples has been randomized. This is done to avoid "expectation errors" in a series of difference tests.

Sensory Difference Test Variations

There are other variations of the above sensory tests, which may be applicable under certain conditions or test circumstances. In one test variation, a control (reference sample) may be presented (as in the duo-trio test), and each panelist is asked to indicate which of a series of subsequent samples are the same as the control and which are different. Unfortunately, as the number of samples in a sensory test increases, the task becomes more difficult for panelists to perform, as well as being tiring, or possibly incurring the effects of flavor adaptation. Hence, this test method may possibly be counterproductive except in visual and tactual (mouthfeel) discrimination.

Two-Sample Tests

"Two-sample tests" also are called "paired comparison tests;" this approach may be used as a discrimination (difference), preference, or acceptance test. The term "preference" in sensory testing refers to those judgments that are rendered by neutral participants in a consumer panel of the products presented to them.

When a paired comparison is used as a sensory difference test, each panelist is instructed to rate the samples on the basis of a selected characteristic such as: (1) flavor intensity; (2) bitterness; (3) viscosity; or (4) any other designated sensory attribute. Typical directions of a ballot (see Fig. 12.2) for a given two-sample test may be: "Taste the two samples in the order indicated below and check the one which you find to have the smoothest texture." The advantage of a two-sample test is the small number of samples to be evaluated, but since there is a 50% chance of selecting one sample over the other, a relatively larger number of judgments (≥ 50) is required to provide statistical validity.

Though two-sample difference tests are commonly used by panels that may have had some sensory training, the key criterion in paired

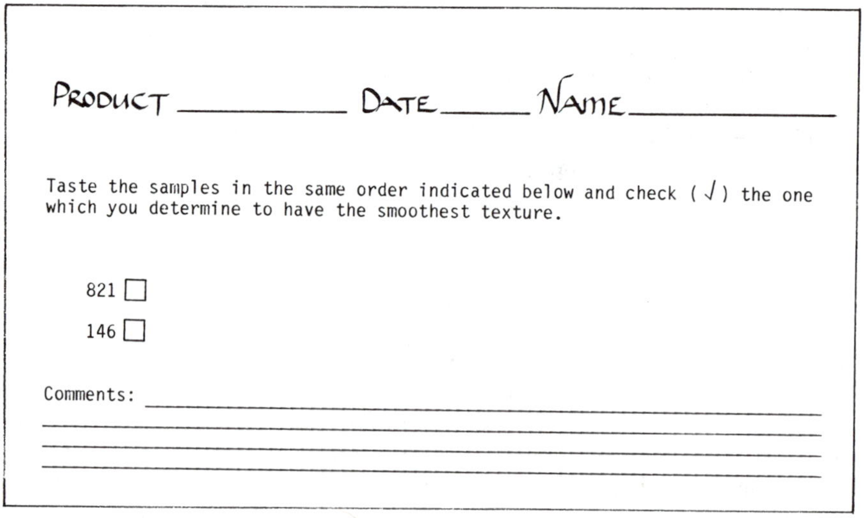

PRODUCT _____ DATE _____ NAME _____

Taste the samples in the same order indicated below and check (√) the one which you determine to have the smoothest texture.

821 ☐

146 ☐

Comments: _____
_____
_____
_____

Fig. 12.2. A sample ballot for a two-sample sensory test.

comparison preference tests is not training, but the necessity that panelists be "representative of the consumer marketplace."

The procedure for the two-sample preference test is similar to the basic difference methods, except that each panelist is asked to indicate a preference for one of the two samples on the basis of either: (1) overall hedonic quality (degree of like or dislike) or (2) merely a singular component of quality such as flavor, texture, degree of sweetness or saltiness, etc. *In a preference test, there are no correct or incorrect responses or answers.* If, for instance, degree of sweetness is the variable in question, the preference test should determine the proportion of people who prefer one level of sweetness in the particular product and the proportion of persons who prefer the other sweetness level. Whether a statistically significant proportion of panelists prefer one sweetness level can be readily determined from tables (Roessler *et al.* 1956, 1978). This approach takes into account the fact that either of the samples may be equally chosen (referred to as a "two-tailed" test).

It should be emphasized that *preference of one sample over another does not, in itself, imply acceptance.* In fact, both samples may be unacceptable. However, if the market performance of one of the compared samples is known, then the preference test can provide either positive or negative signals concerning potential acceptability of the other product sample. This form of sensory test does not provide any indication of the degree to which one sample may be preferred over another. While these technical limitations do not detract from the usefulness of two-sample tests, these constraints should be kept in mind when: (1) selecting a sensory test to meet specific test objectives (e.g., is the fresh sample preferred to a seven-day-old sample?), and (2) when the final test results are interpreted. In many situations, at least two or more sensory tests may need to be used in tandem to provide the required information.

Ranking More than Two Samples

The two-sample approach to sensory testing may be used when more than two samples are to be compared, but each tested sample must be compared against all other samples. This can be time-consuming, tedious and costly; furthermore, interpretation of test results definitely requires the assistance of a statistician. If one of the samples has been designated as the logical control (or reference sample) to which other samples are to be compared, the number of panel tests can be reduced, since a comparison of noncontrol samples to each other is unnecessary. As a rule, difference testing tends to introduce less source of variation than does preference testing. However, a potential problem with the difference test approach is the "anchoring" (preset standard or benchmark) of the reference sample(s).

More than two samples may be compared simultaneously by a procedure called "ranking." In this multisample form of comparison, samples are ranked either on the basis of product preference or on the intensity of a specified characteristic (e.g., lowest to highest flavor intensity, sweetness, carbonation effect, etc.). As in the case of the two-sample test, sample preference may be based on the overall (total) hedonic quality, or on a selected component of sensory quality.

Sample ranking provides a useful and rapid screening method. However, the nature or physical characteristics of the samples impose a practical limit on the number of samples that may be included in a given test. An additional constraint is that sample ranking is the least sensitive sensory method for statistical inference. After some time elapses, the memory capability of the panelists begins to fail. This results in excessive retasting, the onset of panelist fatigue, and possibly questionable results. Three to five samples seem to be the optimal number for taste comparisons. For visually based tests, the number of samples per test may be increased markedly.

Statistical tables are available to assist in interpreting results of ranking tests. These tables provide ranges of rank totals (sum of the ranks assigned to each sample by all panelists), which are not significantly different for various combinations of sample and panel numbers. An experimental sample in which the rank total is outside of the range listed in the table is considered to be significantly different (Kahan et al. 1973). Another statistical interpretation for ranking is an analysis of variance, wherein ranks are converted to scores (Fisher and Yates 1949). Rank-order tables have a provision for the interpretation of results when the set of samples includes a control; for special cases, however, a statistical analysis must be employed appropriate to the type of test design.

In a given test, two or more samples may be compared with a control sample for either sensory difference or preference. To meet this objective, several types of test designs are possible. First, the control sample may simply be one of the blind samples known only to the administrators of the test. A second test possibility is the use of an identified control and a panel that has been instructed to make the sought comparison between the control and each of the other samples. A third test variation is to designate an identified control, but also to include this same sample "blind," as one of the coded test samples. This latter approach is the only valid way for the test administrator to include the control sample in the actual rating, as well as have it serve as the reference sample in the analysis of test results.

Scaling (Quantitation) of Sensory Responses

In order to determine how large the difference is between samples or how much one sample is preferred over another, some type of quantitative scale must be employed. The scale may be numerical, graphical, verbal, pictorial, or a combination of approaches. More specifically, various quantitative scales may take the form of an ordinal, line, category, ratio (magnitude estimation), or a facial expression ("smiley face"). Sensory responses are basically of a psychological nature, rather than physical measurements; hence, for example, a solution containing twice the concentration of sugar does not necessarily taste twice as sweet as a reference sample.

In a way, a psychological quantity is difficult to visualize. Can we really say that one solution is twice as sweet as another? Or that we prefer one sample as much as two, five, or ten times more than another? Psychologists have led the way in providing some answers. Mathematical expressions have been proposed (Dudel 1981) that relate physical or chemical stimuli (in flavor, this is the quantity of the chemical substance that produces the flavor) and sensation (the psychological intensity of flavor). In designing a quantitative sensory experiment, care in the choice of a scale is important. In spite of our best efforts, some compromise may be necessary in certain situations. Basically, this topic is too immense in scope and mathematical detail to discuss within the limitations of this text. Therefore, the reader is advised to consult several of the following authoritative treatments of the subject (ASTM 1968, 1973; Amerine et al. 1965; Kahan et al. 1973; Larmond 1977; Moncrieff 1967; O'Mahony 1986; Roessler et al. 1948, 1956, 1978; Sidel et al. 1975; Stevens 1956; Stone and Sidel 1985).

The Hedonic Scale

The "hedonic scale" has been a most useful tool for the sensory evaluation of food and beverages. Generally, a rating based on the hedonic scale is carried out with nontrained tasters in order to obtain answers that are both spontaneous and subjective. A typical form or chart of the hedonic scale is illustrated in Fig. 12.3. The meaning of the word "hedonic," as provided by Webster, relates to "the psychological range of feelings from pleasant to unpleasant." Instructions that usually ac-

```
                        DAIRY PRODUCT ATTITUDE REPORT

Panel                                            Dairy
member_____  Date_____   product_____
```

Taste the samples in the order listed on this ballot. After tasting each
sample, check (√) in the square to the left that reaction which most nearly
expresses your attitude toward the product. If you have any comments on the
samples, write them in the proper space below.

Code: 425	Code: 317	Code: 701	Code: 532	Code: 675
☐ Like extremely	☐ Like extremely	☐ Like extremely	☐ Like extremely	☐ Like extremely
☐ Like very much	☐ Like very much	☐ Like very much	☐ Like very much	☐ Like very much
☐ Like moderately	☐ Like moderately	☐ Like moderately	☐ Like moderately	☐ Like moderately
☐ Like slightly	☐ Like slightly	☐ Like slightly	☐ Like slightly	☐ Like slightly
☐ Neither like nor dislike	☐ Neither like nor dislike	☐ Neither like nor dislike	☐ Neither like nor dislike	☐ Neither like nor dislike
☐ Dislike slightly	☐ Dislike slightly	☐ Dislike slightly	☐ Dislike slightly	☐ Dislike slightly
☐ Dislike moderately	☐ Dislike moderately	☐ Dislike moderately	☐ Dislike moderately	☐ Dislike moderately
☐ Dislike very much	☐ Dislike very much	☐ Dislike very much	☐ Dislike very much	☐ Dislike very much
☐ Dislike extremely	☐ Dislike extremely	☐ Dislike extremely	☐ Dislike extremely	☐ Dislike extremely
Comments	Comments	Comments	Comments	Comments

Fig. 12.3. Example of a hedonic scale ballot used to record a panel member's attitude towards a given dairy product.

company a hedonic chart are brief. As an example, Peryam and Pilgrim (1957) stated:

"You will be given several servings of food to eat and you are asked to say about each, how much you *like* or *dislike* it. Use the scales to indicate your attitude by checking at the point which best describes your feeling about the food. Keep in mind that you are the judge. You are the only one who can tell what you like. Nobody knows whether these foods should be considered good, bad or indifferent. An honest expression of your personal feeling will help us decide. Take a drink of water after you finish each sample and then wait for the next."

The major advantages of the hedonic scale method, according to Peryam and Pilgrim (1957) are:

"(1) Its simplicity which make it suitable for use with a wide range of populations; (2) subjects can respond meaningfully without previous experience; (3) the data can be handled by the statistics of variables—an advantage inherent in rating scale data; and (4) in contrast to other methods (within broad limits), the results are meaningful for indicating general levels of preference."

As observed from Fig. 12.3, each point on the scale is identified by a verbal description, which should, ideally, be easily understood and have the same meaning to each panel member. Another important requirement is that each descriptor of a categorical scale be equidistant from the preceding or subsequent one in psychological terms. This implies that the distance between "like extremely" and "like very much" should be the same as the distance between "dislike moderately" and "dislike very much." For interpretation of the results of a 9-point hedonic scale, the descriptors are converted to numbers by assigning a value of one (1.0) to "dislike extremely" and progressively advancing upwards to a value of nine (9.0) to represent "like extremely." Sensory evaluators have also used abbreviated five- (5) and seven- (7) point hedonic scales. Interpretation of panel data on the basis of simple arithmetic averages for each sample would not be statistically valid if: (1) the descriptors deviated from an equal partitioning of the linear scale, and (2) the meaning of the various descriptors was not clear to each panelist. Some modifications of the ballot for the hedonic scale (Fig. 12.3) have been employed by sensory specialists. Both the numerical values and the descriptors may be included, or only the numbers or descriptors may be used. The relationship between the descriptors and the numerical values (when used together) must be explained clearly in the instructions.

Another ballot modification is a numerical scale in which only the extremes of the hedonic scale are identified. The instructions for this hedonic test might direct the panelists to score the samples on a scale of from one to nine (1 to 9), where one (1.0) is equivalent to "dislike extremely" and nine (9.0) refers to "like extremely." Graphically, it is permissible to use an unstructured line scale without numbers. Panelists simply place a vertical line at the appropriate point within the continuum that reflects each judge's degree of like or dislike for the sensory attribute(s) in question. For each panelist, the distance of their mark on a line (e.g., 6 in. [15.1 cm] long) is measured and

|⊢—————————————————————————————————⊣

dislike extremely   like extremely

converted to an appropriate numerical value. This is based on the measured inches or centimeters of where the sample was rated on the line. On such an approach, the midpoint of the 6 in. (15.1cm) line may or may not be identified. Scales of more or fewer than nine points may be used, but some precision may be lost when using a shorter numerical scale.

Numerical Intensity Scales

Even when verbal descriptors are used, they must be converted to numbers before they can be analyzed statistically. Linearity (equal distance partitioning of categories) is a definite prerequisite for score averaging. A five-point (5.0) scale is considered quite short for the preferred optimum scale length for sensory tests based on numerical intensity scales. Sensory specialists definitely prefer to use scale lengths of at least seven (7) points; this serves to lend greater precision and often enhances the chance that differences can be found in intensity between samples.

We can consider a hypothetical scale for the intensity of a given flavor note (e.g., an "oxidized" off-flavor in a product following one month of storage). The possible intensity scale could be: absent = 0; threshold level = 1; slight = 2; moderate = 3; strong = 4; and very strong = 5. A pertinent concern of the sensory test coordinator should be: "Is the difference in intensity the same between 'threshold level' and 'slight' as between 'moderate' and 'strong?'" Obviously, an experimenter cannot be absolutely sure; hence, some degree of uncertainty remains even after treating resultant data statistically. On the other hand, if the panel is trained using valid physical examples (i.e., equal interval of intensities) of each descriptor, the data may be reasonably meaningful and trustworthy. The point of emphasis is that if statistical analysis is to be conducted based on mean scores, then the "linearity" of the scale should be kept constant. A category scale (e.g., 1–9) which is identified only at the scale extremes (i.e., "just perceptible"–"extremely strong") should eliminate some of the "linearity" problems for trained panels. Sensory specialists prefer to use what is referred to as "well-anchored scales."

In certain situations for sensory testing, an appropriate standard may be available for comparison against experimental samples for the intensity of a specified characteristic. Trained panelists can be asked to determine: (1) if a difference exists, as well as (2) the extent or degree of difference they can perceive. The panelists can indicate the *degree* of difference as: none, slight, moderate, large, or extreme for the designated characteristic for each sample (midpoints may add sensitivity to the test). Numerical values for this degree of difference may then be assigned as follows: 1 = extremely less intense than standard (e.g., less sweet, softer, coarser, etc.); 5 = same as standard; and 9 = extremely more intense than standard (e.g., more sweet, harder, smoother, etc.). There may be the constant problem of not achieving test objectives due to the misunderstanding by the panelists of the "magnitude of difference" terms and lack of true linearity with respect to the actual line scale or word descriptors.

Whether or not a numerical scale is employed with "verbal anchors," successive points along the line scale must always be of equal distance. A scale length of only a few numbers (or categories) may be more reproducible, but such a short scale is usually less sensitive to smaller differences as perceived and reported by panelists; the opposite result is true of larger or expanded scales.

In essence, the nature of the sample and the type of information desired serve to influence the choice of the scale. Nine- or ten-point scales seem to be most common, but five-point scales have also been employed, as have 25- or 30-point scales. One fact that a sensory test coordinator must keep in mind is that people tend to avoid endpoints of the designated scale. Seven points is usually the minimum number of points recommended by authorities on this subject (Amerine *et al.* 1965; O'Mahony 1986; and Stone and Sidel 1985).

Verbal Scales

According to Stevens (1956) and Sidel *et al.* (1975), a scale that consists of a number of categories, each of which differs in magnitude from another (but not linearly) is called an "ordinal scale." Even if numerical values are assigned to each descriptor, ordinary statistics are not applicable, as a rule. Only "interval scales" which have equal distances between intervals qualify for statistical treatment. Hence, under these circumstances a compromise is often applied to the situation.

Generally, the hedonic scale has been demonstrated to possess essentially equal intervals, but most other tests scales have not been tested as rigorously. Usually, "verbal intensity" scales use descriptors that denote a progression from "lowest to highest" intensity. The following is such an example: absent; just detectable; very slight; slight; slight to moderate; moderate; moderate to strong; strong; strong to very strong; etc. Also the sensory investigator can allow categories in between those just stated, if it is deemed desirable. A scale similar to the 9-point hedonic scale may be used for overall quality assessment (rather than preference). The top anchor "like extremely" becomes "extremely good" or "excellent;" dislike extremely" becomes "extremely poor;" and the midpoint may be designated as "fair." According to Sidel *et al.* (1975), this would be categorized as an ordinal scale. Verbal scales also may be used in descriptive tests, although these applications are usually in combination with numerical or graphical scales.

Graphical Scales

"Graphical scales" as a sensory test method are usually quite unstructured; they merely take the form of a straight horizontal line, with only the extremes identified by verbal anchors. The midpoint of the scale may be identified, also. The panelists are asked to place a vertical mark at the location on the line which corresponds to their assessment of attribute intensity. The distances of the vertical marks from the lowest extreme are then measured and suffice as intensity scores. As indicated earlier, panelists tend to avoid the endpoints of the scale.

Unstructured scales provide a means of overcoming the "unequal interval" problem, as well as minimizing the difficulty of selecting descriptors that convey the same meaning to each panelist. A separate line is used on the form or ballot for each characteristic that is to be evaluated. Figure 12.4 illustrates an example of graphical scaling.

Magnitude Estimation

"Magnitude estimation" is the most common of several methods that can be classified under the heading of "ratio-scaling." This sensory test method (Moskowitz 1977) requires that the first-presented sample be assigned an arbitrary score (e.g., 10 or 50). In some instances, the panelist is instructed to assign a predetermined score to the initial sample; however, the panelists may be allowed to assign their own score, also. The other samples in the test set are now compared to the initial sample that was presented. If the intensity of the designated attribute appears to be twice as high, then the assigned score should be twice that given to the initial sample. If the attribute intensity appears to be half as intense, the assigned score is supposed to be one-half of the initial sample score. There is no upper or lower limit to the scores (no negative numerical values) for the sample set, since statistical analysis involves

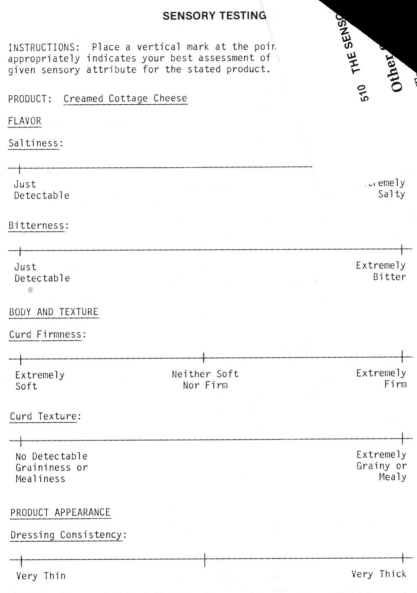

INSTRUCTIONS: Place a vertical mark at the poir.
appropriately indicates your best assessment of
given sensory attribute for the stated product.

PRODUCT: Creamed Cottage Cheese

FLAVOR

Saltiness:

Just   ..remely
Detectable   Salty

Bitterness:

Just   Extremely
Detectable   Bitter

BODY AND TEXTURE

Curd Firmness:

Extremely   Neither Soft   Extremely
Soft   Nor Firm   Firm

Curd Texture:

No Detectable   Extremely
Graininess or   Grainy or
Mealiness   Mealy

PRODUCT APPEARANCE

Dressing Consistency:

Very Thin   Very Thick

Fig. 12.4.   Example of a ballot for the graphical scale method for assessing the intensity of various sensory attributes. Designation of a midpoint
(with or without a descriptive anchor) is optional.

ratios of values. These ratios of the various scores are not simply aver-
aged, but instead are normalized for all panelists and may be handled
as geometric means (usually, "geometric mean normalization" is
employed).

Scales

st situations may arise (e.g., with children) when the use of numerical scales may be inappropriate for sensory evaluation. A sequence of pictures showing facial expressions from a broad smile to a deep frown may be substituted ("facial hedonic scale"). The sequence of facial illustrations should be challenged and evaluated for validity by the test coordinator.

Under some special circumstances, the actual quantities of products consumed by participants may be quantitated and compared. One possible approach is to select an institutional eating place (e.g., college cafeteria) in which consumption levels of a designated product over a period of time are known. Evaluators could determine, for instance, whether substituting a vanilla ice cream with a different source of vanilla flavor changes the level of consumption. Another test approach would be to have two different products "freely available" and monitor the consumption level of each. For example, in determining the relative flavor preferences of animals, such as dogs and cats, the comparable quantities of pet food consumed on a "free choice" basis can be a most useful indicator of sample or product preference.

A "perfect scale" for sensory testing is probably not currently available. If such a scale is within the realm of possibility, it will have to await further research in psychology and psychophysics. However, many of the available methods (discussed above) can be quite useful in providing desired information. To avoid major errors in interpretation of test results, the researcher must be aware of the assumptions undertaken when selecting a scale, since the scales are not perfect (i.e., a lack of true linearity between categories).

DESCRIPTIVE TESTS

The purpose of "descriptive tests" in sensory work is to provide as much information about the flavor and/or texture of a food as can be stated in words and/or numbers. This method of sensory testing helps characterize or articulate sensory properties and product differences in both qualitative and quantitative terms. Members of descriptive panels must be carefully selected and thoroughly trained. This training may require from one to twelve months, depending on the type and depth of descriptive analysis and procedures employed. Flavor has distinct components and properties which must be: (1) recognized; (2) scaled (where applicable); and (3) communicated. The individual flavor notes are described in "meaningful" language and quantitated as to intensity. Other pertinent observations may include: (1) the order in which the flavor notes are perceived (upon tasting and smelling); (2) aftertaste; (3) total flavor intensity; and (4) an overall flavor impression.

Descriptive analysis of foods or beverages may be undertaken by a number of different approaches; some formalized procedures are in common use and will be discussed further. Basically, dairy products evaluation (judging), as treated in previous chapters, may be considered as a form of descriptive analysis in which the observer attempts to identify all undesirable sensory notes that may exist in presented samples. However, descriptive analysis goes beyond that, as illustrated in the following example. Two vanilla ice creams that contain 16% and 10% milkfat, respectively, are found to be free of sensory defects and, therefore, receive so-called "perfect scores," even though they do not taste the same. The sensory differences between these two products may be articulated by descriptive analysis. The best-recognized and widely utilized forms of descriptive tests are: (1) the flavor profile (Caul 1957; Sjostrom *et al.* 1957); (2) the texture profile (Brandt *et al.* 1963); and (3) the Quantitative Descriptive Analysis (QDA) methods (Stone *et al.* 1974).

Flavor Profile

The "flavor profile" procedure (Caul 1957) was developed by the Arthur D. Little Company as early as the 1940s. The aroma and taste dimensions of a sample are evaluated separately. Five or six highly qualified and trained panelists meet as a group and establish the proper terminology for referring to the various and unique flavor notes associated with the product subjected to evaluation. The sensory perception of each note and its verbal description must be perfectly clear to all panelists. When prepared or ready, the panelists independently evaluate and record their observations, one sample at a time. Results of each panelist's observations are reported to the panel leader, who in turn studies these results and holds further discussion with members of the panel. With this approach, problem areas which may have caused variations between panelists are identified. The information which the panelists are asked to provide independently may include the following:

1. Flavor character notes (taste and aroma, separately).
2. The order of appearance of each flavor note.
3. The intensity of each flavor note (using a structured scale).
4. Any perceived aftertaste.
5. Total flavor intensity.
6. Amplitude (an overall flavor impression).

The final results may be analyzed statistically. Also, the descriptive observations are presented usually either in tabular or graphical form. The flavor character notes are usually presented in common or "straightforward" terminology, such as "garlic," "vanilla," "caramel, "boiled milk," etc. These verbal descriptors are listed in the order of their perceived appearance. "Anchored" symbols and numbers are used typically as intensity scale; for instance: 0 = none; 0–1 = just detectable (threshold); 1 = just detectable to slight; 1–2 = slight to moderate; 2 = moderate; 2–3 = moderate to strong; and 3 = strong. Additional midpoints can be added where needed and when appropriate. Sensory differences in products become apparent when: (1) certain notes are absent or of different intensity in aroma, taste, or aftertaste; (2) the order of appearance of flavor notes differs; and (3) total attribute intensity or flavor amplitudes differ.

The Texture Profile

The concept of "texture profile" (Brandt *et al.* 1963) is quite similar to that of flavor profiling, except that various components of texture (Fig. 12.5) are analyzed rather than flavor. First, a carefully selected panel is trained in texture terminology, as well as skills in discrimination of intensity levels of texture (Civille and Szczesniak 1973: Civille and Liska 1975). Although texture characteristics may be defined in a general way, tactual features do not contribute to all products in a uniform manner. The task of the descriptive panel is to select and define the terms which apply to the particular class of product evaluated. Some of the various characteristics that constitute food texture are: (1) mechanical (shear and stress reactions); (2) geometric (shape, size, and particle orientation); and (3) fat and moisture properties (juiciness, saliva reactions, etc.).

According to Civille (1979), intensities of textural features may be expressed numerically, graphically, or by ratio scaling. An example of the order of appearance for textural observations generated during the sequence of selected observations for a particular product are: (1) the surface; (2) partial compression; (3) first bite incisors; (4) first bite molars; (5) during chewing; and (6) after swallowing. Observations on texture amplitude may also be included.

For the dairy products judge, the various terms used to describe texture include all of those descriptors or concepts frequently used to evaluate the body and texture of dairy foods. In addition, certain product attributes which do not necessarily connote a defect (i.e., smoothness, springiness, adhesiveness, moistness, dryness, etc.) are possible descriptive terms for texture profile analysis. Final results may be pro-

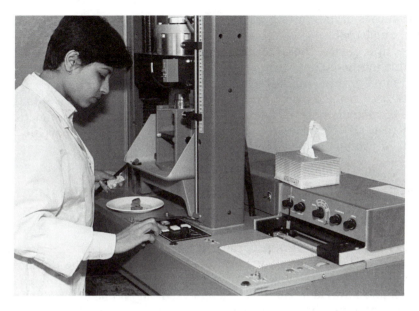

Fig. 12.5. Texture evaluation can be effectively supported by data obtained with an Instron Universal Testing Machine (*Courtesy* Division of Foods and Nutrition, University of Illinois, Urbana-Champaign and Instron Corporation, Canton, Ma.)

vided in tabular or graphical form, and may be analyzed statistically as well.

Quantitative Descriptive Analysis (QDA)

The "QDA" (Stone *et al.* 1974) procedure was developed originally at the Stanford Research Institute, but has found industry application through programs of the Tragon Corporation of Palo Alto, CA. The objectives of Quantitative Descriptive Analysis are essentially the same as those of the just-discussed flavor and texture profile methods. Panels consisting of 10–15 individuals are selected and trained for a specific type of product. Panelists then develop and pretest applicable terminology and an intensity scale that is applicable to each selected term. When ready for testing, each panelist independently evaluates a number of coded sample replicates, one sample at a time. Attribute intensities are recorded on a graphical scale that is six inches in length. With the aid of a plastic overlay, the linear distances of the marked scale are converted to scores in the range of 0 to 60. Usually, a computer that has been programmed to efficiently perform an analysis of variance is used to interpret and summarize the generated data (Stone and Sidel 1985).

At first glance, the QDA method appears to be quite similar to other profile methods, but there are some fundamental differences. The flavor and texture profile techniques depend on sensory experts, who have been trained over a long period of time (usually a year or longer), and are usually able to evaluate several different types of products. By contrast, QDA panels are selected and trained for only a specific type of food product or beverage.

SENSORY TEST SELECTION AND INTERPRETATION

Test Objectives

Selection of the most appropriate sensory test is a key consideration in sensory evaluation of food or beverage products. The decision of whether to use one-sample, two-samples, or multi-sample comparison tests, and how the results will be analyzed, should be determined during the planning stages for the test. *The sensory methods that are selected must be compatible with the test objective* (i.e., they must provide the required information). When a sensory difference test fails to demonstrate an actual "difference," administration of a preference test may be counterproductive. When a difference in product samples is obvious (e.g., in color or appearance), a side-by-side test may easily demonstrate a preference for one of two samples. However, if these products are presented as single samples to separate groups of consumers, both may receive equivalent hedonic ratings (Fig. 12.6). These projections or insights are helpful in planning the entire sequence of sensory testing from product development through marketing and quality control.

Sensory Quality Control

In the dairy industry, substantial reliance is often placed on one or more individual experts to perform sensory evaluation. When performance demonstrates these persons to be qualified and reliable, this system usually works well, provided the method of sample collection is defined as a part of the quality assurance program. Other sensory tests are also applicable for performing quality testing; difference tests (e.g., a triangle test format) may demonstrate day-to-day or batch-to-batch variations. If a highly standardized product is available, other products may be compared to it by the use of a difference test. Descriptive profile tests are applicable for quality assurance purposes, if a trained

Fig. 12.6.   A flavor panel in session in the Division of Foods and Nutrition, University of Illinois (Urbana-Champaign).

panel is available. When competitive products are compared to a company's own products, a descriptive test can articulate subtle and interesting product differences.

Basically, quality control testing should employ the sensory method(s) deemed most appropriate to help insure that raw materials and the finished products conform to the desired and specified sensory characteristics for the given product. The reader is advised to consult recent reviews of the application of sensory evaluation methodology in quality control programs by Nakayama and Wessman (1979), Meroli (1980), Rutenbeck (1985), Skinner (1980), Symons and Wren (1977), and Wolfe (1979).

Objective Tests

Some of the "objective tests" which supplement sensory methodology include chemical tests (both instrumental and noninstrumental) and rheological tests that measure specific physical and mechanical properties of foods by means of instruments. These objective tests are quite useful in both product development and quality control. They cannot replace sensory preference or acceptance tests, but as a quality control tool, various objective tests can help insure that acceptable product properties are maintained. In some instances, certain objective and sensory tests, conducted in concert with each other, may provide complementary and quite useful information on product attributes.

Determination of pH and titratable acidity are among the most common and useful objective procedures. Determination of the Acid Degree Value (ADV), which measure free fatty acids (FFA) in dairy products may be useful on occasion (see Appendix IV). As a chemical test for the extent of hydrolytic rancidity, the ADV is more sensitive than most taste tests. Not all of the free fatty acids that are released by lipase activity contribute to a rancid off-flavor, but all the FFA that may be present react chemically and, hence, are measured in the ADV chemical test. Thus, lipase activity can be monitored somewhat before rancid off-flavor can be detected organoleptically. There are also chemical tests that measure the degree of fat oxidation, such as the peroxide test and the thiobarbituric acid (TBA) test; however, they are not used commonly in the quality control of dairy products. Numerous chromatographic techniques are capable of separating chemical constituents and providing the observer with a profile of many of the chemical components responsible for the flavor characteristics of a given food product. In the dairy industry, these techniques are used primarily in flavor research and product development (Brandt and Arnold 1977; O'Mahony et al. 1979; Pangborn and Dunkley 1964).

Instrumental methods are used to some extent to assess the body or consistency of dairy products. Viscosity of buttermilk, kefir, chocolate milk, and ice cream mix, and the body of sour cream, yogurt, and dips represent some possible applications. Various instruments are available for measuring shear properties, compressibility, cutting characteristics, tensile strength, and other parameters that contribute to product body and texture (e.g., the Instron unit illustrated in Fig. 12.5). However, for dairy products, judgments for these attributes are most commonly made by sensory evaluation. This is also true for the color and appearance aspects of dairy products, despite the fact that the spectrophotometric methods and Munsell Color Notation represent instrumental procedures for potential use.

Interpretation of Test Results

At times, the results of sensory tests are presented as means (average values) in tabular or graphical form, for instance in the flavor and texture profile methods. In nearly all other instances of sensory test methods, some form of statistical analysis is applied in dealing with test results. Table 12.1 provides examples of some of the more commonly applied procedures.

Table 12.1. A List of Common Sensory Tests and the Appropriate Statistical Treatments Associated with Each Method.

Sensory Test	Statistical Treatment
Two-sample preference	Two-tailed, 50% probability tables[a]
Triangle	One-tailed, 33.3% probability tables[a]
Duo-Trio	One-tailed, 50% probability tables[a]
Ranking two or more samples	Rank order tables or analysis of variance (ANOVA)
Ratings of two samples by the same panel	t-test for dependent means
Ratings of two samples each by a different panel	t-test for independent means
Ratings of three or more samples	Analysis of variance and either Duncan's multiple range test or the least significant difference (LSD) test

[a] Available tables round the numbers for attaining statistical significance values to the closest integer; hence, for deployment of exact values, the Chi Square test can be applied.

A completed statistical analysis for a sensory test indicates whether a significant difference exists for a selected level of confidence. If the test results are significant at the 1% level, the inference is that in only 1 case out of 100 would such a result have been obtained by chance alone (in the customary design of statistical analysis). The challenge is to select a probability level which provides adequate confidence; should it be 10%, 5%, 1%, or 0.1%? This is an individual decision that must be made by the administration within the company or organization. However, as a general rule, 5% is chosen as the probability level that provides the lowest level of acceptable confidence. Excellent examples of statistics applied to sensory data (that show stepwise calculations) can be found in the following three references (ASTM 1968, Larmond 1977, and O'Mahony 1986).

MECHANICS OF DISCRIMINATION PANELS

Panelist Selection

In the process of selecting panel members (Fig. 12.7) for discrimination testing, careful consideration should be given to availability, experience, health, taste-smell sensitivity, reliability of judgment, and the personality of the individuals. The emphasis placed on the foregoing factors generally depends upon the test objective(s). Sensory authorities constantly emphasize the point that *data obtained from a taste*

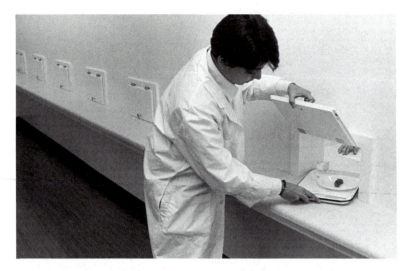

Fig. 12.7. A food sample is presented for sensory evaluation to a panelist in an individual booth in the Division of Foods and Nutrition, University of Illinois (Urbana-Champaign).

panel can be no more reliable than the quality of the panel which provided them.

Since there are no hard-and-fast rules on how to proceed in organizing and training a discrimination panel, the approach used should be appropriate for the existing circumstances. Usually, there will be a need to interview approximately twice as many individuals as will be required for the panel. Since discrimination panels perform specific sensory duties, the panelists need to qualify only for the intended task(s). While some individuals may be disqualified on the basis of some facts that surfaced during the interview, final panel selection is deferred until the training is completed and the performance of each individual has been "checked" or confirmed. A difference test, such as the triangle test, may be used for this purpose. This test is repeated a number of times to obtain a valid indication of both the evaluator's discrimination ability and consistency of performance. Since the quality of performance in a discrimination panel has a greater impact than the quantity of panelists, a group of five to ten panelists, plus one or two alternates, often proves to be satisfactory. If the panel is to perform over a long period of time (more than six months), periodic requalification may be advisable. There is no reason why some individuals cannot qualify for several different panels.

Training of Panelists

Training of panelists should be relevant to the intended task or test objective. Each panelist needs to learn or recognize the sensory notes of interest by repeated exposure to them, both as "knowns" and as "unknowns." Group and private discussions with panel coordinators and others involved in the sensory project help provide the needed guidance and motivation for the project. Final stages of training may include several "dress rehearsals" of the actual tests, followed by instructive discussions or constructive critiques (Amerine *et al.* 1965, O'Mahony 1986, and Stone and Sidel 1985).

Sometimes, it may be desirable to determine each panelists' threshold for the sensory notes of interest. In certain situations, a low threshold is important. Identification of individuals with abnormally high thresholds may help explain their inability to perform effectively. For basic tastes (such as sweet, sour, salty, and bitter), thresholds are easily determined by tasting progressively higher dilutions of a solution of known strength. In the case of off-flavors, this may be more difficult, particularly when the actual flavor-producing substance is not known or is unavailable. When feasible, an alternative is to start with a sample in which the flavor is quite strong and dilute it with the same product free of the defect. The diluent should be a member of a series tested along with about equal numbers of samples that are both below and above the average threshold for the entire group. Several trials are usually required. In some instances, the individual thresholds need not be pinpointed. An alert panel administrator can qualitatively assess the panels' abilities and identify the individuals whose sensitivity may be inadequate for serving as a member of a panel.

Dairy products experts and descriptive test panelists must possess broad qualifications in sensory discrimination (Dawson *et al.* 1963a, 1963b; Hirsh 1975b; O'Mahony 1979). Some individuals may perform poorly on the identification of sensory notes at first, but improve with practice and eventually prove themselves capable; other persons may always struggle. For panel membership, administrators should look for people who possess an excellent memory for taste and odors. Hopefully these same people also possess the other necessary attributes of an effective discriminator of sensory characteristics.

Management of Samples

Except for the sensory property being investigated (e.g., flavor or texture), the products being evaluated should be as uniform as possible in size, form, consistency, color, appearance, and temperature. However, if the study is focusing on changes as the result of cooking or storage (for example), then several attributes might be expected to change (e.g., color, appearance, texture, etc.).

The samples must be presented in identical containers. The identification of each sample should be such that it can be switched or changed easily between sessions of the panel, without compromising the original identity of the samples. Three digit code numbers are highly recommended as the best method of identifying samples without introducing a bias. In so far as possible, the samples examined by the panelists must be identical. The samples are usually presented to panelists in a randomized order.

The amount of sample submitted to each panel member is quite important. The sample size should be adequate, but not excessive. The fear of exhausting all the sample before determining the final judgment is psychologically detrimental to sensory evaluation. Although as little as 8 to 12 ml (1/3 oz) may be an adequate volume for evaluating milk, a serving of 1 oz (28.6 g) or more is preferred. The larger sample size provides the panel member with a sufficient quantity for retasting, if necessary. On the other hand, limiting the sample size is an effective way of preventing retasting, should that be an intended test requirement.

Sample temperature is an important factor in tasting and noting odors. The commonly invoked principle is to use the particular temperature which optimizes sensory discrimination; hence, trained panels usually follow this approach. By contrast, members of preference panels should taste samples that have been tempered to the normal serving or consumption temperature for the product. To optimize the discrimination process for trained panels performing sensory examination of dairy products, it is usually helpful to temper the samples above the typical storage temperature. About 12.8°C (55°F) is the common serving temperature for milk, butter, cheese, and similar dairy products. Ice cream and other hard-frozen desserts should be tempered to about −15°C (5°F), exercising care that body and texture are not altered. Chocolate milk may lose viscosity when warmed, while buttermilk may lose both viscosity and the carbonation effect at higher temperatures. When these particular characteristics are evaluated, normal serving temperatures for the given product should be used. While increasing the temperature facilitates the detection of off-flavors, desirable flavors in some products may be exaggerated simultaneously, to the extent that difficulties may be encountered in assessing their actual intensity level. Judgment on those criteria and sensory attributes must be made as the situation demands.

Fatigue of the taste buds and olfactory center ("sensory fatigue") is a limiting factor in determining the number of samples that can be effectively evaluated in each session. Most sensory investigators recommend that, for a given session, the number of samples to be judged be limited. Experience has indicated that for products of low flavor intensity (such as milk, sweet cream, or cottage cheese), as many as ten samples can be evaluated by the panel without the onset of sensory fatigue. For products of higher flavor intensity, the number of samples should be limited usually to three to five for any one panel session. On the other hand, experienced judges may routinely evaluate as many as 25 to 30 samples. On occasion, perhaps up to 50 samples may be examined, apparently, without losing sensory acuity. In essence, good judgment is required for a determination of the maximum number of samples, based on the type of samples and the performance of available panelists or judges.

All samples must be coded to conceal their true identity or source; however, the code numbers should not suggest to panel members any rank or quality dimension of the samples. Most unfortunately, the first of an alphabetical or numerical series usually suggests "first choice." Consequently, from a psychological standpoint the use of "A, B, C" or "1, 2, 3" or any other logical series of sample identifications is not without an associated bias. The sample code should be such that it may be readily changed between panel sessions, without retaining any relation to the previous sample marking. Commonly, three-digit numbers are recommended, as obtained from tables of random numbers.

Taste Panel Room

Panel evaluations should be performed in a room suitable for the purpose (see Fig. 12.7 and 12.8). The use of separate booths is desirable, particularly in the interest of concentration and independent judgment for each panel member. The room should be air-conditioned, properly ventilated, and free from odors and various distractions (such as human activity, noise, talking, etc.). Smoking must not be tolerated in the vicinity of the taste-panel room. A moderate, comfortable temperature should be maintained. Controlled lighting is desirable, especially where color is a factor in making sensory judgments. Quietness, orderliness, and regularity of sample presentation are conducive to a more accurate evaluation of the dairy product(s) under study. One possible floor plan for such a room (Moser et al. 1950) is shown in Fig. 12.9. (See also Amerine et al. 1965, Larmond 1973, Trichese 1968).

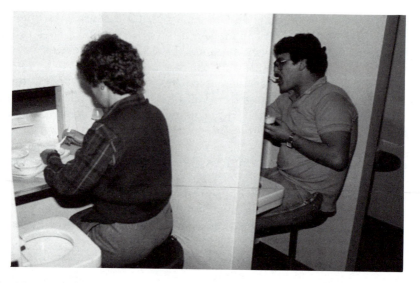

Fig. 12.8. An ice cream flavor panel underway in the Sensory Science Laboratory, Department of Food Science and Technology at Oregon State University (Corvallis).

Equipment and Materials

Care should be exercised in furnishing appropriate equipment and materials for the panel tasting room. All spoons, cups, and dishes used by the panel members, and all dipping or slicing devices (used to prepare samples) should be made of materials that will not influence the flavor (or other sensory properties) of the test product. Cups or dishes that are too small should be avoided. Paper towels or napkins should be available; dentist cuspidors or other suitable means for expectorating samples need to be provided. Specific colored lights or filters to minimize or mask "color bias" are required in any studies where sample color might be a negative or positive factor in determining panelists' judgments.

Tasting Techniques

In discrimination testing, most individuals develop their own styles of tasting. There are distinct variations among persons as to the amount of sample placed into the mouth for tasting. Some persons prefer to swallow a small amount of sample; others meticulously try to avoid swallowing any product. Some judges rinse their mouths with water between samples, others do not do so unless intense-flavored or objec-

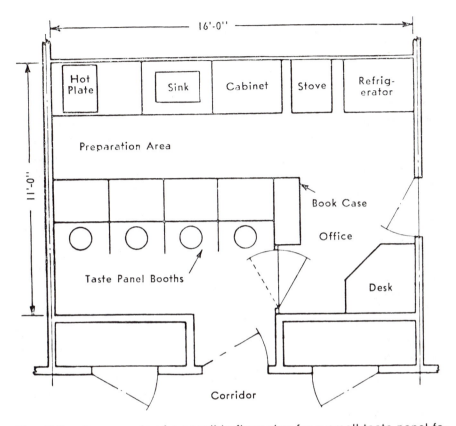

Fig. 12.9.   An example of a possible floor plan for a small taste panel facility.

tionable samples are encountered. The optimum "waiting interval" between trying samples varies among individuals. There appears to be no "right" or "wrong" procedure, as long as meaningful sensory results are obtained. In spite of this panelist individuality, it is sometimes necessary to establish sample size, length of time between sample presentations, etc. for statistical purposes.

Generally, individuals should not be forced to adopt a technique with which they are not comfortable and which adversely affects their performance. For novice judges who have not yet developed their own style, it may be advisable to: (1) place a small amount of sample into the mouth; (2) refrain from swallowing the sample; (3) decide to either rinse or not rinse the mouth (but practice the same routine with all samples); (4) allow about 30 sec between samples (more or less depending on the type of sample) to avoid adaptation; and (5) consciously observe for any aftertaste.

Ballots for Sensory Tests

Preprinted ballots should provide clear, neutral, and unambiguous instructions for the panelists of any sensory test. The ballot for each form of sensory test requires careful planning and design in order to help achieve the test objectives. Ballots are the key device for generating and temporarily storing the all important data. Examples of ballots for various types of sensory tests are illustrated in Figs. 12.1 through 12.4.

Personnel Management

Good rapport is required between panelists, the panel administrator(s), and the particular groups requesting the panel work (e.g., research, quality control and/or marketing). The obvious support of management is most helpful in reinforcing the position that panel participation provides an important contribution to the company's or organization's mission. Panelists should not feel that their participation in the sensory panel is a personal burden, as they take time away from their regular job responsibilities. To help maintain a high level of interest, panelists should be: (1) periodically apprised of the progress being made on the sensory activity, as well as (2) being given an opportunity to contribute their own suggestions. A spirit of teamwork must be fostered, wherein each participant is anxious to help achieve the objectives of the sensory test.

REFERENCES AND BIBLIOGRAPHY

American Society for Testing and Materials. 1968. *Manual on Sensory Testing Methods.* ASTM Special Tech. Public. No. 434. Philadelphia, PA. 77 pp.

American Society for Testing and Materials. 1973. *Sensory Evaluation of Appearance and Materials.* Special Tech. Public. No. 545. Philadelphia, PA. 194 pp.

Amerine, M. A., Pangborn, R. M., and Roessler, E. B. 1965. *Principles of Sensory Evaluation of Food.* Academic Press. New York. 602 pp.

Birch, G. G., Brennan, J. G. and Parker, K. J. 1977. *Sensory Properties of Foods.* Applied Science Publ., Ltd. London, U.K. 326 pp.

Brandt, M. A., Skinner, E. Z., and Coleman, J. A. 1963. Texture profile method. *J. Food Sci. 28:*404.

Brandt, F. I. and Arnold, R. G. 1977. Sensory tests used in food product development. *Food Prod. Dev. 11*(8):56.

Caul, J. F. 1957. The profile method of flavor analysis. *Adv. in Food Res. 7:*1.

Civille, C. V. and Szczesniak, A. S. 1973. Guidelines to training a texture profile panel. *J. Text. Studies* 4:204.

Civille, G. V. and Liska, H. L. 1975. Modifications and applications to food of the General Foods sensory texture profile technique. *J. Text. Studies* 6:19.

Civille, G. V. 1979. Descriptive analysis. In: *Sensory Evaluation Methods for the Practicing Food Technologist.* Johnston, M. R. (Editor). Institute of Food Technologists. Chicago, IL. 6:1–28.

Dawson, E. H., Brogen, J. L., and McManus, S. 1963a. Sensory testing of differences in taste. I. Methods. *Food Technol.* 17(9):45.

Dawson, E. H., Brogen, J. L., and McManus, S. 1963b. Sensory testing of differences in taste. II. Selection of panel members. *Food Technol.* 17(10):39.

Dudel, J. 1981. General sensory physiology, psychophysics. In: *Fundamentals of Sensory Physiology.* R. F. Schmidt (Editor). Springer Verlag. New York. 1–30.

Ellis, B. H. 1969. Acceptance and consumer preference testing. *J. Dairy Sci.* 52:823.

Fisher, R. A. and Yates, F. 1949. *Statistical Tables for Biological, Agricultural and Medical Research.* (3rd Edition). Oliver and Boyd Ltd. London, U.K.

Hirsh, N. L. 1971. Sensory good sense. *Food Prod. Develop.* 5(6):27.

Hirsh, N. L. 1974. Getting fullest value from sensory testing. Part I: Use and misuse of testing methods. *Food Prod. Develop.* 8(10):33.

Hirsh, N. L. 1975a. Getting fullest value from sensory testing. Part II: Considering the test objectives. *Food Prod. Develop.* 9(1):10.

Hirsh, N. L. 1975b. Getting fullest value from sensory testing. Part III: Use and misuse of test panels. *Food Prod. Develop.* 9(2):78.

Jellinek, G. 1962. Flavor testing with the profile method. In: *Recent Advances in Food Science.* Hawthorn, J. and Leitch, J. M. (Editors). Butterworths. London, U.K. 317 pp.

Kahan, G., Cooper, D., Papavasiliou, A. and Kramer, A. 1973. Expanded tables for determining significance of differences for ranked data. *Food Tech.* 27(5):61.

Kramer, A. and Twigg, B. A. 1970. *Quality Control for the Food Industry.* Vol. 1—Fundamentals. AVI Pub. Co. Westport, CT. 541 pp.

Kramer, A. and Twigg, B. A. 1973. *Quality Control for the Food Industry.* Vol. 2—Applications. AVI Pub. Co. Westport, CT.

Larmond, E. 1973. Physical requirements for sensory testing. *Food Technol.* 27(11):28.

Larmond, E. 1977. *Laboratory Methods for Sensory Evaluation of Food.* Public. No. 1637, Research Branch, Canada Dept. Agric. Ottawa, Ontario, CANADA. 74 pp.

Merolli, A. 1980. Sensory evaluation in operations. *Food Technol.* 34(11):65.

Moncrieff, R. W. 1967. *The Chemical Senses.* Chemical Rubber Co. Press. Cleveland, OH. 760 pp.

Moser, H. A., Dutton, H. J., Evans, C. D., and Cowan, J. C. 1950. Conducting a taste panel for the evaluaton of edible oils. *Food Technol.* 4:105.

Moskowitz, H. L. 1977. Magnitude estimation: Notes on what, how, when and why to use it. *J. Food Quality* 3:195.

Nakayama, M. and Wessman, C. 1979. Application of sensory evaluation to the routine maintenance of product quality. *Food Technol.* 33(9):38.

Noble, A. C. 1975. Instrumental analysis of the sensory properties of food. *Food Technol.* 29(12):56.

O'Mahony, M. 1979. Psychophysical aspects of sensory analysis of dairy products: A critique. *J. Dairy Sci.* 62:1954.

O'Mahony, M., Kulp, J., and Wheeler, L. 1979. Sensory detection of off-flavors in milk incorporating short-cut signal detection measures. *J. Dairy Sci.* 62:1857.

O'Mahony, M. and Odbert, N. 1985. A comparison of sensory difference testing procedures: Sequential sensitivity analysis and aspects of taste adaptation. *J. Food Sci.* *50:*1055.

O'Mahony, M. 1986. *Sensory Evaluation of Food. Statistical Methods and Procedures.* Marcel Dekker, Inc. New York. 487 pp.

Pangborn, R. M. and Dunkley, W. L. 1964. Laboratory procedures for evaluating the sensory properties of milk. *Dairy Sci. Abstr. 26:*55.

Peryam, D. R., Pilgrim, F. J., and Petersen, M. S. (Editors). 1954. *Food Acceptance Testing Methodology—A Symposium.* Quartermaster Food and Container Inst. Armed Forces. Natick, MA.

Peryam, D. R. and Pilgrim, F. J. 1957. Hedonic scale method of measuring food preferences. *Symposium Proc. 17th Annual Mtg. Inst. Food Technol.* Pittsburgh, PA.

Prell, P. A. 1976. Preparation of reports and manuscripts which include sensory evaluation data. *Food Technol. 30*(11):40.

Roessler, E. B., Warren, J. and Guymon, J. F. 1948. Significance in triangular taste tests. *Food Res. 13:*503.

Roessler, E. B., Baker, G. A., and Amerine, M. A. 1956. One-tailed and two-tailed tests in organoleptic comparisons. *Food Res. 21:*117.

Roessler, E. B., Pangborn, R. M., Sidel, J. L., and Stone, H. 1978. Expanded statistical tables for estimating significance in paired-preference, paired-difference, duo-trio, and triangle tests. *J. Food Sci. 43:*940.

Rutenbeck, S. K. 1985. Initiating an in-plant quality control/sensory evaluation program. *Food Technol. 40*(11):124.

Sidel, J. L. and Stone, H. 1976. Experimental design and analysis of sensory tests. *Food Technol. 30*(11):32.

Sidel, J., Woolsey, A., and Stone, H. 1975. Sensory analysis: Theory, methodology, and evaluation. In: *Fabricated Foods.* Inglett, G. (Editor). AVI Pub. Co. Westport, CT. pp. 109–126.

Simone, M. and Pangborn, R. 1957. Consumer acceptance methodology: One vs. two samples. *Symposium Proc. 17th Annual Mtg. Inst. Food Technol.* Pittsburgh, PA.

Sjostrom, L. B., Cairncros, S. E. and Caul, Jean F. 1957. Methodology of the flavor profile. *Symposium Proc. 17th Annual Mtg. Inst. Food Technol.* Pittsburgh, PA.

Skinner, E. 1980. Sensory evaluation in distribution. *Food Technol. 34*(11):65.

Stahl, W. H. and Einstein, M. A. 1973. Sensory testing methods. In: *Encyclopedia of Industrial Chemical Analysis.* Vol. 17, P. 608.

Stevens, S. S. 1956. The direct estimation of sensory magnitudes: Loudness. *Am. J. Psychol. 69*(1):1.

Stone, H. and Sidel, J. L. 1985. *Sensory Evaluation Practices.* Academic Press, Inc. New York.

Stone, H., Sidel, J., Oliver, S., Woolsey, A., and Singleton, R. C. 1974. Sensory evaluation by quantitative descriptive analysis. *Food Technol. 28*(11):24.

Symons, H. W. and Wren, J. J. 1977. *Sensory Quality Control: Practical Approaches in Food and Drink Production.* Soc. Chem. Ind. London, U.K.

Szczesniak, A. S., Brandt, M. A., and Friedman, H. H. 1963. Development of standard rating scales for mechanical parameters of texture and correlation between the objective and the sensory methods of texture evaluation. *J. Food Sci. 28:*397.

Trichese, T. 1968. Sensory laboratory testing facilities. *Food Prod. Develop. 2*(2):72.

Wolfe, K. A. 1979. Use of reference standards for sensory evaluation of product quality. *Food Technol. 33*(9):43.

Appendices

I. Basics of Grade A Raw Milk Sampling, Grading, and Transport
II. A Suggested Milk Grading Exercise
III. Federal (FDA/USPHS) Chemical, Bacteriological, and Temperature Standards for Grade A Milk
IV. Measurement of Hydrolytic Rancidity in Milk
V. Measurement of Auto-Oxidation of Milkfat
VI. Copper Sensitivity Test
VII. Test for Whippability
VIII. Grades and Standards for American Cheddar Cheese
IX. A Score Card for Milk
X. Collegiate Contest Milk Score Card
XI. Collegiate Contest Butter Score Card
XII. Collegiate Contest Cheddar Cheese Score Card
XIII. Collegiate Contest Ice Cream Score Card
XIV. Collegiate Contest Cottage Cheese Score Card
XV. Collegiate Contest Swiss-Style Yogurt Score Card
XVI. Names and Addresses of Organizations
XVII. Collegiate Dairy Products Evaluation Contest Grading

I. BASICS OF GRADE A RAW MILK SAMPLING, GRADING, AND TRANSPORT

The licensed milk hauler occupies an important position in the dairy industry. The hauler represents both the seller (producer) and buyer (processor) of milk supplies intended for fluid or manufactured products use. The judgment, actions, and decisions of this key individual impact on: (1) representative sampling; (2) the measurement of both quantity and quality of milk (which have a direct effect upon milk payment); (3) compliance with food safety requirements; and (4) the quality of dairy products made from the milk.

To be competent, a bulk milk hauler should possess several attributes and skills in addition to being able to drive safely and competently handle a truck and trailer. First, it must be recognized that the hauler is a food handler; one's appearance and personal sanitation habits should definitely reflect this role. Cleanliness, neatness, integrity, a "spirit of fairness" and "attention-to-detail" are additional personal attributes that are indicative of a milk hauler's ability and performance.

More specifically, the milk hauler must be able to accurately determine the milk volume (or weight) of bulk milk tanks, and be able to aseptically collect, identify, and properly care for "official" milk samples. Hence, the hauler should be familiar with farm tank installation and the factors which affect the accurate measurement of milk, sampling, milk abuse, and temperature control. It is most helpful if the hauler can identify certain causes of milk quality problems and then offer suggestions to the dairyman for corrective action.

Ideally, milk haulers should possess a keen sense of smell and be able to identify the more objectionable milk off-flavors (odors). This person needs to have an appreciation of the role of cleanliness and sanitation in the handling and protection of milk. Additionally, the hauler must be able to clean and sanitize all equipment and "tools" entrusted to his or her care.

The milk hauler is the most appropriate communication link between the milk producer, the milk marketing cooperative, and the milk processor. A milk hauler's awareness and knowledge of many aspects of farm milk collection and handling, the determinants of good milk quality, and essential sanitation and hygiene requirements are factors that can lead to an increased level of understanding and trust between the milk producer and the dairy processor.

"To accept" or "to reject" a given tank of milk at the farm is essentially an "all-or-none" type of decision; this is undoubtedly the most important form of decisionmaking required of the milk hauler (who is simultaneously a milk sampler/grader). Training and experience play an important role in the development of capable milk haulers. Milk haulers must constantly:

1. Be alert—always observing for details or signs of shortcomings of sanitation or milk quality.
2. Be willing to assume their share of responsibility.
3. Maintain day-to-day positive relationships with milk producers.
4. Demonstrate integrity—measure and sample milk with equity.
5. Exercise a keen sense of smell (and perhaps taste in some instances).

Milk Hauler (Receiver) Defined

A "milk hauler" or "receiver" is defined as a person who, in the course of their employment, accepts bulk milk or milk products (ingredients) from a producer, milk plant, receiving or transfer station, and transports such commodity to a milk or dairy products plant.

Licensing of Milk Haulers

In most states of the United States, all persons involved in the transporting or hauling of bulk milk from producers to processing plants (or transfer stations) must be licensed. Actual licensing arrangements will vary among the states.

The purpose of licensing is: (1) to make certain that a milk hauler is qualified to accept, measure, sample, collect, and transport farm bulk milk samples in such a manner that the results of one's activities are reliable, accurate, and equally fair to the producer and processor and (2) to help insure continued satisfactory performance by the hauler.

Required Equipment and Supplies

Any milk hauler must have the proper equipment and supplies to perform the stipulated requirements of milk sampling, transfer, and transport. This includes:

1. Tank truck with appropriate milk transfer equipment (pump) that complies with the official 3A Standards Committee guidelines and is properly cleaned and sanitized.
2. Appropriate sampling equipment and supplies: Sample transfer device (e.g., dipper, sterile straws, etc.), sample containers (bags, bottles), sanitizing solution, and an insulated sample case with support racks.
3. An accurate ($\pm 2°F[1.2°C]$ nonbreakable dial thermometer with a dial range of $25°F$ to $125°F$ ($-4°$ to $54°C$).
4. Refrigerant (ice) required to keep samples between $32°F$ to $40°F$ ($0°C$–$4.4°C$).
5. Waterproof, indelible marking pen for helping identify samples.
6. A supply of milk weight receipt forms and a pen or pencil to record all required information (milk quantity, temperature, date, etc.).
7. A watch or other timing device.

8. Single-service paper towels and a bright flashlight are also helpful items.
9. A stainless steel or plastic cup of about 2–4 oz (60–120 ml) capacity for tempering (warming to 90°F[32°C]) milk samples for checking the odor of "suspicious" samples.

All milk contact surfaces of the tank truck and transfer equipment must be constructed of approved materials and be readily cleanable, in good condition, and accessible for inspection. The overall sanitary condition of the tank truck, milk contact equipment and other items used by the milk hauler are subject to inspection by a milk sanitarian on a regular basis.

Milk Hauler Responsibilities

Regardless of who washes and sanitizes the tank truck and accessories, the milk hauler is ultimately responsible for equipment cleanliness and sanitization. Generally, facilities for circulation cleaning (CIP) of the tanker and pump are available at the milk plant receiving stations of most milk plants. Many state regulatory agencies require that cleaned tankers be indicated with a designated or "official" tag to reflect the date, time, place, and the person responsible for overseeing or performing the cleaning procedures.

A responsible milk hauler will not assume that automated washing (CIP) systems are foolproof. A hauler must check that: (1) all ports are open on the spray ball or tube, (2) the CIP system cycles solutions correctly, (3) water rinse and detergent solutions are supplied properly, (4) appropriate temperatures are reached and maintained, (5) the "spray" solutions reach all critical or intended surfaces, and (6) the tank interior and transfer equipment are actually "clean." Pump and valve parts that cannot be satisfactorily cleaned-in-place (or cleaned-out-of-place (COP)) must be cleaned manually (by hand).

Stainless steel and other milk contact surfaces are usually considered clean when: (1) rinse water slips (sheets) in a continuous layer from top to bottom of the surface and moisture droplets do not appear, (2) a "squeaky" noise results when clean, dry finger tips are pressed against the dry surface and (3) the dry equipment surface is bright and shiny with no visible waterstone, milkstone, or blue discoloration. By contrast, when water drains by a pattern of distinct "patches" or random moisture droplets appear on the surface, it is most indicative of an unclean surface.

Milk Weights and Other Records

The milk hauler has the responsibility for maintaining an important written record of milk quantity, temperature observations, and other pertinent information with each bulk tank collection. Generally, the milk weight record (receipt) contains provisions for recording the following minimum information:

1. Date of collection
2. Time of pick-up
3. Producer name and/or number
4. Sensory quality (odor, appearance, etc.)
5. Milk temperature
6. "Dipstick" or gauge reading
7. Converted milk weight (lbs)
8. Name of milk buyer
9. Hauler's signature (and license number)
10. Specific reason for milk rejection (if rejected)

Milk Grading

In most state jurisdictions of the United States, the milk hauler is required to grade each collection (pick-up) of Grade A fluid milk with respect to its "sensory quality," within his or her capacity as a licensed milk grader. If a given collection (lot) of milk is determined by the hauler/grader to be "unacceptable," and thus rejected as being unsuitable for Grade A milk use, a designated record form must be completed by the grader (hauler). This official "rejection" form usually identifies: (1) the producer (by name and/or number); (2) the usual marketing agent (cooperative, if appropriate); (3) the date and time; (4) the quantity of milk; and (5) the cause or reasons for rejection as Grade A milk.

As a rule, licensed milk haulers and graders are expected to be somewhat familiar or acquainted with some of the following criteria for the effective determination of milk quality:

1. The official regulatory agency "sampling" and "compliance" periods for Grade A raw milk.
2. Abnormal milk standards (maximum somatic cell count per ml).
3. General methods of examining milk for abnormal composition.
4. Sediment testing procedures and official sediment standards.
5. Test alternatives for bacterial inhibitors (antibiotics and sanitizers).
6. Bacterial standards for Grade A raw milk for pasteurization (for the farm level and commingled milk).
7. Bactericidal treatment (sanitization) of sampling equipment.

8. Importance of personnel cleanliness and health in performance of tasks and duties as a sampler/grader.
9. The general technical and sanitation requirements for milk tanker loading and unloading.
10. Aseptic technique for the collection of "official" milk ("universal") samples.[a]
11. The purpose or intent of various milk quality tests that the cooperative marketing agency or processor may have implemented for specific milk quality incentive (bonus) programs.
12. A need to exhibit knowledge of and the ability to comply with changes in applicable laws and regulations.

II. A SUGGESTED MILK GRADING EXERCISE

A SUGGESTED MILK GRADING EXERCISE
for
MILK FLAVOR CLINICS OR TRAINING SESSIONS

Name _____

Representing _____ Date _____

Procedure: A set of 10 milk samples will be presented for the purpose of grading for their respective flavor (odor and taste) and appearance characteristics (sensory evaluation). For each milk sample, one of three choices or decisions is sought from you:

1.   Generally or unquestionably, in your judgment, if a given milk sample has no off-flavor or appearance problem and you find it acceptable for fluid milk use, then place a √ mark in the ACCEPT column opposite that sample number(s).

---

[a]"Universal" sample. A "universal" milk sample is a representative portion of milk that has been aseptically collected from a given bulk milk tank by a licensed milk hauler/grader by a technique often referred to as the Universal Sampling Plan. This sampling protocol provides a singular milk sample that can be used for all anticipated laboratory analyses (microbiological, inhibitory substances and/or composition). Not all analyses necessarily need be done on the same sample. Some advantages cited for this milk sampling approach include:

1. Eliminates the need for the hauler to collect several different types of samples, hence, it simplifies the sampling process and the necessary equipment.
2. Enables the laboratory to more equitably monitor a given producer's milk quality without requesting "special" samples.
3. The producer is unable to anticipate when bacteria, antibiotics, composition, sediment, added water, and other tests are to be conducted, because the same size sample is "universally" removed from the dairyman's bulk tank at every milk pick-up.
4. If serious or questionable quality concerns occur within the raw milk supply, there are available representative milk samples for each producer that can be analyzed to help pinpoint the specific origin of many problems.

2. If you have some doubt or reservations about the odor, taste and/or appearance characteristics of a given sample(s), then √ mark the QUESTIONABLE QUALITY column.

3. And without much doubt, you determine that a given milk sample(s) is unacceptable for odor, taste and/or appearance for fluid use, then √ mark the REJECT column.

Place a (√) mark in only one column/sample

Sample No.	Accept	Questionable Quality	Reject	List Given Flavor Defect (Optional)
1.	_____	_____	_____	_____
2.	_____	_____	_____	_____
3.	_____	_____	_____	_____
4.	_____	_____	_____	_____
5.	_____	_____	_____	_____
6.	_____	_____	_____	_____
7.	_____	_____	_____	_____
8.	_____	_____	_____	_____
9.	_____	_____	_____	_____
10.	_____	_____	_____	_____

III. FEDERAL (FDA/USPHS) CHEMICAL, BACTERIOLOGICAL, AND TEMPERATURE STANDARDS FOR GRADE A MILK

Grade A Raw Milk for Pasteurization, Ultra-pasteurization, or Aseptic Processing	Temperature	Cooled to 45°F (7°C) or less within two hours after milking, provided that the blend temperature after the first and subsequent milkings ≤50°F (10°C).
	Bacterial limits (SPC)	Individual producer milk ≤100,000/ml prior to commingling with milk from other producers; ≤300,000/ml for commingled milk prior to pasteurization.
	Antibiotics	No zone ≥16 mm by *Bacillus stearothermophilus* disc assay method.

(continued)

	Somatic Cell Count (SCC)	Individual producer milk: $\leq 1,000,000$/ml
Grade A Pasteurized Milk and Milk Products	Temperature	Cooled to $\leq 45\,^{\circ}$F ($7\,^{\circ}$C) and maintained thereat.
	Bacterial limits[a] (SPC)	$\leq 20,000$/ml
	Coliform	≤ 10/ml
	Phosphatase	≤ 1 microgram/ml (Scharer Rapid Method or equivalent)
	Antibiotics	No zone ≥ 16 mm by *B. stearothermophilus* disc assay method
Grade A Aseptically Processed Milk and Milk Products	Temperature	None applicable
	Bacterial limits	No growth by a specified test.
	Antibiotics	No zone ≥ 16 mm with the *B. stearothermophilus* disc assay method

[a] Not applicable to cultured products.

IV. MEASUREMENT OF HYDROLYTIC RANCIDITY IN MILK

The Thomas *et al.* Test (Acid Degree Value)

Apparatus and Reagents.

1. Centrifuge—Babcock centrifuge that will recieve an 18 g, 8% milk test bottle may be used. An unheated centrifuge is satisfactory.
2. Boiling water bath—Water in the bath should be at a depth sufficient to cover the base of the test bottles.
3. Tempering bath—This should be controlled to within a temperature range of 54°C to 60°C (129°F–140°F).
4. Glassware—The glassware required consists of the following: a standard 18g, 8% milk test bottle; a 1 ml tuberculin type syringe with a No. 19 needle; a 50 ml graduated syringe with a No. 15 needle or a standard 17.6 ml milk pipette; a 50 ml Erlenmeyer flask; and a 5 ml microburette.
5. BDI Reagent (Bureau of Dairy Industry Reagent)—Thirty grams of Triton X-100 (a nonionic surface active agent manufactured by Rohm and Haas Co., Philadelphia, PA) and 70 g of sodium tetraphosphate are made up to a 1 l with distilled water.

The reagent is used since it will not change the physical or chemical properties of the milkfat; also it quickly and effectively de-emulsifies milkfat.

6. Alcoholic-KOH—To 50 g of KOH add 50 ml of distilled water (50% KOH solution) and swirl until completely dissolved. Next, mix 1.08 ml of 50% KOH into 500 ml of ethanol and standardize to an 0.02N alcoholic-KOH solution. This solution should be standardized frequently against standard potassium acid phthalate (or other suitable standards).
7. Indicator solution—This is prepared by dissolving 1 g of phenolphthalein in 100 ml of absolute ethanol.
8. Absolute ethanol.
9. Petroleum ether—Boiling point range 30°C to 60°C (86°F–140°F).
10. 50% Aqueous methanol—This reagent consists of equal volumes of chemically pure methanol and distilled water.

Recovery of the Fat.

1. Place 35 ml of the milk sample in an 18 g, 8% milk test bottle. The milk may be transferred to the bottle by either a 50 ml syringe or a standard 17.6 ml milk pipette.
2. Next, add 10 ml of BDI reagent to the milk sample and mix thoroughly. (It is important to use only 10 ml of BDI reagent in the 35 ml of whole milk, since the acid degree value of the sample could decrease with an increased amount of reagent).
3. Place the bottle and contents into a gently boiling water bath. Agitate the bottle contents thoroughly, after 5 min and again after 10 min in the water bath.
4. After 15 min in the boiling water bath, centrifuge the bottles for 1 min. Sufficient 50% aqueous methanol is then added to bring the top of the fat column to the 6% graduation. Centrifuge for an additional 1 min.
5. Place sample bottles in the tempering bath (54°C to 60°C (129°F–140°F)) for 5 min. The water level should be equivalent to the height of the fat column.

Titration.

1. Transfer 1 ml of the tempered milkfat from the test bottle into a 50 ml Erlenmeyer flask (by use of a 1 ml syringe and a No. 19 needle). *Note:* It is preferable to weigh 1 g of milkfat into the flask by use of an appropriate top-loading balance.

2. Dissolve the extracted fat by the addition of 10 ml of petroleum ether and 5 ml absolute ethanol into the flask. Add 10 drops of indicator solution.
3. Titrate to the first definite color change with the standardized alcoholic-KOH solution (0.02N) by use of a 5 ml microburette.
4. Express results in terms of acid degree value (ADV). The ADV is equivalent to the number of ml of 1N base required to titrate 100 g of fat.

Calculations.

1. If milkfat was measured volumetrically (pipetted), determine the actual g of milkfat in the sample titrated by multiplying the ml of milkfat by the specific gravity of tempered milkfat (0.90):

$$(ml\ milkfat \times 0.90 = g\ of\ milkfat)$$

2. Calculate the Acid Degree Value (ADV) as follows:

$$\frac{ml\ of\ KOH\ used\ in\ titration\ -\ ml\ used\ in\ titrating\ ``blank"\ \times\ N.F.^a\ \times\ 100}{Wt.\ (g)\ of\ milkfat\ from\ sample} = ADV$$

"Blank" = the titration value of the fat solvent (in the absence of fat and by use of 5 drops of phenolphthalein indicator). "Blank" determinations should be run on each new batch of solvent, and then retitrated at frequent intervals thereafter.

Example—Assume that 0.75 ml of milkfat required 0.70 ml of alcoholic-KOH (normality factor of 0.0203) and that the blank titration value = 0.03, then:

$$\frac{(0.70-0.03) \times 0.0203 \times 100}{0.75 \times 0.90} = \frac{0.67 \times 0.0203 \times 100}{0.675} =$$

$$\frac{1.36}{0.675} = 2.014 = ADV$$

The ADV should be expressed only to the second decimal place (e.g., ADV = 2.01 for above example).

Interpretation of ADV Results. Milk drawn freshly from "normal" or typical cows should exhibit ADV's ranging from 0.2 to 0.5. Commonly, hydrolytic rancidity may be detected by experienced judges when the

---

[a]N.F = normality factor of the alcoholic-KOH solution.

ADV equals 1.3 to 1.6. Consumer complaints are likely to develop (about milk off-flavor) when ADV results in the region of 2.0 or higher occur.

When ADVs of producer milk samples approach 0.8 to 1.0 (or higher), there is justification for investigation of the conditions related to the harvesting, handling and transporting the milk. Although it is quite unlikely that hydrolytic rancidity will be detected by sensory procedures at this ADV level, the milk in all likelihood is approaching the "threshold of trouble." Any abusive or "triggering" conditions of milk handling should be identified, minimized, or eliminated.

Lipase activation is favored by any milk-handling conditions that produce foam (when milk is within a temperature range of 27°C–32°C (80°F–90°F).

The Shipe *et al.* Modified Copper Soap Solvent Method (CSM)

Reagents.

1. Copper reagent—Consists of a mixture of 5 ml triethanolamine and 10 ml 1 M aqueous Cu $(NO_3)_2 \cdot 3H_2O$, diluted to 100 ml with a saturated NaCl solution. The pH should be adjusted to 8.3 with 1N NaOH. This mixture needs to be stored in the dark at room temperature. Expected stability is about 5 months.
2. Color reagent—a 5% sodium diethyl dithiol-carbamate solution in n-butanol.
3. Solvent chloroform-heptane-methanol (49:49:2).
4. Solubilizing reagent—ethylene diamine tetra acetate (EDTA), disodium salt, 8% wt/vol in distilled water.

Procedure.

1. A 0.1 ml aliquot of 0.7 N HCl is added to 0.5 ml of milk sample in a 16 × 125 mm test tube.
2. This mixture is shaken on a vortex test tube mixer.
3. Then 2 ml of the copper reagent and 6 ml of the solvent are added.
4. The sample and reagents are shaken for 30 min in a rotary Eberbach (or equivalent) shaker at 240 rpm.
5. Next, sample tubes are centrifuged for 10 min at 1940 × g in an International Model HN centrifuge (or equivalent).
6. Then 3.5 ml of the solvent layer are transferred to an acid-washed test tube that contains 0.1 ml of the color reagent.
7. Color development is measured at 440 mμ within 1 hr (after mixing tube contents).

8. Any type of spectrophotometer may be used (e.g., Bausch and Lomb Spectronic 20, 400-3 system, etc.).
9. Results are reported in terms of the absorbance or the concentration of free fatty acids expressed as meq/L of milk. Development of a standard curve based on palmitic acid is suggested for converting absorbance to meq. FFA/L of milk.

References.

1. Shipe, W. F., Senyk, G. F., and Fountain, K. B. 1980. Modified copper soap solvent extraction method for measuring free fatty acids in milk. J. Dairy Sci. 63:193.
2. Thomas, E. L., Nelson A. J., and Olson, J. C., Jr. 1955. Hydrolytic rancidity—A simplified method for estimating the extent of its development. *Am. Milk Rev.* 77(1):50.

V. MEASUREMENT OF AUTO-OXIDATION OF MILK FAT

The Thiobarbituric Acid (TBA) Test

Reagents.

1. Trichloroacetic acid (TCA) solution is prepared by adding 1 g TCA to each 1 ml distilled water.
2. 95% ethanol (in distilled water).
3. TBA solution prepared by dissolving 1.4 g 2-thiobarbituric acid in 95% ethanol and taking to 100 ml volume. Facilitate solubilization of the TBA by heating in a 60°C (140°F) water bath.
   Note: TBA solution undergoes deterioration easily; hence, it should not be stored longer than three days prior to use.

Procedure.

1. Pipet 17.6 ml of a milk sample into a small flask fitted with a glass stopper.
2. Warm flask contents to 30°C (86°F).
3. Add 1.0 ml trichloroacetic (TCA) solution.
4. Add 2.0 ml 95% ethanol.
5. Stopper the flask and shake contents vigorously for 10 sec.
6. After 5 min. filter the flask contents through no. 42 Whatman paper.
7. To 4.0 ml of clear filtrate add 1.0 ml of TBA solution into another flask.

8. Stopper the flask, mix contents, and place into a 60°C (140°F) water bath for 60 min.
9. Cool to room temperature.
10. Determine optical density of 532 mμ (Beckman DU spectrophotometer preferred), using distilled water as a reference (blank).

Interpretation.

King (1962) observed a direct relationship between sensory evaluation of the intensity of an oxidized off-flavor and optical density readings of milk filtrates as follows:

Flavor Score	Flavor Description	Range of Optical Density (532 mμ)
0	No oxidized off-flavor	0.010—0.023
1	Questionable to slight oxidized	0.024—0.029
2	Slight but consistently detectable	0.030—0.040
3	Distinct or strong oxidized	0.041—0.055
4	Very strong oxidized	≥ 0.056

Notes:

1. Strongly oxidized milks will generally yield an optical density in the range of 0.06–0.10.
2. Multiple determinations of a given sample are often reproducible within ± 0.003 optical density units.
3. Contamination of milk samples with copper up to about 0.5 ppm does not affect TBA test results.
4. The TBA test may have some value for monitoring the extent of a light-activated off-flavor of milk.

References.

1. Dunkley, W. L. and Jennings, W. G. 1951. A procedure for the application of the thiobarbituric acid test to milk. *J. Dairy Sci.* *34*:1064.
2. King, R. L. 1962. Oxidation of milk fat globule membrane material. I. Thiobarbituric acid reaction as a measure of oxidized flavor in milk and model systems. *J. Dairy Sci.* *45*:1165.
3. King, R. L. and Dunkley, W. L. 1959. Relation of natural copper in milk to incidence of spontaneous oxidized flavor. *J. Dairy Sci.* *42*:420.
4. Schroder, M. J. A. 1982. Effect of oxygen on the keeping quality of milk. I. Oxidized flavour development and oxygen uptake in milk in relation to oxygen availability. *J. Dairy Res.* *49*:407.

VI. COPPER SENSITIVITY TEST

Procedure for Detecting Milk Susceptible to an Oxidized Off-Flavor

1. Prepare a 0.01 molar solution of copper sulfate by dissolving 2.496 gms of $CuSO_4$ $5H_2O$ in 1000 ml of distilled water. Each 0.1 ml of this solution, when added to 10 ml of milk, is equivalent to 0.1 part per million (ppm) of copper.
2. Prepare a series of four tubes, each containing 10 ml of the milk to be tested, and add 0, 0.2, 0.5, and 1.0 ml of the 0.01 molar $CuSO_4$ solution.
3. Shake and hold overnight at 10 °C (50 °F) and taste (or smell) for the characteristic flavor. Heat the milk to 71 °C (160 °F) and cool just before tasting to protect the taster (from raw milk).

Interpretation

1. If any of the first three test tubes in the test series shows an oxidized off-flavor, it is unlikely that such milk can proceed through processing without picking up sufficient additional copper to cause an off-flavor (e.g., this milk is susceptible to the development of an oxidized off-flavor).
2. If only the last (1.0 ml) tube in the series exhibits the off-flavor, the given milk is on the "threshold" and may present quality problems, depending on how much has been undertaken to eliminate copper from all equipment.
3. If none of the test tubes shows the off-flavor, in all likelihood the milk may be processed without difficulty, assuming that there is little or no copper contamination.
4. If any final products from the plant exhibit a copper-induced oxidized off-flavor (even though the raw milk supply shows no off-flavor when the maximum amount of copper was added (1.0 ppm)), attention must be given to finding and eliminating any copper exposure in the processing plant.
5. If off-flavors are found in the raw milk supply, apply the copper sensitivity test to milk from individual herds to identify specific troublesome sources. When the herd (or herds) are identified, check farm conditions with particular attention to the following: any equipment with potential sources of copper, iron, or manganese; pipe lines of milker units for air leaks; and note the operating temperature and the recovery rate of the bulk tank. Espe-

cially look for the presence of any copper tubing in hot water lines, heating (or heat recovery) coils, or white (monel metal) fittings in CIP cleaning systems.

Source: Milk Industry Foundation. 1967. *Milk Plant Operators Manual.* Washington, D.C. p. 900.

VII. TEST FOR WHIPPABILITY

Whippability, or the capacity to form stable foams with air, is an important functional property of proteins for applications in food products such as angel food and sponge cakes, whipped toppings, meringues, souffles, frozen desserts, and a variety of candy confections. The terms "whippability" and "foamability" are often used interchangeably. However, (1) whippability suggests that "testing" was conducted by use of a household-type mixer with a wire whip; while (2) foamability refers to the intensive shaking/agitation of a protein solution within a cylinder or the bubbling of air through a protein and the resultant "system" is measured under a set of milder (and more controlled) conditions. The so-called "whipping" procedure is simpler and provides more reproducible results from test-to-test. Furthermore, results of whipping tests are more easily translated from the laboratory bench to the pilot or processing plant.

The stability of the whip or foam relates to the ability of the product to retain its maximum volume. It is usually measured by the rate and/or amount of "leakage of fluid" from the initial foam. Both whippability and foam stability are expressed in many different ways in the literature; this makes comparisons between various products and between laboratories somewhat difficult. Adding to the confusion is the fact that whipping properties are affected not only by the method of creating the foam, but by other factors that must be controlled during testing. These include the protein concentration and source, product processing methods (especially preheat treatments) and subsequent effect on physicochemical properties (such as solubility), pH, whipping temperature, the relative concentration of salts, sugars, and lipids, and the extent of shearing that occurs during the whipping action. Combinations of variables that are optimum for one protein system may be quite different for other protein sources. Because of the complexity of variables and corresponding interactions, a statistical approach should be employed to best determine optimum whipping conditions (see Lah *et al.* 1980. *J. Food Sci.* 45(6):1720).

Procedure for Cream Whippability Test.

1. All testing steps should be conducted in a 40°F (4°C) cold room (or chamber).
2. Transfer the cream to be tested into a weight tared 50 ml beaker. Fill the beaker to "full" capacity and weigh. Pour the cream back into the original container.
3. Transfer 300 ml cream into a chilled mixing bowl. Whip the cream at the highest mixer speed. Record the time required to achieve a "stiff" whipped cream.
4. Fill the same 50 ml beaker (used in step 2 (above)) to "full" capacity with whipped cream and reweigh the container.
5. Transfer the remainder of the whipped cream to a funnel that is placed on top of a 100 ml calibrated cylinder. Place the funnel, cylinder, and contents in an area at room temperature. After 30 min, observe any separated fluid that may have accumulated in the cylinder. A stable foam will show little or no separated liquid (serum).
6. Record whipping test results as follows:

   (a) Whipping time as elapsed seconds of time.
   (b) Calculate % overrun of whipped cream =

   $$\frac{\text{Wt. of unwhipped cream} - \text{Wt. of whipped cream}}{\text{Wt. of whipped cream}} \times 100.$$

   (c) Stability of whipped cream =
       ml. of volume of drained liquid (serum).

VIII. GRADES AND STANDARDS FOR AMERICAN CHEDDAR CHEESE

Federal Definition and Standards

Cheddar cheese is defined by Federal regulations (21 CFR 133) as cheese made by the Cheddar process *or* by another procedure which produces a finished cheese having the same physical and chemical properties as that produced by the Cheddar process. It is made from cow's milk with or without the addition of coloring matter. Common salt may be added (usually is). The cheese shall contain not more than 39% moisture, and in the water-free substance, not less that 50% milk fat (fat-in-dry-matter). Such cheese must also comply with the regulations of the Federal Food, Drug and Cosmetic act. The United States standards for grades of Cheddar cheese are contained in the Code of

Federal Regulations, Title 7, Part 58, Subpart K. paragraphs 58.2501 through 58.2506. The following eight tables provide details of the various grade requirements. They have been reproduced from the 1987 edition of CFR, paragraph 58.2505.

State Standards

Since a high percentage of American Cheddar cheese is made in Wisconsin, a cheese grader should be familiar with the specific standards and grades established by the Wisconsin legislature for this product as well as those standards set by the United States Department of Agriculture.

Wisconsin Grades. The state of Wisconsin recognizes two brands of Cheddar cheese. These are described in the following paragraphs.

Wisconsin State (Brand). Cheese labeled or sold as Wisconsin State Brand shall conform to the following standards:

Flavor. The flavor shall be pleasing. The cheese shall be free from undesirable flavors and odors. It may be lacking in flavor development or may possess characteristic cheese flavor. It may possess very slight acid or slight feed off-flavors.

Body and Texture. A plug drawn from the cheese should appear reasonably solid, compact, smooth and close, and should be slightly translucent. It may have a few mechanical openings if not large and connecting, and may have not more than 2 sweet holes per plug, but should be free from other gas holes. It may be definitely curdy or partially broken down if the cheese is more than 3 weeks old.

Color. It may be uncolored or of any degree of color recognized in the markets. The color shall be uniform, not dull or faded, and practically free from white lines or seams. The cheese may have numerous tiny white specks associated with aged cheese.

Wisconsin Junior (Brand). Any cheese labeled or sold as Wisconsin Junior shall conform to the following standards:

Flavor. It shall have fairly pleasing, characteristic cheese flavor, but may possess the following off-flavors to a slight degree: acid, flat, bitter, utensil, fruity, whey-taint, yeasty, malty, old milk, weedy, barny, and lipase; and feed flavor to a definite degree.

Body and Texture. Body and texture may be slightly defective. A plug drawn from the cheese may possess the following characteristics to a slight degree: coarse, short, mealy, weak, pasty, crumbly, gassy, slitty, and corky; and the following characteristics to a definite degree: curdy, open, and sweet holes.

Table I. Detailed Specifications for U.S. Grade AA Cheddar Cheese

	Fresh or Current	Medium Cured	Cured or Aged
(a) *Flavor:*	Fine and highly pleasing. May be lacking in flavor development or may possess slight characteristic cheddar cheese flavor. May possess a very slight feed flavor, but shall be free from any undesirable flavors and odors.	Fine and highly pleasing. Possesses a moderate degree of characteristic cheddar cheese flavor. May possess a very slight feed flavor but shall be free from any undesirable flavors and odors.	Fine and highly pleasing characteristic cheddar cheese flavor showing moderate to well-developed degrees of flavor or sharpness. May possess a very slight feed flavor but shall be free from any undesirable flavors and odors.
(a) *Body and texture:*	A plug drawn from the cheese shall be firm, appear smooth, compact, close and should be slightly translucent, but may have a few small mechanical openings. The texture may be definitely curdy or may be partially broken down if more than 3 weeks old. Shall be free from sweet holes, yeast holes and gas holes of any kind.	A plug drawn from the cheese shall be firm, appear smooth, waxy, compact, close, flexible and translucent, but may have a few mechanical openings if not large and connecting. May be slightly or not entirely broken down. May possess not more than one sweet hole per plug but shall be free from other gas holes.	A plug drawn from the cheese shall be firm, appear smooth, waxy, compact, close and translucent but may have a few mechanical openings if not large and connecting. Should be free from curdiness and possess a cohesive velvet-like texture. May possess not more than one sweet hole per plug but shall be free from other gas holes.
(c) *Color:*	Shall have a uniform, bright attractive appearance; practically free from white lines or seams. May be colored or uncolored but if colored it should be a medium yellow-orange.	Shall have a uniform, bright attractive appearance; practically free from white lines or seams. May be colored or uncolored, but if colored it should be a medium yellow-orange.	Shall have a uniform, bright attractive appearance; practially free from white lines or seams. May show numerous tiny white specks. May be colored or uncolored, but if colored it should be a medium yellow-orange.

544

(d) *Finish and appearance:*

Bandaged and paraffin-dipped. Shall possess a sound, firm rind with a smooth bandage and paraffin coating adhering tightly but may possess soiled surface to a very slight degree. The cheese shall be even and uniform in shape.

Rindless. The wrapper or covering shall be practically smooth, properly sealed with adequate overlapping at the seams or by any other satisfactory type of closure. The wrapper or covering shall be neat and adequately and securely envelop the cheese. May be slightly wrinkled but shall be of such character as to fully protect the surface of the cheese and not detract from its initial quality. Shall be free from mold under wrapper or covering and shall not be huffed or lopsided.

Bandaged and paraffin dipped. Shall possess a sound, firm rind with a smooth bandage and paraffin coating adhering tightly but may possess very slight mold under bandage and paraffin, and the following other characteristics to a slight degree; soiled surface and surface mold. The cheese shall be even and uniform in shape.

Rindless. Same as for current, except very slight mold under wrapper or covering permitted.

Bandaged and paraffin dipped. Shall possess a sound, firm rind with a smooth bandage and paraffin coating adhering tightly but may possess the following characteristics to a slight degree; soiled surface and mold under bandage and paraffin; and surface mold to a definite degree. The cheese shall be even and uniform in shape.

Rindless. Same as for medium.

Table II. Detailed Specifications for U.S. Grade A Cheddar Cheese

	Fresh or Current	Medium Cured	Cured or Aged
(a) *Flavor:*	Shall possess a pleasing flavor. May be lacking in flavor development or may possess slight characteristic cheddar cheese flavor. May possess very slight acid, slight feed but shall not possess any undesirable flavors and odors.	Shall possess a pleasing characteristic cheddar cheese flavor and aroma. May possess a very slight bitter flavor and the following flavors to a slight degree; feed and acid.	Shall possess a pleasing characteristic cheddar cheese flavor and aroma with moderate to well developed degrees of flavor or sharpness. May possess the following flavors to a slight degree; bitter, feed and acid.
(b) *Body and Texture:*	A plug drawn from the cheese shall be firm, appear smooth, compact, close and should be slightly translucent but may have a few mechanical openings if not large and connecting. May possess not more than two sweet holes per plug but shall be free from other gas holes. May be definitely curdy or partially broken down if more than 3 weeks old.	A plug drawn from the cheese shall be reasonably firm, appear reasonably smooth, waxy, fairly close and translucent but may have a few mechanical openings if not large and connecting. May be slightly curdy or not entirely broken down. May possess not more than two sweet holes per plug but shall be free from other gas holes. May possess the following other characteristics to a slight degree; mealy, short and weak.	A plug drawn from the cheese should be fairly firm, appear smooth, waxy, fairly close and translucent but may have a few mechanical openings. Should be free from curdiness. May possess not more than two sweet holes per plug but shall be free from other gas holes. May possess the following other characteristics to a slight degree; crumbly, mealy, short, weak and pasty.
(c) *Color:*	Shall have a fairly uniform, bright attractive appearance. May have slight white lines or seams or be very slightly wavy. May be colored or uncolored but if colored, it should be a medium yellow-orange.	Shall have a uniform, bright attractive appearance. May have slight white lines or seams. May be colored or uncolored but if colored, it should be a medium yellow-orange.	Shall have a uniform, bright attractive appearance. May have slight white lines or seams and numerous tiny white specks. May be colored or uncolored, but if colored, it should be a medium yellow-orange.

(d) *Finish and appearance:*

Bandaged and paraffin dipped. Shall possess a sound, firm rind with the bandage and paraffin coating adhering tightly, but may possess the following characteristics to a very slight degree; soiled surface and surface mold; and to a slight degree; rough surface, irregular bandaging, lopsided and high edges.

Rindless. The wrapper or covering shall be practically smooth, properly sealed with adequate overlapping at the seams or by any other satisfactory type of closure. The wrapper or covering shall be neat and adequately and securely envelop the cheese. May be slightly wrinkled but shall be of such character as to fully protect the surface of the cheese and not detract from its initial quality. Shall be free from mold under the wrapper or covering and shall not be huffed but may be slightly lopsided.

Bandaged and paraffin dipped. Shall possess a sound, firm rind with the bandage and paraffin coating adhering tightly but may possess very slight mold under bandage and paraffin and the following other characteristics to a slight degree; soiled surface, surface mold, rough surface, irregular bandaging, lopsided and high edges.

Rindless. Same as for current, except very slight mold under wrapper or covering permitted.

Bandaged and paraffin dipped. Shall possess a sound, firm rind with the bandage and paraffin coating adhering tightly but may possess the following characteristics to a slight degree; soiled surface, rough surface, mold under bandage and paraffin, irregular bandaging, lopsided and high edges; and surface mold to a definite degree.

Rindless. Same as for medium.

Table III. Detailed Specifications for U.S. Grade B Cheddar Cheese

	Fresh or Current	Medium Cured	Cured or Aged
(a) Flavor:	Should possess a fairly pleasing characteristic cheddar cheese flavor, but may possess very slight onion and the following flavors to a slight degree: Acid, flat, bitter, fruity, utensil, whey-taint, yeasty, malty, old milk, weedy, barny and lipase; feed flavor to a definite degree.	Should possess a fairly pleasing characteristic cheddar cheese flavor and aroma. May possess very slight onion and the following flavors to a slight degree; flat, yeasty, malty, old milk, weedy, barny and lipase; the following to a definite degree; feed, acid, bitter, fruity, utensil, and whey-taint.	Should possess a fairly pleasing characteristic cheddar cheese flavor and aroma, with moderate to well developed degrees of flavor or sharpness. May possess very slight onion and the following flavors to a slight degree; flat, yeasty, malty, old milk, weedy, barny, lipase and sulfide; the following to a definite degree: Feed, acid, bitter, fruity, utensil, and whey-taint.
(b) Body and texture:	A plug drawn from the cheese may possess the following characteristics to a slight degree: Coarse, short, mealy, weak, pasty, crumbly, gassy, slitty and corky; the following to a definite degree: Curdy open, and sweet holes.	A plug drawn from the cheese may possess the following characteristics to a slight degree: Curdy, coarse, gassy, slitty, and corky; the following to a definite degree: Open, short, mealy, weak, pasty, crumbly, and sweet holes.	A plug drawn from the cheese may possess the following characteristics to a slight degree; gassy and slitty; the following to a definite degree: Open, sweet holes, short, mealy, weak, pasty and crumbly.
(c) Color:	May possess the following characteristics to a slight degree: Wavy, acid cut, mottled, salt spots, dull or faded and definitely seamy. May be colored or uncolored but if colored, may be slightly unnatural.	May possess a very slight bleached surface; and the following characteristics to a slight degree: Wavy, acid-cut, mottled, salt spots, dull or faded; and definitely seamy. May be colored or uncolored but if colored, may be slightly unnatural.	May possess the following characteristics to a slight degree: Wavy, acid-cut, mottled, salt spots, dull or faded and bleached surface; and definitely seamy. May be colored or uncolored but if colored, may be slightly unnatural.

(d) *Finish and appearance:*

Bandaged and paraffin dipped. Shall possess a reasonably firm sound rind, but may possess very slight mold under bandage and paraffin. The following characteristics to a slight degree: Soiled surface, surface mold, defective coating, checked rind, huffed, weak rind, and sour rind; and to a definite degree: Rough surface, irregular bandaging, lopsided and high edges.

Rindless. The wrapper or covering shall be fairly smooth and properly sealed with adequate overlapping at the seams or by other satisfactory type of closure. The wrapper or covering shall be fairly neat and adequately and securely envelop the cheese. May be definitely wrinkled but shall be of such character as to protect the surface of the cheese and not detract from its initial quality. Shall be free from mold under wrapper or covering but may be slightly huffed and slightly lopsided.

Bandaged and paraffin dipped. Shall possess a reasonably firm sound rind, but may possess the following characteristics to a slight degree; surface mold, mold under bandage and paraffin, checked rind, huffed, weak rind, and sour rind; the following to a definite degree: Soiled surface, rough surface, irregular bandaging, lopsided, high edges and defective coating.

Rindless. Same as for current, except slight mold underwrapper or covering permitted.

Bandaged and paraffin dipped. Shall possess a reasonably firm sound rind, but may possess the following characteristics to a slight degree; checked rind, huffed, weak rind, and sour rind; the following to a definite degree: Soiled surface, surface mold, mold under bandage and paraffin, rough surface, irregular bandaging, lopsided, high edges and defective coating.

Rindless. Same as for medium.

Table IV. Detailed Specifications for U.S. Grade C Cheddar Cheese

Fresh or Current	Medium Cured	Cured or Aged
(a) *Flavor:*		
May possess the following flavors to a slight degree: Sour, metallic, onion; and to a definite degree: Acid, flat, bitter, fruity, utensil, whey-taint, yeasty, malty, old milk, weedy, barny, and lipase; feed flavor to a pronounced degree.	May possess the following flavors to a slight degree: Onion, and sulfide; and to a definite degree; Flat, sour, metallic, yeasty, malty, old milk, weedy, barny and lipage; and to a pronounced degree: Feed, acid, bitter, fruity, utensil, and whey-taint.	May possess slight onion and the following flavors to a definite degree: Flat, sour, metallic, yeasty, malty, old milk, weedy, barny, lipase and sulfide; and to a pronounced degree: Feed, acid, bitter, fruity, utensil and whey-taint.
(b) *Body and texture:*		
A plug drawn from the cheese may possess the following characteristics to a definite degree: Curdy, coarse, corky, crumbly, mealy, short, weak, pasty, gassy, slitty, pinny; and to a pronounced degree; open and sweet holes. The cheese shall be sufficiently compact to permit the drawing of a plug.	A plug drawn from the cheese may be slightly curdy and may possess the following other characteristics to a definite degree: Coarse, corky, gassy, slitty and pinny; and to a pronounced degree: Open, sweet holes, short, weak, pasty, crumbly and mealy. The cheese shall be sufficiently compact to permit the drawing of a plug.	A plug drawn from the cheese may possess the following characteristics to a definite degree: Gassy, slitty, pinny; and to a pronounced degree: open, sweet holes, short, weak, pasty, crumbly and mealy. The cheese shall be sufficiently compact to permit the drawing of a plug.
(c) *Color:*		
May have a slight bleached surface and possess the following other characteristics to a definite degree: Wavy, acid-cut, mottled, salt spots, dull or faded; and seamy to a pronounced degree. May be colored or uncolored but if colored, may be definitely unnatural. The color shall not be particularly unattractive.	May possess the following characteristics to a definite degree: Wavy, acid-cut, mottled, salt spots, bleached surface, dull or faded; and seamy to a pronounced degree. May be colored or uncolored but if colored may be definitely unnatural. The color shall not be particularly unattractive.	Same as for medium.

550

(d) *Finish and appearance:*

Bandaged and paraffin dipped. May possess the following characteristics to a slight degree: Cracks in rind, soft spots and wet rind; and mold under bandage and paraffin; and to a definite degree: Soiled surface, surface mold, defective coating, checked rind, weak rind, sour rind, and huffed; and to a pronounced degree: Rough surface, irregular bandaging, lopsided and high edges.

Rindless. The wrapper or covering shall be fairly smooth and properly sealed with adequate overlapping at the seams or by other satisfactory type of closure. The wrapper or covering shall adequately and securely envelop the cheese. May be definitely soiled and wrinkled but shall be of such character as to protect the surface of the cheese and not detract from its initial quality. May have slight mold under the wrapper or covering and may be definitely huffed and lopsided.

Bandaged and paraffin dipped. May possess very slight rind rot and the following other characteristics to a slight degree: Cracks in rind, soft spots and wet rind; and to a definite degree: Surface mold, mold under bandage and paraffin, checked rind, weak rind, sour rind and huffed; and to a pronounced degree: Soiled surface, rough surface, defective coating, irregular bandaging, lopsided and high edges.

Rindless. Same as for current, except definite mold under the wrapper or covering permitted.

Bandaged and paraffin dipped. May possess the following characteristics to a slight degree: Rind rot, cracks in rind; and to a definite degree; checked rind, weak rind, sour rind, wet rind, soft spots and huffed; and to a pronounced degree: Rough surface, soiled surface, surface mold, mold under bandage and paraffin, defective coating, irregular bandaging, lopsided and high edges.

Rindless: Same as for medium.

Table V. Classification of Flavor According to Degree of Curing (Cheddar Cheese)

Identification of flavor characteristics	AA			A			B			C		
	Fresh or Current	Medium Cured	Cured or Aged	Fresh or Current	Medium Cured	Cured or Aged	Fresh or Current	Medium Cured	Cured or Aged	Fresh or Current	Medium Cured	Cured or Aged
Feed	VS	VS	VS	S	S	S	D	D	D	P	P	P
Acid				VS	S	S	S	D	D	D	P	P
Flat							S	S	S	D	D	D
Bitter					VS	S	S	D	D	D	P	P
Fruity							S	D	D	D	P	P
Utensil							S	D	D	D	P	P
Metallic										S	D	D
Sour										S	D	D
Whey-Taint							S	D	D	D	P	P
Yeasty							S	S	S	D	D	D
Malty							S	S	S	D	D	D
Old Milk							S	S	S	D	D	D
Weedy							S	S	S	D	D	D
Onion							VS	VS	VS	S	S	S
Barny							S	S	S	D	D	D
Lipase							S	S	S	D	D	D
Sulfide									S		S	D

VS—Very Slight.   S—Slight.   D—Definite.   P—Pronounced.

Table VI. Classification of Body and Texture According to Degree of Curing (Cheddar Cheese)

Identification of body and texture characteristics	AA			A			B			C		
	Fresh or Current	Medium Cured	Cured or Aged	Fresh or Current	Medium Cured	Cured or Aged	Fresh or Current	Medium Cured	Cured or Aged	Fresh or Current	Medium Cured	Cured or Aged
Curdy	D	S		D	S		D	S		D	S	
Coarse	VS	S	S				S	S		D	D	
Open		VS	VS	S	S	S	D	D	D	P	P	P
Sweet holes				S	S	S	D	D	D	P	P	P
Short					S	S	S	D	D	D	P	P
Mealy					S	S	S	D	D	D	P	P
Weak					S	S	S	D	D	D	P	P
Pasty						S	S	D	D	D	P	P
Crumbly						S	S	D	D	D	P	P
Gassy							S	S	S	D	D	D
Slitty							S	S	S	D	D	D
Corky							S			D	D	
Pinny										D	D	D

VS—Very slight. S—Slight. D—Definite. P—Pronounced.

Table VII. Classification of Color According to Degree of Curing (Cheddar Cheese)

Identification of color characteristics	AA			A			B			C		
	Fresh or Current	Medium Cured	Cured or Aged	Fresh or Current	Medium Cured	Cured or Aged	Fresh or Current	Medium Cured	Cured or Aged	Fresh or Current	Medium Cured	Cured or Aged
Seamy	VS	VS	VS	S	S	S	D	D	D	P	P	P
Wavy				VS			S	S	S	D	D	D
Acid-cut							S	S	S	D	D	D
Unnatural							S	S	S	D	D	D
Mottled							S	S	S	D	D	D
Salt spots							S	S	S	D	D	D
Dull or faded							S	S	S	D	D	D
Bleached surfaced (rindless)								VS	S	S	D	D

VS—Very Slight.   S—Slight.   D—Definite.   P—Pronounced.

Table VIII. Classification of Finish and Appearance According to Degree of Curing (Cheddar Cheese)

Identification of finish and appearance characteristics	AA			A			B			C		
	Fresh or Current	Medium Cured	Cured or Aged	Fresh or Current	Medium Cured	Cured or Aged	Fresh or Current	Medium Cured	Cured or Aged	Fresh or Current	Medium Cured	Cured or Aged
Soiled surface	VS	S	S	VS	S	S	S	D	D	D	P	P
Surface mold		S	D	VS	S	D	S	S	D	D	D	P
Mold under bandage and paraffin		VS	S		VS	S	VS	S	D	S	D	P
Mold under wrapper or covering (rindless)		VS	VS		VS	VS		S	S	S	D	D
Rough surface				S	S	VS	D	D	D	P	P	P
Irregular bandaging (uneven, wrinkled and overlapping)						S	D	D	D	P	P	P
Lopsided				S	S	S	D	D	D	P	P	P
Lopsided (rindless)				S	S	S	S	S	S	P	P	D
High edges				S	S	S	D	D	D	D	D	P
Huffed				S	S	S	S	S	S	D	D	D
Defective coating (scaly, blistered and checked)							S		D	D	S	P
Cracks in rind								S		S	S	S
Checked rind							S		S	D	D	D
Soft spots										D	D	D
Weak rind							S	S	S	D	D	D
Sour rind							S	S	S	D	D	D
Wet rind										S	S	D
Rind rot										S	VS	S

VS—Very Slight. S—Slight. D—Definite. P—Pronounced.

Wisconsin Junior (Brand) Cheese (cont.)

Color. It may be either uncolored or any degree of color that is recognized or accepted in the marketplace. The color may be slightly defective and may possess the following defects to a slight degree: wavy, acid-cut, dull, or faded.

Finish and Appearance. The surface shall be sound; it may be slightly weak, but shall be free from soft spots, rind rot, cracks and openings of any kind. . . . In the case of rindless cheese, the covering or wrapper shall adequately and securely envelop the cheese, be neat and unbroken, and protect the cheese, but may be wrinkled. Rindless cheese shall be free from mold under the wrapper or covering, but may be slightly lopsided.

IX. A SCORE CARD FOR MILK

A SCORE CARD FOR MILK

NAME OF EVALUATOR: _____ DATE _____

DIRECTIONS: Rate each milk sample for the general characteristics listed (I-IX), by use of the flavor rating scale immediately following:

0 - Not perceptible	2 - Slight level
1 - Just barely perceptible	3 - Distinct level
(threshold)	4 - Pronounced level

In addition, place a check mark opposite the sub-category, if any, that helps to identify the cause of the flavor.

SAMPLE NUMBER	1	2	3	4	5	6	7	8	9	10
I. HEATED										
Cooked										
Scorched										
II. LIGHT ACTIVATED										
III. LACKS FINE FLAVOR										
Lacks flavor										
Unnatural blend										
IV. LIPOLYZED (Hydrolyzed rancidity)										
V. MICROBIAL										
Acid										
Bitter										
Fruity										
Malty										
Unclean										
VI. MISCELLANEOUS										
Astringent										
Foreign										
Salty										
VII. OXIDIZED										
Papery-cardboardy										
Metallic										
Oily										
VIII. STORED (Lacks freshness)										
IX. TRANSMITTED										
Feed										
Cowy										
FLAVOR CLASS SCORE[a]										

[a] (Excellent = 5) (Good = 4) (Fair = 3) (Very poor = 1)

X. COLLEGIATE CONTEST MILK SCORE CARD

MARKING INSTRUCTIONS

USE NO. 2 PENCIL ONLY

IMPROPER MARKS ⊗ ⊙ ◐ ⊙   PROPER MARK ●

- ERASE CHANGES CLEANLY AND COMPLETELY
- DO NOT MAKE ANY STRAY MARKS

MILK

NCS Trans-Optic® MP30-73529-321   A2400

PR CONTESTANT NO.   DATE

CRITICISMS		SAMPLE NUMBER							
		1	2	3	4	5	6	7	8
FLAVOR		①②③④⑤⑥⑦⑧⑨⑩	①②③④⑤⑥⑦⑧⑨⑩	①②③④⑤⑥⑦⑧⑨⑩	①②③④⑤⑥⑦⑧⑨⑩	①②③④⑤⑥⑦⑧⑨⑩	①②③④⑤⑥⑦⑧⑨⑩	①②③④⑤⑥⑦⑧⑨⑩	①②③④⑤⑥⑦⑧⑨⑩
NO CRITICISM 10 NORMAL RANGE 1-10	ACID	○	○	○	○	○	○	○	○
	BITTER	○	○	○	○	○	○	○	○
	COOKED	○	○	○	○	○	○	○	○
	FEED	○	○	○	○	○	○	○	○
	FERMENTED/FRUITY	○	○	○	○	○	○	○	○
	FLAT	○	○	○	○	○	○	○	○
	FOREIGN	○	○	○	○	○	○	○	○
	GARLIC/ONION	○	○	○	○	○	○	○	○
	LACKS FRESHNESS	○	○	○	○	○	○	○	○
	MALTY	○	○	○	○	○	○	○	○
	OXIDIZED - LIGHT INDUCED	○	○	○	○	○	○	○	○
	OXIDIZED - METAL INDUCED	○	○	○	○	○	○	○	○
	RANCID	○	○	○	○	○	○	○	○
	SALTY	○	○	○	○	○	○	○	○
	UNCLEAN	○	○	○	○	○	○	○	○
		○	○	○	○	○	○	○	○
		○	○	○	○	○	○	○	○
		○	○	○	○	○	○	○	○
		○	○	○	○	○	○	○	○
		○	○	○	○	○	○	○	○

BODY AND TEXTURE	1	2	3	4	5	6	7	8
	①②③④⑤	①②③④⑤	①②③④⑤	①②③④⑤	①②③④⑤	①②③④⑤	①②③④⑤	①②③④⑤
NO CRITICISM 5	○	○	○	○	○	○	○	○
	○	○	○	○	○	○	○	○
	○	○	○	○	○	○	○	○
NORMAL RANGE 1-5	○	○	○	○	○	○	○	○
	○	○	○	○	○	○	○	○
	○	○	○	○	○	○	○	○

APPEARANCE AND COLOR	1	2	3	4	5	6	7	8
	①②③④⑤	①②③④⑤	①②③④⑤	①②③④⑤	①②③④⑤	①②③④⑤	①②③④⑤	①②③④⑤
NO CRITICISM 5	○	○	○	○	○	○	○	○
	○	○	○	○	○	○	○	○
	○	○	○	○	○	○	○	○
NORMAL RANGE 1-5	○	○	○	○	○	○	○	○
	○	○	○	○	○	○	○	○
	○	○	○	○	○	○	○	○

XI. COLLEGIATE CONTEST BUTTER SCORE CARD

MARKING INSTRUCTIONS

USE NO. 2 PENCIL ONLY

IMPROPER MARKS	PROPER MARK
⊗ ⊘ ⦿ ⊙	●

- ERASE CHANGES CLEANLY AND COMPLETELY
- DO NOT MAKE ANY STRAY MARKS

NCS Trans-Optic® MP30-73532-321   A2400

BUTTER

PR | CONTESTANT NO. | DATE

CRITICISMS

SAMPLE NUMBER

CRITICISMS	1	2	3	4	5	6	7	8

FLAVOR

	ACID
NO	BITTER
CRITICISM	CHEESY
10	COARSE
	FEED
NORMAL	FLAT
RANGE	GARLIC/ONION
1-10	HIGH SALT
	METALLIC
	MUSTY
	NEUTRALIZER
	OLD CREAM
	OXIDIZED
	RANCID
	SCORCHED
	STORAGE
	UNCLEAN/UTENSIL
	WHEY
	YEASTY

BODY AND TEXTURE

NO CRITICISM 5

NORMAL RANGE 1-5

APPEARANCE AND COLOR

NO CRITICISM 5

NORMAL RANGE 1-5

XII. COLLEGIATE CONTEST CHEDDAR CHEESE SCORE CARD

MARKING INSTRUCTIONS

USE NO. 2 PENCIL ONLY

IMPROPER MARKS

PROPER MARK

- ERASE CHANGES CLEANLY AND COMPLETELY
- DO NOT MAKE ANY STRAY MARKS

CHEDDAR CHEESE

NCS Trans-Optic® MP30-73530-321   A2400

PR CONTESTANT NO.   DATE

CRITICISMS		SAMPLE NUMBER							
		1	2	3	4	5	6	7	8

FLAVOR

		1	2	3	4	5	6	7	8
NO CRITICISM 10	BITTER								
	FEED								
	FERMENTED/FRUITY								
	FLAT/LACKS FLAVOR								
	GARLIC/ONION								
	HEATED								
NORMAL RANGE 1-10	HIGH ACID								
	MOLDY								
	RANCID								
	SULFIDE								
	UNCLEAN								
	WHEY TAINT								
	YEASTY								

BODY AND TEXTURE

		1	2	3	4	5	6	7	8
NO CRITICISM 5	CORKY								
	CRUMBLY								
	CURDY								
	GASSY								
	MEALY								
NORMAL RANGE 1-5	OPEN								
	PASTY								
	SHORT								
	WEAK								

APPEARANCE AND COLOR

	1	2	3	4	5	6	7	8
NO CRITICISM 5								
NORMAL RANGE 1-5								

XIII. COLLEGIATE CONTEST ICE CREAM SCORE CARD

MARKING INSTRUCTIONS

USE NO. 2 PENCIL ONLY

IMPROPER MARKS	PROPER MARK

⊗ ⊘ ⊙ ⊙   ●

- ERASE CHANGES CLEANLY AND COMPLETELY
- DO NOT MAKE ANY STRAY MARKS

ICE CREAM

NCS Trans-Optic® MP30-73533-321   A2400

PR CONTESTANT NO.   DATE

SAMPLE NUMBER

CRITICISMS

		1	2	3	4	5	6	7	8

FLAVOR

		1 2 3 4 5	1 2 3 4 5	1 2 3 4 5	1 2 3 4 5	1 2 3 4 5	1 2 3 4 5	1 2 3 4 5	1 2 3 4 5
		6 7 8 9 10	6 7 8 9 10	6 7 8 9 10	6 7 8 9 10	6 7 8 9 10	6 7 8 9 10	6 7 8 9 10	6 7 8 9 10
	ACID								
NO	COOKED								
CRITICISM	LACKS FINE FLAVOR								
10	LACKS FLAVORING								
	LACKS FRESHNESS								
	LACKS SWEETNESS								
NORMAL	METALLIC								
RANGE	OLD INGREDIENT								
1-10	OXIDIZED								
	RANCID								
	SALTY								
	STORAGE								
	SYRUP FLAVOR								
	TOO HIGH FLAVOR								
	TOO SWEET								
	UNNATURAL FLAVOR								
	WHEY								

BODY AND TEXTURE

		1 2 3 4 5	1 2 3 4 5	1 2 3 4 5	1 2 3 4 5	1 2 3 4 5	1 2 3 4 5	1 2 3 4 5	1 2 3 4 5
		1	2	3	4	5	6	7	8
	COARSE/ICY								
NO	CRUMBLY								
CRITICISM	FLUFFY								
5	GUMMY								
	SANDY								
	SOGGY								
NORMAL	WEAK								
RANGE									
1-5									

APPEARANCE AND COLOR

		1 2 3 4 5	1 2 3 4 5	1 2 3 4 5	1 2 3 4 5	1 2 3 4 5	1 2 3 4 5	1 2 3 4 5	1 2 3 4 5
		1	2	3	4	5	6	7	8
NO									
CRITICISM									
5									
NORMAL									
RANGE									
1-5									

XIV. COLLEGIATE CONTEST COTTAGE CHEESE SCORE CARD

MARKING INSTRUCTIONS

USE NO. 2 PENCIL ONLY

IMPROPER MARKS   PROPER MARK

- ERASE CHANGES CLEANLY AND COMPLETELY
- DO NOT MAKE ANY STRAY MARKS

COTTAGE CHEESE

NCS Trans-Optic® MP30-73535-321   A2400

PR   CONTESTANT NO.   DATE

CRITICISMS	SAMPLE NUMBER

		1	2	3	4	5	6	7	8
FLAVOR		①②③④⑤⑥⑦⑧⑨⑩	①②③④⑤⑥⑦⑧⑨⑩	①②③④⑤⑥⑦⑧⑨⑩	①②③④⑤⑥⑦⑧⑨⑩	①②③④⑤⑥⑦⑧⑨⑩	①②③④⑤⑥⑦⑧⑨⑩	①②③④⑤⑥⑦⑧⑨⑩	①②③④⑤⑥⑦⑧⑨⑩
	BITTER								
NO	COOKED								
CRITICISM	DIACETYL								
10	FEED								
	FERMENTED/FRUITY								
	FLAT								
NORMAL	FOREIGN								
RANGE	HIGH ACID								
1-10	HIGH SALT								
	LACKS FINE FLAVOR								
	LACKS FRESHNESS								
	MALTY								
	METALLIC								
	MUSTY								
	OXIDIZED								
	RANCID								
	UNCLEAN								
	YEASTY								

BODY AND TEXTURE		1	2	3	4	5	6	7	8
		①②③④⑤	①②③④⑤	①②③④⑤	①②③④⑤	①②③④⑤	①②③④⑤	①②③④⑤	①②③④⑤
NO	FIRM/RUBBERY								
CRITICISM	GELATINOUS								
5	MEALY/GRAINY								
	OVERSTABILIZED								
NORMAL	PASTY								
RANGE	WEAK/SOFT								
1-5									

APPEARANCE AND COLOR		1	2	3	4	5	6	7	8
		①②③④⑤	①②③④⑤	①②③④⑤	①②③④⑤	①②③④⑤	①②③④⑤	①②③④⑤	①②③④⑤
NO	FREE CREAM								
CRITICISM	FREE WHEY								
5	LACKS CREAM								
	MATTED								
NORMAL	SHATTERED CURD								
RANGE									
1-5									

XV. COLLEGIATE CONTEST SWISS-STYLE YOGURT SCORE CARD

MARKING INSTRUCTIONS

USE NO. 2 PENCIL ONLY

IMPROPER MARKS

PROPER MARK

- ERASE CHANGES CLEANLY AND COMPLETELY
- DO NOT MAKE ANY STRAY MARKS

SWISS STYLE YOGURT

NCS Trans-Optic · MP30-73528-321   A2400

PR CONTESTANT NO.   DATE

CRITICISMS	SAMPLE NUMBER							
	1	2	3	4	5	6	7	8

FLAVOR

NO CRITICISM 10

NORMAL RANGE 1-10

	ACETALDEHYDE
	BITTER
	COOKED
	FOREIGN
	HIGH ACID
	LACKS FINE FLAVOR
	LACKS FLAVORING
	LACKS FRESHNESS
	LACKS SWEETNESS
	LOW ACID
	OLD INGREDIENT
	OXIDIZED
	RANCID
	TOO HIGH FLAVORING
	TOO SWEET
	UNNATURAL FLAVORING
	UNCLEAN
	YEASTY

BODY AND TEXTURE

NO CRITICISM 5

NORMAL RANGE 1-5

	GEL-LIKE
	GRAINY
	ROPY
	TOO FIRM
	WEAK

APPEARANCE AND COLOR

NO CRITICISM 5

NORMAL RANGE 1-5

	ATYPICAL COLOR
	COLOR LEACHING
	ENTRAPPED GAS
	EXCESS FRUIT
	FREE WHEY
	LACKS FRUIT
	LUMPY
	SHRUNKEN

XVI. NAMES AND ADDRESSES
OF ORGANIZATIONS

Secretariat, Collegiate Dairy Products Evaluation Contest
Dairy and Food Industries Supply Association
6245 Executive Blvd.
Rockville, MD 20852 (301) 984–1414

Source: Collegiate Dairy Products Evaluation Contest information

Superintendent, Collegiate Dairy Products Evaluation Contest
Dairy Division, Room 2750-S
Agricultural Marketing Service
United States Department of Agriculture
Washington, D.C. 20250 (202) 382–9383

Source: Collegiate Dairy Products Evaluation Contest official rules

Chairman, Committee on Sensory Evaluation of Dairy Products
American Dairy Science Association
309 W. Clark St.
Champaign, IL 61820 (217) 356–3182

Source: Official dairy product score cards (ADSA)

Milk Safety Branch
Public Health Service/Food and Drug Administration
U.S. Department of Health and Human Services
Washington, D.C. 20204 (202) 485–0138

Source: Grade A Pasteurized Milk Ordinance

Standardization Branch—Dairy Division
Agricultural Marketing Service
United States Department of Agriculture
Washington, D.C. 20250 (202) 447–3245

Source: Sediment standard charts and discs

American Cultured Dairy Products Institute
888 16th St. N.W.
Washington, D.C. 20006 (202) 223–1931

Source: Quality assurance manual for cultured dairy foods

Associate Manager of Contests and Awards
National FFA Center
P. O. Box 15160
Alexandra, VA 22309   (703) 360-3600

Source: FFA Milk quality contest rules and regulations

XVII. COLLEGIATE DAIRY PRODUCTS EVALUATION CONTEST GRADING

Grading of Contestants' Score Cards

The grading of contestants' score cards in the annual National Collegiate Contest is under the direct supervision of a member of the Committee on Sensory Evaluation of Dairy Products, American Dairy Science Association. The grading is conducted as described in the following material.

Scoring/Grading of Product Samples. A contestant's score for each item on the score card is assigned a grade, expressed as the difference between his/her score and the official score (except as indicated below). For example, if a contestant (for a given sample) scores "flavor" as 7 but the judges' or official score is 5, the contestant receives a grade of *2 points*. If, however, a contestant recognizes: (1) that the quality factor (item) scores "perfect," but fails to indicate the appropriate maximum score on his/her score card, or (2) records any score that is outside the score range for the quality factor (item), or (3) indicates the score by a dash (–), the contestant receives a grade equivalent to the maximum "point cut" for that quality item (score card sector). For example, the normal score range for body and texture of cottage cheese is "1–5" and represents a maximum "cut" of 5 points. The contestant's grade, therefore, is 5 when he/she fails to record any numerical score for that given quality item. This rule holds regardless of the official score assigned to the sample.

Grading of Contestant Criticisms

The grading of criticisms, which is independent of the grading of contestant scores, is based on the contestant's proficiency in recognizing the specific quality merits and defects of the various samples as noted by the official judge. Each criticism marked by the contestant on his/her score card is involved in the grading scheme. The contestant's grade for determining sample criticism(s) for a single quality factor is ascertained as follows:

(1) Perfect, or *0*, when the contestant:
    (a) Checks precisely the same defect(s) as the official judge(s) and checks only those made by the official; or
    (b) Recognizes, along with the official judge, that the given quality item is "above criticism," under which situation no designation mark is required (for the given sample).
(2) The maximum "point cut," i.e., *2.0* points, when the contestant:
    (a) Fails to check any of the defect(s) that were noted by the official judge;
    (b) Checks a defect(s) when the given sample was determined by the official judge as being "above criticism;"
    (c) Fails to check a criticism(s) when the official judge scores the sample within the "criticizable range" (although the contestant may have recorded a score for the sample as "above criticism"); or
    (d) Fails to check a criticism(s) when his/her recorded score indicates he/she should have (although the official scored the sample "above criticism").
(3) Subsequently, less than *2* points, i.e., *0.50, 0.67, 1.00, 1.33, 1.50,* or *1.60*, etc. (as an assigned grade), which reflects the percentage of the total criticisms recorded (marked) correctly on the score card, is given in the absence of either of the two previous conditions.

Examples: (a) The official judge(s) check 3 different criticisms for a quality factor, but the contestant checks only 1, but which is identical with one of the official criticisms. The assigned grade for the contestant is 1.33 for this instance. (b) The official judge checks 1 criticism, but the contestant checks 3 different ones, one of which coincides with that of the official judge. The contestant's grade is 1.33. (c) The official judge checks 3 different criticisms, and the contestant checks 3, one of which is identical with one of the official judges' criticisms. The contestant grade is 1.60. (A total of 5 different criticisms has been involved between the official judge and the contestant. The contestant and official judge agree on but one of them. Thus, the contestant is only 1/5 correct (or 4/5 incorrect), thus meriting a grade of 1.60 [80% of 2]).

Grading Guide for Contestant Score Cards.

Number of "Wrong" Criticisms/Item[a]	Total "Right" and "Wrong" Marks Involved	Contestants Grade Per Quality Item[a]
1	2	1.00
1	3	0.67
2	3	1.33
1	4	0.5
2	4	1.0
3	4	1.5
1	5	0.4
2	5	0.8
3	5	1.2
4	5	1.6
1	6	0.33
3	6	0.67
3	6	1.00
4	6	1.33
5	6	1.67

[a] Item refers to each quality factor (item) that undergoes sensory evaluation, i.e., flavor, body and texture, and/or color and appearance for a given sample.

Final Contestant Grades/Sample

A contestant's grade for a given product sample is the sum of his/her grades (deducted points) for both the "scoring" and "criticisms" aspects of that sample. In turn, the contestant's grade for a dairy product is the sum of his/her grades on all eight samples of that product.

Correspondingly, a team grade for each product is the sum of the grades for the three team members.

Since "grade" in the Collegiate Contest is synonymous with "points lost," the contestant scoring the lowest "grade" is deemed to be the winner for the given product; and the team scoring the lowest "grade" is deemed to be the winning team for that product.

Index

Acceptance tests, 494–95, 501
Acetaldehyde ("green") off-flavor
 concentration in yogurt, 255, 273
 in products
  Bulgarian buttermilk, 246
  cottage cheese, 233, 292–93
  cultured buttermilk, 233, 240, 246
  cultured milk products, 231, 233
  kefir, 297
  sour cream, cultured, 233
  yogurt, 233, 252, 253, 255, 272–73,
   297
 See also "Green" off-flavor
Acetone (in ketosis), 66
Acid off-flavor, 66, 68
 in butter, 379, 381, 384, 385, 405
 in Cheddar cheese, 311, 339
 in cottage cheese, 280, 291, 478
 in cream cheese, 372
 in cultured milk products, 231, 233, 236
 in dry milk products, 460, 462
 in evaporated milk, 426, 427
 in ice cream, 171, 173, 174, 175, 204,
  210–11
 in Italian cheese varieties, 370
 in Limburger cheese, 362
 in milk, 66, 68, 124, 140, 141–42, 476
 in yogurt, 261, 271
 See also High acid off-flavor
Acid-cut (Cheddar cheese), 316, 320
Acid, low, off-flavor
 cultured milk products, 236–37, 240,
  242
 lactic cultures, 236–37
 yogurt, 261, 272
Acidified milk products, 238
Acidity
 in buttermilk, 242
 in cultured milk products, 231, 235, 236
 in yogurt, 255

Acidophilus milks, 246–47
Adaptation (time), 18–19, 499–500
 to products
  butter, 404
  ice cream, 198
  milk, 147
  salt (in cottage cheese), 295
Aged flavor (butter), 381, 405
All-natural products, 169, 183
American Dairy Products Institute
  (ADPI), 451, 468–71
American Dairy Science Association
  (ADSA)
 Committee on Flavor Nomenclature
  and Reference Standards, 59,
  143
 Committee on Sensory Evaluation of
  Dairy Products, 56, 124, 140,
  172, 229, 259, 278, 345, 406, 492
 dry milk products evaluation, 460
 score card approval, 44, 124
 sponsor of Collegiate Contest, 54–57
Ammoniacal off-flavor, (cheese)
 blue-veined types, 364
 brick, 361
 Camembert and Brie, 366
 Limburger, 362
Anaerobic environment (cheese), 361
Anesthesia (numbness), 23, 172, 179, 197
Antioxidants, 70, 71
 alpha-tocopherol (acetate), 71
 high temperature heat treatment, 70,
  72, 408
 in dry cream, 457
 natural forms of, 70, 144
 sulfhydryl compounds, 72, 408
 vitamins (C and E) as, 71
Appearance
 Cheddar cheese, 100, 313
 cultured products, 101

Appearance (*cont.*)
  cultures, 237
  dry milk products, 466
  flavored products, 100
  unflavored products, 100
Appearance and finish of Cheddar cheese
  cheese mites, 329
  huffed (bloated), 329
  paraffin defects, 326
  pliable wax-coated, 326
  rindless defects, 325, 326
  surface of cheese, 328
  workmanship, 327
Aroma (properties and qualities), 11, 22,
    38, 59–62, 179, 249, 318
  See also Odors and/or Olfaction
Aromatic flavor, 228
  in cultured milk products, 233
  in soft, mold-ripened cheeses, 366
Astringency, 88, 89
Astringent defect, 23, 72, 88–89, 92–93
  in butter, 412–13
  in concentrated milks, 427
  in cottage cheese, 293
  in cultured milk products, 231, 233
  in dry milk products, 119, 460, 462
  in evaporated milk, 427
  in ice cream, 97, 206–7
  in milk, 124, 140, 142, 150
  in skim milk, 119
  tactile properties, related to, 93
  See also Chalky defect and Puckery
    mouthfeel
Atypical color defect
  as an index of workmanship, 99
  in Cheddar cheese, 321
  in ice cream, 180–83
  specks in Cheddar cheese, 320, 321
  in yogurt, 266
Auto-oxidation of lipids, 62, 73, 210, 294
  See also Oxidation of lipids; Oxidized
    off-flavor; and Metallic off-flavor

Bacterial content, 68, 204, 227, 253, 453,
    454
  score guides, 139
  scoring schemes
    in ice cream, 172
    in milk, 137–39
  USPHA/FDA P.M.O. standards, 138
Bacterial lipases, 80

Ballots for sensory testing panels, 498,
    500, 504, 509 (figs.), 524
Barny off-flavor, 66
  in milk, 124, 140, 142
  in Italian cheese, 370
Basic tastes, 179
  See also Primary tastes
Bifidobacterium sp., 253
Bite resistance (ice cream), 93, 96, 188,
    192
Bitter off-taste, 23, 68, 87
  causes of, 23, 62, 68, 87
    lipid hydrolysis (rancidity), 76–81, 87
    lipid oxidation, 72
    proteolytic activity, 65–66, 87
    psychrotrophic bacteria, 65–66, 87
    weed ingestion by cows, 65–66, 87,
      146
  in blue-veined cheese varieties, 364
  in Brie cheese, 366
  in butter, 379, 381, 384, 406
  in Camembert cheese, 366
  in Cheddar cheese, 340–44
  in chocolate flavored products, 90, 213–
    14
  in Colby/Monterey Jack cheese, 351
  in cottage cheese, 280, 287–88, 291–92,
    294, 478
  in cultured milk products, 231–32
  in dry buttermilk, 451
  in dry milk products, 460, 462
  in dry whey, 452
  in Italian cheese, 370
  in Limburger cheese, 362
  in milk, 124, 140, 142, 150, 477
  in nonfat dry milk, 445
  in Swiss cheese, 353
  in yogurt, 273
"Blind" defect
  Colby/Monterey Jack cheese, 351–52
  Swiss cheese, 354–55, 358
Bloat defect (Swiss cheese), 354
Blue-veined cheese varieties, 98, 101, 301,
    303
  body and texture, 363–64
  desired sensory qualities, 364
  flavor, 362–65
  ketones in, 364
  mold growth, 364
  Penicillium mold in, 101
  related types, 362–63
  ripening conditions, 363–64

score card, 363-64
undesirable properties, 364
Body and texture of
  ice cream, 178, 188, 189
  brick cheese, 359
  butter, 394, 395
  buttermilk, 240-42
  buttermilk, flake type, 244, 245
  Camembert cheese, 368
  Cheddar cheese, 315, 332-37
  chocolate milk products, 154-55
  cottage cheese, 285, 290
  cream cheese, 372
  sour cream, 249-51
  yogurt, 263, 268-71
Body of cultured milk products, 234-35
Bound water, 169, 251
Brands and trademarks, 2-4
Brevibacterium linens, 359, 361
Brick cheese, 98, 303
  body and texture, 359
  color and appearance, 360
  flavor, 359
  salt content of, 360
  score card, 359
  sensory evaluation of, 360
Brie cheese, 98
  body and texture, 365-66, 368
  desired flavor characteristics, 366-68
  score card, 366-67
  tempering of samples, 368
Briny off-flavor (butter), 94, 384, 386,
      390-91, 393, 406
  See also Salt defects (butter)
Browned defect (dry whole milk), 442
Bulky-flavored frozen desserts
  body and texture, 212
  quality criteria, 214, 217, 219
"Burnt" off-flavor, 121
Butter
  body and texture, 94, 394-95
  color and appearance, 393-94
  composition, 376
  cultured cream type, 232, 278
  definition (CFR), 376
  diacetyl (buttery) aroma, 232, 278
  flavor, 5, 402
  frequency of sensory defects, 401
  grading milk and cream for buttermak-
      ing, 378, 380
  historical developments, 2-4, 376-77
  ingredients, 377

melting point, 94
packaging, 390
plug (core for sample), 95, 389
regional sensory preferences, 376
salt in, 94, 390-91, 393, 406
salted, 378
sampling, 388
Scandinavian style, 232
sequence of observations, 389-90
sour cream, 377
spreadability, 94, 396-97, 399-401
sweet cream type, 377-78
tempering of samples, 385-86
triers for, 95, 387
triglycerides in, 94
unsalted, 378
U.S. Grades of, 377, 381
whey cream type, 415
whipped, 378
Buttermilk, cultured, 93
  acidity (pH), 236, 242-43
  body and texture, 240-42, 244-45
  butter granules, 243-45
  carbonation effect, 235
  color and appearance, 242, 245
  containers and closures, 243
  flavor attributes, 240
  flavor defects, 240, 244
  sensory evaluation of (scoring), 243-45
  tactile properties, 93, 97
Buttermilk, Bulgarian-style, 240, 246
Buttermilk, flake-type
  body and texture, 244, 245
  flavor, 244
Buttery defect
  in evaporated milk, 427
  in ice cream, 190, 194
  in soft-serve, 224
"Buttons" (sweetened condensed milk),
      433

Caked defect, (dry whole milk), 442
Calcium lactate (Cheddar cheese), 323-24
  See also White specks (Cheddar cheese)
Camembert cheese, 303, 365-66
  body and texture, 368
  color and appearance, 368
  composition, 366
  continuous manufacture, 365
  score card, 367
  sensory characteristics, 366
  tempering of samples, 368

Caramel off-flavor, 62, 63, 119, 121, 143, 169, 204, 205
  in acidophilus milk, 247
  in evaporated milk, 119, 406-7
  in dry whole milk, 441
  in frozen dairy desserts, 169, 204-5, 210
  in sweetened condensed milk, 430, 433
  in UHT/ultrapasteurized products, 121
Carbonyl compounds, 60-63, 73
Carbonation effect
  in buttermilk, 235
  in cultured milk products, 235
  in kefir, 297-98
  in lactic cultures, 235
  in sour cream, cultured, 249
Carbon dioxide (CO_2), in Swiss cheese, 353
Casein. *See* Edible dry casein
"Catty" off-flavor (Cheddar cheese), 62, 345
Chalky defect, 93, 119
  in concentrated milks, 427
  in cottage cheese, 293
  in cultured milk products, 231, 233
  in dry milk products, 460, 462-63
  in ice cream, 97, 206-7
  in nonfat dry milk, 445
  in skim milk, 119
  tactile properties, relationship to, 93
  See also Astringent defect and Puckery mouthfeel
Checked appearance defect (Swiss cheese), 357
Cheddar cheese
  barrel cheese, 306
  body and texture, 95, 315, 331
    defect remedies, 316
    defects of, 332-37
      corky, 307, 316, 333
      crumbly, 315-16, 318, 333
      curdy, 307, 315-16, 318, 333
      gassy (gas holes), 96, 329, 331, 335-37
      mealy/grainy, 316, 318, 333, 335
      open (openness), 315, 330, 337
      pasty (sticky), 315, 318, 336
      short, 315, 332
      weak, 315-16, 330, 336
    defined, 330
  Cheddaring step, 306, 331
  color and appearance, 99-100, 313

  defect remedies, 316-317
  defects of, 316-24
    acid-cut, 316, 320-21, 339
    color, too high, 321
    moldy, 317
    mottled, 317, 322
    seamy, 317, 322-23
    white specks, 317, 321, 323-24, 335
  composition, 307
  curdiness, 95, 307, 315-16, 333-34
  definitions (CFR), 307
  "finish" (and/or package), 100, 313, 323, 325-29
    damaged wrappers, 325
    high edges, 327
    loosened wrappers, 325
    lopsided (misshapen), 327
    mold growth, 325-26
    paraffin defects, 326-27
    rindless defects, 325
    soiled wrappers, 326
    surface defects, 328-29
    uneven edges and sizes, 327
    workmanship, 327-328
  flavor, 307, 338, 345-48
    Balanced Component Theory of, 344, 347
    defects of
      bitter, 340, 342
      "catty", 62, 345
      fermented/fruity, 336, 342
      flat (lacks flavor), 307, 333-34, 342
      garlic/onion, 342-43
      "green" (flat, fresh, immature), 333-34
      heated (cooked), 343
      high acid (sour), 95-96, 339, 340, 342
      malty, 343
      metallic (oxidized), 343
      moldy (musty), 343
      rancid (lipase), 343
      whey taint, 340, 344-45
      yeasty, 345
  flavor and body, correlated, 95, 96, 305
  flavor compounds, 344-47
  form, (style), 308-10
  manufacturing process, 306
  packaging materials, 313
  quality determinants, 308
  reduction-oxidation (redox) potential, 72-73, 343

ripening considerations, 307–8
score card, 312
"sharp" (mature) flavor, 342
slicing properties, 330–32, 337
U.S. grades, 349–51, 542–55
texture, 96, 333
yields (product), 307
"Cheddar" flavor, 307, 338, 345–48
Cheese
  background and historical, 300–301
  calcium in, 304
  calories in, 304
  cheesemaking steps, 301
  classification systems, 300–301, 306
  composition, 302–4, 307
  contests and clinics, 51, 53
  critique of samples, 53
  curd types, 305
  definition of, 300
  determinants of cheese type, 301–2
  flavor development, 307
  grading, 53, 305, 347–51, 542–55
  keeping quality, 301, 349
  manufacturing steps, 301
  moisture content, 301
  nutritive values, 302–4
  pest infestation, 329
  physical properties, 301–2, 305
  reduction-oxidation (redox) potential,
    72–73, 343
  ripening considerations, 306, 338, 346–
    48
  sample size, 53
  varieties of
    blue-veined, 361–65
    brick, 358–61
    Brie, 365–69
    Camembert, 365–68
    Colby, 349, 351–52
    cream, 371–73
    Italian types, 369–71
    Limburger, 361–62
    Monterey Jack, 349, 351–52
    other miscellaneous types, 373–74
    pasteurized processed American,
      303, 304
    soft, mold-ripened, 365–69
    Swiss, 352–58
Cheesy off-flavor
  in butter, 68, 232, 379, 384, 386, 406
  in buttermilk, cultured, 232, 234
  in cultured milk products, 232, 234

in ice cream, 210
in sour cream, cultured, 232, 234
Chemical receptors, 10
Chemical senses, 14–15, 23
  pain, 23
  stimuli, 503
Chemoreceptor sites
  role of molecules, 15, 17–19, 20, 28
    electrophilic, 28
    nucleophilic, 28
  sense of smell, 17–21, 28–29
  sense of taste, 14–17
Chlorophenol off-flavor, 83–84
Chocolate
  ice cream, 213–15
  milk, 154, 155, 489
    cocoa sedimentaion, 155
    color and appearance, 154
    consumer preferences (regional), 155,
      490
    masking of off-flavors, 154
    sensory properties, 154–55
    viscosity parameters, 154–55
Cholesterol, 303
Churned fat defect
  in cultured milk products, 231
  in dried milk products, 466, 467
  in ice cream, 190, 194
Clarification (of milk), 117
Coarse/icy texture defect (frozen des-
    serts), 194–96, 216, 221
Coarse off-flavor
  in butter, 407–8
  in cultured milk products, 233
Coffee (cereal) cream. See Half-and-half
Colby cheese, 98, 301, 303
  age, 351
  body and texture, 351
  flavor characteristics, 349
Collegiate Dairy Products Evaluation
    Contest, 1–2, 54–55
  all-products, 488
  awards, 57–58
  butter, 403
  Cheddar cheese, 312
  ice cream, 172
  milk, 474
  objectives, 55
  official judges, 56–57
  organization and structure, 55–57
  rules, 56
  sponsors, 54–56

Collegiate Dairy Products (*cont.*)
  yogurt, 259
Color and appearance of dairy products,
    21-22
  added colorants, 99
  aesthetic aspects of colorants, 98-99
  characteristics and defects
    in brick cheese, 360
    in butter, 393-94
    in buttermilk, cultured, 242, 245
    in Cheddar cheese, 99-100, 313-14,
      316
    in chocolate ice cream, 214
    in cultured milk products, 101, 234,
      237, 242, 251
    in dry milk products, 442, 448
    in evaporated milk, 423
    in ice cream, 180-81, 183
      fruit type, 216
      nut type, 217
      variegated types, 219-20
    in sherbet, 222
    in sour cream, cultured, 251
    in yogurt, 265
  color specks in butter, 382, 385, 393
  color variations due to milk source, 99
  effect on color of
    casein, 99
    freezing, 99, 117
    homogenization, 99
    milkfat emulsion, 99
  extraneous matter in products, 100,
    116
  guidelines for colorants, 99-100
  lack of color uniformity
    in dry milk, 466
    in ice cream, 181
  reflectance properties (color), 99
  stratification/settling of flavorants, 100
  white specks in Cheddar cheese, 317,
    321, 323-24, 335
Color leaching (yogurt), 266
Competitive samples, 49
Component Balance Theory (cheese fla-
    vor), 344-47, 370
Concentrated milks, 119-21
  beverage quality, 419
  body and texture, 424, 427-30
  color and appearance, 422-23, 429-30
  composition, 419
  container (package) condition, 424-25
  defined (CFR), 418-19, 420

examination procedures, physical, 423-
    26
  Federal Standards of Identity, 418-19
  flavor characteristics, 420
  frozen concentrates, 119, 419
  manufacturing procedures, 421
  product types, 418-22
  reconstitution of, 420, 425
  sample preparation, 420
  sterile milk concentrates, 119, 418, 420
  tactile properties, 420
Condensed milk
  defined, 420-22
  flow diagram (process), 421
  quality parameters, 422-30
Conditioned (psychic) reflex, 33
Consumer acceptance, 491
Consumer (sensory) panels, 490, 494
  number of participants, 495
  number of samples for, 497
  precautions of use, 496
  questionnaires, 496
  when and where employed, 495
Consumer (sensory) impressions, 48
  of butter, 378, 391
  of buttermilk, cultured, 240-41
  of Cheddar cheese, 344
  of cottage cheese, 285
  of concentrated milk products, 418,
    425, 427
  of cooked flavor, 81-82
  of cultured milk products, 98, 228
  of direct-draw milk shakes, 224
  of dry milk products, 435
  of ice cream
    body and texture, 192
    color and appearance, 183
    chocolate flavor, 214
    flavor, 207
  of milk, 81-82
  of salt in butter, 378, 391-92
  of sour cream, cultured, 249, 251
  of Swiss cheese, 354
  of yogurt, 254-56, 265, 271-72
  regional preferences (product character-
    istics), 46, 228, 240-44, 285,
    344-47, 376, 391-93
Containers and closures, 126-31
  buttermilk, cultured, 243
  cultured milk products, 231, 243
  ice cream, 178
  milk and cream products

functions, 126–27
PHS/FDA P.M.O. requirements,
    127–28
See also Milk and cream containers
Contamination of milk and milk products
    in butter, 394
    chemical compounds, 83–84
    copper compounds, 72, 147, 207
    extraneous matter, 100, 136
    foreign off-flavor, cause of, 146
    iron (rust), 72, 147, 207
    sanitizing and cleaning agents, 83
Contests and clinics (exhibitions), 48–49
    for butter, 405
    for cheese, 51, 53
    Collegiate Dairy Products Evaluation
        Contest, 54–56
    consumer impressions, 48
    "cuttings" of products, 48
    educational value of, 51
    for ice cream, 54
    kinds of, 50–51, 54
    objectives/purposes of, 48–54
    for other dairy products, 54
    use of score cards in, 46–47
    for youth (4-H and FFA), 50, 55, 57
Continuous manufacturing methods
    of butter, 377, 401
    of soft, mold-ripened cheeses, 365
Cooked (heated) flavor, 62, 81–83
    antioxidant properties of, 70, 82, 408
        butter, 70, 82, 381, 384, 386, 408
        buttermilk, cultured, 240
        Cheddar cheese, 343
    caramel (scorched) off-flavor, 82
    consumer reactions to, 82
    in cultured milk products, 231
    descriptors of, 81
    desirable properties, 70, 81–82
    dissipation of, 82
    in dry milk products, 462–63
    in evaporated milk, 119, 427
    in ice cream (mixes), 83, 171, 173–75,
        204, 210, 216
    impact on shelf-life of products, 82
    Maillard reaction, relationship of, 62–
        64, 81, 83, 119, 169
    masking effect on flavor, 82, 205, 408–
        9
    in milk, 62, 81–83, 118
    in nonfat dry milk, 445
    richness attribute, 81–82, 197, 204–5

in UHT/ultrapasteurized products, 121
in yogurt, 261, 273
See also Caramel and Scorched off-
    flavors
Copper
    content of dry whole milk, 439
    in cream, 414
    in milk, 69
    role in milkfat oxidation, 64, 69, 71–72,
        147, 207, 294
    sensitivity index, 71, 540–41
Corky (dry, hard, tough) defect of Ched-
    dar cheese, 311, 316, 333
Corn syrup, 166, 168–69, 202, 204, 222
Cottage cheese
    body and texture, 96, 285–90
    color and appearance, 101, 282–84
    consumer (sensory) impressions, 285
    creaming methods, 281, 289
    curd size, 281, 285
    "cutting pH," 288
    defect causes and remedies, 284, 290
    defined (CFR), 277–78
    desirable product characteristics, 282,
        285
    dry curd, product type, 239, 281, 285
    examination of curd characteristics,
        281
    flavor characteristics and defects, 278,
        281–82, 291–95
    lowfat, product type, 239
    manufacturing steps, 281
    package examination, 282
    processing techniques, 283–84, 288
    product types, 278, 281, 285, 589
    psychrotrophic bacteria in, 278, 287,
        292, 294–95
    quality attributes, 282
    regional preferences and variations,
        228, 285
    score cards for, 278–79
    scoring guides for, 280
    sensory evaluation techniques, 281–95
    shelf-life of, 281, 292–95
    tactile properties, 96, 285–90
Cowy off-flavor, 66
    in Italian cheese, 370
    in milk, 124, 140, 144
Cranial nerves, 16–17
Cream
    body and texture defects
        cream plug (fat ring), 100, 117, 128,
        139, 158–60

Cream (*cont.*)
  creaming-off, 160
  "feathering," 158–59
  leathery plug, 160
  oiling-off, 159–60
  off-color, 159
  ropiness, 160
  separation (into layers), 160
  thin (weak) body, 160–61
  defined (CFR), 376
  dry, product type, 457
  farm separated, 377
  freezing, effects of, 159
  grading of, 377–80
  heavy, product type, 161–62
  light, product type, 159–61
  overrun, 162
  product types, 120
  sensory characteristics, 120
  whippability, 120, 541–42
  whipped cream (topping), 162
  whipping, product type, 161–62
  whipping properties, 156, 541–42
Cream cheese, 108, 120, 160–61, 371–73
  body and texture, 372
  composition, 371
  direct acidification manufacturing
      method, 371
  flavor, 372
  flavored type, 373
  manufacturing steps, 371
Cream dressing (cottage cheese), 281–83,
    285–87, 289
Cream plug (fat ring), 100, 117, 128, 139,
    158–60
Crumbly (brittle, friable) defect
  in blue-veined cheese varieties, 364
  in butter, 382, 385, 396–97
  in Cheddar cheese, 311, 315–16, 318,
      333, 336
  in cream cheese, 372
  in ice cream, 190–91, 221
Cultured buttermilk. *See* Buttermilk, cul-
    tured
Cultured half-and-half. *See* Sour cream,
    cultured
Cultured lowfat milk, 238
Cultured milk products, 101, 227–29
  body and texture, 234
  consumer acceptance, 227–28
  definitions (CFR), 238–39
  direct acidified, product types, 238–39

extended shelf-life of, 227–28
  flavor characteristics, 232
  historical development of, 227–28
  kefir, 295–98
  nutritional/health aspects, 227–28
  off-flavors, 232–34
  quality criteria, 229
  score cards, 229–30
  scoring guides, 230
  sensory attributes, 229
  sensory evaluation, role of, 228–29
Cultured skim milk, 238
Curd
  acid coagulated, 302, 305
  rennet (chymosin) coagulated, 302, 305
Curd particles
  butter, 393
  cottage cheese, 283, 285–86
Curdy (curdiness) defect
  Brie and Camembert cheese, 368
  buttermilk, cultured, 241
  Cheddar cheese, 95, 307, 311, 315–16,
      333–34
  concentrated milks, 428
  cultured milk products, 234–35
  evaporated milk, 427–28
  ice cream meltdown, 185, 187–88
  sour cream, cultured, 249
Custard-like flavor. *See* Cooked or Rich
    flavor

Dairy and Food Industries Supply Asso-
    ciation (DFISA), 54–56
Dairy products sensory evaluation (judg-
    ing, scoring), 1, 5
Dark particles defect (dry milk products),
    445, 467
Deamination, 63
Decarboxylation, 63
Descriptive tests (sensory panels)
  components of, 510
  difference from scoring, 511
Dextrose, 168, 222
Dextrose Equivalent (DE), 89, 166, 168–
    69
Diacetyl (buttery aroma)
  of butter, 232, 378, 404–5
  of buttermilk, cultured, 232, 240, 244
  of cottage cheese, 282, 289, 292
  of cultured milk products, 232, 240,
      244, 247–49
  of kefir, 297

of lactic cultures, 232
of sour cream, cultured, 247–49
of yogurt, 254
Diacetyl defect (cottage cheese), 280, 292, 479
Difference tests (sensory panels), 499
Diffusion (aroma innervation), 20
Directions (for sensory panelists), 505
Discrimination (sensory) panels
  qualification of panelists, 493
  tasks assigned, 493
  training of, 492, 518–19
Discrimination test (sensory panels), 500
Dry buttermilk
  defined (CFR), 449
  USDA grades and standards, 449–50
Dry casein. See Edible dry casein
Dry cream
  defined (CFR), 457
  sensory properties, 458
Dry curd cottage cheese, 239, 289–90
Dry ice cream (milk) mix, 458
Dry malted milk, 458–59
Dry milk products
  appearance, 466–68
  drying methods, 436–38
  flavor characteristics and defects, 460–66
  laboratory tests, quality determination, 468–69
  product types, 438
  reconstitution methods, 469
  scorched particles in, 468
  score card, 461
  scoring guide, 462
Dry whey
  acidity classification, 452
  defined (CFR), 451
  flavor characteristics, 453
  laboratory tests, quality determination, 452
  physical properties, 452
  USDA grades and standards, 451–53
Dry whole milk
  color of, 442
  defined, 439
  flavor characteristics, 441–42
  grading of, 439
  physical properties, 442–43
  tactile properties, 442
  USDA grades and standards, 439–41
Dull appearance defect

cultured milk products, 231, 251
sour cream, cultured, 251
Dull defect (Swiss cheese), 354–55
Duo-trio test (sensory panels)
  description of, 499
  statistical significance of, 499

Edam cheese, 303, 304
Edible dry casein
  bacterial content of, 453–54
  defined (CFR), 453
  flavor characteristics, 453–54
  laboratory tests, quality determination, 453–54
Egg flavor
  in ice cream, 208, 209, 212
  in eggnog, 162
Eggnog, 162
  manufacturing steps, 163
  sensory characteristics, 162–63
  viscosity, 97
Emmental or Emmentaler cheese. See Swiss cheese
Emulsifiers
  in dry cream, 457
  in ice cream, 169, 194, 209
  lecithin, 169, 209
  mono- and diglycerides, 169, 209
  Polysorbate-65, 169, 209
  Polysorbate-80, 169, 209
Enzymatic activity (cheese)
  in blue-veined varieties, 364–65
  in brick, 360–61
  in Brie, 366–69
  in Camembert, 365–66
  in Cheddar, 305, 307, 338, 344, 346, 349
  in Limburger, 361
  in soft, mold-ripened types, 365–68
Enzymes
  bacterial, 346
  chymosin (rennin), 346
  lipase, 76–79, 80–81, 343
  sulfhydryl oxidase, 121
  xanthine oxidase, 72
Evaporated milk
  body and texture, 425, 427
  can (container) condition, 423–24
  color and appearance, 423, 429
  defined (CFR), 420–22
  desirable sensory properties, 422
  examination (observations) sequence, 423

Evaporated milk (*cont.*)
  flavor characteristics, 119, 426
  flow diagram (process), 421
  quality parameters, 422–30
  reaction(s) in coffee, 425
  sterilization of, 119, 425–26
Excess fruit defect (yogurt), 266
Expert sensory evaluator (judge), 489,
      491–92
  qualifications of, 492
  tasks assigned, 492
  training of, 492

Fabricated foods, 489
Fat separation defect, 433, 434
Fatty acids, 149, 27, 302, 303, 364
Fatigue (sensory, tastebuds), 502, 521
"Feathering" defect, 158, 427, 428
  of concentrated milks, 424, 425, 428
  of cream products, 156, 158–59
  reaction in coffee, 156, 158, 427, 428
Federal (USDA) grades and standards
      (dairy products), 3–4
  butter, 376
  Cheddar cheese, 347–49, 351, 542–55
  dry milk products, 449–51
  dry whey, 451
  sediment content, 136–37
Federal Standards of Identity (dairy
      products) for
  acidified milk products, 238–39
  butter, 378
  Cheddar cheese, 307
  concentrated milk products, 419
  cream, 379–80
  cultured milk products, 238–39
  ice cream, 166–68
  milk and cream, 108–12, 114, 117–20
  sherbet, 223
  yogurt, 252–53
Feed off-flavor
  in butter, 408–9
  control of, 64
  descriptors, 144
  in dry milk products, 462–63
  in Italian cheese varieties, 370
  in milk (and cream), 124, 140, 144–45,
      152, 476
  in nonfat dry milk (NFDM), 445
  preparing samples of (training), 476
  transmission mode, 65
  vacuum process, removal by, 65

Fermented/fruity off-flavor, 68
  in Cheddar cheese, 336, 347
  in cottage cheese, 280, 292, 479
  in dry milk products, 462, 464
  in dry whey, 452
  in ice cream, 210
  in Italian cheese, 370
  in milk (and cream), 67–68, 124, 140,
      145, 477
  sample preparation (training), 477
  in sweetening agents, 169
  vinegar-like defect type, 231
Finish (package workmanship)
  butter, 390–91
  Cheddar cheese, 325–29
Firm (too) body defect
  in cottage cheese, 280
  in cream cheese, 172
  in cultured milk products, 231, 235
  in sour cream, cultured, 250
  in yogurt, 259, 261, 270
"Fish eyes" defect (Cheddar cheese), 311,
      335
Fishy off-flavor (butter), 384, 386, 409,
      412
  See also Oxidized off-flavor
Fissures defect (Cheddar cheese), 335, 337
Flake style (granulated) buttermilk
  body and texture, 244–45
  color and appearance, 245
  description of, 244
  flavor characteristics, 244
Flaky defect (ice cream), 190, 196
Flat flavor defect
  in butter, 381, 384, 386, 409
  in Cheddar cheese, 307, 311, 339–40,
      342
  in cheese, 88
  in cottage cheese, 280, 292
  in cream cheese, 372
  in cultured milk products, 88, 233, 237
  in dry milk products, 445, 462–64
  in flavored dairy products, 88
  in ice cream, 200–201, 219
  in Italian cheese, 370
  in milk (and cream), 88, 124, 140, 145–
      46, 476
  in yogurt, 274
  See also Lacks flavor defect
Flavor
  balance(d), 197–98, 201
  Balanced Component Theory (cheese),
      344–47

complexity of sensation, 8
components (elements) of, 5, 179, 264–65
compounds, 60–62
concept of, 23
"language," 59
masking of, 82, 205, 256, 268, 273, 277, 408–409
"memory," of expert evaluators, 29, 156
mental attitude, role in sensory evaluation, 29
perception sequence, 40
"release," 249
research techniques, 346
"richness," 81–82, 197, 204–5
stability, 69, 116
threshold (rancidity), 78–79
unnatural, 200–202
Flavor chemistry, 60–62, 345
Flavor profile (analysis), 200
  methodology and components, 511
  scaling and descriptors, 512
Flavor too high (excessive) defect, 90–91, 170–71, 174–75, 200
  in ice cream, 170, 200
  in yogurt, 261, 274
Flavorings (flavor agents)
  artificial, 170
  defects derived from, 90–91
  descriptors, 91
  desirable criteria, 91
  excessive (too high), 90–91, 170
  guidelines for evaluation, 91
  ingredient listing, 100
  insufficient amount (lacking), 90, 200
  lacking fine balance (character), 91
  natural, 170
  types, 91
  unnatural, 91, 201
Flavoring systems
  chocolate, 154–55
  defects in
    frozen dairy desserts, 170, 198–202
    yogurt, 273–74
  described, 91
  natural, 91, 169, 183
  objectives of, 91
  strawberry, 215–16, 266–67
  unnatural, 91
  vanilla, 170
Fluffy defect (ice cream), 190, 196

Foam drying, 437
Food and Drug Administration (FDA)
  processing guidelines for food colors, 99–100
  PHS/FDA Pasteurized Milk Ordinance (P.M.O.), 108–11, 144–15, 117–20, 127–28, 130–32, 138
Foreign off-flavor
  burnt plastic-like, 84
  in butter, 379, 384, 386, 410
  from chemical contaminants, 83–84
    chlorophenols, 83
    medicinal-like substances, 83
    sanitizing and cleaning agents, 83
    "steam" source, 84
  in cottage cheese, 280, 292
  in cultured milk products, 231–32
  in ice cream, 171, 173–75, 211
  in Italian cheese, 370
  in milk, 124, 140, 146, 152, 477
  modes of transmission, 83–84, 85
  sample preparation (training), 477
  in yogurt, 261
Four-H (4-H) contests, 55, 474
Free cream defect
  in cottage cheese, 280, 283
  sample preparation (training), 480
Free fatty acids (FFA)
  in blue-veined cheese varieties, 364
  in Cheddar cheese, 343, 347
  pK$_a$ value, 79
  rancid off-flavor, role in, 76, 78–80, 343
  salts of FFAs, 80
  sensory characteristics, 79
Free radicals, 62, 73
Free whey defect
  in cottage cheese, 280, 284
  in cultured milk products, 98, 101
  sample preparation (training), 480
  in yogurt, 256, 266–67
  See also Wheyed-off defect
Freeze drying, 437
Freezing of milk, 117
Freezing point (ice cream mix), 168–69
French custard, 209
French vanilla ice cream, 208–9, 212
"Frog mouth" defect (Swiss cheese), 357–58
Frozen custard, 167, 209, 212
Frozen dairy desserts
  anesthesia, due to, 172, 179
  body and texture, 178, 188–97

Frozen dairy desserts (*cont.*)
  bound water, 169
  color and appearance, 178, 180–84
  composition differences, 167
  corn syrup in, 166, 168–69
  defined (CFR), 166
  Federal Standards of Identity, 167,
    208, 212
  flavor characteristics, 197–211
  flavor labeling requirements, 203
  "heat shock" problems, 169, 187, 192,
    195, 197, 219–21
  nonmilk food solids, 209
  optional milk ingredients, 167–68, 209
  product types
    candy flavored, 218
    chocolate flavored, 213–15
    direct-draw milk shakes, 224–25
    French vanilla (custard), 208–9, 212
    frozen novelties, 220
    frozen yogurt, 222
    fruit flavored, 215–16
    nut flavored, 217–18
    sherbet, 220–23
    soft-serve, 223–24
    vanilla, 180–211
    variegated, 218–20
    water ices, 222–23
  sensory evaluation of, 170–225
  shrinkage defect, 102, 191–93, 197
  stabilizers/emulsifiers, 169–70, 191,
    192, 195, 196, 209
  sweetening agents, 166, 168–69
  See also Ice cream and Sensory evalua-
    tion of ice cream
Frozen novelties. *See* Frozen dairy des-
    serts
Fructose, 168
Fruit
  excess defect (yogurt), 261, 266
  in frozen dairy desserts, 215–16
  lacking defect (yogurt), 261, 266–67
  quality criteria, 216
Fruity off-flavor, 68
  See also Fermented/fruity off-flavor
Future Farmers of America (FFA) con-
    tests, 55, 474

Garlic/onion off-flavor, 65–66, 146
  in butter, 379, 384, 386, 410
  in Cheddar cheese, 311, 339, 342
  in cottage cheese, 280

  in milk, 65–66, 124, 140, 146, 152, 476
Gassy defect
  in brick cheese, 359–60
  in buttermilk, cultured, 235
  in Cheddar cheese, 311, 335
  in concentrated milks, 428
  in cultured milk products, 231, 235
  in evaporated milk, 427–28
  in Italian cheese varieties, 371
  in lactic cultures, 235
  in Limburger cheese, 362
  in sour cream, cultured, 249
  in sweetened condensed milk, 434
  in yogurt, 265
Geometric mean normalization, 508
Geotrichum candidum (kefir), 296
Gelatinous defect (cottage cheese), 280,
    287
Gel-like defect, 98
  in sour cream, cultured, 249–50
  in yogurt, 257–58, 268–70
Glaesler defect (Swiss cheese), 354, 356
Glucono delta lactone (GDL), 372
Gluey off-flavor
  in dry milk products, 462
  in edible dry casein, 456
Grades and grading (dairy products), 1, 3
  of butter, 380–83
  of Cheddar cheese, 347–49, 544–55
  of cream, 378–80
  disratings scheme (butter), 382–83
  disratings scheme (Cheddar cheese),
    552–55
  of dry milk products, 439–41
  quality categories of cheese, 348
  validity of performing, 492
  workmanship faults (butter), 382
"Grain" defect (evaporated milk), 424
Grainy (gritty) defect
  in buttermilk, cultured, 241
  in Cheddar cheese, 311, 323, 335
  in cottage cheese, 280
  in cream cheese, 172
  in cultured milk products, 231
  in dry milk products, 466, 468
  in evaporated milk, 427–28
  in nonfat dry milk, 445
  in sour cream, cultured, 249
  in yogurt, 258, 261, 270
Greasy defect
  in butter, 396–97
  in Cheddar cheese, 311, 333–34

"Green" (fresh) cheese, 347
"Green" off-flavor
    Bulgarian buttermilk, 246
    cottage cheese, 233, 292–93
    cultured buttermilk, 233, 240, 246
    cultured milk products, 231, 233, 246
    kefir, 297
    sour cream, cultured, 233
    yogurt, 233, 252, 253, 255, 272–73, 297
    See also Acetaldehyde off-flavor
Gummy (pasty, sticky) defect
    association with syrup flavor (ice
        cream), 204
    in brick cheese, 359
    in butter, 382, 385, 396–97, 400
    in Cheddar cheese, 336
    in cottage cheese, 288
    in ice cream, 178, 190–91, 204
    in Limburger cheese, 362
    in Swiss cheese, 357
    See also Pasty defect

Half-and-half
    coffee whitening properties, 120, 156
    defined (CFR), 120, 157
    feathering defect (in coffee), 156, 158–
        59
    oiling-off defect, 159
    optional ingredients, 158
    sensory characteristics, 157–58
    viscosity, 159
Heated (cooked) flavor
    in Cheddar cheese, 311, 339, 343
    in milk, 60, 143
Heat treatment (of milk and milk prod-
        ucts)
    direct vs. indirect process, 112, 114
    effects on
        cooked flavor, 81–83, 143–44
        laboratory pasteurization, 110
        oxidized off-flavor, prevention of, 70
        product viscosity, 92
Heavy body defect
    in buttermilk, cultured, 240
    in ice cream, 190, 192
    in sherbet, 222
    See also Too firm body defect
Hedonic scale, 503–7
High acid off-flavor
    in buttermilk, cultured, 233, 242–43
    in Cheddar cheese, 311, 339, 340, 342
    in Colby cheese, 349–50

    in cottage cheese, 233, 291
    in cultured milk products, 233, 236
    in Monterey Jack cheese, 349–50
    in sour cream, cultured, 233, 236
    in yogurt, 271–72
    See also Acid off-flavor
High salt defect
    in butter, 384, 386, 406
    in cottage cheese, 294–95
History of sensory evaluation (dairy prod-
        ucts), 1–3
Homogenization
    of chocolate milk, 155
    of cream, 158–59
    effects on
        feathering defect (cream), 158–59
        ice cream texture, 194–96
        light-induced off-flavor, 74, 149
        oiling-off defect, 159
        oxidized off-flavor, 69–70, 120, 149
        product viscosity, 92
        sour cream body and texture, 247,
            250
    of reconstituted milk, 120
Homogenized milk, 114–15
    cooked flavor in, 115
    cream plug (fat ring) defect in, 117
    defined (CFR), 114
    light-induced off-flavor in, 116, 149
    oxidized off-flavor in, 116, 149
    rancid off-flavor in, 116
    sedimentation problems in, 116–17
    sensory properties of, 115–17
    watery appearance of, 117
Human senses, 9, 11, 198
Human sensory reactions, 9–12
Hydrolytic rancidity. See Rancid off-
        flavor

Ice cream
    adaptation (time), 198
    air cells in, 97
    bite resistance, 93, 96, 188, 192
    body and texture of, 188–97
        defects, 190–97
        determinants of, 96–97
        freezing point, effect of, 97
        "heat shock," effect of, 169, 187, 192,
            195, 197
        heat treatment, effect of, 96
        homogenization, effect of, 96, 194–96
        ice crystals, 97, 169, 193, 194–96

Ice cream (*cont.*)
  sandiness, 97
  bound water in, 169
  color and appearance of, 180–81, 183–88
    atypical level (colorants), 181, 183–84
    meltdown properties, 176, 184–88
    shrinkage, 102, 191–93, 197
  contests and clinics, 54
  flavor considerations, 179, 197
  flavor defects
    acid (sour), 204, 210–11
    cooked, 204–5, 210
    eggy, 208–9
    foreign, 211
    lacks fine flavor, 201
    lacks flavor (flat), 200–201
    lacks freshness, 210
    lacks sweetness, 202
    old ingredient, 206
    oxidized (metallic), 206–7
    rancid, 207
    salty, 207–8
    syrup flavor, 202, 204
    too high flavor, 201
    too sweet, 202
    unnatural flavor, 201–2
    whey, 208
  flavor defects due to, 197–202, 203–11
    chemical changes, 210
    dairy products (ingredients), 204–8
    flavoring system, 199–202
    microbial growth, 210–11
    other ingredients, 208–9
    processing operations, 210
    sweetening agents, 202, 204
  flavoring systems, 199–202
  freezing point of mix, 97, 168
  ingredients, importance of, 168, 199–202, 204–9
  melting properties, 176, 184, 186
  precautions in sensory evaluation, 197–98
  product types, 167
    bulky flavored, 212–13
    candy flavored, 218
    chocolate, 213–15
    frozen novelties, 220
    fruit flavored, 215–16
    nut flavored, 217–18
    soft-serve, 223–24
    vanilla, 180–211

    variegated, 218–20
  scoops for dipping, 176–77
  score cards, 45, 170–72, 174, 561
  scoring guides, 172–73, 175, 182, 186, 189
  sequence of sensory observations, 177–80
  storage cabinets (freezers), 175–76
  sweeteners in, 166, 202, 204
  tactile properties, 96–97
  tempering of samples, 97, 172, 174–76, 188
  See also Frozen dairy desserts
Ice milk, 167, 209, 211–12, 223–24
  See also Frozen dairy desserts
Icy (iciness). *See* Coarse/icy defect and Ice cream (ice crystals)
"Ideal' product, 489
  blue-veined cheese varieties, 364
  brick cheese, 360–61
  butter
    body and texture, 94–95, 395
    flavor, 404–5
  Cheddar cheese
    body and texture, 95, 318, 330–32, 336–37, 348–49
    flavor, 307, 338, 348–49
  cottage cheese, 96, 282, 285
  cultured milk products, 237–38
  ice cream
    body and texture, 188
    flavor, 200
    melting properties, 184–85
  Swiss cheese, 352, 57
  yogurt
    body and texture, 256, 263, 268–70
    color and appearance, 263, 265
    flavor, 264, 271
"Ideal" sensory (quality) characteristics, 37
  "fix in mind," 40, 198
  mental standards of, 40, 198, 200, 238
Infrared spectra, 346
Ingredient quality (importance), 5, 170
"In-house" sensory panels, 493–94
  size (number of participants), 494
  tasks assigned, 494
  value of, 494
Instantizing process (NDM), 436–37
Instrumental (objective) sensory techniques, 94, 512–13, 515–16
International Ice Cream Association, (IICM), 54

Introspection (mental concentration), 36–37, 41, 45, 126, 491
for expert judges, 32
as related to evaluation of
butter, 389, 401
cultured milk products, 238
dry milk products, 470
ice cream, 180
technique of, 41
value of, 28, 126, 180, 491
Iron content of dry whole milk, 439
Irradiation, 75
Italian cheese varieties, 370–71
body and texture, 371
classification (type), 370
finish and package, 371
flavor characteristics, 370
mold growth on, 371

Judging. *See* Sensory evaluation of dairy products

Keeping quality
butter, 377–78
Cheddar cheese, 301
cheese, 301, 349, 394, 413
cultured milk products, 227–28
Limburger cheese, 362
milk and cream, 141–42
soft, mold-ripened cheese, 365–66, 368
Kefir, 295–98
characteristics, 296–98
grains, 296
microorganisms, 296
process of manufacture, 296
Ketosis (cowy off-flavor), 66

Laboratory pasteurization, 110, 373
Lacks cream defect
in cottage cheese, 280
sample preparation (training), 480
Lacks desirable flavor (Swiss cheese), 353
Lacks fine flavor
in butter (as "coarse" flavor), 407
in cottage cheese, 280, 292, 479
in cultured milk products, 233
in ice cream, 171, 173–75, 201, 217
chocolate flavored, 214
fruit flavored, 216
variegated type, 219
sample preparation (training), 479
in yogurt, 261, 273

See also Coarse off-flavor
Lacks flavor (flat) defect, 88, 90, 200
causes of, 88
descriptors of, 88
in products
butter, 409–10
Cheddar cheese, 307, 333–34, 342
cottage cheese, 292
cultured milk products, 233, 237
ice cream, 200–201, 119
milk, 88, 145–46
yogurt, 261, 273
See also Flat flavor defect
Lacks freshness (stale) off-flavor, 84–86
in butter (aged), 405–6
in concentrated milk products, 420, 426–27
in cottage cheese, 280, 293, 479
in cultured milk products, 231
in dry milk products, 441
in dry whey, 453, 447–48
in fruit ice cream, 216
in ice cream, 171, 173–75, 205, 210
in milk, 124, 140, 146–47, 477
sample preparation (training), 477, 479
in skim milk, 118–19
in yogurt, 261, 275–76
Lacks fruit (yogurt), 267
Lacks malt flavor (malted milk), 459
Lacks sweetness
in ice cream, 171, 173–75, 179, 202
in yogurt, 261, 274
Lacks uniformity
cottage cheese, curd appearance, 284
cultured milk products, appearance, 231 .
dry whole milk, color of, 443
sour cream, cultured, appearance, 251
Lactic (starter) cultures, 229–38, 306
acidity of, 236–37
body and texture, 234–38
color and appearance, 234–38
curdiness defect, 234–35
flavor characteristics, 232–34
function of, 232
microbial contamination of, 232–35
sensory evaluation of, 234–38
types (forms) of
concentrates, 232
frozen, 232
lyophilized (freeze dried), 232
"mother" cultures, 229

Lactic (starter) cultures (*cont.*)
Lactobacilli sp. (cultures), 240, 252–53, 271, 272, 353
Lactobacillus acidolphillus, 246, 296
Lactobacillus bulgaricus, 246, 253, 294
Lactose, 63–64, 168, 303–4
   crystallization in ice cream, 178, 196, 224
   flavor contributions, 63–64
   See also Sandy (sandiness) defect
Lactose-reduced milk, 120
Late gas defect
   in brick cheese, 359
   in Cheddar cheese, 331
Leaky defect (butter), 23, 382, 385, 389, 393–95, 397–98
Leuconostoc sp. (cultures), 247, 278, 296
Light cream, 108, 160
Light-induced (activated) off-flavor, 62, 64, 70–71
   antioxidants, role of, 74
   descriptor, 70–71, 147–49
   homogenization, role of, 70–71, 74–75
   in ice cream, 207
   impact on nutrient value, 74
   irradiated milk, response of, 75
   light source, role of, 74–76
   mechanism of formation, 75
   methionine, role of, 62, 75–76
   prevention and control, 76
   products affected, 75–76
   proteins, role of, 62
   riboflavin, role of, 74
   ultra-sound treatment, response to, 75
Limburger cheese, 301, 361–62
   ammoniacal-like flavor characteristics, 362
   anaerobic environment, 361
   keeping quality of, 362
   manufacturing process, 361
   score card, 361
   sensory characteristics, 361
   surface growth, 361
   undesirable sensory properties, 362
Lipid oxidation. *See* Oxidation and/or Oxidized off-flavor
Lipids (fat), 303–4
Lipolytic (or Lipase) off-flavor. *See* Rancid off-flavor
Loop test (evaporated milk), 424
Lowfat dry milk, 456–57
   defined (CFR), 457

   sensory properties, 457
Lowfat milk, 108, 118, 157
   optional ingredients, 118
   sensory characteristics
     flavor and off-flavors, 118, 157
     mouthfeel properties, 118
     stabilizers/emulsifiers in, 118
     vitamin fortification, 118
Lowfat spreads, 378
Low-sodium milk, 120
Low viscosity (concentrated milks), 427–29
Lumpy defect
   in buttermilk, cultured, 230, 235
   in cultured milk products, 231, 235
   in dry milk products, 466, 467, 468
   in dry whole milk, 442
   in ice cream, 190
   in lactic (starter) cultures, 230, 235
   in nonfat dry milk, 445
   in sour cream, cultured, 230, 235
   in sweetened condensed milk, 433–34
   in yogurt, 260–61

Magnitude estimation method (sensory panels), 508
Maillard reaction, 62–64, 81, 83, 119, 169, 429, 456, 457, 465, 467
Malted milk, 459
Maltose, 168
Malty off-flavor, 61, 68, 204
   in butter, 381, 384, 386, 410
   in Cheddar cheese, 311, 339, 343
   in cottage cheese, 280, 293, 479
   in Italian cheese, 370
   in milk and cream, 61, 68, 147, 204
Margarine, 378
Market milk
   certified product forms, 111
   Grade A pasteurized, 111
   grades of, 110–12, 114–21
   whole milk, 111–12, 114–18
   raw form of, 109–11
Matted curd defect
   in cottage cheese, 280, 284
   sample preparation (training), 480
Mealy/grainy (gritty) defect
   in Brie cheese, 368
   in butter, 382, 385, 396, 399
   in Cheddar cheese, 23, 311, 333, 335
   in concentrated milks, 428
   in cottage cheese, 287–88

in cream cheese, 372
Meaty-like texture, 96
  in Cheddar cheese, 330, 337
  in cottage cheese, 278, 282, 285
Mechanical holes. *See* Open defect
Medicinal off-flavor. *See* Foreign off-
    flavor
Mellorine, 167, 212
  See also Frozen dairy desserts
Melting quality (meltdown) defect (ice
    cream), 176, 180, 184-88
Mendel's laws, 31
Mesityl oxide, 345
Metallic (oxidized) off-flavor, 67, 72
  in butter, 70, 384, 386, 410-12
  in Cheddar cheese, 311, 339, 343
  in cottage cheese, 293-94
  in cultured milk products, 231, 233
  in cream cheese, 372
  in dry milk products, 462, 464
  in ice cream, 171, 173-75, 206-7, 210,
    217
  in Italian cheese, 370
  in milk and cream, 67, 69-76, 147-49
  in sweetened condensed milk, 432
  in yogurt, 261, 277
  See also Oxidized off-flavor
Microbial contamination
  of brick cheese, 359
  of Brie cheese, 369
  of butter, 394, 406-7, 410-11, 414-16
  of Cheddar cheese, 331, 335, 340, 343,
    345
  of concentrated milks, 426
  of cottage cheese, 281, 287, 292-95
  of cultured milk products, 232, 234
  of ice cream, 204, 206
  of Italian cheese, 371
  of Limburger cheese, 362
  of milk and cream, 66-68, 141-42, 147,
    150
  of soft, mold-ripened cheese, 369
  of yogurt, 267, 273, 276-77
Microbial caused off-flavors, 66-68
  in butter, 410-12
  in buttermilk, cultured, 232, 234, 242-
    43
  in Cheddar cheese, 331, 335, 340
  in cottage cheese, 277, 287, 293-94
  in cultured milk products, 232, 234,
    242-43, 246
  in ice cream, 204, 210-11

in milk and cream, 68, 141-42
in sour cream, cultured, 242-43, 249
in yogurt, 267-69, 273
Microorganisms, 66-67, 305
  osmophilic type, 433
  psychrotrophs, 66-68
  See also Lactic (starter) cultures and/or
      Psychrotrophic (spoilage) bac-
      teria
Milk
  bacteria in, 68, 137-39, 141-42
  classes (categories) of, 108-9
  containers and closures, 126-30
  extraneous matter in, 116
  flavor, 128
    categories, 139
    from individual cows, 153-54
  grades of, 108-14
    Grade A, 108-14
  manufacturing, 108-9
  homogenized, defects of, 116-17
  "ideal" sensory characteristics, 139,
    489
  laboratory pasteurization, 110, 373
  preparation of samples (training), 125,
    489
  raw, evaluation precautions for, 109-10
  score cards for, 121-23, 141, 558
  scoring guides, 124, 131, 137-139
  sediment content of, 125-26
  sensory evaluation of, 139-50
  tempering of samples, 128
  ultrapasteurized, 111
  vitamin fortified, 117
Milk (and cream) containers, 126, 128-30
  bottles, 128-29
  "burnt plastic" off-flavor, 128
  closures, 129-30
  correct fill level, 128, 130
  possible defects of, 131
  scoring
    guide for, 131
    procedure for, 129-32
  types of
    glass, 128
    multiuse, 128
    paperboard, 128
    tamperproof, 130
Milkfat (lipid), 60-62, 210
  auto-oxidation, 67, 69-72
  hydrolysis of, 60, 148-50
  phospholipids, 62

Milkfat (lipid) (*cont.*)
  short-chain fatty acids, 60
  solvent for flavor compounds, 60
  role in flavor of
    butter, 410–11
    cheese, 338
    cultured milk products, 240
    ice cream, 179, 197–98
    milk and cream, 60, 62
Milk haulers, 492, 527–32
Milk Industry Foundation, 54
Milk off-flavors, 141–50
  dilution effect, 153
  distinguishing characteristics, 151–52
  dry lot feeding, effect of, 153
  individual cows, effect of, 153
  seasonal occurrence of, 151–52
  stage of lactation, 153
  troubleshooting, for cause of, 151–54
Milk producers (dairyman), 5–6, 528
Milk proteins, 62
Milk score cards
  modified ADSA form, 122
  official version (ADSA), 121, 124, 558
  USDA original version of, 122, 124
  use of, 125
  version of, 557
Milk shakes, 224–25
Milk sugar (lactose), 63–64, 168, 303–4
Minerals (of milk)
  copper, 64, 69, 71
  flavor contribution by, 64
  iron, 64
  lactose-mineral balance, 64, 86–87
  as oxidation catalysts, 64, 69, 71
  salty taste, attributed to, 64
Modality, 9, 11
Mold discoloration (butter), 385, 393
Mold growth (cheese), 325–26, 328, 343, 363
Moldy off-flavor, 66, 68, 317, 343
  in Cheddar cheese, 311, 339, 343
Mold toxins (mycotoxins)
  aflatoxins, 328
  allergenic properties, 343
  in Italian cheese varieties, 371
Mottled color defect
  in butter, 382, 385, 393
  in Cheddar cheese, 317, 320, 322
Monterey Jack cheese, 98, 304, 349
Mouthfeel
  gel-like, 268

  puckery, 148
  "slick," 268, 289
  See also Astringent, Puckery mouth-
    feel, and/or Tactile properties
Mozzarella cheese, 301, 303
  See also Italian cheese varieties
Mushroom-like flavor, 61
  in Brie cheese, 366, 368
  in Camembert cheese, 366, 368
Musty off-flavor
  in blue-veined cheese varieties, 364
  in butter, 379, 381, 384, 386, 411
  in Cheddar cheese, 343
  in cottage cheese, 280, 294

Natural foods, 489
Neufchatel cheese, 303, 372
Neutralizer off-flavor
  in butter, 381, 384, 386, 411–12
  in dry milk products, 462, 464
  in ice cream, 171, 173, 211
Niszler defect (Swiss cheese), 356
Nonenzymatic browning. *See* Maillard re-
    action
Nonfat dry milk (NDM)
  color of, 448
  defined (CFR), 443
  flavor characteristics, 447
  flow diagram (process), 444
  laboratory tests, quality determination,
    446
  miscellaneous dry milk products, 459–
    60
  physical characteristics, 448
  sensory properties, 447–48
  USDA grades and standards, 444–46
Nutty-like flavor. *See* Cooked or Rich
    flavor

Objective tests (sensory panels), 515–16
Odor classifications, 26–29
  comparison of, 27
  primary classes, 28–29
  proposed classes of odor, 24–29
    Boring, 27
    Bains, 26
    Crocker and Henderson, 26–28
    Henning, 26–27
    Moncrieff, 27–28
    Rimmel, 26
    Zwaademaker, 26
Odor perception, 17–22, 24–29

"smellability," 18–19
stimualtion, 24–26
volatile molecules, 18–19
Odor receptors, 17–19, 28
Off-flavors (lists of), 59–61, 63–64, 80, 84
in butter, 68, 404
in Cheddar cheese, 339
chemical compounds responsible for, 60–62
in cottage cheese, 291
in cultured milk products, 232
in dry milk products, 460
in ice cream, 199
in milk (and cream), 62–68, 84, 140
in yogurt, 271
"Official" scores (student contests), 46–47
Oiling-off (cream), 159–60
Oily off-flavor. *See* Oxidized off-flavor and Oxidation of lipids
Old cream defect (butter), 68, 379, 381, 384, 386, 412–13
Old ingredient defect
in ice cream, 68, 171, 173–75, 205–6
in yogurt, 68, 261, 276
Olfaction
physiology of, 14, 20, 24
theories of, 19–20, 24–26, 29
Amoore's stereochemical, 24–25, 29
Beet's profile-functional group, 26
Davies' penetration and puncture, 19–20, 25
enzyme action, 24
intermolecular interaction, 20
molecular shape (and size), 20, 24
Wright's vibrational, 20, 24, 25
Olfactory prism, 26–27
Olfactory stimulation, 24–26
Open (openness) defect
in brick cheese, 360
in Cheddar cheese, 311, 315, 330, 335, 337
in Swiss cheese, 357
Overrun
in direct-draw shakes, 224
gradient (ice cream), 219
in sherbet, 221–22
in whipping cream, 162
Overset defect (Swiss cheese), 354, 356
Overstabilized dressing defect (cottage cheese), 289
Oxidation of lipids

catalysts of, 64, 69, 71–72, 148
occurrence in products
butter, 70, 410–12
buttermilk, cultured, 239
Cheddar cheese, 343
cottage cheese, 293–94
cultured milk products, 233–34
dry milk products, 442, 448, 464
ice cream, 206–7, 210, 217
milk and cream, 67, 69–76, 147–49
yogurt, 277
peroxide radicals, relation to, 73
resistance/susceptibility of milk (cream), 69–72
role of
chlorine, 72, 148
fat globule membrane, 69
homogenization, 69–70
phospholipids, 69–70, 72
sanitizing agents, 72
sulfhydryl compounds, 72
unsaturated fatty acids, 70, 72
xanthine oxidase, association with, 72
Oxidized off-flavor (light-induced)
descriptors of, 147–49
effect of
homogenization, 69, 116
light source (wavelength), 149
riboflavin, 149
mechanism, 74
in milk (and cream), 124, 140, 147–49, 152, 476
sample preparation (training), 476
Oxidized off-flavor (metal-induced), 60, 67, 69–73
adaptation (time), 147
antioxidants, 70–71
auto-oxidation mechanism, 73
catalysts, 64, 69, 71–72, 148
comparison to light-induced and rancid off-flavors, 148
descriptors of, 67, 147
cardboardy, 67, 73, 412
metallic, 67, 69–70, 147–48, 239, 293, 206–7, 410–11
puckery mouthfeel, 72, 147–48, 206–7, 233–34, 246
occurrence in products
butter, 70, 384, 386, 410–12
buttermilk, cultured, 239
Cheddar cheese, 343
cottage cheese, 280, 293–94, 479

Oxidized off-flavor (metal-induced) (*cont.*)
   cultured milk products, 231, 233–34
   dry milk products, 442, 445, 447–48,
      451, 462, 464
   "fat-free" products, 72–73
   ice cream, 171, 173–75, 206–7, 210,
      217
   milk (and cream), 67, 69–76, 124, 140,
      147–49, 152, 476
   preparation of samples (training), 476,
      479
   prevention and control measures, 70–
      72, 148
   role of
      carbonyl compounds, 73, 148
      catalysts, 64, 69, 71–72, 147
      chlorine, 72, 148
      copper, 64, 69, 71–72, 147–48, 207
      dry lot feeding, 153
      homogenization, 69–70
      iron, 148
      manganese, 148
      milkfat, 67, 69–70, 148
      milk from individual cows, 153
      phospholipids, 69–70, 72
      reduction-oxidation (redox) potential,
         72, 343
      storage time, 85
      surface oxidation, 85
      vitamins C and E, 71

Packages and packaging (dairy products)
   aesthetics of, 101
   consumer appeal, 101
   contamination of, 84
   criteria for selection/use, 101
   exterior appearance, 21–22
   functions, 101
   multiuse forms of, 84
   sensory quality, impact on, 101
   See also Milk containers and closures
Panel evaluation. *See* Sensory testing
      panels
Panelist (sensory), selection of, 517–19
Papillae, 16–17, 21
Painty off-flavor. *See* Oxidized (metal-
      induced) off-flavor and Oxidation
      of lipids
Parmesan cheese, 304
   See also Italian cheese varieties
Partial churning defect (cream), 159
Pasteurization

defined (CFR), 111–12
   HTST method, 112
   of ice cream mix, 205
   laboratory conditions for, 110
   methods and systems, 111–13
   minimum conditions, 112
   UHT/ultrapasteurized, 112, 120–21
Pasteurized Milk Ordinance and Code
      (P.M.O.)
   as related to
      concentrated milk products, 418–19
      milk (and cream), 108–12, 114, 120,
         128, 130–32, 138
Pasty (gummy, sticky) defect
   in brick cheese, 359
   in butter, 400
   in Cheddar cheese, 280, 288
   in cream cheese, 372
   in Limburger cheese, 362
   in Swiss cheese, 357
   See also Gummy defect
Pavlov's experiments (salivation), 33
Peel flavor (fruit sherbet), 221
Penicillium sp. (molds) in cheese, 343,
      362–63, 369
   P. camemberti, 369
   P. candidum, 369
   P. roqueforti, 362–63
Personnel management (sensory panels),
      524
pH (hydrogen ion concentration)
   buttermilk, cultured, 243–45
   cultured milk products, 236, 243–45
   yogurt, 254, 266, 267, 272
Penyl-thiolcarbamide (P.T.C.), 31
Phospholipids, 62, 69, 148, 239
Physical senses, 15, 21–23
   primary (basic) senses, 21–23
      aroma (smell), 22
      sight, 21–22
      sound, 23
      taste, 22
      touch, 22–23
Physical stimuli, 15, 21–23, 503
Physiology of aroma and taste, 14–32
   adaptation (time), 18–19
   "centers" of olfaction, 20
   chemical senses, 14–15
   diffusion, 20
   odor
      perception, 17–22, 24–29
      receptors, 15–17

sensitivity, 15
olfaction theories, 19–20, 24–26, 29
taste
  receptors, 15–17
  sensitivity, 15
"Picks and checks" defect (Swiss cheese), 357
Plug (core) samples
  from brick cheese, 359–60
  from butter, 388–90, 404
  from Cheddar cheese, 95–96, 314–19, 330–34, 337
  from Swiss cheese, 354, 356, 358
Powdery (chalky) defect (Brie cheese), 368
Preference tests (sensory panels), 500–501
Pressler defect (Swiss cheese), 356–57
"Premium" quality dairy products, 169, 255
Primary (basic) tastes, 11, 21, 23, 59
  in butter, 405, 406
  in Cheddar cheese, 339, 342
  in milk (and cream), 141–43
  stimuli and response interactions, 90
Product uniformity (concept of), 6
Protein "break" (concentrated milks), 424
Proteins (milk), 61–62, 70, 196, 210, 302–4
  as precursors of off-flavor, 61–62
Proteolysis (in cheese)
  blue-veined varieties, 364–65
  brick, 360–61
  Brie, 366–69
  Camembert, 365–66
  Cheddar, 305, 307, 348
  Limburger, 361
  soft, mold-ripened, 365
Propionibacterium shermanii, 278, 281, 352–53, 356
Provolone cheese, 303
Psuedomonas taetrolens, 294, 411
Psychophysics, 12, 13, 510
Psychrotrophic (spoilage) bacteria, 66–67, 87, 150, 204
  in butter, 406–7, 410–11, 414–16
  in Cheddar cheese, 331, 335, 340, 344
  in cottage cheese, 278, 287, 292, 294–95
  in dry milks, 465
  flavor defects, list of, 68
  in ice cream, 204, 206
  in milk, 66–67, 150
  post-pasteurization contamination, 67

sporeforming, heat resistant bacteria, 67
in yogurt, 273
Puckery mouthfeel, 148
  of Bulgarian buttermilk, 246
  of cottage cheese, 293–94
  of cultured milk products, 233–34, 272
  of ice cream, 206–7
  of milk (and cream), 72, 147–48
  of yogurt, 272
Putrid off-flavor, 68, 370

Quality
  "potential," 53
  "relative degree of excellence," 6, 491
  "search for excellence," 5–6
  of sensory perception, 9, 11
Quantitative Descriptive Analysis (Q.D.A. [sensory panels]), 513–14

Ragged-boring defect (butter), 382, 385, 396, 399, 401
Rancid off-flavor, 60, 68, 76–79, 87, 148–50
  Acid Degree Value (ADV), 78, 534–39
  activation of lipase, 76, 80–81
  aftertaste characteristic of, 150
  comparison to oxidized off-flavor(s), 148
  "desirable" aspects of defect, 80–81
  development factors, 78
  effect of
    free fatty acids (and salts of), 76, 79
    homogenization, 76, 81
    milk from individual cows, 153–54
    stage of lactation, 153
  flavor threshold for, 78–79, 149, 530
  measurement of defect intensity, 78, 534–39
  mechanism of formation, 76–79, 148
  occurrence in products
    blue-veined cheese varieties, 364
    butter, 413
    Cheddar cheese, 343–44
    cottage cheese, 294
    dry milk products, 464
    ice cream, 207, 217
      walnut-flavored, 217
    Italian cheese, 370
    milk, 76–81, 149–50
    Swiss cheese, 353

Rancid off-flavor (*cont.*)
  yogurt, 277
  prevention and control measures, 81
  resistance and susceptibility of various
    milks, 77–78
  taste blindness (deficiency) to, 78–79
Ranking tests (sensory panels), 501–3
Raw milk (impact on sensory quality), 5,
    132, 136–37, 170
Reconstituted milks, 119, 467, 469
Reduction-oxidation (redox) potential,
    72–73, 343
Representative product samples, 38, 125
Relative quality, 6, 491
Rich (richness) flavor attribute
  cooked "note," 81–82
  in ice cream, 197, 204–5, 210
  in milk, 81–82, 118
Ricotta cheese, 303
Rinsing of mouth (sensory technique), 40
Roller drying, atmospheric, 436
Romano cheese, 98, 304
Ropy (ropiness) body defect
  in Bulgarian buttermilk, 246
  in buttermilk, cultured, 97, 235, 246
  in cultured milk products, 235, 246
  in milk (and cream), 93
  in yogurt, 258–59, 270–71
Roquefort cheese, 363

Saliva
  composition of, 33
  flow stimulation (secretion), 33
  function of, 32–33
  glands, 32–33
  properties, 32–33
  types of, 32–33
Salmonella sp., 453
Salt, 23, 391
Salty defect (dairy products), 64, 86–87
  in butter, 87, 390–91, 393, 406
  as "briny" defect, 22, 87, 391, 393
  in cheese, 87
  in cottage cheese, 280, 289, 294–95, 479
  in cultured milk products, 231
  in dry milk products, 464
  in ice cream, 171, 173–75, 207–8, 217
  in milk, 124, 140, 150
  preparation of samples (training), 479
Sample preparation
  butter, 94, 478
  Cheddar cheese, 95, 312–14

cottage cheese, 478–79
  cultured milk products, 238, 481
  ice cream, 172, 174–77, 481
  milk, 125–26, 128–29, 474–77
  tempering, 38, 94–95, 128, 172, 174–76,
    188, 238, 260, 312–13
  yogurt, 259–63
Sample tempering. *See* Tempering of
    samples
Samples (dairy products)
  arrangement of, 53
  collection of, 50
  for contests and clinics, 53
  critique, 53
  identification methods, 125–26
  preparation of known defects, 474–79
    "doctored" samples, 473, 475, 479
    precautions for tasters, 475, 477, 479
  size of, 39–40, 53, 129
  See also Representative product sam-
    ples and Tempering of samples
Sampling
  of butter, 388–90, 404
  Cheddar cheese, 95–96, 314–19, 330–
    34, 337
  ice cream, 176
  Swiss cheese, 354, 356, 358
Salvy body defect (butter), 399
Sandy (sandiness) defect
  in ice cream, 22, 178, 196–97
  in sweetened condensed milk, 434
Sapid substances, 21, 32
Saturated fatty acids, 303
Sauerkraut-like off-flavor (cultured milk
    products), 231
Scales and scaling. *See* Sensory scales
Schweizer or Sweitzer. *See* Swiss cheese
Scorched off-flavor, 121, 143, 205
  in dry buttermilk, 451
  in dry milk products, 462, 464
  in dry whole milk, 441, 443
  in nonfat dry milk, 445, 447–48
Score cards, 37, 43–44
  brick cheese, 359
  butter, 382, 384, 387, 398, 559
  Cheddar cheese, 310–11, 313, 319, 320,
    560
  cottage cheese, 278–280, 562
  cultured milk products, 230–31
  description of, 43
  dry milk products, 461
  example, proper completion, 45

functions and purposes, 43–44
ice cream, 45, 170–72, 174, 561
milk, 121, 123–25, 557, 558
"quality factors," 43–44, 47–48
scoring techniques (use of), 44–47
summary format, 51
yogurt, 259–60, 563
Scoring
of bacteria (ice cream), 172
of bacteria (milk), 137–39
drills (self-test), 482–88
Scoring guides (dairy products), 47–48,
492
for butter
body and texture, 385
color and appearance, 385
finish, packaging and salt, 392
for Cheddar cheese, 311, 319, 338
for cottage cheese, 280
for cultured milk products, 231
for dry milk
flavor, 462
package, 469
physical appearance, 466
for ice cream
bacteria, 199
body and texture, 189
color, appearance and package, 182
flavor, 173, 175
meltdown, 186
for yogurt, 261
Seamy (seaminess) defect (Cheddar
cheese), 317, 320, 322, 333, 344
Sedimentaion (precipitation)
in concentrated milks, 429
cellular debris, 100, 116
freezing, due to, 116–17
somatic cells, 116
stratification/settling of flavorants, 100
in sweetened condensed milk, 431
Sediment examination, 125–26
of cream, 380
of evaporated milk, 427, 429
of processed milk, 125–26, 137–38
of raw milk, 133–37
via sediment discs, 125, 134–35
standard charts, 126–27, 135
visual observation, 134
Sensation intensity (quantity), 12–14
absolute threshold, 12
detection (stimulus) threshold, 12–13
difference threshold, 13

recognition threshold, 13–14
terminal (threshold), 13
threshold values, 12–14
Senses (human), 14, 21, 23
Sensitivity (to aroma and taste), 20, 21,
28
Sensory defects (of dairy products)
categories, 59–60
defined, 59
quantitative dimensions, 491
reference standards, 59
verbal description, 59
Sensory evaluation (judging, scoring), 6,
8, 29–30
art and science of, 8
checking of scoring, 42
emotional stress, 33
expert judges, 32
facilities, 36
guidelines, 36
health and physical conditions, 29–30
honesty, 42
introspection (mental concentration),
36
practical aspects, 36–42
proficiency of judgments, 46
scoring (judging) performance, 47–48
sensory acuity, 31
training, 48–49, 473–88
vacillating judgments (guesswork), 42
Sensory evaluation (testing) panels
acceptance tests, 494–96
"anchoring" points, 502, 507
bias, 494, 521–22
ballots, 498, 500, 504, 509, 524
blind samples, 502
color bias, 522
components of analysis, 510–11
consumer panels, 490, 494–96
correlation with, consumer acceptance,
489, 491
standards of quality, 489
descriptive evaluation (tests), 494–96,
510–17
descriptive tests, 510–17
components of analysis, 510–11
flavor profile, 511–12
Quantitative Descriptive Analysis
(QDA), 511, 513–14
quantitative vs. qualitative terms,
510
texture profile, 510, 512–13

Sensory evaluation (testing) panels (*cont.*)
  difference tests, 493–94, 497
  discrimination panels, 49
  duo-trio tests, 499
  "facial" hedonic scale, 508
  flavor profile analysis, 511–12
  graphical scales, 508, 510
  hedonic aspects, 490, 493–94, 501–4
  hedonic scales, 503–6
  individual experts (judges), 489, 491–92
  intensity scales, 491
  interpretation of tests, 514–17
  mechanics of testing, 517
  methods selection, 514
  objectives of panel tests, 495–96, 514
  objective tests, 515–16
    Acid Degree Value (ADV), 516
    infrared analysis, 516
    Instron (rheological measurement),
      512, 516
    instrumental methods, 515–16
    Munsel color notation, 516
    peroxide test, 516
    thiobarbituric acid test (TBA), 516
    titratable acidity of pH, 516
    viscosity, 516
  panel size, 494–95
  personal interviews, 496
  personnel management, 524
  preference evaluation, 494
  preference scales, 496
  role in
    product development, 490–91
    quality control, 514–15
  room and facilities, 521–22
  sensory description, 489
  sensory difference tests, 497–510
    duo-trio test, 499
    magnitude estimation, 508
    numerical intensity scales, 505–7
    ranking, 501–3
    scaling (quantitation), 503
    triangle test, 497–99
    two-sample (paired comparison) test,
      500–501
    two tailed, 501
    verbal scale, 507
  statistical inference, 495–96, 500, 502,
    505, 508, 513, 517, 523
  three digit code numbers, 520–21
  trained discrimination panels, 492–93
  training of panelists, 519
  untrained (in-house) panels, 493–94
Sensory evaluation techniques (judging,
    scoring), 32, 38–40, 43–47
  aroma observations, 38–39
  of butter, 382–417
    body and texture, 394–402
    color and appearance, 393–94
    flavor, 402–17
    frequency of sensory defects, 401–2
    "ideal" sensory characteristics, 395,
      404–5
    room and facilities, 384–85
    salt, 390–92
    sampling, 388–89
    score cards, 382, 384, 387, 398
    scoring guides, 384–86
    sequence of observations, 389–90
    tempering of samples, 385–86
    triers, use of, 386, 388
  of buttermilk, 243
  of Cheddar cheese, 310–15, 318–19,
    330–39
  of cheese, 305, 310–12
    aroma, 318–19, 338–39
    body and texture, 330–37
    color and appearance, 313–15
    facilities and equipment, 313
    flavor, 318–19, 338–45
    sample preparation, 313
    sensory characteristics, 305, 330–33,
      338–39
    sequence of observations, 313–19
    tempering of samples, 95, 312–13
  of concentrated milk products, 419–20,
    423–26
  of cottage cheese, 281–95
  of cultured milk products, 237–38
    buttermilk, 243–45
    sour cream, cultured, 247–51
    yogurt, 259–65
  of dry milk products, 460–71
  of ice cream, 170–80, 197–99
    anesthesia, 172, 179
    body and texture, 178, 188–97
    color and appearance, 178, 180–84
    equipment and supplies, 176–77
    flavor, 179–80, 190, 197–99
    melting properties, 180, 184–88
    sampling, 176–77, 197–98
    sandiness, 178
    score cards, 45, 170–72, 174
    scoring guides, 172–73, 175

sequence of observations, 177–80
sweetness, 179
tempering of samples, 172, 174–76, 188
laboratory facilities and equipment, 43
of milk and cream
flavor, 128–29, 140–50
score cards, 121–25, 141, 557–58
scoring guides, 124, 140
sequence of observations, 126
tempering of samples, 132–33
psychological dimensions of, 503
sample quantity (volume), 39–40
score cards, use of, 43–44
sequence of observations (in scoring), 40, 45–57
tasting techinques, 522–24
tracing sources of off-flavors (trouble-shooting), 150–54
Sensory evaluation training
for butter, 478, 480–81
for Cheddar cheese, 481
for cottage cheese, 478–80
for ice cream, 481
for milk (and cream), 473–78
safety and health considerations, 475
self-test sheets (scoring guides), 482–88
Sensory excellence, 5–6
Sensory impression, 9–11
quality, 9, 11
modality, 10–11
stimuli, 9–11
Sensory perception, 11–13, 31
acuity of, 31
intensity (quantity), 11
intensity (measurement), 12–13
quality, 11
sensation, 11
space, 11
threshold values, 12
time, 11
Sensory performance, 518–19
Sensory physiology, 9–34
human senses, 9–12
"mapping" concept, 12
olfaction, 17–21
psychophysics, 12–14
stimuli, 9–12
subjective vs. objective, 11–13
taste, 15–17
Sensory scales and scaling, 503–8
categories, 503

graphical type, 503, 505–6, 508
intensity, 506
interval, 507
numerical type, 503, 505–7
Sensory tests (panels), 497
interpretation, 516–17
mechanics of, 517–22
objective format, 515–16
objectives, 514
personnel management, 524
selection of, 514
tasting techniques, 522–24
types of, 490–91
consumer panels, 494–99
descriptive, 510–514
individual experts, 491–92
sensory difference, 497–510
trained discrimination panels, 492–93
untrained laboratory (in-house) panels, 493–94
Settled defect (sweetened condensed milk), 433, 434
Shakes (milk), direct-draw, 224
Shattered curd defect (cottage cheese), 280, 284
preparation of samples (training), 480
Sherbet, 167, 221–22
Short body defect
in butter, 382, 385, 396, 399
in Cheddar cheese, 311, 333–34
Shrinkage (shrunken) defect
in ice cream, 102, 190–93, 197
in yogurt, 258, 261, 266–67
Simulation of sensory defects
body and texture, 478
color and appearance, 480–81
flavor, 473–78
of milk, 474–78
reference standards, 59–60, 473–74
safety and health considerations, 475
Skim milk, 108, 118, 157
defined (CFR), 157
sensory characteristics, 157
"Slick" defect (cream cheese), 172
Slimy defect (cottage cheese), 280
Slit-open (slits) defect
brick cheese, 359
Cheddar cheese, 311, 331, 335
Small (too) eyes defect (Swiss cheese), 354, 356
"Sniffing" ("whiffing") technique (for aroma), 20, 22, 28, 129, 233, 318–19

Soft, mold-ripened cheeses, 365–69
  Brie, 366–69
  Camembert, 365–66
  color and appearance, 368–69
  salt content, 366
  score card, 367
  sensory characteristics, 366
Soft-serve frozen desserts, 223
  See also Ice milk
Somesthetic (body) receptors, 10
Sour cream, cultured (and half-and-half),
      98, 239, 247–51
  acidified type, 239
  body and texture, 249–51
  color and appearance, 101, 251
  gloss (sheen), 101, 251
  manufacture of, 247
  sensory attributes, 247–48
  sensory defects, 248–51
  sensory evaluation, 248–50
  tactile properties, 98
  uses of, 247
Spangling (container condition), 424–25
Spongy defect (Cheddar cheese), 333–34
Spray drying, 436–37
Stabilizer/emulsifier off-flavor
  in frozen dairy desserts, 171, 173, 209
  in sour cream, cultured, 248–49
  in yogurt, 275
Stabilizers and emulsifiers (in dairy prod-
      ucts)
  in concentrated milks, 427
  in cottage cheese (dressing), 286, 289
  in dry cream, 457
  in dry whole milk, 439
  in eggnog, 163
  in frozen dairy desserts, 169–70, 191,
      192, 194–95, 206, 209
  meltdown (ice cream), effect on, 186–87
  in milk and cream products, 118
  off-flavors, derived from, 171, 173, 209,
      248–49, 275
  in sherbet, 222
  in sour cream, cultured, 248–51
  in yogurt, 265, 268, 270
Stale off-flavor, 85, 169, 205, 210
  in butter, 379, 416
  in dry buttermilk, 451
  in dry milk products, 462, 464–65
  in dry whey, 452–53
  in dry whole milk, 441
  in edible dry casein, 456

  in evaporated milk, 119, 426–27
  in frozen dairy desserts, 169, 205–6,
      210
  in nonfat dry milk (NDM), 445, 447–48
  sensory properties of, 465
  in skim milk, 119
  in sterile milk, 121
Staphylococci sp., 453
Statistical analysis (sensory panels), 495–
      96, 500, 502, 505–8, 513, 517,
      523
Sticky (gummy, pasty) defect
  in butter, 382, 385, 396, 400
  in cream cheese, 172
Sterile milk, 120
  See also Concentrated milks
Stimuli (sensory), 9–11
"Stinker" off-flavor (Swiss cheese), 353
Stirred (granular) cheese curd, 352
Storage off-flavor, 84–86
  in butter, 381, 384, 386, 413
  descriptors for, 85
  in dry buttermilk, 451
  in dry milk products, 462, 465
  in dry whey, 452
  in ice cream, 210
  as "limited oxidation," form of, 85
  in milk and cream, 86
  mechanism of formation, 85
  in nonfat dry milk (NDM), 445
  in skim milk, 119, 157
  as "staling" action, 85–86
  as "surface oxidation," 85
Streaky color defect (butter), 382, 385,
      393–94
Streptococcus cremoris (culture), 141,
      240, 247, 296
Streptococcus lactis (culture), 141, 240,
      247, 296
  subsp. diacetylactis, 233, 240, 247, 278,
      281, 296, 336
  var. maltigenes, 147, 293, 343
Streptococcus thermophilous, 240, 246,
      252–53, 353
Sucrose (sugar). See Sweetening agents
Sulfhydryl oxidase, 121
Sulfide (skunky, stinker) off-flavor
  in Cheddar cheese, 311, 339, 344
  in Swiss cheese, 353
Superheated condensed skim milk, 435
Surface discolored defect (cottage cheese),
      280

Surface faded color defect (butter), 385, 393
Surface growth (mold and/or yeast)
  in cultured milk products, 231
  in yogurt, 258, 261, 266–68
Surface taint defect, 85, 102, 412, 414
Sweet-curd holes defect (Cheddar cheese), 311, 335–37
Sweetened condensed milk, 430–35
  body and texture, 431, 433–34
  color and appearance, 431
  defined (CFR), 430
  flavor characteristics, 432–33
  flow diagram (process), 421
  optional ingredients, 430
  sedimentation problems, 431
  sensory quality, 432
  sequence of observations, 431
Sweetened condensed skim milk, 435
Sweetening agents (sucrose, sugar, syrups)
  carmelization effects, 89–90
  consumer preferences, 89
  corn syrup, 166, 168–69·
  fructose, 168
  functions of,·89
  sensory defects, derived from, 90, 169
  sucrose equivalent, 168
  sweetening (level) power, 168, 213–14, 215–16
  types of, 89, 166, 168–69
  use in dairy products
    chocolate flavored, 154–55, 213–14
    frozen dairy desserts, 166, 168–69, 213–14
    fruit-flavored ice creams, 215–16
    sweetened condensed milk, 430
  yeast contamination, 90
Sweetness/acidity "balance," 223
Swiss cheese, 98, 301–4, 352–58
  body and texture, 356–57
  "domestic" Swiss, 352
  "eye" size and distribution, 23, 101, 354–57
  finish and appearance, 357
  flavor characteristics, 352–54
  salt content, 357
  score card, 352
  styles (types), 352, 357–58
  technique for examination, 358
Syneresis (wheyed-off) defect
  in buttermilk, cultured, 235–36, 241, 243
  in cottage cheese, 283–84
  in cultured milk products, 231, 235–36
  in ice cream, 186–87
  in lactic cultures, 235–36
  in sour cream, cultured, 251
  in yogurt, 261
  See also Free whey and Wheyed-off defects
Syrup off-flavor, 89, 171, 173–75, 179, 202

Tactile properties (dairy products), 90–98, 442
  of butter, 93–95, 394–95
  of Cheddar cheese, 93, 95–96, 330–37
  consumer acceptance, 92
  correlation with flavor, 92, 95, 98
  of cottage cheese, 96
  of cultured milk products, 240–42, 244–45, 249–51
  of ice cream, 93, 188–97
  instrumental measurement (objective methods), 94, 512, 515–16, 541–42
  of lowfat milk, 118
  of other dairy products, 97–98
  physical parameters, 94
  principle criteria, 92
  puckery mouthfeel, 72, 147, 206–7
  of skim milk, 157
  of yogurt, 256–69
Tallowy off-flavor, 67, 294
  in butter, 384, 386, 412–13
  in dry milk products, 462, 464
  in dry whole milk, 442
  in malted milk, 459
  in nonfat dry milk (NDM), 448
  in sweetened condensed milk, 432
Tartness flavor defect (sherbet), 211
Taste, 14–17, 20–24, 30–32
  adaptation (time), 18–19
  blindness (deficiency), 29, 31–32, 78–79
  classification, 23
  detection thresholds, 30
  fatigue, 9, 197–98
  perception, 15–17, 20–22
  primary (basic or "classic"), 11, 17, 21, 23–24, 59, 90
    butter, 11, 17, 21, 23–24
    salt, 17, 23–24
    sour (acid), 17, 23–24
    sweet, 17, 23–24

Taste (*cont.*)
  others, 23
    alkaline, 23
    meaty, 23
    metallic, 23
    watery, 23
  qualities (mechanoreceptors), 11, 21, 23
  receptors (taste buds), 15–17, 20–21
  sensitivity, 28–29
  tastebud conditioners (sensitizers), 40–41
Taste receptors (taste buds), 15–17, 20–21, 40–41
Tasting techniques (sensory panels), 522
Tempering of samples, 38, 420
  of butter, 94, 385–86
  of Cheddar cheese, 95, 312–13
  of cultured milk products, 238
  of ice cream, 38, 172, 174–76, 188
  of milk (and cream), 128
  of soft, mold-ripened cheeses, 368
  of yogurt, 260
Texture profile analysis, 512–13
Thickened defect (sweetened condensed milk), 433–34
Thin (weak) body defect
  buttermilk, cultured, 241
  sour cream, cultured, 250
Threshold values, 12–14
  absolute, 12–13
  detection (stimulus), 12
  difference, 12–13
  individual response, 14, 29
  recognition (identification), 12, 14
  suprathreshold stimulus, 14
  terminal, 12
Thresholds (sensory), 12, 29, 519
Titratable acidity (% T.A.)
  of buttermilk, cultured, 236–37
  of cultured milk products, 236–37, 242–43
  of lactic cultures, 236–37
  of sour cream, cultured, 248
  of yogurt, 254, 272
Tobacco products, constraints of, 37, 384–85
Too firm body defect
  of Brie and Camembert cheese, 368
  of buttermilk, cultured, 241
  of cottage cheese, 286–87
  of cream cheese, 372
  of cultured milk products, 235

  of sour cream, cultured, 249–51
  of yogurt, 259, 270
Too sweet flavor defect
  in ice cream, 171, 173–75, 179, 202
  yogurt, 261, 275
Torula lactis condensi, 434
Touch (mechanoreception) qualities, 11
Trademarks and brands (dairy products), 2–4
Translucent appearance defect (cottage cheese), 280
Triangle test (sensory panels), 497–98
Triers (core sampling devices)
  for butter, 386, 388
  for cheese, 313–14
Troubleshooting (tracing) sensory defects
  of ice cream, 199
  of milk (and cream), 148, 151–54
  of other frozen dairy desserts, 224–25
Two-sample test (sensory panels), 500
Two-tailed test (sensory panels), 501

UHT sterilized milk, 120
Ultrapasteurization, 205
Unclean off-flavor, 66–67, 124, 140, 150
  bacterial cause (origin), 66–67
  in butter, 379, 381, 384, 386, 414
  in Cheddar cheese, 344, 347
  in cottage cheese, 295
  in cream cheese, 372
  in cultured milk products, 231
  descriptors for, 66–67, 150
    as "barny/cowy," 66
    as "cheesy," 68
    as "putrid," 66, 68
    as "utensil," 68, 414
  in ice cream, 210
  in milk (and cream), 124, 140, 150
  in Swiss cheese, 353–54
  in yogurt, 261, 276
  See also Utensil off-flavor
Unconditioned reflex, 33
"Undesirable" defect (dry milk products), 466
Unhomogenized milk, 115, 149
United States Department of Agriculture (USDA)
  butter grades/standards, 70, 380–83
  Cheddar cheese grades/standards, 347–49, 544–55
  as coordinator of Collegiate Contest, 54–57

dry milk grades/standards, 439–41, 469
federal consumer grades, 3–4, 491
grading and inspection shields, 4
laboratory analysis, dry milk products,
469
manufacturing grade milk standards,
109
milk score card (original), 122, 124
sediment content (of milk) standards,
136
Unnatural (atypical) color defect
in Cheddar cheese, 320–21
in cottage cheese, 280
in cultured milk products, 231, 245
in dry milk products, 466–67
in ice cream, 181
in nonfat dry milk, 445
in sour cream, cultured, 251
in yogurt, 265–66
Unnatural (atypical) flavor defect
ice cream, 171, 173–75, 201–2, 217
in sherbet, 221
in Swiss cheese, 353–54
in yogurt, 261, 274
Unnatural sweetness. See Syrup off-
flavor
Unsaturated fatty acids, 148, 303
Utensil off-flavor
in butter, 68, 379, 381, 384, 386, 414,
417
in cheese, 68
in dry buttermilk, 451
in dry milk products, 462, 465
in dry whey, 452
in nonfat dry milk, 445
See also Unclean off-flavor

Vacuum drum drying, 436
Vacuum treatment (milk), 64–65
Vanilla (ice cream), 170, 180, 203
Variegated frozen dairy desserts, 218
Verbal anchors (points), 505, 507
Viscosity (dairy products), 92–93
of chocolate milk/drinks, 98
causes of variation, 92–93
control of, 93
in cultured milk products, 98
descriptors for, 93
desirable level, 92
determination of, 92
in eggnog, 97–98
heat treatment, effect of, 92

regional preferences, 92, 97
stabilization, achieved by, 93
Vitamin(s)
contribution to off-flavor (milks), 64
fortification, 117
of concentrated milks, 119, 419, 421–
22
of dry milk, 439
fortification (of products)
off-flavors caused by, vitamin A,
(hay-like), 64, 117
with vitamin A, 117–18, 422
with vitamin D, 117–18, 119, 419,
421
of lowfat milk, 118
of milk, 117–18
riboflavin, 64
vitamin A, 117–18, 422, 439, 441, 443
vitamin C
antioxidant properties, 64, 70
as oxidizing agent, 64
vitamin D, 117–18, 119, 419, 421, 439,
441, 443
vitamin E (antioxidant), 64, 70
Vitamin (caused) off-flavor, 157
Volatile fatty acids, 76–80, 343, 347, 353
Volatile substances (compounds)
from feed (cow consumed), 64–65
in flavor(s), 60
in garlic/onion off-flavor, 65–66
removal, vacuum processing, 64–65
from weeds, 65–66

Water ices, 167, 222, 233
See also Frozen dairy desserts
Wavy color defect (butter), 382, 385, 393–
94
Waxy (waxiness) body (cheese), 330, 331–
32
Weak (thin, watery) body defect
in butter, 382, 385, 396, 400
in buttermilk, cultured, 235, 241
in Cheddar cheese, 311, 316, 333–34
in Colby cheese, 351
in cottage cheese, 288–89
in cream cheese, 372
in cultured milk products, 231, 235
in ice cream, 190, 192, 193
in Italian cheese varieties, 371
in lactic cultures, 235
in Limburger cheese, 362
in Monterey Jack cheese, 351

Weak (thin, watery) body defect (*cont.*)
  in soft-serve (ice milk), 224
  in sour cream, cultured, 251
  in Swiss cheese, 358
  in yogurt, 259, 261, 270
Weber's rule, 13-14
Weedy off-flavor, 65
  in butter, 379, 381, 384, 386, 415
  in dry milk products, 462, 466
  in dry whey, 452
  in milk (and cream), 65
Wheyed-off (syneresis) appearance defect
  in buttermilk, cultured, 235-36, 241,
    243
  in cottage cheese, 283-84
  in cultured milk products, 231, 235-36
  in ice cream, 186-87
  in lactic cultures, 235-36
  in sour cream, cultured, 186-87
  in yogurt, 261
  See also Free whey and Syneresis de-
    fects
Whey off-flavor
  in butter, 381, 384, 386, 415
  in dry whey, 453
  in ice cream, 171, 173-75, 208, 453
Whey taint
  in Cheddar cheese, 311, 339, 341, 344
  in Italian cheese varieties, 370
  in Swiss cheese, 356
Whipping cream, 108, 161-62
  body defects, 161-62
  mouthfeel, 162
  in pressurized containers, 162
  product types, 161-62
    whipped cream (topping), 162
  whipping properties, 156, 161-62, 541-
    42
White specks (Cheddar cheese), 335
"Woody" off-flavor (butter), 416

Yeasts
  as contaminants of
    butter, 379, 384, 386, 415
    Cheddar cheese, 335, 345

    cottage cheese, 234, 295
    cream cheese, 372
    cultured milk products, 232, 234
    Swiss cheese, 354
    yogurt, 277
  as cultures for
    kefir, 296
    Limburger cheese, 361
    soft, mold-ripened cheeses, 365
Yeasty off-flavor, 68
  in butter, 379, 384, 386, 415
  in Cheddar cheese, 311, 339, 345
  in cottage cheese, 280, 295
  in cream cheese, 372
  in cultured milk products, 231, 234
Yogurt
  acidity (% T.A.), 254, 266, 267, 272
  body and texture, 252, 255-56, 268-71
  color and appearance, 263, 265-68, 269
  defined (CFR), 252-53
  flavor characteristics, 253-56, 264-65,
    271-77
  flavor defects, 271-77
  historical developments, 251-52
  "ideal" or desirable sensory properties,
    256
  nutritional and health aspects, 252-53
  optional dairy ingredients in, 252, 253
  sample preparation, 259-60, 262-63
  score card for, 259-60
  scoring guide for, 259, 261
  sequence of observations (evaluation),
    263-65
  Standards of Identity (federal), 252-
    53
  styles (types) of product
    drinkable, 252
    frozen, 254-55
    plain, 253-59
    novelties, 255
    sundae (fruit-on-bottom), 254-55
    Swiss (prestirred), 254-55
  sweeteners in, 253
  tactile properties, 98